29. $\displaystyle\int \frac{dx}{\sin ax} = \frac{1}{a} \ln \left| \tan \left(\frac{ax}{2}\right)\right| = \frac{1}{a} \ln |\csc ax - \cot ax|$

30. $\displaystyle\int \frac{dx}{\cos ax} = \frac{1}{a} \ln |\tan ax + \sec ax|$

31. $\displaystyle\int \sin ax \sin bx \, dx$
$= \frac{1}{2} \left[\frac{\sin (a-b)x}{(a-b)} - \frac{\sin (a+b)x}{(a+b)}\right] \quad (a^2 \neq b^2)$

32. $\displaystyle\int \sin ax \cos bx \, dx$
$= -\frac{1}{2} \left[\frac{\cos (a-b)x}{(a-b)} + \frac{\cos (a+b)x}{(a+b)}\right] \quad (a^2 \neq b^2)$

33. $\displaystyle\int \cos ax \cos bx \, dx$
$= \frac{1}{2} \left[\frac{\sin (a-b)x}{(a-b)} + \frac{\sin (a+b)x}{(a+b)}\right] \quad (a^2 \neq b^2)$

34. $\displaystyle\int \sin^n u \, du = \frac{-\sin^{n-1} u \cos u}{n} + \frac{n-1}{n} \int \sin^{n-2} u \, du$

35. $\displaystyle\int \cos^n u \, du = \frac{\cos^{n-1} u \sin u}{n} + \frac{n-1}{n} \int \cos^{n-2} u \, du$

36. $\displaystyle\int \tan^n u \, du = \frac{\tan^{n-1} u}{n-1} - \int \tan^{n-2} u \, du$

37. $\displaystyle\int \cot^n u \, du = \frac{-\cot^{n-1} u}{n-1} - \int \cot^{n-2} u \, du$

38. $\displaystyle\int \sec^n u \, du$
$= \frac{\tan u \sec^{n-2} u}{n-1} + \frac{n-2}{n-1} \int \sec^{n-2} u \, du \quad (n \neq 1)$

39. $\displaystyle\int \csc^n u \, du$
$= \frac{-\cot u \csc^{n-2} u}{n-1} + \frac{n-2}{n-1} \int \csc^{n-2} u \, du \quad (n \neq 1)$

40. $\displaystyle\int x^n \sin ax \, dx = -\frac{1}{a} x^n \cos ax + \frac{n}{a} \int x^{n-1} \cos ax \, dx$

41. $\displaystyle\int x^n \cos ax \, dx = \frac{1}{a} x^n \sin ax - \frac{n}{a} \int x^{n-1} \sin ax \, dx$

42. $\displaystyle\int e^u \, du = e^u$

43. $\displaystyle\int x^n e^{ax} \, dx = \frac{1}{a} x^n e^{ax} - \frac{n}{a} \int x^{n-1} e^{ax} \, dx \quad (n > 0)$

44. $\displaystyle\int \frac{e^{ax}}{x^n} \, dx = -\frac{e^{ax}}{(n-1)x^{n-1}} + \frac{a}{n-1} \int \frac{e^{ax}}{x^{n-1}} \, dx \quad (n > 0)$

45. $\displaystyle\int a^u \, du = \frac{a^u}{\ln a}$

46. $\displaystyle\int x^n a^{bx} \, dx = \frac{x^n a^{bx}}{b \ln a} - \frac{n}{b \ln a} \int x^{n-1} a^{bx} \, dx \quad (n > 0)$

47. $\displaystyle\int \ln u \, du = u \ln u - u$

48. $\displaystyle\int \frac{du}{u \ln u} = \ln |\ln u|$

49. $\displaystyle\int x^n \ln (ax) \, dx$
$= x^{n+1} \left[\frac{\ln (ax)}{n+1} - \frac{1}{(n+1)^2}\right] \quad (n \neq -1)$

50. $\displaystyle\int x^n (\ln ax)^m \, dx$
$= \frac{x^{n+1}}{n+1} (\ln ax)^m - \frac{m}{n+1} \int x^n [\ln (ax)]^{m-1} \, dx$
$\qquad\qquad\qquad\qquad\qquad (n \neq -1)$

51. $\displaystyle\int x e^{ax} \, dx = \frac{ax-1}{a^2} e^{ax}$

52. $\displaystyle\int_0^\infty e^{-ax^2} \, dx = \frac{1}{2} \sqrt{\frac{\pi}{a}} \quad (a > 0)$

53. $\displaystyle\int_0^\infty x^{n-1} e^{-x} \, dx = (n-1)! \quad (n > 0)$

54. $\displaystyle\int e^{ax} \sin (bx) \, dx = \frac{e^{ax}}{a^2 + b^2} [a \sin (bx) - b \cos (bx)]$

55. $\displaystyle\int e^{ax} \cos (bx) \, dx = \frac{e^{ax}}{a^2 + b^2} [a \cos (bx) + b \sin (bx)]$

56. $\displaystyle\int \sin^{-1} ax \, dx = x \sin^{-1} ax + \frac{1}{a} \sqrt{1 - a^2 x^2}$

57. $\displaystyle\int \tan^{-1} ax \, dx = x \tan^{-1} ax - \frac{1}{2a} \ln (1 + a^2 x^2)$

Calculus for Business, Biology, and the Social Sciences
Third Edition

Calculus for Business, Biology, and the Social Sciences

Third Edition

David G. Crowdis
Sacramento City College

Susanne M. Shelley
Sacramento City College

Brandon W. Wheeler
Sacramento City College

GLENCOE PUBLISHING CO., INC.
Encino, California
Collier Macmillan Publishers
London

Glencoe Publishing Co., Inc.
17337 Ventura Boulevard
Encino, California 91316
Collier Macmillan Canada, Ltd.

Library of Congress Catalog Card Number: 77-094765

1 2 3 4 5 6 7 8 9 10   83 82 81 80 79

ISBN 0-02-472050-X

Contents

Preface

This text is designed for a one-semester, two-quarter, or full-year course in calculus for business, biology, and social science majors. The main difference between the material presented here and the more traditional approach to calculus lies in the kinds of applications chosen to illustrate the mathematical concepts, rather than in the mathematics itself. Whenever possible, applications are taken from the social, management, and biological sciences rather than from engineering and physics. For example, the derivative is interpreted as marginal cost—an economic concept—and the area under a curve is interpreted as a probability—a concept with applications in the social sciences.

It should be pointed out that this book is intended as a mathematics text, not as an introduction to economics, biology, or sociology. As such, it deals primarily with the basic concepts of a first course in calculus. The approach is intuitive rather than formal; however, this does not mean it is a totally mechanical approach. Rather, it means that ideas are discussed and their underlying principles are examined for "reasonableness" without attempting to structure formal proofs.

Organization. The first seven chapters of this text present the calculus of a single variable. The first four chapters constitute an introduction to the derivative and the integral. These chapters are designed so that the student will receive a solid introduction to these concepts in the first semester. In addition, Chapter 1 includes a review of important precalculus concepts and a basic, noncalculus introduction to logarithmic and exponential functions. Chapters 5–7 cover the calculus of trigonometric functions, the calculus of logarithmic and exponential functions, and techniques and applications of integration. Chapters 8 and 9 present multivariate calculus and series and sequences.

Features of the Third Edition.   In this edition we have incorporated many changes suggested by users of previous editions in the hope that they will make the subject more meaningful and teachable. These changes include the following:

The first four chapters as well as the first three sections of Chapter 6 have no trigonometry-related exercises or examples. This permits instructors to present a one-term course which is trigonometry-free but which includes the calculus of logarithmic and exponential functions. Furthermore, the instructor may select applications from Chapter 7 as time permits. Such a course meets the needs of many business and social science students.

Trigonometry is reviewed in Chapter 5 and, with the exception of the first three sections of Chapter 6 (which deal with logarithmic and exponential functions), is incorporated throughout subsequent chapters as it was in the previous editions. Thus, the second term of a two-term course will give biology and other science students the basic background in the calculus of trigonometric functions that they need.

In addition, the entire text has been reviewed for accuracy and clarity and many minor adjustments have been made in writing and organization. The review of algebraic concepts in Chapter 1 has been considerably expanded, and Chapter 4 has been reorganized so that antiderivatives are treated before the definite integral.

Pedagogical Aids.   Each chapter contains numerous worked-out examples and exercises. The applications from business, biology, and the social sciences contained in these examples and exercises are marked in the margins of the text with the symbols 🝰, 🙂, and 👀 , respectively. Every chapter has a set of review exercises in addition to the exercises following each section. The answers to the odd-numbered exercises and to all review exercises are supplied in the back of the book. The text also contains a separate index of applications in addition to the regular index. This allows the reader easy access to particular kinds of applications within the text.

A separate Answer and Test Booklet contains the answers to the even-numbered text exercises and two tests for each chapter, along with two mid year tests and an end-of-year test.

Course Outlines.   The following flowchart indicates three possible course sequences for presenting the material in this text.

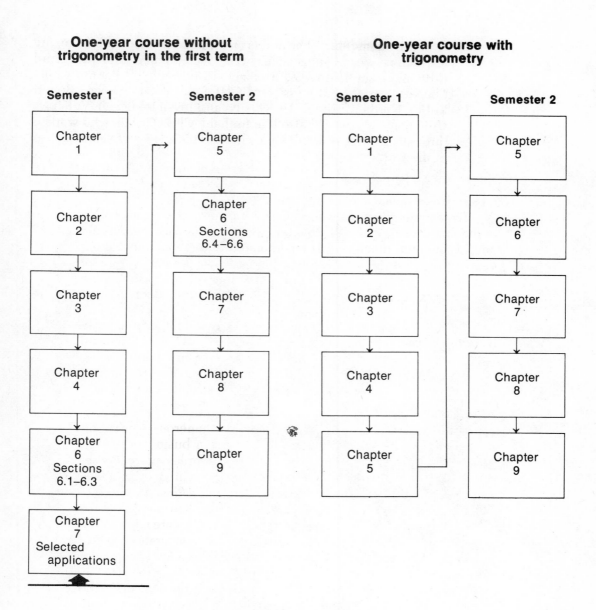

One-year course without trigonometry in the first term

Semester 1

Chapter 1

Chapter 2

Chapter 3

Chapter 4

Chapter 6
Sections
6.1–6.3

Chapter 7
Selected
applications

Semester 2

Chapter 5

Chapter 6
Sections
6.4–6.6

Chapter 7

Chapter 8

Chapter 9

One-year course with trigonometry

Semester 1

Chapter 1

Chapter 2

Chapter 3

Chapter 4

Chapter 5

Semester 2

Chapter 5

Chapter 6

Chapter 7

Chapter 8

Chapter 9

One-semester course without trigonometry, with logarithmic and exponential functions

Acknowledgments. The authors would like to thank the following instructors whose suggestions were incorporated into the Third Edition: Joseph Angelo, Indiana University of Pennsylvania; Thomas Enright, University of California at San Diego; Robert Kuller, Northern Illinois University; and Joyce Longman, Villanova University. We would also like to thank Chris Cagan, who worked through all of the examples and checked the answers to all exercises in the text.

Calculus for Business, Biology, and the Social Sciences
Third Edition

1

Operations and Functions

Calculus is built upon two great concepts, the derivative, which examines rates of change, and the integral, which deals with sums and, initially, areas. Both of these concepts have been known and worked with in various forms for many centuries. It remained for two of the greatest minds of the seventeenth century, Issac Newton and G. W. Leibniz, to see that these two concepts were not separate but fit together to form one unified discipline. The study of these two concepts, their interrelation, and their application, forms the subject we call *calculus.*

It would be interesting to jump right in and examine how the rate of change, or derivative, relates economic concepts like the rate of change of profit to a change in sales, or how the rate of radioactive decay of carbon-14 helps the anthropologist date finds, or how the sum concept of the integral relates to ideas of supply and demand or population growth. But first we must study the basic concepts of functions and limits.

Mathematicians know that the key to understanding the derivative and the integral lies in the limit. The limit concept is used for the examination of the behavior of one variable quantity as a second related variable approaches but does not reach a selected value. To develop the concept of a limit it will be necessary to examine the function concept, which formalizes relationships between variables. One final word: Calculus can be and is applied to complex numbers; however, that application is beyond the scope of this text, which will restrict its discussion to real numbers.

In preparation for the study of the concepts of calculus, we will examine functions and their inverses and introduce the idea of a limit. The concept of a *function* should not be a new idea to students who have studied precalculus mathematics. It is in many ways only the formalization of the ideas of formulas we have encountered as early as the second or third grade. In this chapter we will review the use of formulas and use this review to examine the nature of mathematical operations. We will also introduce some new symbolism and notation, present definitions for a function and its inverse, and review the logarithmic function. This will enable us to examine the limit concept, which is the key to calculus.

1.1   FORMULAS AND THE ORDER OF OPERATIONS

Many important quantities are related by formulas.

Example 1   The Celsius temperature reading, °C, is related to the Fahrenheit temperature reading, °F, by the formula

$$°C = \frac{5}{9}(°F - 32)$$

Example 2   Find the Celsius reading corresponding to 104°F.

Solution   Since $°C = \frac{5}{9}(°F - 32)$, when $°F = 104$,

$$°C = \frac{5}{9}(104 - 32)$$
$$= \frac{5}{9}(72)$$
$$= \frac{5}{9} \cdot \frac{72}{1} = 5 \cdot 8 = 40°C$$

The above problem is a very simple one. It does, however, illustrate some extremely important points. First, the formula is given in such a way that each unique value of °F corresponds to a unique value of °C. We will have much more to say about this later. Second, the operations of arithmetic (and thus also algebra) are always performed in a very specific order:

1. Operations within grouping symbols
2. Exponentiation (raising to powers)
3. Ungrouped multiplications and divisions (from left to right as they occur)
4. Ungrouped additions and subtractions (from left to right)

Example 3   What does 3×2^4 equal?

Solution   Since there are no grouping symbols, exponentiation is done first:

$$3 \times 2^4 = 3 \times 16 = 48$$

Example 4   What does $(3 \times 2)^4$ equal?

Solution   Here the grouping symbol takes precedence. Thus

$$(3 \times 2)^4 = 6^4 = 1296$$

Example 5   Evaluate the following expressions:

a. $3 \cdot 4 + 8 \div 2$

b. $3 + 4 \div 2 + 8$

c. $27 - 7 \cdot 3 + 6 \div 2$

d. $[(27 - 7) \cdot 3 + 6] \div 2$

Solution

a. $3 \cdot 4 + 8 \div 2 = 12 + 4 = 16$

b. $3 + 4 \div 2 + 8 = 3 + 2 + 8 = 5 + 8 = 13$

c. $27 - 7 \cdot 3 + 6 \div 2 = 27 - 21 + 3 = 6 + 3 = 9$

d. $[(27 - 7) \cdot 3 + 6] \div 2 = [20 \cdot 3 + 6] \div 2$
$$= [60 + 6] \div 2 = 66 \div 2 = 33$$

When we apply the order of operations to formulas, we are usually trying to find the specific value for one variable quantity when a specific value is assigned to a related quantity.

Example 6   Suppose two variable quantities, S and t, are related by the formula

$$S = 16t^2 - 8t + 144$$

What value does S have if $t = 4$?

Solution
$$
\begin{aligned}
S &= 16t^2 - 8t + 144 \\
&= 16(4^2) - 8(4) + 144 \\
&= 16(16) - 8(4) + 144 \\
&= 256 - 32 + 144 \\
&= 224 + 144 = 368
\end{aligned}
$$

Example 7   Suppose in a certain experiment that a population of bacteria P is given by

$$P = \frac{2(t^3 - 3t^2 + 4t)}{\sqrt{t^2 + 2}}$$

where t is time in hours from the start of the experiment. What is the population 12 hours after the start of the experiment?

Solution $P = \dfrac{2[12^3 - 3(12)^2 + 4(12)]}{\sqrt{12^2 + 2}}$

$= \dfrac{2[1728 - 3(144) + 4(12)]}{\sqrt{144 + 2}}$

$= \dfrac{2[1728 - 432 + 48]}{\sqrt{146}}$

$= \dfrac{2(1344)}{\sqrt{146}} \approx 222.46$

(Here a calculator was used to approximate the answer.)

Example 8   An object is thrown upward from the top of a 120 meter tall building. The distance S, in meters, of the object from the ground is given by

$$S = -4.9t^2 + 30t + 120$$

where t is time measured in seconds. Make a table showing various S values for $t = 0, 1, 2, 3, 4, 5, 6, 7, 8, 9,$ and 10. Use this table to estimate the time it takes the object to strike the ground.

Solution   At $t = 0$ $S = -4.9(0)^2 + 30(0) + 120 = 120$ meters
At $t = 1$ $S = -4.9(1)^2 + 30(1) + 120$
$= -4.9 + 30 + 120 = 145.1$ meters
At $t = 2$ $S = -4.9(2)^2 + 30(2) + 120$
$= -4.9(4) + 60 + 120 = 160.4$ meters

The remainder of the table can be constructed in the same manner. The result is given here.

t	S
0	120
1	145.1
2	160.4
3	165.9
4	161.6
5	147.5
6	123.6
7	89.9
8	46.4
9	−6.9
10	−70.0

The object strikes the ground somewhere between $t = 8$ and $t = 9$. The values for S when $t = 9$ and $t = 10$ are nonsense. Clearly the formula no longer applies.

Often, the value we want to put into a formula is not a specific numerical value but another symbol instead.

Example 9   Assume that, as in the previous example, an object is thrown upward, but this time from a balloon 4 kilometers above the ground. Assume further that the distance S, in meters, the object is above the ground t seconds after it is thrown is given by

$$S = -4.9t^2 + 100t + 4000$$

Find a formula that gives S for any given value of T, time in minutes.

Solution   A time of T minutes translates into $60T$ seconds. Thus $t = 60T$. By replacing t in the formula with $60T$ we will have a new formula giving S in terms of T. Thus

$$
\begin{aligned}
S &= -4.9t^2 + 100t + 4000 \\
&= -4.9(60T)^2 + 100(60T) + 4000 \\
&= -4.9(60)^2 T^2 + 6000T + 4000 \\
&= -4.9(3600)T^2 + 6000T + 4000 \\
&= -17{,}640T^2 + 6000T + 4000
\end{aligned}
$$

Example 10   Suppose that two variables x and y are related by the formula

$$y = \frac{1}{x^2 - 36} - \frac{4}{x^2 - 4}$$

Further suppose that x is to be replaced by $w + 2$. Find the resulting formula relating y and w algebraically. Simplify the result.

Solution First consider the denominators of the fractions. The expression $x^2 - 36$ becomes $(w + 2)^2 - 36$ and

$$(w + 2)^2 - 36 = w^2 + 4w + 4 - 36$$
$$= w^2 + 4w - 32$$
$$= (w - 4)(w + 8)$$

Similarly, $x^2 - 4$ becomes $(w + 2)^2 - 4$ and

$$(w + 2)^2 - 4 = w^2 + 4w + 4 - 4$$
$$= w^2 + 4w$$
$$= w(w + 4)$$

Then y becomes

$$y = \frac{1}{(w - 4)(w + 8)} - \frac{4}{w(w + 4)}$$
$$= \frac{w(w + 4)}{w(w + 4)(w - 4)(w + 8)} - \frac{4(w - 4)(w + 8)}{w(w + 4)(w - 4)(w + 8)}$$
$$= \frac{w^2 + 4w - 4w^2 - 16w + 128}{w(w + 4)(w - 4)(w + 8)}$$
$$= \frac{-3w^2 - 12w + 128}{w(w + 4)(w - 4)(w + 8)}$$

One key feature of all the formulas used above should be kept in mind. In each case there was an *independent* variable. We were free to choose any value for that variable. Once that value was picked, the formula permitted us to calculate a unique value for a second, or *dependent*, variable. In other words, a value of the independent variable was always paired with a unique value of the dependent variable. In our study of calculus we will make extensive use of relationships defined by formulas that have this unique value feature.

1.1 EXERCISES

1. Use the formula $°C = \frac{5}{9}(°F - 32)$ to find the Celsius temperature corresponding to $10°F$, $50°F$, $72°F$, $500°F$, and $-32°F$.

2. The formula relating Fahrenheit to Celsius temperature is $°F = \frac{9}{5}°C + 32$. Find the Fahrenheit temperatures corresponding to 40°C, 20°C, 18°C, 0°C, and −40°C.

3. Find the temperature that has the same reading (that is, numerical value) on both scales.

In Exercises 4–10, use the order of operations to evaluate the expressions.

4. $-(-37) + (8 + 2) \times 5 - 4$

5. $8^2 - 2^2 + 3^3 \times 4$

6. $(7 + 5^2 - 4) \times 2 + (8 - 2^3) \times 5$

7. $\left(\frac{1}{2} + \frac{1}{4}\right) \times \frac{1}{2} \times 2$

8. $\frac{1}{3} + \frac{1}{6} - 3\left(\frac{1}{12} + \frac{3}{24}\right)$

9. $(4 + 6 - 2 \times 3)^2$

10. $\dfrac{5 - 2^2 - 3^3 + (4 + 7) \cdot 2}{(8^2 - 36) \times 2}$

In Exercises 11–20, evaluate the formulas for the value of the independent variable given.

11. $T = (7H + 4 - H^2)(10 + 3H)$, at $H = 2$

12. $y = (x - 2)^2(x + 2)^2$, at $x = 7$

13. $z = (x^2 - 2x + 4)(x^2 + 2x + 1)$, at $x = -3$

14. $v = (u^2 - 2u)(3u^3 - 2u)$, at $u = -1$

15. $k = \dfrac{(L - 2)^2}{(L + 2)^2}$, at $L = -4$

16. $M = -4.9t^2 + 100t - 1200$, at $t = 3$

17. $M = \dfrac{-32t + 16}{t^3}$, at $t = -1$

18. $S = \dfrac{-16t^2 + 100}{t - 2}$, at $t = 4$

19. $y = \dfrac{(3x^3 - 2x)x + 2}{x(3 - x) + 4}$, at $x = 3$

20. $y = \{[(x - 4)x + 3]x - 2\}x + 5$, at $x = 4$

In Exercises 21–30, for each formula replace x by the indicated expression in z and simplify the result.

21. $y = x + 3$, $x = z + 7$

22. $y = (x + 3)^2$, $x = z - 2$

23. $y = x^2 + 3$, $x = z - 2$

24. $y = x^2 + 3$, $x = (z + 2)^2$

25. $y = \dfrac{1}{x} + \dfrac{1}{x^2}$, $x = \dfrac{1}{z}$

26. $y = \dfrac{1}{x - 2} + \dfrac{1}{x + 2}$, $x = (z + 2)^2$

27. $y = \dfrac{1}{x^2 - 3} + \dfrac{1}{x^2 + 3}, \quad x = z - 3$

28. $y = \dfrac{x}{x + 2} - \dfrac{x + 2}{x}, \quad x = z - 2$

29. $y = \{[(x + 2)x + 3]x + 7\}x, \quad x = \dfrac{z}{2}$

30. $y = \dfrac{1}{[(x + 2)x + 3]x}, \quad x = z - 2$

Exercises 31–60 represent a general review of the algebraic skills a student should have before starting calculus. In Exercises 31–34, find the indicated quotients.

31. $(3x^2 + 13x + 12) \div (3x + 4)$ **32.** $(8a^2 - 14a - 15) \div (4a + 3)$

33. $\dfrac{6x^3 - 7x^2 - 17x + 4}{2x^2 - 5x + 1}$ **34.** $\dfrac{4a^3 + 5a^2 + 6a - 9}{4a - 3}$

In Exercises 35–38, find the indicated products.

35. $\left(\dfrac{x}{2} + 8y\right)\left(\dfrac{x}{2} + 8y\right)$ **36.** $(w - 2)^2(w + 2)^2$

37. $(x^2 - 2x + 4)(x^2 + 2x + 1)$ **38.** $(u^2 + v^2)(u^4 - u^2v^2 + v^4)$

In Exercises 39–42, factor the expressions.

39. $10x^2 - 5x - 5$ **40.** $8x^2 - 18x - 18$

41. $9(x - y)^2 + 12w(x - y) + 4w^2$ **42.** $5t^2 + 14t - 3$

In Exercises 43–46, find the indicated products or quotients.

43. $\dfrac{x^4 - 81}{x} \cdot \dfrac{7x}{x^2 - 9}$ **44.** $\dfrac{4x^2}{x^2 - y^2} \cdot \dfrac{y^2 - x^2}{x}$

45. $\dfrac{x^2 + 10x + 21}{x^2 - 2x - 15} \div (x^2 + 2x - 35)$ **46.** $\dfrac{a + b}{a^4 - b^4} \div \dfrac{(a + b)^2}{a^2 - b^2}$

In Exercises 47–54, find the indicated sums.

47. $\dfrac{x}{4} + \dfrac{2x}{3}$ **48.** $\dfrac{1}{3} - \dfrac{5}{x}$

49. $\dfrac{x}{3} + 5$ **50.** $\dfrac{x - 3}{4} - \dfrac{5}{10}$

51. $\dfrac{x}{x^3 + 1} + \dfrac{x - y}{y}$ **52.** $\dfrac{10}{x^2 + x - 6} + \dfrac{3x}{x^2 - 4x + 4}$

53. $\dfrac{1}{x^2 - y^2} + \dfrac{1}{x^2 + 2xy + y^2}$ **54.** $\dfrac{3}{x + 5} + \dfrac{2}{x - 5} - 1$

In Exercises 55–58, solve the systems of equations.

55. $3x + 3y = 4$
$3x - 3y = -4$

56. $4x + 2y = 2$
$6x - 3y = 3$

57. $5x - 2y = 8$
$x + y = 7$

58. $3x + y - z = 2$
$x + 4y + z = 1$
$3x - y + z = 10$

In Exercises 59 and 60, use the quadratic formula or the method of completing the square to solve the equations.

59. $3x^2 = 14x - 8$

60. $\dfrac{x^2}{4} - 4x + 10 = 0$

1.2 FORMULAS AND FUNCTIONAL NOTATION

In this section we will examine a notational device that will make much of our work with calculus easier.

Instead of formulas of the form $y = $ (Expression in x) we often use the notation $f(x) = $ (Expression in x). In this form the corresponding values of the variables are easy to identify. Consider

$$f(x) = x^2 + 2x - 3$$

The idea of the notation is that if one replaces x by a new numerical or symbolic value on the left side then the same replacement is made on the right side. For example, if $f(x) = x^2 + 2x - 3$, then

$$f(3) = 3^2 + 2(3) - 3$$
$$= 9 + 6 - 3 = 12$$
$$f(-2) = (-2)^2 + 2(-2) - 3$$
$$= 4 - 4 - 3 = -3$$
$$f(10) = 10^2 + 2(10) - 3$$
$$= 100 + 20 - 3 = 117$$
$$f(-4) = (-4)^2 + 2(-4) - 3$$
$$= 16 - 8 - 3 = 5$$

The replacements need not be numerical. For example,

$$f(z) = z^2 + 2z - 3$$
$$f(k) = k^2 + 2k - 3$$
$$f(-w) = (-w)^2 + 2(-w) - 3 = w^2 - 2w - 3$$

Notice in the last example parentheses were used to assure correct substitution. This is even more important when the substitution is a polynomial. For example,

$$
\begin{aligned}
f(z + 2) &= (z + 2)^2 + 2(z + 2) - 3 \\
&= z^2 + 4z + 4 + 2z + 4 - 3 \\
&= z^2 + 6z + 5 \\
f(w - y) &= (w - y)^2 + 2(w - y) - 3 \\
&= w^2 - 2wy + y^2 + 2w - 2y - 3
\end{aligned}
$$

The substitution can even contain x. For example, if

$$f(x) = x^2 + 2x - 3$$

then

$$
\begin{aligned}
f(x + 3) &= (x + 3)^2 + 2(x + 3) - 3 \\
&= x^2 + 6x + 9 + 2x + 6 - 3 \\
&= x^2 + 8x + 12 \\
f(x^2 + 4) &= (x^2 + 4)^2 + 2(x^2 + 4) - 3 \\
&= x^4 + 8x^2 + 16 + 2x^2 + 8 - 3 \\
&= x^4 + 10x^2 + 21
\end{aligned}
$$

Naturally the letters used in the notation need not be f or x.

Notation   The functional notation for formulas:

$$g(a) = \text{Expression using placeholder variable}$$

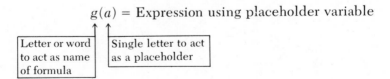

| Letter or word to act as name of formula | Single letter to act as a placeholder |

Example 1   The expression

$$h(k) = \frac{k}{k - 2}$$

defines a formula. For reference the formula's name is h. The placeholder, a dummy variable, is k. Note that

$$h(x) = \frac{x}{x-2} \qquad h(y) = \frac{y}{y-2} \qquad h(t) = \frac{t}{t-2}$$

all define the same formula.

Example 2

Assuming h is a formula as in Example 1,

$$h(x) = \frac{x}{x-2}$$

find $h(1)$, $h(\tfrac{3}{7})$, $h(-7)$, and $h(2)$.

Solution

$$h(1) = \frac{1}{1-2} = \frac{1}{-1} = -1$$

$$h(\tfrac{3}{7}) = \frac{3/7}{(3/7)-2} = \frac{3/7}{(3/7)-(14/7)}$$

$$= \frac{3/7}{-11/7}$$

$$= \frac{3}{7} \times \frac{7}{-11} = \frac{-3}{11}$$

$$h(-7) = \frac{-7}{-7-2} = \frac{-7}{-9} = \frac{7}{9}$$

$$h(2) = \frac{2}{2-2} = \frac{2}{0} \qquad \text{and is undefined}$$

Example 3

If

$$h(k) = \frac{k}{k-2}$$

find $h(t)$, $h(t^2 - 4)$, $h(x^2)$, and $h(x - 2)$.

Solution

$$h(t) = \frac{t}{t-2}$$

$$h(t^2 - 4) = \frac{(t^2 - 4)}{(t^2 - 4) - 2} = \frac{t^2 - 4}{t^2 - 6}$$

$$h(x^2) = \frac{x^2}{x^2 - 2}$$

$$h(x - 2) = \frac{(x-2)}{(x-2) - 2} = \frac{x-2}{x-4}$$

Example 4   Let

$$t(z) = \frac{z^2}{z^2 - 4}$$

Find $t(3)$, $t(-5)$, $t(x^2)$, and $t(x^3 - 2x)$.

Solution

$$t(3) = \frac{3^2}{3^2 - 4} = \frac{9}{9 - 4} = \frac{9}{5}$$

$$t(-5) = \frac{(-5)^2}{(-5)^2 - 4} = \frac{25}{25 - 4} = \frac{25}{21}$$

$$t(x^2) = \frac{(x^2)^2}{(x^2)^2 - 4} = \frac{x^4}{x^4 - 4}$$

$$t(x^3 - 2x) = \frac{(x^3 - 2x)^2}{(x^3 - 2x)^2 - 4}$$

$$= \frac{x^6 - 4x^4 + 4x^2}{x^6 - 4x^4 + 4x^2 - 4}$$

Example 5   If f is the name of a formula such that $f(x) = 2x + 3$, find a value for a such that $f(x + a) = 2x$.

Solution $f(x + a) = 2(x + a) + 3$
$\qquad\qquad = 2x + 2a + 3$

If this expression is to equal $2x$, then

$$2x + 2a + 3 = 2x$$
$$2a + 3 = 0$$
$$a = -\frac{3}{2}$$

We can develop more complex problems than the one above.

Example 6   Let h be a name of a formula such that $h(x) = ax + b$ where a and b are constants. Find specific values for a and b such that $h(3) = 4$ and $h(-1) = -2$.

Solution   If $h(3) = 4$, we also know that

$$h(3) = a(3) + b = 3a + b$$

Thus $3a + b = 4$. If $h(-1) = -2$, we also have

$$h(-1) = a(-1) + b = -a + b$$

Thus $-a + b = -2$.

Solving these as two linear equations with two unknowns (a and b),

$$3a + b = 4$$
$$-a + b = -2 \quad \text{Subtracting the second equation from the first}$$
$$4a = 6$$
$$a = \frac{6}{4} = \frac{3}{2}$$

Substituting for a in the second equation,

$$-\left(\frac{3}{2}\right) + b = -2$$
$$b = -2 + \frac{3}{2} = -\frac{1}{2}$$

Therefore, $a = \frac{3}{2}$, $b = -\frac{1}{2}$, and

$$h(x) = \frac{3}{2}x - \frac{1}{2}$$

Checking our answers,

$$h(x) = \frac{3}{2}x - \frac{1}{2}$$
$$h(3) = \frac{3}{2}(3) - \frac{1}{2}$$
$$= \frac{9}{2} - \frac{1}{2}$$
$$= \frac{8}{2} = 4$$
$$h(-1) = \frac{3}{2}(-1) - \frac{1}{2}$$
$$= \frac{-3}{2} - \frac{1}{2}$$
$$= \frac{-4}{2} = -2$$

The solution checks.

Example 7   Let f be a name of a formula such that $f(x) = x^2 - 2$. Find and simplify $f(x + h)$ and

$$\frac{f(x + h) - f(x)}{h}$$

Solution
$$f(x) = x^2 - 2$$
$$f(x + h) = (x + h)^2 - 2 = x^2 + 2xh + h^2 - 2$$
$$f(x + h) - f(x) = x^2 + 2xh + h^2 - 2 - (x^2 - 2)$$
$$= x^2 + 2xh + h^2 - 2 - x^2 + 2$$
$$= 2xh + h^2$$

Therefore

$$\frac{f(x + h) - f(x)}{h} = \frac{2xh + h^2}{h} = \frac{h(2x + h)}{h}$$
$$= 2x + h$$

1.2   EXERCISES

In Exercises 1–5, let f be a formula defined by $f(x) = x^2 - 4x + 4$. Find:

1. $f(1)$ and $f(-1)$ **2.** $f(2)$ and $f(-2)$

3. $f(\frac{7}{2})$ and $f(-\frac{7}{2})$ **4.** $f(\frac{1}{3})$ and $f(-\frac{1}{3})$

5. $f(z)$ and $f(-z)$

In Exercises 6–10, let h be a formula defined by $h(k) = \dfrac{k}{k - 5}$. Find:

6. $h(0)$ and $h(2)$ **7.** $h(-3)$ and $h(\frac{3}{2})$

8. $h(5)$ and $h(-5)$ **9.** $h(x^2)$ and $h(x^3)$

10. $h(k + 1)$ and $h(k - 1)$

In Exercises 11–15, let $g(x) = \sqrt{x - 1}$. Find:

11. $g(x + 1)$ and $g(x - 1)$ **12.** $g(x^2)$ and $g(-x^2)$

13. $[g(-x)]^2$ and $g(x - 3)$ **14.** $g(x^2 + 2x + 1)$ and $g\left(\dfrac{1}{x}\right)$

15. $g\left(\dfrac{x}{x - 5}\right)$ and $g\left(\dfrac{x - 5}{x}\right)$

In Exercises 16–20, let g be a formula defined by $g(y) = \dfrac{1}{2(y - 1)}$.

Find:

16. $g(3)$ and $g(-3)$

17. $g(y + 2)$ and $g(y - 2)$

18. $g(x)$ and $g(-x)$

19. $g\left(\dfrac{1}{x - 1}\right)$ and $g(y - 1)$

20. $g\left(\dfrac{1}{2(y - 1)}\right)$ and $g(2(y - 1))$

In Exercises 21–25, let h be a formula defined by $h(x) = -x - 3$. Find:

21. $h(x + 1)$

22. $h(x^2 + \frac{1}{2})$

23. $h(x + d)$

24. $\dfrac{h(x + d) - h(x)}{d}$

25. $\dfrac{h(6 - x) - h(6)}{x}$

In Exercises 26 and 27, if g is a formula defined by $g(k) = ak + b$, and a and b are constants, find:

26. a and b such that $g(0) = 2$ and $g(1) = 2$.

27. a and b such that $g(0) = 0$ and $g(-5) = -5$.

In Exercises 28–33, let f be a formula given by $f(x) = \dfrac{1}{2x + 1}$ and g be a second formula given by $g(x) = \sqrt{1 - x}$.

28. Find $g(-8)$ and $f(g(-8))$.

29. Find $f(-\frac{5}{2})$ and $g(f(-\frac{5}{2}))$.

30. Using the symbolic pattern, find $g(z)$ and $f(g(z))$.

31. Find $f(y)$ and $g(f(y))$.

32. Find $f(x^2 + 2)$ and $g(f(x^2 + 2))$.

33. Find $g(x^2 + 2)$ and $f(g(x^2 + 2))$.

In Exercises 34 and 35, consider formulas h, j, and s given by $h(k) = k^2, j(l) = l + 2,$ and $s(t) = \sqrt{t^2 + 1}$.

34. Find $h(j(s(3)))$.

35. Find $j(h(s(-1)))$.

1.3 FUNCTIONS

It is reasonable to assume that any model attempting to describe the profit made from the sales of an item should reflect in some manner the relationship between profit and the number of items produced. Suppose, on one hand, the more items produced, the cheaper it is to make each individual item and, on the other hand, the more of the item on the market, the lower the price consumers are willing to pay. In mathematical terms one would say that profit is a *function* of the number of items produced. In a similar way, a mathematical model attempting to describe voter willingness to pass a school bond may be found to depend on the current tax rate. Again in mathematical terms, the size of the favorable vote may be viewed as a *function* of the tax rate. The idea of two or more variables being dependent on each other is fundamental to any application of mathematical concepts to real life.

Although the concept of a function is fundamental to the study of calculus, there are several equivalent ways in which the term *function* may be defined. We will use the following.

Definition

A *function* is a *rule of correspondence* that associates each number from a set of numbers called the *domain* with one and only one number from a set called the *range*.

In other words, if we use x and y as symbols for the numbers involved, with x the independent number and y the corresponding dependent value, the set of values that may be substituted for x in the functional rule is called the *domain*. The corresponding set of values of y that these substitutions create is called the *range* of the function.

Example 1

If x is a number taken from the domain and y is the corresponding range value, are the following equations suitable as functional rules?

$$y = 2x - 6 \qquad y = \pm \sqrt{x}$$

Solution

The equation $y = 2x - 6$ defines a function, because for any real number, x, there is produced one and only one real number, y.

The equation $y = \pm \sqrt{x}$ is not a function, since for any positive x there are two corresponding y values; y is not unique as it would have to be for a function. For instance, if $x = 16$ then $y = +4$ or $y = -4$.

Functions may be defined by sets of ordered pairs where it is understood that the first elements of the ordered pairs are from the domain of the function and the second from its range.

Example 2	Which of the sets of ordered pairs are functions?

$$R = \{(2, 4), (3, 6), (2, 5)\} \qquad G = \{(1, 5), (2, 5), (3, 5)\}$$

Solution   The set R is not a functional rule because two different elements in the range (4 and 5) correspond to the element 2 in the domain.

The set G does define a function because every element in the domain corresponds to one and only one element in the range.

Definition

Domain elements are called *arguments* of a function; range elements are called *images*.

When a function is defined by an equation, the domain and range of the function should be specified separately. However, unless specified otherwise, we will assume that the domain of a function defined by an equation, such as $y = 2x + 1$, is the set of all real numbers for which there is a meaningful application of the equation to produce a real number image. The range will consist of all such real number images.

An alternate definition of a function can be given in terms of ordered pairs.

Definition

A *function* is a set of ordered pairs such that no two distinct ordered pairs have the same first component. The set of all elements used as first components is called the *domain* of the function, while the set of all elements used as second components is its *range*.

This is an equivalent definition and presents a useful alternate view of a function.

Example 3   Can $y = \dfrac{1}{x - 3}$ be used to define a function?

Solution   Yes, the domain of the function is all real numbers x, such that $x \neq 3$ (3 must be excluded, because replacement of x by 3 would result in division by zero). The range is all real numbers y such that $y \neq 0$.

From this point on references to the fact that the numbers involved are real numbers will be dropped. *It is understood that only real numbers are being considered.*

Example 4   Consider $y = \sqrt{x - 4}$. Here the domain is x such that $x \geqslant 4$; this is required to produce real number images. The range is y such that $y \geqslant 0$.

Notation of the form $f(x)$, $g(x)$, or $h(x)$ is commonly used to denote the element in the range which corresponds to the element x in the domain of the function.

This is the same point we made about this symbolism in the previous section. The function name can be f or g or h or anything else, and $f(x)$, $g(x)$, and $h(x)$ each stands for a single number in the range of the function. The notation $f(x)$ stands for the number in the range that the function named f associates with the number x in the domain. The symbol $f(x)$ is read "f of x"; it denotes the image of the argument x under the function f.

Example 5   Let x and y be real numbers. Let $y = 2x^2 + 3x + 1$ or $y = f(x)$. Then

$$f(x) = 2x^2 + 3x + 1$$
$$f(1) = 2(1)^2 + 3(1) + 1 = 6$$
$$f(2) = 2(2)^2 + 3(2) + 1 = 15$$
$$f(-2) = 2(-2)^2 + 3(-2) + 1 = 3$$
$$f(x + 2) = 2(x + 2)^2 + 3(x + 2) + 1 = 2x^2 + 11x + 15$$
$$f(x + h) = 2(x + h)^2 + 3(x + h) + 1$$
$$= 2x^2 + 4xh + 2h^2 + 3x + 3h + 1$$

Note that the value of y, or $f(x)$, depends on the value assigned to x. The variable x is, therefore, called the *independent* variable, and y is the *dependent* variable.

A graph is a useful representation of functions and nonfunctions. The graph in Figure 1.1 represents the function $y = x^2 + 2$, or $f(x) = x^2 + 2$. It is easily seen that for every x there exists a unique y, and the graph can be recognized as a function.

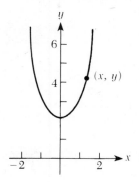

Figure 1.1   Graph of $y = x^2 + 2$

Figure 1.2 is the graph of a curve whose equation is *not* a function, since the points $(4, 2), (4, -2), (1, 1)$, and $(1, -1)$ satisfy the equation. The equation $x = y^2$ may be defined as *two* functions: $y = \sqrt{x}$, the portion of the graph on or above the x axis, and $y = -\sqrt{x}$, the portion of the graph on or below the x axis. The point $(0, 0)$ belongs to both functions. (See Figure 1.3.)

It is reasonable to study complicated functional relationships, such as one might find in constructing a mathematical model of profit or sales, by breaking them down into their simpler compo-

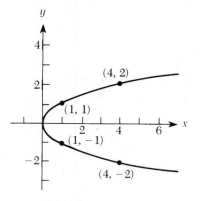

Figure 1.2   Graph of $x = y^2$

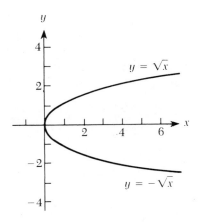

Figure 1.3　Graph of $y = \sqrt{x}$ and $y = -\sqrt{x}$

nents. To see how this approach would work, it will be necessary to examine how complicated functions can be built up from basic functions.

There are five operations defined on functions. They can be used to build up simple functions into more complicated functions. Four of these operations are analogous to addition, subtraction, multiplication, and division as defined on numbers. The fifth operation, the *composition* of two functions, is somewhat unique and does not correspond to an operation on real numbers.

Definition

Let f and g be functions in x. Then

Image		Operation	Function name
$(f + g)(x) = f(x) + g(x)$		Sum	$f + g$
$(f - g)(x) = f(x) - g(x)$		Difference	$f - g$
$(f \cdot g)(x) = f(x) \cdot g(x)$		Product	$f \cdot g$
$(f/g)(x) = f(x)/g(x)$	$g(x) \neq 0$	Quotient	f/g

The domain of each of these functions is the intersection of the domain of f and the domain of g, the numbers common to both domains. The domain of the quotient function excludes values which result in a zero denominator.

Example 6

Let $f(x) = \dfrac{1}{2x + 1}$ and $g(x) = \sqrt{1 - x}$ define functions f and g. Find $f + g$, $f - g$, $f \cdot g$, and f/g, and examine the domains.

Solution

$$(f + g)(x) = \frac{1}{2x + 1} + \sqrt{1 - x}$$

$$(f - g)(x) = \frac{1}{2x + 1} - \sqrt{1 - x}$$

$$(f \cdot g)(x) = \frac{\sqrt{1 - x}}{2x + 1}$$

$$\left(\frac{f}{g}\right)(x) = \frac{1}{(2x + 1)\sqrt{1 - x}}$$

The domain of f is $D_f = \{x \mid x \neq -\frac{1}{2}\}$. The domain of g is $D_g = \{x \mid x \leq 1\}$. The domain of $f + g$, $f - g$, and $f \cdot g$, is $\{x \mid x \leq 1$ and $x \neq -\frac{1}{2}\}$. The domain of f/g is $\{x \mid x < 1$ and $x \neq -\frac{1}{2}\}$.

Definition

Let f and g be functions in x. Then the *composite function*, $f \circ g$ (read "f circle g"), is defined by

$$(f \circ g)(x) = f(g(x))$$

The domain of $f \circ g$ consists of all values of x in the domain of g for which $g(x)$ is in the domain of f.

Example 7

If $f(x) = \dfrac{1}{2x + 1}$ and $g(x) = \sqrt{1 - x}$, then

$$(f \circ g)(x) = f(g(x)) = \frac{1}{2\sqrt{1 - x} + 1}$$

The domain of $f \circ g$ is $\{x \mid x \leq 1\}$.

Example 8

Let $f(x) = 2x^2 + 3x + 1$ and $g(x) = x + 2$. Evaluate $f(g(2))$, $g(f(2))$, $f(g(-2))$, and $g(f(-2))$.

Solution

$$g(2) = 4$$

$$f(g(2)) = f(4) = 2(4)^2 + 3(4)$$
$$+ 1 = 45$$

$$f(2) = 2(2)^2 + 3(2) + 1 = 15$$

$$g(f(2)) = g(15) = 15 + 2$$
$$= 17$$

$$g(-2) = -2 + 2 = 0$$

$$f(g(-2)) = f(0) = 1$$

$$f(-2) = 2(-2)^2 + 3(-2) + 1 = 3$$

$$g(f(-2)) = g(3) = 3 + 2 = 5$$

1.3 EXERCISES

In Exercises 1–9, state the domain of each of the given functions.

1. $y = 3x + 6$ **2.** $y = x^2$ **3.** $f(x) = 3x^3 + 4x^2$

4. $g(x) = \sqrt{x}$ **5.** $g(x) = \sqrt{x - 6}$ **6.** $h(x) = \dfrac{1}{x}$

7. $h(x) = \dfrac{1}{x + 2}$ **8.** $y = \dfrac{1}{\sqrt{x + 2}}$ **9.** $y = \dfrac{x}{\sqrt{x - 5}}$

In Exercises 10–15, determine which of the following sets or equations define functions. Give the domain and range for each function. If an expression is not a function, justify your decision.

10. $f = \{(1, 2), (3, 4), (5, 6), (7, 8)\}$

11. $g = \{(2, 2), (3, 3), (4, 4), (-1, -1)\}$

12. $y = 2x + 1$ **13.** $y = 2x^2 - 4x + 1$

14. $f(x) = \sqrt{x^2 - 3}$ **15.** $f(x) = \pm \sqrt{16 - x^2}$

In Exercises 16–21, if $f(x) = 3x^2 - 2x + 1$, find:

16. $f(2)$ **17.** $f(-1)$ **18.** $f(-3)$

19. $f(a)$ **20.** $f(x + h)$ **21.** $f(x + h) - f(x)$

In Exercises 22–27, if $f(x) = \frac{1}{3}x^3 - 2x + 4$, find:

22. $f(-3)$ **23.** $f(0)$ **24.** $f(1)$

25. $f(x + 1)$ **26.** $f(x + h)$ **27.** $\dfrac{f(x + h) - f(x)}{h}$

In Exercises 28–35, graph each of the following equations and state whether they can be used to define functional rules.

28. $y = 2x + 1$ **29.** $y = (x - \pi)^2 + 2$

30. $y = |x + 1|$

31. $y = \dfrac{1}{x^2}$

32. $x^2 + y^2 = 1$

33. $y = \sqrt{x^2 + 4}$

34. $f(x) = \begin{cases} 1 & \text{when } x > 0 \\ 0 & \text{when } x = 0 \\ -1 & \text{when } x < 0 \end{cases}$

35. $f(x) = \dfrac{x^2 - 4}{x + 2}$ when $x \neq -2$, and $f(-2) = -4$

36. A box manufacturing company wishes to design a lidless box of maximum volume from a 5 by 7 inch rectangular piece of cardboard by cutting equal-sized squares from each corner and folding the edges up. Express the volume of the box as a function of its depth.

37. A certain item manufactured by the Little Profit Manufacturing Company yields a profit expressed by the function rule

$$f(x) = -500 + 0.35x$$

where x is the number of items manufactured. Determine the profit when $x = 1000$, 10,000, or 100,000. For what value of x is $f(x) = 0$? How would you interpret the statement $f(x) = 0$ in terms of profit and the number of items manufactured?

The revenue produced from the sale of an item is a function of the number of items produced. In Exercises 38–41, you are given an expression relating revenue, R, to the number of items produced, x. In each exercise, define the domain of the function based on the equation given and its real-world domain, and sketch the function.

38. The revenue in thousands of dollars from sales of a talking doll is given by $R(x) = 2.1x^2 - 2\sqrt{x^4 + 3}$, where x is in hundreds of dolls.

39. The revenue from vodka sales in Moscow is given in thousands of rubles by $R(x) = \frac{1}{3}x^2 - 2x + 4$, where x is in kiloliters.

40. From the sale of x 10-pound tins of tobacco, the revenue of a store in London is given in pounds sterling by

$$R(x) = 4(x - \sqrt{x})$$

41. The revenue in dollars from the sales of x Volkswagens in New York is given by $R(x) = 895x + 895\sqrt{x^2 - 400}$, where $x \geq 20$.

In Exercises 42–46, the sketch of a possible function is given. Based on the information given by the sketch determine if y has a functional relationship with x.

42.

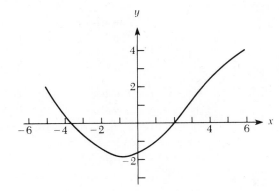

43.

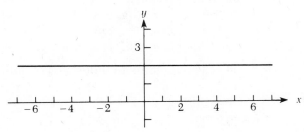

44.

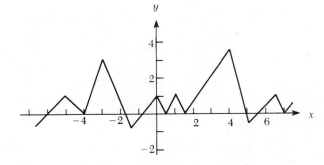

45.

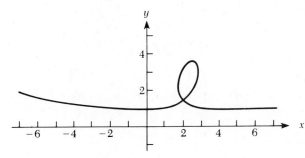

46.

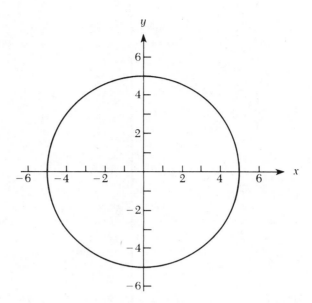

In Exercises 47–51, find the sum, difference, product, and quotient of the functions f and g defined with the given rules, and state the domains of the resulting functions.

47. $f(x) = x^2 + 3x$ $\qquad$ $g(x) = 2x + 1$

48. $f(x) = \dfrac{1}{1 - x}$ $\qquad$ $g(x) = \sqrt{x}$

49. $f(x) = \sqrt{x^2 - 1}$ $\qquad$ $g(x) = x + 2$

50. $f(x) = x^2$ $\qquad$ $g(x) = 1 + \sqrt{x}$

51. $f(x) = 1 - \sqrt{x}$ $\qquad$ $g(x) = x^3$

As Exercises 52–56 find $f \circ g$ for the functions defined in Exercises 47–51, and state the domain of $f \circ g$ for each exercise.

As Exercises 57–61, find $g \circ f$ for the functions defined in Exercises 47–51, and state the domain of $g \circ f$ for each exercise. Compare the answers with those of Exercises 52–56. Is the composition operation commutative?

In Exercises 62–68, let $f(x) = 1 - \sqrt{x}$ and $g(x) = 1/x$, and evaluate each one.

62. $f(g(4))$ $\qquad$ **63.** $g(f(4))$

64. $f(g(1))$ $\qquad$ **65.** $g(f(1))$

66. $f(g(x + h))$, $h > 0$ $\qquad$ **67.** $g(f(x + h))$, $h > 0$

68. $g(f(0))$

69. The function f defined by $f(x) = [x]$ is called the greatest integer function, and the $[\ ]$ mean that $f(x)$ is the greatest integer less than or equal to x. Thus $f(1) = 1, f(\frac{1}{2}) = 0, f(-\frac{1}{2}) = -1$, and $f(2.999) = 2$. Graph this function for $-2 \leqslant x \leqslant 5$.

70. If $f(x) = [x]$, and $g(x) = [x + 1]$, find the value of:

    a. $(f + g)(2.3)$ **b.** $(f - g)(2.3)$

    c. $(f \cdot g)(2.3)$ **d.** $\left(\dfrac{f}{g}\right)(2.3)$

    e. $f(g(2.3))$

$[\$]$ 71. The greatest integer function can be used to solve this problem: A taxi charges 25 cents for the first mile of fare, and 10 cents for each subsequent half mile or part thereof. Graph this function for a journey of five miles, and from your graph evaluate the fare for the first 4.3 miles of travel.

1.4 FUNCTIONS AND INTERVAL NOTATION

Functional rules need not be in the form of a single equation. In fact, they need not be in the form of equations at all. The classic example of functions defined with other than an equation are the trigonometric functions. They will be considered in Chapter 5. In this section we will consider functions whose defining equations change, depending on where we take the argument.

Example 1   Let f be a function defined by the following rule:

$$f(x) = \begin{cases} x & \text{if } x \text{ is less than or equal to zero} \\ x^2 + 2 & \text{if } x \text{ is greater than zero but less than or equal to 4} \\ -x^2 & \text{if } x \text{ is greater than 4} \end{cases}$$

What is the domain of this function? Also find $f(-2), f(3)$, and $f(5)$.

Solution   While the functional rule uses three equations, it is possible to find an image for *any* real number. Therefore, the domain is all real numbers.

$f(-2)$ is calculated using the first part of the rule since -2 is less than zero

$f(-2)$ $= -2$

$f(3)$ is calculated using the second part of the rule since 3 is between zero and 4

$f(3)$ $= 3^2 + 2 = 9 + 2 = 11$

$f(5)$ is calculated using the third part of the rule since 5 is greater than 4

$f(5)$ $= -(5)^2 = -25$

It is important to note that f is a *single function;* only the functional rule has more than one part.

Before examining more functions with multipart rules it will be very useful to develop some notation that can be used to make the descriptions of such functions more compact.

Definitions

$[a, b]$ will be used to denote all the real numbers between a and b including a and b; the notation $[a, b]$ is called a *closed interval*

(a, b) will be used to denote all the real numbers between a and b *not* including a or b; the notation (a, b) is called an *open interval*

$[a, b)$ will be used to denote all the real numbers between a and b including a but not b; the notation $[a, b)$ is called a *half-open* or a *half-closed interval*

$(a, b]$ is defined in a manner similar to $[a, b)$ but including b but not a; it is also called a half open or half-closed interval

---

Note that the open interval notation is the same as the notation normally used for ordered pairs in analytic geometry (that is, points in a plane and coordinates). It is usually clear from the context which is which.

Example 2   The real numbers between -2 and 3 could be described as

$[-2, 3]$ if -2 and 3 are to be included

$[-2, 3)$ if -2 is to be included and 3 not included

$(-2, 3]$ if 3 is to be included and -2 not included

$(-2, 3)$ if neither -2 nor 3 is to be included

Notation $x \in [a, b]$ is equivalent to $a \leq x \leq b$

$x \in [a, b)$ is equivalent to $a \leq x < b$

$x \in (a, b]$ is equivalent to $a < x \leq b$

$x \in (a, b)$ is equivalent to $a < x < b$

(a, ∞) denotes all real numbers greater than a (∞ is read "infinity")

$[a, \infty)$ denotes all real numbers greater than or equal to a

$(-\infty, a)$ denotes all real numbers less than a

$(-\infty, a]$ denotes all real numbers less than or equal to a

Note that the symbol $(a, \infty]$ is nonsense, because it indicates that ∞ is included in the set, but infinity is not a real number and cannot be included in any set of real numbers.

A geometric picture of a real number interval is the number line. Figure 1.4 illustrates the first four types of intervals. The solid dots indicate that the point at the end is to be included. The open circles indicate that the point is *not* to be included.

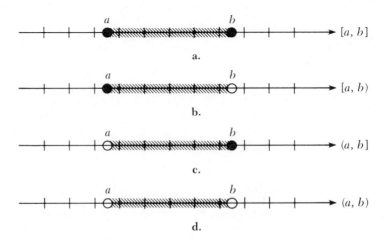

Figure 1.4   The four types of intervals illustrated on a number line

Example 3	The functional rule described in Example 1 can be restated in either of the following ways: f is a function whose rule is given by

$$f(x) = \begin{cases} x & \text{if } x \in (-\infty, 0] \\ x^2 + 2 & \text{if } x \in (0, 4] \\ -x^2 & \text{if } x \in (4, \infty) \end{cases}$$

or we might define f as follows

$$f(x) = \begin{cases} x & \text{if } x \leq 0 \\ x^2 + 2 & \text{if } 0 < x \leq 4 \\ -x^2 & \text{if } x > 4 \end{cases}$$

Example 4	Consider the function g, whose rule is given by

$$g(x) = \begin{cases} -x & \text{if } x \in (-\infty, 0] \\ x & \text{if } x \in (0, \infty) \end{cases}$$

Find $g(1)$, $g(15)$, $g(-3)$, and $g(-5)$.

Solution
$$g(1) \ = 1 \qquad \qquad \text{since } 1 \in (0, \infty)$$
$$g(15) \ = 15 \qquad \quad \text{since } 15 \in (0, \infty)$$
$$g(-3) = -(-3) = 3 \quad \text{since } -3 \in (-\infty, 0]$$
$$g(-5) = -(-5) = 5 \quad \text{since } -5 \in (-\infty, 0]$$

Interval notation can be used to describe the domain of a function.

Example 5	Let h be a function such that

$$h(x) = \sqrt{x^2 + 2} \quad \text{for } x \in [0, 7]$$

Find $h(6)$, $h(\sqrt{3})$, and $h(10)$.

Solution
$$h(6) \ = \sqrt{6^2 + 2} = \sqrt{36 + 2} = \sqrt{38} \approx 6.16$$
$$h(\sqrt{3}) = \sqrt{(\sqrt{3})^2 + 2} = \sqrt{3 + 2} = \sqrt{5} \approx 2.24$$

$h(10)$ does not exist! While we could calculate $\sqrt{10^2 + 2}$, the definition of h is valid only for numbers in the $[0, 7]$ interval, and 10 is not between 0 and 7.

Example 6	Consider the function g, whose rule is given by $g(x) = \sqrt{6 + x - x^2}$. Describe the domain of this function using interval notation.
Solution	For a real number x to be in the domain of g,

$$6 + x - x^2 \geqslant 0$$

Factoring,

$$(3 - x)(2 + x) \geqslant 0$$

$3 - x$ is positive or zero if x is 3 or less, or $x \in (-\infty, 3]$

$2 + x$ is positive or zero if x is -2 or greater, or $x \in [-2, \infty)$

The factors will agree in sign and thus the product will be positive or zero only if $x \in [-2, 3]$. (See Figure 1.5.)

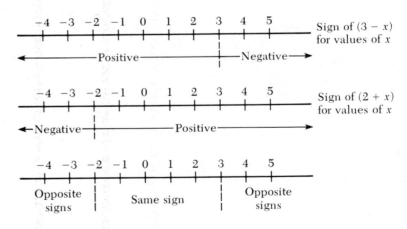

Figure 1.5

We can make up complicated problems, and give the answer using intervals and interval notation.

Example 7	Let f be a function given by $f(x) = \sqrt{6 + x - x^2}$ and g be a function given by $g(x) = 3x + 2$. Find the domain of $f \circ g$.

Solution First let us try a specific example.

$$g(-1) = 3(-1) + 2 = -1$$
$$f(g(-1)) = f(-1) = \sqrt{6 + (-1) - (-1)^2} = \sqrt{4} = 2$$

Not all numbers in the domain of g produce images in the domain of f. For example,

$$g(2) = 3 \cdot 2 + 2 = 8$$
$$f(g(2)) = f(8)$$

which is undefined; in other words, 8 is not in the domain of f.

The composite function requires restrictions on the original domain of g. We have to restrict ourselves to numbers in the domain of g that have images in $[-2, 3]$, the domain of f.

Specifically, x as an argument of g must produce an image such that $(3x + 2) \in [-2, 3]$, or translating into inequalities,

$$-2 \leqslant 3x + 2 \leqslant 3$$
$$-4 \leqslant 3x \leqslant 1$$
$$-\frac{4}{3} \leqslant x \leqslant \frac{1}{3} \quad \text{or} \quad x \in \left[-\frac{4}{3}, \frac{1}{3} \right]$$

This is the domain of $f \circ g$.

We can sketch the graphs of functions with multiple equation definitions without too much difficulty.

Example 8 Consider f a function defined by

$$f(x) = \begin{cases} x & x \in (-\infty, 0] \\ x^2 + 2 & x \in (0, 4] \\ -x^2 & x \in (4, \infty) \end{cases}$$

Graph this function.

Solution Figure 1.6 shows the graph of three equations, $y = x$, $y = x^2 + 2$, and $y = -x^2$. The graph of f matches that of $y = x$ if $x \in (-\infty, 0]$. It matches that of $y = x^2 + 2$ if $x \in (0, 4]$ and it matches that of $y = -x^2$ if $x \in (4, \infty)$. These portions have been shaded in Figure 1.6. The three graphs have been combined in Figure 1.7. The closed and open dots show included or not included points respectively.

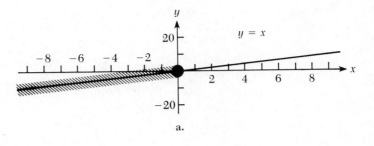

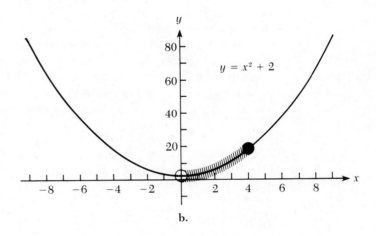

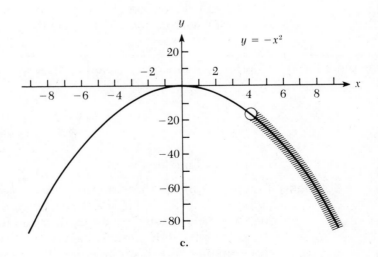

Figure 1.6   The graphs of **a.** $y = x$; **b.** $y = x^2 + 2$; and **c.** $y = -x^2$

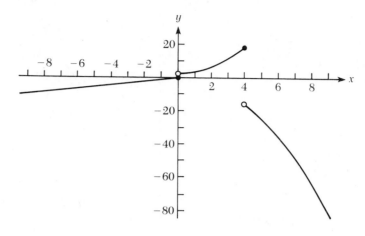

Figure 1.7 The graph of $f(x) = \begin{cases} x & x \in (-\infty, 0] \\ x^2 + 2 & x \in (0, 4] \\ -x^2 & x \in (4, \infty) \end{cases}$

1.4 EXERCISES

In Exercises 1–5, let f be a function defined by

$$f(x) = \begin{cases} -x - 2 & \text{for } x < -1 \\ -x & \text{for } -1 \leq x < 2 \\ x + 2 & \text{for } x \geq 2 \end{cases}$$

1. Restate the definition of this function in interval notation.
2. Find $f(-1), f(-2)$, and $f(-3)$.
3. Find $f(2)$ and $f(6)$.
4. Find $f(a)$ if $a \geq 10, f(b)$ if $b \leq -10$.
5. Find $f(-3), f(3), f((-3)^2)$, and $f((3)^2)$.

In Exercises 6–10, write the interval notation for:

6. The real numbers between 3 and 7 excluding both 3 and 7.
7. The real numbers between -4 and -5 including both numbers.
8. The real numbers between 100 and 200 including the smaller value, but not the larger.

9.   The real numbers between $-\frac{1}{2}$ and $\frac{1}{2}$, a half-open interval excluding the upper value.

10.   An open interval 5 units wide centered on -3.

As Exercises 11–15, sketch on a number line the intervals described in Exercises 6–10.

In Exercises 16–20, sketch the intervals on a number line and state their meaning in words.

16. $[3, 4)$ **17.** $(-4, 8)$ **18.** $(-\frac{1}{2}, 7]$
19. $[-3, 3]$ **20.** $(-1, 10)$

In Exercises 21–25, use the infinity symbol and interval notation to write:

21.   All real numbers x such that $x \geqslant -1$.

22.   All real numbers y such that $y \leqslant 3$.

23.   All positive real numbers.

24.   All nonnegative real numbers.

25.   All nonpositive real numbers.

In Exercises 26–30, translate into interval notation:

26. $-1 < x \leqslant 3$ **27.** $-1 \leqslant x < 3$ **28.** $-1 < x < 3$
29. $-1 \leqslant x \leqslant 3$ **30.** $-1 < x \leqslant \frac{5}{2}$

In Exercises 31–35, translate into inequality notation:

31. $x \in [-1, 5]$ **32.** $x \in [-1, 5)$ **33.** $x \in (-1, 5]$
34. $x \in (-1, 5)$ **35.** $x \in [-\frac{1}{2}, \frac{1}{2}]$ but $x \neq 0$

In Exercises 36–40, there is something about each exercise that does not make sense. What is it?

36. $x \in [-2, -4]$ **37.** $x \in (-2, \infty]$ **38.** $x \in [-\infty, 0)$
39. $x \in (2, -1)$ **40.** $x \in (1, -4)$

In Exercises 41–46, let f and g be defined as follows:

$$f(x) = \begin{cases} -x & \text{for } x \in (-\infty, 0] \\ x & \text{for } x \in (0, \infty) \end{cases}$$

$$g(x) = \begin{cases} x - 2 & \text{for } x \in [1, 5] \\ x + 2 & \text{for } x \in [-1, 1) \end{cases}$$

41.   Find $f(3), f(-3),$ and $f(-x^2)$.

42.   Find $g(3), g(-3),$ and $g(5)$.

43. For what values of x is $f(x^2)$ defined?

44. For what values of x is $g(x^2)$ defined?

45. For what values of x is $f(g(x))$ defined?

46. For what values of x is $g(f(x))$ defined?

In Exercises 47–50, sketch the functions.

47. f defined by $f(x) = \begin{cases} x^2 & \text{for } x \geqslant 0 \\ -x^2 & \text{for } x < 0 \end{cases}$

48. g defined by $g(x) = \begin{cases} 2 - x & \text{for } x \in (-\infty, 2] \\ x - 2 & \text{for } x \in (2, \infty) \end{cases}$

49. h defined by $h(x) = \begin{cases} \sqrt{x^2 - 9} & \text{for } x \in (-\infty, -3) \\ \sqrt{9 - x^2} & \text{for } x \in [-3, 3] \\ \sqrt{x^2 - 9} & \text{for } x \in (3, \infty) \end{cases}$

50. k defined by $k(x) = \begin{cases} x + 3 & \text{for } x \in (-\infty, -3] \\ \sqrt{9 - x^2} & \text{for } x \in (-3, 3) \\ x - 3 & \text{for } x \in [3, \infty) \end{cases}$

In Exercises 51 and 52, consider the function f defined by

$$f(x) = \begin{cases} \frac{1}{2}x + 2 & \text{for } x \in [-5, -2] \\ x + 1 & \text{for } x \in (-2, 4) \\ x & \text{for } x \in [4, 10] \end{cases}$$

51. Sketch f.

52. Find the values of $f(0), f(7)$, and $f(11)$.

In Exercises 53–56, consider the function g defined as

$$g(x) = \begin{cases} \vdots \\ -4 & \text{for } x \in [-4, -3) \\ -3 & \text{for } x \in [-3, -2) \\ -2 & \text{for } x \in [-2, -1) \\ -1 & \text{for } x \in [-1, 0) \\ 0 & \text{for } x \in [0, 1) \\ 1 & \text{for } x \in [1, 2) \\ 2 & \text{for } x \in [2, 3) \\ \vdots \end{cases}$$

The dots indicate the pattern is repeated endlessly.

53. Find $g(\frac{3}{2})$, $g(\sqrt{3})$, $g(-\sqrt{3})$, and $g(-\pi)$.

54. Sketch g.

In Exercises 55 and 56, let f be a function defined by

$$f(x) = \begin{cases} -x & \text{if } x \in (-\infty, 0] \\ x & \text{if } x \in (0, \infty) \end{cases}$$

55. Sketch $f \circ g$ **56.** Sketch $g \circ f$

1.5 INVERSE FUNCTIONS

The functional example of Celsius and Fahrenheit temperature scales considered in Section 1.1 illustrates the key idea of this section. Consider these formulas:

$$°C = \frac{5}{9} (°F - 32) \qquad °F = \frac{9}{5} °C + 32$$

They both define the same pairings. In fact if we were to replace °C by y and °F by x in the first formula we would have $y = \frac{5}{9}(x - 32)$. Suppose, then, we interchange x and y in this formula (after all, we like x to be our independent variable and y our dependent variable). We would then have

$$x = \frac{5}{9} (y - 32)$$

"Solving" this for y,

$$9x = 5(y - 32)$$
$$\frac{9x}{5} = y - 32$$
$$y = \frac{9x}{5} + 32$$

which is what we get when we replace °F by y and °C by x in the second formula. In summary, we could think of the two formulas and the corresponding functions as really being defined by a single formula of the form

$$\Box = \frac{5}{9} (\bigcirc - 32)$$

If we put y in the box and x in the circle we get a functional relationship. If we put x in the box and y in the circle (and then "solve" for y) we get a second functional relationship. This is the idea of an *inverse*.

For any particular function there is a second relationship formed by interchanging the domain and range elements in the defining functional rule. The new relationship thus generated is called the *inverse* of the original function. For example, the inverse of the functional rule $y = 3x + 6$ is $x = 3y + 6$. If this second rule is also a functional rule then the corresponding function is called the *inverse function* of the original. One can interchange the domain and range roles in any functional rule, but one does not always get a functional rule as a result.

Definition

If $y = f(x)$ and $x = f(y)$ both define functional rules (with x the argument and y the image), then the corresponding functions are the *inverse* functions of each other.

Example 1

Since $y = x^3$ *defines* a function as does $x = y^3$ then the two functions involved are the inverse of each other; that is, $x = y^3$ (or $y = \sqrt[3]{x}$) and $y = x^3$ are inverse functional rules.

Example 2

The inverse functional rule of $y = -9x + 4$ is $x = -9y + 4$.

Note that the essential move in creating an inverse functional rule is the interchanging of the positions of x and y in the defining rule. The variable y is then usually solved for in terms of x.

Example 3

Suppose that we have a function f such that $y = f(x) = 2x + 4$. Find the inverse functional rule.

Solution

First we will interchange x and y. That is $y = 2x + 4$ becomes $x = 2y + 4$. Then solve for y.

$$x - 4 = 2y$$
$$y = \frac{x - 4}{2}$$

This is the equation we want; it defines the inverse function.

Notation   Inverse function pairs are denoted as f and f^{-1} (read "f inverse"). The corresponding images are $f(x)$ and $f^{-1}(x)$.

Often we get careless and talk about "the function $f(x) =$ Expression in x." Actually it is important to note that this means the function whose name is f and whose functional rule is given by $f(x) =$ Expression in x. It is usually clear which is meant from the context of the discussion.

The -1 in f^{-1} is a notational device and does not mean the function is to be raised to the -1 power. The notation of $f(x)$ to the negative first power will be written $[f(x)]^{-1}$.

If a function is given as a set of ordered pairs, its inverse is found by interchanging the coordinates of the pairs.

Example 4   If $R = \{(1, 3), (6, 2), (5, 4), (0, 8)\}$, find R^{-1}.

Solution $R^{-1} = \{(3, 1), (2, 6), (4, 5), (8, 0)\}$

Example 5   Find $f^{-1}(x)$ for $y = f(x) = x^3$.

Solution   We interchange x and y in $y = x^3$ and solve for y.

$$x = y^3$$
$$y = \sqrt[3]{x}$$

Since $y = \sqrt[3]{x}$ is a function,

$$f^{-1}(x) = \sqrt[3]{x}$$

Not all inverses are inverse functions.

Example 6   Does $\{(1, 3), (5, 6), (8, 3), (4, 6)\}$ have an inverse function?

Solution   No. The set formed by switching the coordinates of each pair is $\{(3, 1), (6, 5), (3, 8), (6, 4)\}$; it is not a function because the domain element 3 is paired with two different range elements in $(3, 1)$ and $(3, 8)$.

Example 7   Does the function defined by $y = x^2$ have an inverse function?

Solution   Switching x and y and solving for y gives:

$$x = y^2$$
$$y = \pm \sqrt{x}$$

But $y = \pm \sqrt{x}$ is not a functional rule since a domain element such as 4 could be paired with either $+2$ or -2. Therefore $y = f(x) = x^2$ has no inverse function.

In Example 7, $f(x) = x^2$ had no inverse function. However, an inverse function can be found for a restricted version of this function. If

$$f(x) = x^2 \qquad x \geqslant 0$$

then

$$y = f^{-1}(x) = \sqrt{x} \qquad y \geqslant 0$$

The equation $f^{-1}(x) = \sqrt{x}$ results from interchanging x and y in $f(x)$. In the last example, we found that the inverse of $f(x) = x^2$ is $y = f^{-1}(x) = \pm\sqrt{x}$. The restriction $y \geqslant 0$ makes the $\pm$ sign unnecessary because y must be nonnegative, and f^{-1} is now a function.

The concept of restricting the domain of a function to create the inverse function will be used often with the trigonometric functions. In the study of calculus, it is often necessary that each domain element be paired with a unique image; in short, that f^{-1} be a function.

The graphs of a function and its inverse are symmetric about the line $y = x$. This can be seen in Figure 1.8, which contains the graphs of $y = f(x) = 3x + 4$ and its inverse function

$$y = f^{-1}(x) = \frac{x - 4}{3}$$

In the general case, if the point (a, b) is an element of f then (b, a) is the corresponding element of f^{-1}.

If we can show that these two points are symmetric about the line $y = x$, then every pair of such points must be; and, thus, the entire graph of a function and its inverse must be symmetric about the line $y = x$. To show that two points are symmetric about a line it is sufficient to show that the line is the perpendicular bisector of the line joining the two points. In Figure 1.9 the points $A(a, b)$ and $B(b, a)$ are graphed. To prove they are symmetric about the line $y = x$, we must show that $y = x$ is the perpendicular bisector of $\overline{AB}$. Triangles $\triangle BCD$ and $\triangle ACD$ are congruent because they have side CD in common, angles α and β are both $45°$, and the lengths of sides BC and AC are $a - b$. Therefore, the line $y = x$ is the perpendicular bisector of $\overline{AB}$. This shows that a function and its inverse are symmetric about the line $y = x$.

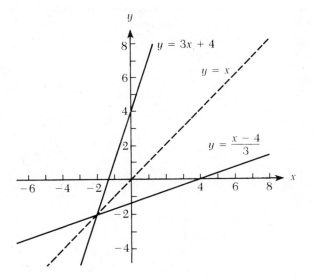

Figure 1.8 Graphs of $f(x) = 3x + 4$, $f^{-1}(x) = \dfrac{x - 4}{3}$, and $y = x$

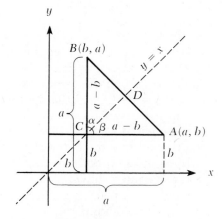

Figure 1.9 The points (a, b) and (b, a) are symmetric about the line $y = x$

Example 8

Consider the function g defined by $g(x) = (2x - 3)/(4x - 1)$. Find $g^{-1}(x)$ and sketch both g and g^{-1}.

Solution

The defining equation is given by

$$y = \frac{2x - 3}{4x - 1}$$

Interchanging x and y,

$$x = \frac{2y - 3}{4y - 1}$$

Solving for y,

$$(4y - 1)x = 2y - 3$$
$$4yx - x = 2y - 3$$
$$4yx - 2y = x - 3$$
$$2y(2x - 1) = x - 3$$
$$y = \frac{x - 3}{2(2x - 1)}$$

or

$$g^{-1}(x) = \frac{x - 3}{2(2x - 1)}$$

The domain of g is all real numbers *except* $\frac{1}{4}$. The domain of g^{-1} is all real numbers *except* $\frac{1}{2}$.

The graph of g is shown in Figure 1.10a. It was constructed with the aid of a calculator (to evaluate values of y for given x values) and noting that no value of y exists when $x = \frac{1}{4}$. The line $x = \frac{1}{4}$ is called an *asymptote* and more will be said about this later. In Figure 1.10b *just* the line $x = \frac{1}{4}$ has been rotated about the 45° line. (Note the change of scale.) This produces the line $y = \frac{1}{4}$. Figure 1.10c shows the graph of $y = g^{-1}(x) = (x - 3)/2(2x - 1)$. The points on this graph can be verified by calculating values using the formula for g^{-1}.

Notice that the vertical asymptote in the graph of g has become a horizontal one in the graph of g^{-1}. Since g^{-1} has a vertical asymptote at $x = \frac{1}{2}$, we can conclude that g has a horizontal asymptote at $y = \frac{1}{2}$.

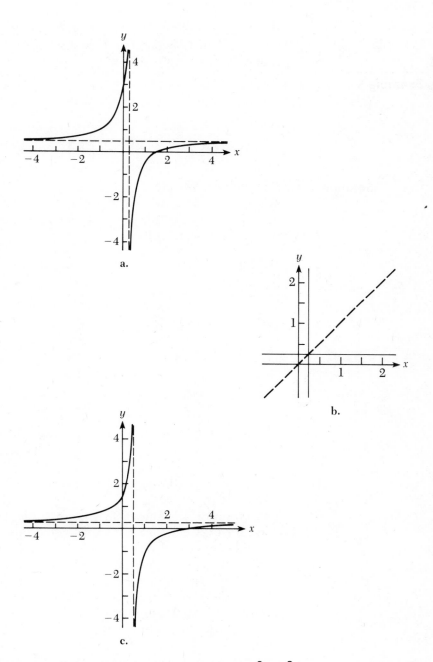

Figure 1.10   Graphs of **a.** g, where $g(x) = \dfrac{2x - 3}{4x - 1}$; **b.** $y = \frac{1}{4}$ and $x = \frac{1}{4}$; and **c.** $y = g^{-1}(x) = \dfrac{x - 3}{2(2x - 1)}$

Functions defined with other than a single formula can still have inverses. Example 9 illustrates just such a case.

Example 9   Let h be a function defined by

$$h(x) = \begin{cases} x & \text{for } x \in (-\infty, 0] \\ x^2 & \text{for } x \in (0, \infty) \end{cases}$$

Find $h^{-1}(x)$ if it exists and sketch the graph of both h and h^{-1}.

Solution   The graph of h is shown in Figure 1.11a. By rotating this graph about the line $x = y$, we get the graph of h^{-1}. We can tell it is a function. Any vertical line intersects the graph only once. In terms of formulas h^{-1} is given by

$$h^{-1}(x) = \begin{cases} x & \text{for } x \in (-\infty, 0] \\ \sqrt{x} & \text{for } x \in (0, \infty) \end{cases}$$

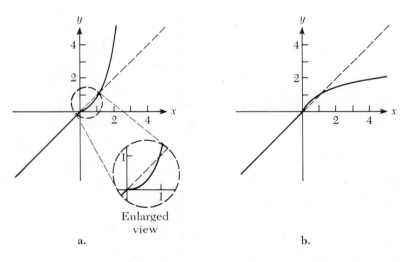

Enlarged
view

a.   b.

Figure 1.11   Graphs of **a.** $y = h(x) = \begin{cases} x & \text{for } x \in (-\infty, 0] \\ x^2 & \text{for } x \in (0, \infty) \end{cases}$ and **b.** $y = h^{-1}(x)$

We know that a formula can be used as a functional rule if its graph intersects any vertical line only once. The graph of the inverse relationship (perhaps not a function) can be found by rotating the graph of the original about the $x = y$ line. Since that graph can tell us if the inverse rule represents a function by examining to see if vertical lines intersect it only once, we can deduce the following guideline about inverses: If any *horizontal* line intersects the graph of a given functional rule only once, at most, then the inverse rule also defines a function.

Example 10   Based on the graphs of functions f_1, f_2, f_3, and f_4 given in Figure 1.12, determine which of these functions have inverse functions.

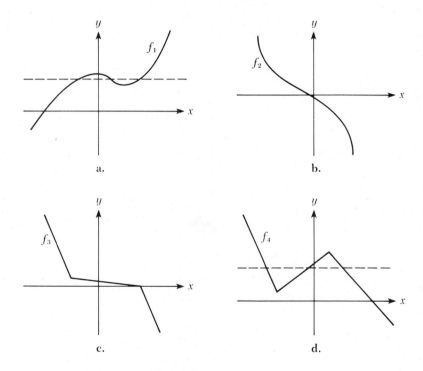

Figure 1.12   Graphs of f_1, f_2, f_3, and f_4

Solution   Since there are horizontal lines that intersect the graphs of f_1 and f_4 more than once, *neither* of these functions have inverse functions. Functions f_2 and f_3 do have inverse functions.

1.5   EXERCISES

In Exercises 1–12, if the formulas define functions, find the inverse functions.

1. $y = x$

2. $y = x + 8$

3. $y = 4x + 5$

4. $y = 7x - 13$

5. $y = -6x + 2$

6. $y = -\dfrac{x}{3} + 9$

7. $y = \dfrac{2x - 1}{x}$

8. $y = \dfrac{x - 3}{2x}$

9. $y = \dfrac{4 - x}{2 - x}$

10. $y = \dfrac{(x/2) - 1}{x/3}$

11. $y = \dfrac{x + 4}{x + 3} + 1$

12. $y = \pm \sqrt{x^4}$

13.   Find the inverse function for $y = \sqrt{x + 25}$, if it exists.

In Exercises 14–18, find the inverse functions if they exist.

14. $y = \sqrt{3 - x}$

15. $y = x^{3/2}$

16. $y = (x + 4)^2$

17. $y = (x - 5)^2$

18. $y = x^5$

In Exercises 19–22, the functions defined do not have inverse functions. Define new functions that use the same rules but with restricted domains so that the inverse is also a function. Note that the answers are not unique.

19. $y = (x + 2)^2$

20. $y = |x - 3|$

21. $y = x^4$

22. $y = x^2 - x - 6$

In Exercises 23–30, sketch the functions and their inverses.

23. f: $f(x) = 2x - 1$ [*Read:* The function f, such that $f(x) = 2x - 1$.]

24. f: $f(x) = \dfrac{x}{2} - 3$

25. f: $f(x) = \dfrac{x - 3}{x + 1}$

26. f: $f(x) = \dfrac{3 - x}{x}$

27. g: $g(x) = \dfrac{2x - 3}{x}$

28. g: $g(x) = \sqrt{5 - x}$

29. h: $h(x) = \sqrt{-x^2}$

30. h: $h(x) = \sqrt{6 - x - x^2}$

In Exercises 31–36, find the inverse, if it is a function. If possible determine the domain of the inverse.

31. f: $f(x) = \begin{cases} x & \text{for } x \in [-4, 0] \\ x^2 & \text{for } x \in (0, 4] \end{cases}$

32. g: $g(x) = \begin{cases} -x^2 & \text{for } x \in (-\infty, -1] \\ x & \text{for } x \in (-1, 1) \\ x^2 & \text{for } x \in [1, \infty) \end{cases}$

33. h: $h(x) = \begin{cases} -x & \text{for } x \in (-\infty, 0) \\ -x^2 & \text{for } x \in [0, \infty) \end{cases}$

34. k: $k(x) = \begin{cases} -x & \text{for } x \leq 0 \\ x & \text{for } x > 0 \end{cases}$

35. f_2: $f_2(x) = \begin{cases} x^2 - x - 6 & \text{for } x \leq -2 \\ x & \text{for } -2 < x \leq 0 \\ -3 & \text{for } x > 0 \end{cases}$

36. m: $m(x) = \begin{cases} x + 2 & \text{for } x \in [-1, 0) \\ x & \text{for } x \in [0, 1) \\ x - 2 & \text{for } x \in [1, 2) \end{cases}$

In Exercises 37–40, based on the graphs, which functions have inverse functions?

37.

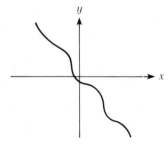

38.

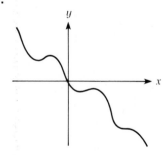

39.

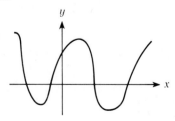

40.

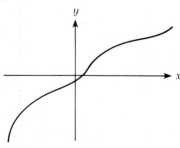

In Exercises 41 and 42, find f^{-1}; then find $f(f^{-1}(x))$ and $f^{-1}(f(x))$.

41. $f: \quad y = f(x) = 3x + 2$ **42.** $f: \quad y = f(x) = 4x - 1$

1.6 LOGARITHMIC AND EXPONENTIAL FUNCTIONS

An important pair of functions, that are inverses of each other, are the logarithmic and exponential functions.

Definition

The exponential function with base b ($b > 0$ and $b \neq 1$) E_b is defined by

$$E_b\, x = b^x$$

It is assumed that the domain of this function is the set of real numbers.

Example 1   Find $E_2\, 1$, $E_2(-7)$, $E_2\, 3.2$, and $E_2\, \pi$.

Solution
$$E_2\, 1 = 2^1 = 2$$
$$E_2(-7) = 2^{-7} = \frac{1}{2^7} = \frac{1}{128}$$
$$E_2\, 3.2 = 2^{3.2} = 2^{32/10} = \sqrt[10]{2^{32}} \approx 9.19 \quad \text{(Using a calculator)}$$
$$E_2\, \pi = 2^\pi \approx 2^{3.14159} \approx 8.82$$

The last example illustrates an important point. In algebra, b^x would only be defined for x as a rational number. Values like 2^π are undefined. There is a way to extend the definition for b^x to all real numbers. Figure 1.13 is the graph of $y = 2^x$ for selected rational values of x. If these dots are connected with a smooth curve as in Figure 1.14 we may assume that 2^x is defined for irrational numbers in such a way as to smooth in the curve.

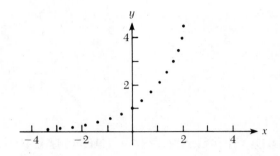

Figure 1.13 Graph of selected values of $y = E_2\, x = 2^x$

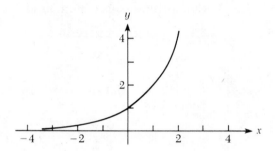

Figure 1.14 Graph of $y = E_2\, x = 2^x$

Example 2   Find $E_3\, 2$, $E_3(-1)$, $E_{10}\, 1000$, $E_{1/2}\, 3$, and $E_{10}\, 1$.

Solution
$$E_3\, 2 = 3^2 = 9$$
$$E_3(-1) = 3^{-1} = \tfrac{1}{3}$$
$$E_{10}\, 1000 = 10^{1000}$$
$$E_{1/2}\, 3 = (\tfrac{1}{2})^3 = \tfrac{1}{8}$$
$$E_{10}\, 1 = 10^1 = 10$$

Definition

The logarithmic function base b, $\log_b$ ($b \neq 1, b > 0$) is defined by

$$y = \log_b x \quad \text{if and only if } b^y = x$$

The logarithmic function base b is the inverse of the exponential function base b; that is, $\log_b x = E_b^{-1} x$ in the inverse function notation.

Example 3   Find $\log_2 8$, $\log_{1/2} 8$, $\log_{10} 10,000$, and $\log_{10} 0.001$.

Solution

$$\log_2 8 = 3 \quad \text{because } 2^3 = 8$$
$$\log_{1/2} 8 = -3 \quad \text{because } (\tfrac{1}{2})^{-3} = 1/2^{-3} = 2^3 = 8$$
$$\log_{10} 10,000 = 4 \quad \text{because } 10^4 = 10,000$$
$$\log_{10} 0.001 = -3 \quad \text{because } 10^{-3} = 0.001$$

A logarithm is an exponent. As an exponent it obeys the same rules as any exponent. The graph of a logarithmic function can be found by reflecting the graph of the corresponding exponential function about the line $y = x$. Figure 1.15 shows the graph of $y = \log_2 x$.

Properties of Logarithms

1. $\log_b 1 = 0$
2. $\log_b b^t = t$
3. $b^{\log_b t} = t$
4. $\log_b xy = \log_b x + \log_b y$
5. $\log_b (x/y) = \log_b x - \log_b y$
6. $\log_b x^y = y \log_b x$
7. $\log_b x = \dfrac{\log_a x}{\log_a b}$

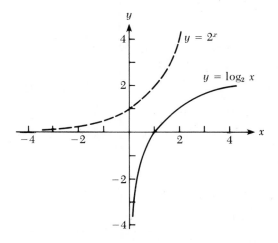

Figure 1.15   Graph of E_2^{-1} or $\log_2$, with all
y such that $x = 2^y$

Each of the following properties is the logarithmic form of a basic rule for exponents:

1. $\log_b 1 = 0$ if and only if $b^0 = 1$
2. $\log_b b^t = t$ if and only if $b^t = b^t$
3. $b^{\log_b t} = t$ if and only if $\log_b t = \log_b t$

Each of these is nothing more than filling in the $\bigcirc$ and $\square$ appropriately in the relationship

$$\log_b \square = \bigcirc \quad \text{if and only if} \quad b^{\bigcirc} = \square$$

4. $\log_b xy = \log_b x + \log_b y$

Let
$$
\begin{array}{lll}
k = \log_b xy & \text{or} & b^k = xy \\
m = \log_b x & \text{or} & b^m = x \\
n = \log_b y & \text{or} & b^n = y
\end{array}
$$

Therefore,

$$b^k = xy = b^m \cdot b^n = b^{m+n}$$

or

$$k = m + n$$

and

$$\log_b xy = \log_b x + \log_b y$$

5. $\log_b (x/y) = \log_b x - \log_b y$

$$
\begin{aligned}
0 = \log_b 1 &= \log_b (y/y) \\
&= \log_b y + \log_b (1/y) \qquad \text{by property 4}
\end{aligned}
$$

Therefore,

$$\log_b (1/y) = -\log_b y$$

and

$$\log_b (x/y) = \log_b x + \log_b (1/y) = \log_b x - \log_b y$$

6. $\log_b x^y = y \log_b x$

Let $\quad l = \log_b x \qquad$ then $b^l = x$
$\quad m = \log_b x^y \qquad$ then $b^m = x^y$

However,

$$x^y = (b^l)^y = b^{ly} = b^m$$

Thus

$$m = ly = yl$$

and

$$\log_b x^y = y \log_b x$$

7. $\log_b x = \dfrac{\log_a x}{\log_a b}$

Consider $\quad \log_b x = l \qquad$ then $b^l = x$

$$\log_a b^l = \log_a x$$
$$l \log_a b = \log_a x \qquad \text{by property 6}$$
$$l = \frac{\log_a x}{\log_a b}$$

or

$$\log_b x = \frac{\log_a x}{\log_a b}$$

Each of the properties is illustrated below.

Example 4

$\log_{10} 1 = 0$
$\log_3 1 = 0$

Example 5

$\log_{10} 10^4 = 4$
$\log_2 2^{7.1} = 7.1$
$\log_{1/2} (\frac{1}{2})^{-3.72} = -3.72$

Example 6

$2^{\log_2 18} = 18$
$10^{\log_{10} 7.1} = 7.1$
$82^{\log_{82} \frac{1}{2}} = \frac{1}{2}$

Example 7	$\log_{10} (20 \times 50) = \log_{10} 20 + \log_{10} 50$ $\log_2 (7 \cdot 5) = \log_2 7 + \log_2 5$

Example 8	$\log_7 3^5 = 5 \log_7 3$ $\log_{10} 10^4 = 4 \log_{10} 10 = 4$ $\log_3 \sqrt{7} = \log_3 7^{1/2} = \frac{1}{2} \log_3 7$ $\log_{10} \sqrt[8]{21} = \frac{1}{8} \log_{10} 21$

Example 9	$\log_2 3.2 = \dfrac{\log_{10} 3.2}{\log_{10} 2} = \dfrac{0.5051}{0.3010} \approx 1.6781$

The values above were computed with a calculator—one could also use the tables provided in the Appendix.

Notation The notation $\log_{10}$ and $\log$ will be used to denote the same logarithmic function.

Example 10	$\log_{10} 1000 = \log 1000 = \log 10^3 = 3$

Notation The symbol ln will be used to denote $\log_e$ where $e \approx 2.718281828$; this is known as the *natural* logarithmic function.

The specific definitions for e and ln will be considered later.

1.6 EXERCISES

In Exercises 1–20, find the images of the logarithmic or exponential function involved. A calculator should not be used.

1.	$E_3 \, 2$	**2.**	$E_3 \, \frac{1}{2}$	**3.**	$E_{1/2} \, \frac{1}{2}$
4.	$E_{1/2} \, 1.5$	**5.**	$E_{81} \, \frac{1}{2}$	**6.**	$E_9 \, \frac{5}{2}$
7.	$E_{10} \, 4$	**8.**	$E_{100} \, 2$	**9.**	$E_{1/10}(-3)$
10.	$E_{1/10} \, 10$	**11.**	$\log_2 8$	**12.**	$\log_3 \frac{1}{27}$
13.	$\log_{10} 10^{10}$	**14.**	$\log 10^{10}$	**15.**	$\log_2 \frac{1}{32}$
16.	$\log_{1/2} \frac{1}{32}$	**17.**	$\log_{64} \frac{1}{4}$	**18.**	$\log_{64} 1024$
19.	$\log 0.00001$	**20.**	$\log_{10} 0.00001$		

In Exercises 21–30, considering the definition of logarithms, find x or y in each case, without using a calculator.

21. $\log_3 9 = y$ **22.** $\log_{1/2} x = -5$ **23.** $\log x = 3$

24. $\log_x 0.1 = 2$ **25.** $\log_x b^2 = 2$ **26.** $\log_5 5 = x$

27. $\log_{10} y = -3$ **28.** $\log_x 8 = -3$ **29.** $\log_x 100 = -2$

30. $\log_y \dfrac{1}{y^4} = -4$

Each of Exercises 31–40 is a specific example of one of the properties of logarithms. Identify the property.

31. $\log_7 49 = 2$

32. $\log \frac{3}{10} = \log 3 - \log 10$

33. $\log_5 \sqrt[4]{81} = \frac{1}{4} \log_5 81$

34. $\log_3 (6^{5/3} \cdot 2^{8/3}) = \log_3 6^{5/3} + \log_3 2^{8/3}$

35. $\log_5 64 = \log_5 2 + \log_5 32$

36. $\log_{15} \frac{3}{7} = \log_{15} 3 - \log_{15} 7$

37. $\log_{12} 12 = 1$

38. $\log_{12} 144 = 2$

39. $\log_{12} \frac{1}{1728} = -3$

40. $\log_{144} 12 = \frac{1}{2}$

In Exercises 41–45, assume that $\log_2 3 = 1.58$ and $\log_2 5 = 2.32$, and use the properties of logarithms to find:

41. $\log_2 15$ **42.** $\log_2 10$ **43.** $\log_2 \sqrt{5}$

44. $\log_2 \sqrt[5]{30}$ **45.** $\log_2 400$

In Exercises 46–50, use property 7 to write in terms of $\log_{10}$.

46. $\log_7 21$ **47.** $\log_2 8.1$ **48.** $\log_6 18$

49. $\ln 8$ **50.** $\ln 100$

1.7   INCREMENTS AND RATE OF CHANGE

We would be remiss if this introductory chapter did not at least begin the study of calculus. Our first step will be to examine the idea of a rate of change.

If an auto maker increases production by 1000 cars a month, a corresponding change in costs, revenue, and, hopefully, profits will be expected. A manufacturer is especially interested in the rate of change in these quantities; that is, how much profits will increase (or decrease) with each additional unit produced. Along a similar line, a doctor might be interested in a patient's blood pressure drop in response to a given dose of a new drug. In a classic problem of this type, auto experts want to know the increase in stopping distance for a given increase in speed. To a student of calculus, all of these problems are part of the more general study of *rates of change of a function*, the change of a function's image per unit change in the function's argument, and the related concept of a limit. These are the mathematical ideas which will be examined.

The concept of change in a function image $f(x)$ with a change in x is basic to calculus. In fact, it is one of the primary differences between calculus and other types of mathematics. This means, for example, we will consider not only the cost of production as a function of the number of units produced, but also how the cost per unit changes as the number of units produced changes.

Suppose that $x_1 = 4$ and $x_2 = 5.7$. Then the change in x is 1.7. The Greek letter Δ (delta) is used to denote change in a variable. In this case, "the change in x is 1.7" is abbreviated to $\Delta x = 1.7$.

Definition

The change in x from x_1 to x_2 will be called an *increment* of x and be denoted by Δx, where

$$\Delta x = x_2 - x_1$$

Example 1

If x changes from 7 to 9, then $\Delta x = 2$. If y changes from 1 to -2, then $\Delta y = -2 - 1 = -3$. The increment of $f(x)$, as $f(x)$ changes from 1.11 to 1.13, is $\Delta f(x) = 0.02$. If $x_2 = x + h$ and $x_1 = x$, then

$$\Delta x = x_2 - x_1 = x + h - x = h$$

If $y = f(x)$, that is, y is a function f of x, and an increment, Δx, is added to x, then y is changed by a corresponding increment Δy. Let $y_1 = f(x_1)$ and $y_2 = f(x_2)$. Then $\Delta y = y_2 - y_1 = f(x_2) - f(x_1)$. Solving $\Delta x = x_2 - x_1$ for x_2 gives $x_2 = x_1 + \Delta x$. Substituting $x_1 + \Delta x$ for x_2 gives $\Delta y = f(x_1 + \Delta x) - f(x_1)$. Since x_1 can be any element in the domain of f, the subscript becomes superfluous and the formula for the increment of y can be written:

$$\Delta y = f(x + \Delta x) - f(x)$$

Example 2	If $f(x) = x^2 + 3$, find Δy when $x = 2$ and $\Delta x = 0.5$.
Solution	Substituting the values specified for x and Δx in $\Delta y = f(x + \Delta x) - f(x)$ gives

$$\Delta y = f(2 + 0.5) - f(2) = f(2.5) - f(2)$$

Substituting in $f(x)$,

$$\Delta y = [(2.5)^2 + 3] - (2^2 + 3)$$
$$= 6.25 + 3 - 4 - 3 = 2.25$$

In this example, a 0.5 change in x resulted in a 2.25 change in y.

Example 3	A painted sphere has a radius of exactly 6 inches. The old paint on the sphere is then ground off and it is repainted, with a net loss in radius of 0.1 inch. How much was the volume changed?
Solution	The volume of a sphere is $V = \frac{4}{3}\pi r^3$. Let $r = 6$ and $\Delta r = -0.1$; Δr is negative because the radius was reduced. To find ΔV, evaluate the volume for $r + \Delta r$ and for r and find the difference between the volumes:

$$r + \Delta r = 6 - 0.1 = 5.9$$
$$\Delta V = \tfrac{4}{3}\pi(5.9)^3 - \tfrac{4}{3}\pi 6^3 = \tfrac{4}{3}(205.379 - 216)\pi$$
$$= -\tfrac{4}{3}(10.621)\pi \approx -45 \text{ cubic inches}$$

Change stated in absolute terms as in the examples discussed above or in more familiar phrases such as "I lost 20 pounds" or "My income has increased \$48" may not be as informative as rates of change; a rate of change is stated in terms such as "I lost 20 pounds over the last six months" or "My income has increased \$48 per week." From the statement "I lost 20 pounds in the last six months" it is possible to compute the average change in weight per month. It is the change in weight divided by the number of months (change in time) or $\frac{20}{6} = 3\frac{1}{3}$ pounds per month. In Example 2, it was shown that $\Delta y = 2.25$ when $\Delta x = 0.5$. In this case the average change in y as x changes from 2 to 2.5 is $\Delta y/\Delta x = 2.25/0.5 = 4.5$. This means that, for the indicated interval of x, the average rate of change of y is 4.5 units for each one unit change in x. In Example 3, a change of -0.1 inch in the radius produced a -45 cubic inches change in the volume. The average rate of change of volume, as the radius goes from 6 to 5.9, is $\Delta V/\Delta r = (-45)/(-0.1) = 450$ cubic inches/inch. Thus, the volume is changing at the average rate of 450 cubic inches per one inch change in radius.

Definition

The *average rate of change* of a function f per unit change in x over an interval of its domain, x to $x + \Delta x$, is:

$$\frac{\Delta y}{\Delta x} = \frac{f(x + \Delta x) - f(x)}{\Delta x}$$

Example 4   Find the average rate of change of $f(x) = x^3$ as x changes from 4 to 6.

Solution $\dfrac{\Delta y}{\Delta x} = \dfrac{f(6) - f(4)}{2} = \dfrac{6^3 - 4^3}{2} = \dfrac{216 - 64}{2} = \dfrac{152}{2} = 76$

Example 5   The distance S, in feet, a body will fall (from rest) in a vacuum after t seconds is given by $S = 16t^2$. Find the average speed of such a falling body between times $t = 5$ and $t = 5\frac{1}{2}$ seconds.

Solution   The average speed is the change in distance with respect to change in time, $\dfrac{\Delta S}{\Delta t}$.

$$\frac{\Delta S}{\Delta t} = \frac{16(5\frac{1}{2})^2 - 16(5)^2}{\frac{1}{2}} = \frac{484 - 400}{\frac{1}{2}} = \frac{84}{\frac{1}{2}} = 168 \text{ feet/second}$$

Thus for the interval of time from 5 to $5\frac{1}{2}$ seconds the body has an average speed of 168 feet per second.

1.7 EXERCISES

In Exercises 1–10, find the increment of each function for the given interval of its domain.

1. $y = 2x + 5$, $x = 3, \Delta x = 0.1$
2. $y = x^2$, $x = 2$ to $x = 2.01$
3. $y = x - 5x^2$, $x = -3$ to $x = -2.5$
4. $g(t) = 1 - 3t^2$, $t = 0, \Delta t = 0.4$
5. $f(x) = x^2$, $x = a$, $\Delta x = h$

6. $h(x) = x^2 + 3x + 5, \quad x \text{ to } x + \Delta x$

7. $f(x) = 3x^2 + 2x, \quad x \text{ to } x + \Delta x$

8. $y = 2x - 3x^2, \quad x \text{ to } x + \Delta x$

9. $y = x^3, \quad x \text{ to } x + h$

10. $f(x) = x^3 + 3x^2, \quad a \text{ to } a + h$

In Exercises 11–17, find the average rate of change of each function for the indicated interval of its domain.

11. $y = 2x + 5, \quad x = 3, \Delta x = 0.1$

12. $f(x) = 4x^2, \quad x = -2, \Delta x = \frac{1}{2}$

13. $f(x) = \sqrt{x}, \quad x = 4, \Delta x = 0.25$

14. $y = x^2 + 5x + 6, \quad x \text{ to } x + \Delta x$

15. $y = ax^2 + bx + c, \quad x \text{ to } x + h$

16. $y = \dfrac{4}{x}, \quad x = 2 \text{ to } x = 2 + \Delta x$

17. $y = 2x^3, \quad a \text{ to } a + h$

In Exercises 18–21, find the average speed of the body for the given interval of time; given that the distance, S, a body initially at rest falls in a vacuum in t seconds is given by $S = 16t^2$.

18. $t = 0 \text{ to } t = 1$ **19.** $t = 3 \text{ to } t = 6$

20. $t = 10 \text{ to } t = 11$ **21.** $t = 3 \text{ to } t = 3.001$

22. The cost, C, to produce x units is given by $C = 100x + (1/x)$. What is the average cost per unit (also known as the average marginal cost) for the units from $x = 16$ to $x = 20$?

23. A growing culture of bacteria weighs $10e^{t/3}$ grams at time t. Find the average rate of growth during the hours $t = 3$ to $t = 6$, where $e \approx 2.7$.

24. The price elasticity of demand for an item is defined to be

$$PE = -\frac{\text{Percentage change in the quantity of the item demanded}}{\text{Percentage change in the price of the item}}$$

Discuss how this relates to increments and rates of change.

25. Using the expression given in Exercise 24, calculate the price elasticity if the sales of an item change from 10 to 15 units and the price goes from \$22,500 to \$21,800 per item.

26. The average cost of the production of an item is defined to be the total cost divided by the total number of items produced.

Discuss how the average cost of production relates to the idea of increments and rates of change.

[$] **27.** The total cost for producing 20 television sets is $900. What is the average cost (defined in Exercise 26) of production?

[$] **28.** Referring to Exercises 26 and 27, the marginal cost is defined to be the additional cost attributable to the addition of one unit produced. That is, the marginal cost for $x = 16$ is the cost of the 16th item. If the total cost of producing 15 of the television sets described in Exercise 27 was $600, what was the marginal cost of the 16th set? Discuss the assumptions which had to be made in order to apply the definition of marginal cost.

[$] **29.** Describe marginal cost in terms of rates of change as considered in this section.

NEW TERMS

Argument	Image
Average rate of change	Increment
Closed interval	Independent variable
Composition	Inverse
Dependent variable	Logarithm
Domain	Natural logarithm
Function	Open interval
Half-closed interval	Range
Half-open interval	Rate of change

REVIEW

In Exercises 1 and 2, use the formulas that relate the Fahrenheit and Celsius temperature scales to find the temperatures called for.

1. The corresponding Fahrenheit temperatures if °C = 37°, °C = 18°, °C = −20°, °C = 0°, and °C = 212°.

2. The corresponding Celsius temperatures if °F = 37°, °F = 18°, °F = −20°, °F = 0°, and °F = 212°.

In Exercises 3–6, simplify the expressions.

3. $-(18) + 2[4 - 6(3 + 7)]$

4. $\left(\dfrac{1}{2} + \dfrac{1}{6}\right) \cdot \dfrac{1}{8} \div \dfrac{1}{12}$

5. $\dfrac{5 + 2 \cdot 3 - 4 \div 2 + 6}{6 + 2 \cdot 4}$

6. $7 + 2^5 - 16 \div 3$

In Exercises 7–10, evaluate the following formulas at the indicated values.

7. $T = 18s^2 + 2s - 6, \quad s = 3$

8. $T = [s(3.7s - 4) + 2]s, \quad s = -1$

9. $y = \dfrac{x^3 - 4x^2 + 2x - 3}{[x(x - 4) + 2]x - 3}, \quad x = 10$

10. $y = \dfrac{x^3 - 4x^2 + 2x - 3}{[x(x - 4) + 2]x - 3}, \quad x = -7$

In Exercises 11–14, in each formula, replace x by the given expression in z and simplify the result.

11. $y = (x + 3)^2, \quad x = z - 3$

12. $y = (x + 3)^2 - 2(x + 3) + 2, \quad x = z^2 - 1$

13. $y = \dfrac{1}{x} + \dfrac{1}{x + 2}, \quad x = z - 1$

14. $y = \dfrac{x + 2}{x}, \quad x = z^2 - z - 6$

In Exercises 15–20, let h and g be defined as $h(x) = \dfrac{x}{x + 1}$; $g(x) = x^2 - x - 6$. Evaluate the expressions.

15. $h(1), h(0),$ and $h(-2)$

16. $g(1), g(0),$ and $g(-2)$

17. $h(x + 2), h(x - 2),$ and $h(x^2 + x)$

18. $g(x + 2), g(x - 2),$ and $g(x^2 + x)$

19. $g(3), g(x^2), h(g(3)),$ and $h(g(x^2))$

20. $h(3), h(x^2), g(h(3)),$ and $g(h(x^2))$

In Exercises 21–24, determine which of the expressions can be used to define functions and state the domain.

21. $y = \dfrac{2x + 1}{x}$

22. $f(x) = \sqrt{x^2 + 4}$

23. $h(x) = \dfrac{1}{\sqrt{9 - x^2}}$

24. y such that $x^4 + y^2 = 2x^2$

In Exercises 25–28, graph each equation. Based on the graph, determine whether the equation could be used to define a function.

25. $y = x^3$

26. $y = x^3 - 2x$

27. $y^2 = x^2 - x^4$

28. $x^3 + y^3 = 1$

In Exercises 29 and 30, consider the function f defined by

$$f(x) = \begin{cases} 2 - x & \text{for } x \in (-\infty, 2] \\ x^2 & \text{for } x \in (2, 4] \\ x^3 & \text{for } x \in (4, 5) \\ x^0 & \text{for } x \in [5, \infty) \end{cases}$$

29. Find $f(1), f(-1), f(3)$, and $f(-3)$.

30. Find $f(-10), f(10)$, and $f(y^2)$ if $y \geq \sqrt{10}$.

In Exercises 31 and 32, consider the possible functional rule for a function named g

$$g(x) = \begin{cases} -x^0 & \text{if } x \in [-2, 0) \\ x^0 & \text{if } x \in (0, 2] \end{cases}$$

31. Is g indeed a function, and if so, what is its domain? Sketch it on a number line. What is its range?

32. Sketch $y = g(x)$ whether or not it is a function.

33. Let $y = 7$. Does this equation represent a potential functional rule? What domain? What range?

34. Let $h(x) = 7$ and find: $h(-1)$, $h(2)$, $h(\pi)$, $h(x^2)$, and $h(z^2 - 2z + 2)$.

In Exercises 35–37, if f is a function given by $f(x) = x^3 - 2x + 4$, find:

35. $f(2)$ and $f(-1)$

36. $f(\sqrt{x})$ and $f(x + h)$

37. $\dfrac{f(x + h) - f(x)}{h}$ (Simplify.)

In Exercises 38–45, let f be a function given by $f(x) = 3x^2 - x - 2$. Let g be a function given by $g(x) = 3x + 2$. Find the domain and an expression for each image.

38. $f + g$

39. $f - g$

40. $f \cdot g$

41. f/g

42. $f \circ g$

43. $g \circ f$

44. Find f^{-1} if it is a function. **45.** Find g^{-1} if it is a function.

46. Does $y = \dfrac{2x - 3}{x}$ define a function? Does it have an inverse function? Find the appropriate formula if it does.

47. Repeat Exercise 46 for the formula $y = 3x - 7$.

In Exercises 48 and 49, sketch the functions and their inverses and determine whether their inverses are functions.

48. f: $f(x) = \dfrac{3 - x}{x}$ **49.** g: $g(x) = x^2 + 2$

50. Consider h, a function defined by

$$h(x) = \begin{cases} x^2 + 2 & \text{for } x \geqslant 0 \\ -x^2 + 2 & \text{for } x < 0 \end{cases}$$

Sketch this function and h^{-1} (if it is a function).

51. Consider k, a function given by

$$k(x) = \begin{cases} x & \text{for } x > 2 \\ x - 2 & \text{for } x < 2 \\ 1 & \text{for } x = 2 \end{cases}$$

Sketch this function and k^{-1}.

In Exercises 52–60, find the indicated values.

52. $E_{1/2}\, 8$ **53.** $E_8\, \tfrac{1}{2}$ **54.** $E_{10}\, 10^3$

55. $E_{1/10}(-3)$ **56.** $\log_{1/2} 8$ **57.** $\log_8 \tfrac{1}{2}$

58. $\log_{10} 10^3$ **59.** $\log_{1/10} (-3)$ **60.** $\log 10^{10}$

In Exercises 61 and 62, find the increment of the function for the given interval of its domain.

61. f: $f(x) = x^2 - 3x + 7$, x to $x + \Delta x$

62. $y = x^3$, x to $x + 1$

In Exercises 63 and 64, find the average rate of change for the indicated interval.

63. $y = 3x - 6$, $x = 3, \Delta x = 0.1$

64. $y = -4x^2$, $x = b$ to $x = b + h$

2

The Derivative

In Chapter 1 we examined the concept of a function and looked at some notations and examples. In this chapter we will show how the function idea is used in the development of the concept of a derivative. The derivative is an extension of the average rate of change idea we examined in Chapter 1. In this chapter we will consider average rates of change over shorter and shorter time periods. These shorter averages will lead us to the concept of the limit of the average rate of change. This concept, the derivative, will have many applications in the real world.

2.1 LIMITS

To refine the idea of an average change it will be necessary to introduce the concept of a limit. This basic idea can be shown by a simple example: Suppose that you are sitting on one end of a bench and the object of your affections is sitting on the other end. Further suppose that your strategy is to halve the distance between you, then halve it again, and so on, over and over. A mathematician would say that the object of your affections is the limit of your moves, no matter how many you make. He or she would also realize that you would never actually reach your love goal, which is an important part of the concept of a *a limit*.

The average rate of change of a function f for the interval a to $a + \Delta x$ is $[f(a + \Delta x) - f(a)]/\Delta x$. As Δx is successively assigned smaller and smaller values, the average rate of change of the function approaches the "instantaneous rate of change" of the function at a. In order to create simple techniques for computing such instantaneous rates of change we introduce *limits*.

The phrase, "The limit, as x approaches a, of $f(x)$ is L," will be denoted by $\lim\limits_{x \to a} f(x) = L$. Intuitively, this means that as x gets closer and closer to a, $f(x)$ approaches L. Thus $\lim\limits_{x \to 6} (2x - 3) = 9$ says that as x approaches 6, $2x - 3$ approaches 9. Intuitively this seems correct, since $2(6) - 3 = 9$.

Statements such as "closer and closer" and "approaches" are too imprecise to be used to prove that certain values are limits of functions or to prove theorems about limits. However, the idea of closer-and-closer can be denoted mathematically by examination of the absolute values of differences. If two numbers are close to the same value, their difference will be small. Thus, x approaches a will mean that $|x - a|$ is less than some small positive number δ. Absolute values are used around $|x - a|$, because the algebraic sign of the difference is not important. In the same manner the phrase "$f(x)$ approaches L" will mean that $|f(x) - L| < \varepsilon$ for any positive number ε no matter how small. Therefore we have the following definition:

Definition

$\lim\limits_{x \to a} f(x) = L$ if and only if for any positive number ε (no matter how small) there is a positive number δ (dependent on ε) such that

$$|f(x) - L| < \varepsilon \qquad \text{whenever} \qquad 0 < |x - a| < \delta$$

This means that if the statement $\lim\limits_{x \to a} f(x) = L$ is true, then there is a δ such that $f(x)$ can be made as close to L as we please (that is, within ε of L), by substituting any value of x within a distance δ of a into $f(x)$. Notice that the definition of limits and the intuitive discussion about limits say nothing about $f(x)$ at a. That is, the function may not be defined at a, and in fact many important ones are not, but these functions can still have a limit as x approaches a. The statement $0 < |x - a| < \delta$ in the definition, requires that f be defined for all x close to a, but it need not be defined at a. The definition is mathematically correct, but not useful on a practical level.

Example 1   Find $\lim\limits_{x \to 6} (2x - 3)$ and prove that it is the limit using the definition.

Solution   It is reasonable to conclude as we did above that the closer x gets to 6, the closer $2x$ will get to 12, and thus the closer $(2x - 3)$ will get to $(12 - 3)$ or 9. Hence $\lim\limits_{x \to 6} (2x - 3) = 9$. To prove this is the limit we must find δ, such that for any positive ε no matter how small:

$$\left|(2x - 3) - 9\right| < \varepsilon \qquad \text{whenever} \qquad 0 < \left|x - 6\right| < \delta$$

To show this, we will write $\left|(2x - 3) - 9\right|$ as a function of $\left|x - 6\right|$.

$$\left|(2x - 3) - 9\right| < \varepsilon$$

if and only if

$$\left|2x - 12\right| < \varepsilon$$

which is equivalent to

$$\left|2(x - 6)\right| < \varepsilon$$

Thus,

$$\left|x - 6\right| < \frac{\varepsilon}{2}$$

Therefore, $\left|(2x - 3) - 9\right| < \varepsilon$ whenever $\left|x - 6\right| < \varepsilon/2$. So take $\delta \leq \varepsilon/2$, and the proof is complete. It shows that $2x - 3$ will be within ε of 9 whenever x is within $\frac{1}{2}\varepsilon$ of 6.

     Using the definition to prove limits is usually very complicated unless, as was the case in Example 1, we have a very simple function. In addition we will be much more interested in finding limits rather than proving values using the definition.

     The definition of a limit is not useful in finding specific limits. It is, however, useful in proving theorems which can then be used to find limit values. The following limit theorems are just that: rules that can be used to find limits. Their proofs make use of the

$\delta - \varepsilon$ definition of limits and can be found in most Engineering/Mathematics-oriented calculus texts.

Theorem 2.1 $\lim\limits_{x \to a} x = a$

In words: "The limit of the function x is a as x tends toward a."

Example 2 $\lim\limits_{x \to 3} x = 3$

Theorem 2.2 $\lim\limits_{x \to a} [f(x) \pm g(x)] = \lim\limits_{x \to a} f(x) \pm \lim\limits_{x \to a} g(x)$, provided $\lim\limits_{x \to a} f(x)$ and $\lim\limits_{x \to a} g(x)$ exist

In words: "The limit of a sum or difference is the sum or difference of the limits, provided the limits exist."

Example 3 If $\lim\limits_{x \to 3} x^3 = 27$ and $\lim\limits_{x \to 3} (1/x) = \frac{1}{3}$, then

$$\lim_{x \to 3} \left(x^3 + \frac{1}{x} \right) = \lim_{x \to 3} x^3 + \lim_{x \to 3} \frac{1}{x} = 27 + \frac{1}{3} = \frac{81}{3} + \frac{1}{3} = \frac{82}{3}$$

Theorem 2.3 $\lim\limits_{x \to a} [f(x) \cdot g(x)] = [\lim\limits_{x \to a} f(x)][\lim\limits_{x \to a} g(x)]$, provided $\lim\limits_{x \to a} f(x)$ and $\lim\limits_{x \to a} g(x)$ exist

In words: "The limit of a product is the product of the limits, provided the limits exist."

Example 4 If $\lim\limits_{x \to 3} x^3 = 27$, and $\lim\limits_{x \to 3} (1/x) = \frac{1}{3}$, then find $\lim\limits_{x \to 3} x^3(1/x)$.

Solution $\lim\limits_{x \to 3} x^3 \left(\frac{1}{x} \right) = \left(\lim\limits_{x \to 3} x^3 \right) \left(\lim\limits_{x \to 3} \frac{1}{x} \right) = 27 \left(\frac{1}{3} \right) = 9$

Theorem 2.4 $\lim\limits_{x \to a} [f(x)^n] = [\lim\limits_{x \to a} f(x)]^n$ where n is a rational number and $\lim\limits_{x \to a} f(x)$ exists

In words: "The limit of a function raised to a rational power is the power of the limit of the function, provided the limit exists."

Example 5   Find $\lim\limits_{x\to 3} x^{5/7}$

Solution $\lim\limits_{x\to 3} x^{5/7} = (\lim\limits_{x\to 3} x)^{5/7} = (3)^{5/7} = 3^{5/7}$

Theorem 2.5 $\lim\limits_{x\to a} \dfrac{f(x)}{g(x)} = \dfrac{\lim\limits_{x\to a} f(x)}{\lim\limits_{x\to a} g(x)}$, provided $\lim\limits_{x\to a} f(x)$ and $\lim\limits_{x\to a} g(x)$ exist and $\lim\limits_{x\to a} g(x) \neq 0$

In words: "The limit of a quotient is the quotient of the limits, provided the limits exist and the limit of the denominator is not zero."

Example 6   Find $\lim\limits_{x\to 3} \dfrac{x^3}{x^2 + x}$.

Solution $\lim\limits_{x\to 3} \dfrac{x^3}{x^2 + x} = \dfrac{\lim\limits_{x\to 3} x^3}{\lim\limits_{x\to 3} (x^2 + x)} = \dfrac{\lim\limits_{x\to 3} x^3}{\lim\limits_{x\to 3} x^2 + \lim\limits_{x\to 3} x}$

$\dfrac{(\lim\limits_{x\to 3} x)^3}{(\lim\limits_{x\to 3} x)^2 + 3} = \dfrac{3^3}{3^2 + 3} = \dfrac{27}{9 + 3} = \dfrac{27}{12}$

Theorem 2.6 $\lim\limits_{x\to a} c = c$, for c constant

In words: "The limit of a constant is the constant itself."

Example 7   Find $\lim\limits_{x\to 3} 2$.

Solution $\lim\limits_{x\to 3} 2 = 2$

Theorems 2.3 and 2.6 can be combined to form Theorem 2.7.

Theorem 2.7 $\lim\limits_{x\to a} [cf(x)] = c \lim\limits_{x\to a} f(x)$, for c a constant

In words: "The limit of a constant times a function is that constant times the limit of the function."

Example 8   Use the above theorems to find $\lim_{x \to 5} (x^2 - 3x + 4)$.

Solution $\lim_{x \to 5} (x^2 - 3x + 4) = \lim_{x \to 5} x^2 - \lim_{x \to 5} 3x + \lim_{x \to 5} 4$ by Theorem 2.2

$\lim_{x \to 5} x^2 = 5^2 = 25$ by Theorems 2.4 and 2.1

$\lim_{x \to 5} 3x = (\lim_{x \to 5} 3)(\lim_{x \to 5} x)$ by Theorem 2.3

$= 3 \cdot \lim_{x \to 5} x$ by Theorem 2.6

$= 3(5) = 15$ by Theorem 2.1

$\lim_{x \to 5} 4 = 4$ by Theorem 2.6

$\therefore \lim_{x \to 5} (x^2 - 3x + 4) = 25 - 15 + 4 = 14.$

In practice the theorems are applied mentally. We will write them in Example 9 to show two applications that Example 8 does not contain.

Example 9   Use limit theorems to find $\lim_{x \to -3} \dfrac{\sqrt{x^2 + 16}}{9x}$.

Solution $\lim_{x \to -3} \dfrac{\sqrt{x^2 + 16}}{9x} = \dfrac{\lim_{x \to -3} \sqrt{x^2 + 16}}{\lim_{x \to -3} 9x}$ by Theorem 2.5

$= \dfrac{\sqrt{\lim_{x \to -3} (x^2 + 16)}}{\lim_{x \to -3} 9x}$ by Theorem 2.4

$= \dfrac{\sqrt{\lim_{x \to -3} x^2 + \lim_{x \to -3} 16}}{\lim_{x \to -3} 9 \cdot \lim_{x \to -3} x}$ by Theorem 2.2
by Theorem 2.3

$= \dfrac{\sqrt{9 + 16}}{9(-3)}$ by Theorems 2.1 and 2.6
by Theorems 2.6 and 2.1

$= \dfrac{5}{-27}$

Example 10   Find $\lim_{x \to 4} \dfrac{2x^2 - 7x - 4}{x - 4}$.

Solution   Theorem 2.5 cannot be applied because the limit of the denominator is 0. From the definition of limits we are only concerned with values of x close to but *different* from 4. Therefore the numerator may be divided by $x - 4$ since it is a small but nonzero number.

$$\lim_{x \to 4} \frac{2x^2 - 7x - 4}{x - 4} = \lim_{x \to 4} \frac{(2x + 1)(x - 4)}{x - 4} = \lim_{x \to 4} (2x + 1) = 9$$

Example 11 Find $\lim\limits_{x \to 1} \dfrac{x - 1}{\sqrt{x} - 1}$.

Solution Once more Theorem 2.5 cannot be applied since the denominator tends to 0. Nor can the numerator be easily factored and divided by the denominator as in the last example. For limits involving radicals, rationalize.

$$\lim_{x \to 1} \frac{x - 1}{\sqrt{x} - 1} = \lim_{x \to 1} \frac{(x - 1)}{(\sqrt{x} - 1)} \frac{(\sqrt{x} + 1)}{(\sqrt{x} + 1)}$$
$$= \lim_{x \to 1} \frac{(x - 1)(\sqrt{x} + 1)}{x - 1}$$
$$= \lim_{x \to 1} (\sqrt{x} + 1) = 2$$

Example 12 For $f(x) = x^2$, find $\lim\limits_{\Delta x \to 0} (\Delta f/\Delta x)$ for the interval x to $x + \Delta x$, where Δf is defined by:

$$\Delta f = f(x + \Delta x) - f(x)$$

Solution $\lim\limits_{\Delta x \to 0} \dfrac{\Delta f}{\Delta x} = \lim\limits_{\Delta x \to 0} \dfrac{f(x + \Delta x) - f(x)}{\Delta x}$
Substituting in $f(x)$ yields:

$$\lim_{\Delta x \to 0} \frac{\Delta f}{\Delta x} = \lim_{\Delta x \to 0} \frac{(x + \Delta x)^2 - x^2}{\Delta x}$$
$$= \lim_{\Delta x \to 0} \frac{x^2 + 2x \, \Delta x + (\Delta x)^2 - x^2}{\Delta x}$$
$$= \lim_{\Delta x \to 0} \frac{2x \, \Delta x + (\Delta x)^2}{\Delta x}$$
$$= \lim_{\Delta x \to 0} (2x + \Delta x) = 2x$$

The definition of limits implies that if a limit exists it is unique. If, as x approaches a, $f(x)$ approaches two different values, the limit, $\lim\limits_{x \to a} f(x)$, does not exist.

Example 13	Find $\lim\limits_{x \to 0} (|x|/x)$.
Solution	The function is not defined for $x = 0$. For x negative $|x|/x = -1$; for x positive $|x|/x = +1$. Therefore if x approaches 0 from the negative side, $|x|/x = -1$; if it approaches 0 from the positive side, $|x|/x = +1$. So $|x|/x$ does not approach a unique number as x approaches zero. The limit does not exist.

While Example 13 illustrates that for the limit to exist it must be the same from both directions (that is x approaching a from above and x approaching a from below), it also suggests the following idea of "one-sided limits."

Definition

$\lim\limits_{x \to a^+} f(x) = L$ if and only if for any positive number ε there is a positive number δ such that

$$|f(x) - L| < \varepsilon \qquad \text{whenever} \qquad 0 < x - a < \delta$$

L is called the *right-hand limit* or limit from the right of f at a.

Definition

$\lim\limits_{x \to a^-} f(x) = L$ if and only if for any positive number ε there is a positive number δ such that

$$|f(x) - L| < \varepsilon \qquad \text{whenever} \qquad 0 < a - x < \delta$$

L is called the *left-hand limit* or limit from the left of f at a.

Right-hand and left-hand limits provide a second way of defining regular limits.

Theorem 2.8 $\lim\limits_{x \to a} f(x) = L$ if and only if $\lim\limits_{x \to a^+} f(x) = L = \lim\limits_{x \to a^-} f(x).$

In other words, for the limit of a function to exist at a point, the two one-sided limits must exist and be equal.

As with the regular limit definition, right- and left-hand limits are easier to find than their definitions suggest.

Example 14   Find $\lim\limits_{x\to0^+} (|x|/x)$ and $\lim\limits_{x\to0^-} (|x|/x)$.

Solution $\lim\limits_{x\to0^+} (|x|/x) = 1$, since here we are insisting that x approach 0 from the right, that is where $x > 0$. Since $x > 0$, $|x| = x$ and $|x|/x = 1$.

$$\lim_{x\to0^-} \frac{|x|}{x} = -1 \qquad \text{by similar reasoning}$$

Example 15   For f defined by

$$f(x) = \begin{cases} x^2 & \text{for } x \le 2 \\ x + 1 & \text{for } x > 2 \end{cases}$$

find $\lim\limits_{x\to2^-} f(x)$, $\lim\limits_{x\to2^+} f(x)$, and $\lim\limits_{x\to2} f(x)$.

Solution   The graph of f is shown in Figure 2.1. Based on the graph, the function is *discontinuous*, that is, broken at $x = 2$. (A formal definition of a continuous function will be given in a later section.)

$$\lim_{x\to2^-} f(x) = \lim_{x\to2^-} x^2 = 2^2 = 4$$

because the closer x gets to 2 "from below" or from the left side, the nearer $f(x)$ gets to 4.

$$\lim_{x\to2^+} f(x) = \lim_{x\to2} (x + 1) = 2 + 1 = 3$$

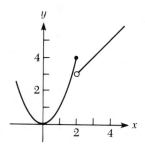

Figure 2.1 $f(x) = \begin{cases} x^2 & x \le 2 \\ x + 1 & x > 2 \end{cases}$

Since $x > 2$, then $f(x) = x + 1$. Finally, $\lim\limits_{x \to 2} f(x)$ does not exist. For this limit to exist the right- and left-hand limits must exist and be equal.

2.1 EXERCISES

1. Calculate $f(x) = 2x - 5$ for $x = 2.7, 2.8, 2.9, 3.2,$ and 3.1.

2. Calculate $f(x) = x^2 + 2x$ for $x = 1,$ $0.5,$ $0.1,$ $0.01,$ $0.001,$ $0.00001,$ $-1,$ $-0.5,$ $-0.1,$ $-0.01,$ $-0.001,$ and -0.00001.

3. What is the limit implied in Exercise 1?

4. What is the limit implied in Exercise 2?

In Exercises 5–23, find the indicated limits.

5. $\lim\limits_{x \to 3} (x - 3)$

6. $\lim\limits_{x \to 5} (x^2 + 2x)$

7. $\lim\limits_{x \to -2} \dfrac{x^2 + 5x + 4}{x + 1}$

8. $\lim\limits_{x \to -3} \dfrac{x^2 + 5x}{\sqrt{x^2 + 16}}$

9. $\lim\limits_{x \to 6} (3x^2 + 5x - 6)$

10. $\lim\limits_{x \to -a} \dfrac{x^2 - a^2}{x - a}$

11. $\lim\limits_{x \to 1} \dfrac{x^2 - 1}{x - 1}$

12. $\lim\limits_{x \to 5} \dfrac{x^2 - 25}{x - 5}$

13. $\lim\limits_{x \to -1} \dfrac{x^2 + x}{x + 1}$

14. $\lim\limits_{x \to 2} \dfrac{x^2 + 3x - 10}{x - 2}$

15. $\lim\limits_{x \to -8} \dfrac{x^2 + 9x + 8}{x + 8}$

16. $\lim\limits_{x \to 1} \dfrac{x^3 - 1}{x - 1}$

17. $\lim\limits_{x \to 2} \dfrac{x^3 - 8}{x - 2}$

18. $\lim\limits_{x \to 0} \dfrac{x^2 + 3x + 7}{x}$

19. $\lim\limits_{x \to 0} \dfrac{x^3 - 5x^2 + 7x + 4}{x}$

20. $\lim\limits_{x \to 4} \dfrac{x - 4}{\sqrt{x} - 2}$

21. $\lim\limits_{x \to a} \dfrac{\sqrt{x} - a}{x - a^2}$

22. $\lim\limits_{\Delta x \to 0} \dfrac{\sqrt{2 + \Delta x} - \sqrt{2}}{\Delta x}$

23. $\lim\limits_{\Delta x \to 0} \dfrac{\Delta x}{\sqrt{a + \Delta x} - \sqrt{a}}$

In Exercises 24–30, a function f and a value for a are given. Compute $\lim\limits_{\Delta x \to 0} \dfrac{f(a + \Delta x) - f(a)}{\Delta x}$.

24. $f(x) = 2x^2,$ $a = 1$

25. $f(x) = x^2 + 5x,$ $a = 3$

26. $f(x) = 3x^2 - 2x,$ $a = -2$

27. $f(x) = 3x^2 + 6x - 7,$ $a = 0$

28. $f(x) = 3x^2 + 9x, \quad a = x$ **29.** $f(x) = x^3, \quad a = x$

30. $f(x) = x^3 + 3x^2, \quad a = x$

In Exercises 31–38, consider the functions whose rules are given. Are these functions continuous, that is, unbroken graphs at the indicated values?

31. $f(x) = x - 3, \quad$ at $x = 3$

32. $f(x) = x^2 + 2x, \quad$ at $x = 5$

33. $f(x) = \dfrac{x^2 + 5x}{\sqrt{x^2 + 16}}, \quad$ at $x = -3$

34. $f(x) = \dfrac{x^2 - 1}{x - 1}, \quad$ at $x = 1$

35. $f(x) = \dfrac{x^2 + 3x - 10}{x - 2}, \quad$ at $x = 2$

36. $f(x) = \dfrac{x^2 + 3x + 7}{x}, \quad$ at $x = 0$

37. $f(x) = \dfrac{x^3 - 8}{x - 2}, \quad$ at $x = 2$

38. $f(x) = \dfrac{x^2 + 2x + 3}{x - 4}, \quad$ at $x = 2$

In Exercises 39–41, find the one-sided limits.

39. $\displaystyle\lim_{x \to 2^+} \dfrac{|x - 2|}{x - 2}$ **40.** $\displaystyle\lim_{x \to 2^+} \dfrac{|x - 2|}{2 - x}$

41. $\displaystyle\lim_{x \to 3^+} \dfrac{x^2 - 9}{x - 3}$

In Exercises 42 and 43, if the function f is defined by

$$f(x) = \begin{cases} x^2 & for\ x \in (-\infty, -2) \\ x + 1 & for\ x \in [-2, 1] \\ x^3 & for\ x \in (1, \infty) \end{cases}$$

42.   Find $\displaystyle\lim_{x \to -2^-} f(x)$ and $\displaystyle\lim_{x \to -2^+} f(x)$.

43.   Find $\displaystyle\lim_{x \to 1^+} f(x)$ and $\displaystyle\lim_{x \to 1^-} f(x)$.

2.2   LIMITS AS x TENDS TO INFINITY

Now that we have examined the limits of functions as the independent variables tend to finite values, the next step is to examine the behavior of functions as the argument becomes infinitely large.

Consider, for example, the function f such that $f(x) = 1/x$. The value of $f(x) = 1/x$ becomes progressively smaller for successively greater values of x. For example, if x has values $4, 5, 10, 100,$ and $1000, f(x)$ has corresponding values $\frac{1}{4}, \frac{1}{5}, \frac{1}{10}, \frac{1}{100},$ and $\frac{1}{1000}$. As x becomes infinitely large, $f(x)$ approaches zero, denoted

$$\lim_{x \to \infty} \frac{1}{x} = 0$$

Intuitively this limit seems obvious. In fact, intuition will serve very well to find most limits as x increases without bound, but it fails miserably in the proof of theorems. In order to prove theorems for functions as x becomes great without bound, we need the following definition.

Definition

If f is a function defined for all positive real numbers, then $f(x)$ approaches L as x tends to infinity, written

$$\lim_{x \to \infty} f(x) = L$$

if and only if for any positive number ε, no matter how small, there is a number N such that $|f(x) - L| < \varepsilon$ whenever $x > N$.

This means that if L is the limit of $f(x)$ as $x \to \infty$, then $f(x)$ can be made as close to L as we please by substituting any value for x sufficiently large (greater than N). Using this definition it is possible to prove that limits of functions of x as x tends to infinity, have many of the same properties as limits taken as x approaches a constant; that is, provided the limits exist. The reader should compare the following theorems with Theorems 2.3–2.6.

Theorem 2.9 $\quad \lim_{x \to \infty} [f(x) \pm g(x)] = \lim_{x \to \infty} f(x) \pm \lim_{x \to \infty} g(x)$

Theorem 2.10 $\quad \lim_{x \to \infty} [f(x) \cdot g(x)] = \lim_{x \to \infty} f(x) \cdot \lim_{x \to \infty} g(x)$

Theorem 2.11 $\quad \lim_{x \to \infty} [f(x)]^n = [\lim_{x \to \infty} f(x)]^n,$ n a rational number

Theorem 2.12 $\lim\limits_{x\to\infty} \dfrac{f(x)}{g(x)} = \dfrac{\lim\limits_{x\to\infty} f(x)}{\lim\limits_{x\to\infty} g(x)}$, provided $\lim\limits_{x\to\infty} g(x) \neq 0$

Theorem 2.13 $\lim\limits_{x\to\infty} c = c$ for c constant

Infinite limits involving the ratio of polynomials can often be evaluated by first dividing numerator and denominator by the highest power of the variable that appears in the denominator.

Example 1   Find $\lim\limits_{x\to\infty} \dfrac{3x + 5}{x}$.

Solution   Dividing numerator and denominator by x gives

$$\lim_{x\to\infty} \frac{3x + 5}{x} = \lim_{x\to\infty} \left(3 + \frac{5}{x} \right)$$

$$= \lim_{x\to\infty} 3 + 5 \lim_{x\to\infty} \frac{1}{x}$$

$$= 3 + 5(0) = 3$$

To solve the next example, first divide the numerator and denominator by x^2, since it is the highest power of x in the denominator.

Example 2   Find $\lim\limits_{x\to\infty} \dfrac{5x^2 + 9}{3x^2 - 7}$.

Solution $\lim\limits_{x\to\infty} \dfrac{5x^2 + 9}{3x^2 - 7} = \lim\limits_{x\to\infty} \dfrac{5 + (9/x^2)}{3 - (7/x^2)} = \dfrac{\lim\limits_{x\to\infty} 5 + \lim\limits_{x\to\infty} (9/x^2)}{\lim\limits_{x\to\infty} 3 - \lim\limits_{x\to\infty} (7/x^2)}$

$$= \frac{5 + 0}{3 - 0} = \frac{5}{3}$$

Example 3   Find $\lim\limits_{x\to\infty} \dfrac{5x + 9}{3x^2 - 7}$.

Solution   Divide numerator and denominator by x^2.

$$\lim_{x\to\infty} \frac{5x + 9}{3x^2 - 7} = \lim_{x\to\infty} \frac{(5/x) + (9/x^2)}{3 - (7/x^2)}$$

$$= \frac{0 + 0}{3 - 0} = 0$$

Example 4   Find $\lim\limits_{x\to\infty} \dfrac{5x^2 + 9}{3x - 7}$.

    Solution   Divide numerator and denominator by x.

$$\lim_{x\to\infty} \frac{5x^2 + 9}{3x - 7} = \lim_{x\to\infty} \frac{5x + (9/x)}{3 - (7/x)}$$

Since the numerator tends to $+\infty$ and the denominator tends to 3,

$$\lim_{x\to\infty} \frac{5x + (9/x)}{3 - (7/x)} = \infty$$

Example 5   Find $\lim\limits_{x\to\infty} \dfrac{5x^3}{\sqrt{x^6 + 1}}$.

    Solution   This example does not contain a ratio of polynomials and will therefore require a slightly different technique from the previous examples. The first step is to divide numerator and denominator by x^3. We can assume $x > 0$ (it approaches ∞); therefore $x^3 = \sqrt{x^6}$.

$$\lim_{x\to\infty} \frac{5x^3}{\sqrt{x^6 + 1}} = \lim_{x\to\infty} \frac{5}{\dfrac{\sqrt{x^6 + 1}}{x^3}}$$

$$= \lim_{x\to\infty} \frac{5}{\dfrac{\sqrt{x^6 + 1}}{\sqrt{x^6}}} = \lim_{x\to\infty} \frac{5}{\sqrt{\dfrac{x^6 + 1}{x^6}}}$$

$$= \lim_{x\to\infty} \frac{5}{\sqrt{1 + \dfrac{1}{x^6}}} = 5$$

Limits at negative infinity are treated in the same manner as those at positive infinity. The limit

$$\lim_{x\to-\infty} \frac{1}{x} = 0$$

is zero because the value of $f(x) = 1/x$ approaches zero from the left as negative numbers with greater and greater absolute value are substituted for x. If x is replaced by -10, -1000, -5000, $-1,000,000$, then $f(x)$ has corresponding values $-\dfrac{1}{10}$, $-\dfrac{1}{1000}$,

$-\dfrac{1}{5000}$, $-\dfrac{1}{1,000,000}$. Theorems 2.9–2.13 are equally true when ∞ is replaced by $-\infty$.

Example 6

Find $\displaystyle\lim_{x\to-\infty} \dfrac{3x^2 + 8x + 6}{4x^2 - 2x - 50}$.

Solution   Divide numerator and denominator by x^2.

$$\lim_{x\to-\infty} \frac{3x^2 + 8x + 6}{4x^2 - 2x - 50} = \lim_{x\to-\infty} \frac{3 + \dfrac{8}{x} + \dfrac{6}{x^2}}{4 - \dfrac{2}{x} - \dfrac{50}{x^2}}$$

$$= \frac{3 + 0 + 0}{4 - 0 - 0} = \frac{3}{4}$$

2.2   EXERCISES

In Exercises 1–28, find the indicated limits.

1. $\displaystyle\lim_{x\to\infty} \dfrac{1}{x^4}$

2. $\displaystyle\lim_{x\to\infty} \dfrac{5}{x^3}$

3. $\displaystyle\lim_{x\to\infty} \dfrac{4x + 7}{8x + 3}$

4. $\displaystyle\lim_{x\to\infty} \dfrac{-x + 4}{x - 7}$

5. $\displaystyle\lim_{x\to\infty} \dfrac{3x^2 + 5x + 2}{5x^2 + 2x + 7}$

6. $\displaystyle\lim_{x\to\infty} \dfrac{ax^2 + bx + c}{cx^2 + bx + a}$,
$a, b,$ and c nonzero constants

7. $\displaystyle\lim_{x\to\infty} \dfrac{-3x^2 + 2x - 1}{x^2 - 3x}$

8. $\displaystyle\lim_{x\to\infty} \dfrac{x^2 - x}{-2x^2 + 5}$

9. $\displaystyle\lim_{x\to\infty} \dfrac{2x^2 + 3x - 100}{4x + 9}$

10. $\displaystyle\lim_{x\to\infty} \dfrac{4x + 9}{2x^2 + 3x - 100}$

11. $\displaystyle\lim_{x\to\infty} \dfrac{13 - 4x^2}{5x^3 - 8x^2}$

12. $\displaystyle\lim_{x\to\infty} \dfrac{-7x^4 + 5x^3}{3x^3 + 5x^2}$

13. $\displaystyle\lim_{x\to\infty} \dfrac{-4x^5 + 5x^4 - 8}{-3x^4 + 12x^5 - 9}$

14. $\displaystyle\lim_{x\to\infty} \dfrac{2x^2 - 3x^3 + x}{x - 2x^2 - x^3}$

15. $\displaystyle\lim_{x\to\infty} \dfrac{\sqrt{4x^2 + 5}}{x}$

16. $\displaystyle\lim_{x\to\infty} \dfrac{\sqrt{3x^2 + 2}}{x - 8}$

17. $\displaystyle\lim_{x\to\infty} \dfrac{-5x^2 + x}{\sqrt{16x^4 - 8}}$

18. $\displaystyle\lim_{x\to\infty} \dfrac{x^4 + x^3 + 5x^2}{\sqrt{x^6 - 2}}$

19. $\lim\limits_{x \to \infty} \dfrac{\sqrt{x + 2}}{\sqrt{x^2 - 5}}$

20. $\lim\limits_{x \to \infty} \dfrac{(x - 2)(x + 2)(x - 1)}{x^3 - 3x^2 + 3x + 1}$

21. $\lim\limits_{x \to -\infty} \dfrac{-2x^2 - 8}{5x^2 + 25}$

22. $\lim\limits_{x \to -\infty} \dfrac{x}{\sqrt{x^2 - 1}}$

23. $\lim\limits_{x \to -\infty} \dfrac{x^3 - 5x^2}{\sqrt{x}}$

24. $\lim\limits_{x \to -\infty} \dfrac{x^3}{\sqrt{1 - x^6}}$

25. $\lim\limits_{x \to -\infty} \dfrac{x^2 - 2x + 1}{-\sqrt{x^4 + 5x}}$

26. $\lim\limits_{x \to -\infty} \dfrac{\sqrt{x^3 + 80}}{x^2}$

27. $\lim\limits_{x \to -\infty} \dfrac{(x - 2)(x + 2)(x - 1)}{1 - 3x^2 + 2x + x^3}$

28. $\lim\limits_{x \to -\infty} \dfrac{(x - 2)(2 - x)}{x^2 + 2x + 1}$

2.3 SLOPE AND MARGINAL COST

The average rate of change has been used to describe the growth rate of a population of bacteria and, earlier, to calculate speed. It can also be interpreted as marginal cost or, geometrically, as the slope of a line.

Definition

The *slope, m*, of the line segment between points (x_1, y_1) and (x_2, y_2) is

$$m = \frac{y_2 - y_1}{x_2 - x_1}$$

If $x_1 = x_2$, the slope is undefined.

Recall that in Chapter 1 we found it convenient to use the relations $\Delta y = y_2 - y_1$ and $\Delta x = x_2 - x_1$. Thus, the slope is also the average rate of change of y as x changes from x_1 to x_2, since

$$m = \frac{y_2 - y_1}{x_2 - x_1} = \frac{\Delta y}{\Delta x}$$

When computing the slope of a line segment, either point may be designated as (x_2, y_2), for the order of subtraction is not important. This may be shown as follows:

$$m = \frac{y_2 - y_1}{x_2 - x_1}$$

Multiplying numerator and denominator by -1,

$$m = \frac{-1(y_2 - y_1)}{-1(x_2 - x_1)} = \frac{-y_2 + y_1}{-x_2 + x_1}$$
$$= \frac{y_1 - y_2}{x_1 \quad x_2}$$

Example 1   Find the slope of a line segment between points $(-5, 6)$ and $(3, -2)$. See Figure 2.2.

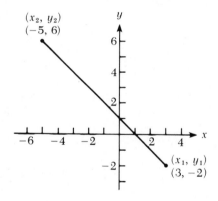

Figure 2.2

Solution   Arbitrarily let $(-5, 6)$ be (x_2, y_2), $(3, -2)$ be (x_1, y_1), and apply the slope formula.

$$m = \frac{y_2 - y_1}{x_2 \quad x_1} = \frac{6 - (-2)}{-5 - 3}$$
$$= \frac{6 + 2}{-8} = \frac{8}{-8} = -1$$

The slope of a straight line is, of course, constant. If three or more points are on a single line, the slope computed by using any two of them will be the same as that computed by using any other two.

Example 2   In Figure 2.3 points $(2, 5)$, $(0, 1)$, and $(-2, -3)$ are all on the same line. Compute the slope of the line.

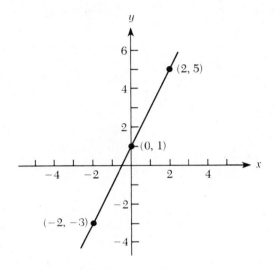

Figure 2.3   Computation of the slope of a line through three points

Solution   If the slope is computed using $(2, 5)$ and $(-2, -3)$ it is:

$$m = \frac{5 - (-3)}{2 - (-2)} = \frac{8}{4} = 2$$

If the slope is computed using $(2, 5)$ and $(0, 1)$ it is:

$$m = \frac{5 - 1}{2 - 0} = \frac{4}{2} = 2$$

If the slope m of a line and a point (x_1, y_1) on the line are known, the equation of the line can easily be found. If (x, y) is any other point on the line, then

$$\frac{y - y_1}{x - x_1} = m$$

and

$$y - y_1 = m(x - x_1)$$

This is called the *point–slope form* of a linear equation.

Example 3

Find an equation of the line that has slope 6 and passes through $(4, -8)$.

Solution

Substituting into the point–slope form of a linear equation:

$$y - y_1 = m(x - x_1)$$
$$y - (-8) = 6(x - 4)$$
$$y + 8 = 6x - 24$$
$$y = 6x - 32$$

If a line has slope m and intersects the y axis at $(0, b)$, we can use the point–slope form to find its equation:

$$y - y_1 = m(x - x_1)$$
$$y - b = m(x - 0)$$
$$y - b = mx$$
$$y = mx + b$$

$y = mx + b$ is called the *slope–intercept form* of a linear equation.

Example 4

Find the slope and y intercept of the graph of $3x - 9y = 5$.

Solution

Put the equation in the slope–intercept form. The coefficient of x will be the slope and the constant term the y intercept.

$$3x - 9y = 5$$
$$-9y = -3x + 5$$
$$y = \frac{-3x}{-9} + \frac{5}{-9}$$
$$y = \frac{1}{3} x - \frac{5}{9}$$

Therefore the slope is $\frac{1}{3}$, and the line crosses the y axis at $-\frac{5}{9}$.

If a linear equation represents the total cost of manufacturing a certain number of items, in terms of the number of items, then the economist's term for the slope of the graph of the equation is *marginal* cost per item. One further restriction is required to interpret a linear equation of cost. The domain must be restricted to the nonnegative numbers, because it is unrealistic to talk about a negative number of items. The marginal cost when the equation is nonlinear will be considered later.

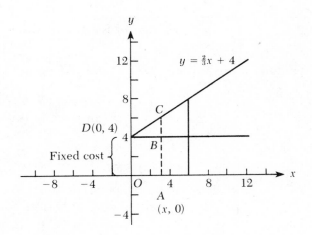

Figure 2.4 Graph of $y = \frac{2}{3}x + 4$

The equation $y = \frac{2}{3}x + 4$ is graphed in Figure 2.4. Suppose y represents the cost in dollars of manufacturing x number of items. The point D $(0, 4)$, on the line means that the cost of producing 0 items is $4. The reason that there is some cost before any items are produced is that there is some capital outlay before production can begin, for such things as supplies. Thus, the line segment OD, which represents $4, is called *fixed cost*. The length of line segment AC represents the cost of manufacturing x items. The length of AB, which is equal to that of OD, also represents the fixed cost, and that of BC is called the *variable cost for x items*. Thus, if $x = 3$, then the cost, $y = \frac{2}{3}(3) + 4 = \$6$; $4 of the $6 is fixed cost, and $2 is the variable cost for 3 items. The ratio of BC to DB is the marginal cost per item (slope). In this case it is $\frac{2}{3}$ of a dollar.

The average cost per item is simply the total cost divided by the number of items produced. The average cost of producing 12 items is the total cost, $12, divided by 12, or $1 per item. The average cost of producing 30 items is the total cost, $24, divided by 30, or $0.80. The average cost per item for 30 items is lower than that for 12 items because the fixed cost is spread over a greater number of items.

2.3 EXERCISES

In Exercises 1–6, find the slope of the line segments between the given points:

1. $(0, 0)$ and $(4, 5)$ **2.** $(-1, -1)$ and $(3, 3)$

3. $(-5, 4)$ and $(2, -1)$ **4.** $(4, 3)$ and $(-2, 3)$

5. $(-1, -2)$ and $(-1, 6)$ **6.** $(4, 3)$ and $(4, 17)$

In Exercises 7–12, find the slope and y intercept.

7. $y = -3x + 5$ **8.** $2x = 5y + 7$

9. $4x - 3y + 7 = 0$ **10.** $x = 4$

11. $y = 4$ **12.** $5x - 19y = 28$

In Exercises 13–18, find an equation of the line with the given slope m, that contains the given point.

13. $m = 1, \quad (3, 3)$ **14.** $m = 5, \quad (-2, 4)$

15. $m = \frac{3}{4}, \quad (1, 4)$ **16.** $m = -2, \quad (-3, -5)$

17. $m = 0, \quad (-3, 2)$ **18.** $m = -\frac{5}{8}, \quad (-3, 7)$

In Exercises 19–26, find an equation of the line that contains the given points.

19. $(0, 0), (2, 2)$ **20.** $(3, 4), (-1, 3)$

21. $(1, -2), (-3, 5)$ **22.** $(4, 2), (-8, 2)$

23. $(6, -5), (0, -5)$ **24.** $(1, 0), (1, 4)$

25. $(-5, 2), (-5, -2)$ **26.** $(2, 3), (7, -2)$

Ⓢ 27. What are the economist's terms for m and b in the equation $y = mx + b$?

Ⓢ *In Exercises 28–33, y represents the cost in dollars of manufacturing x items. The number of items produced is given with each equation. Find the following for each equation:*

   a. Total cost   **b.** Average cost per item

   c. Fixed cost   **d.** Variable cost

   e. Marginal cost per item

28. $y = 5x + 40, \quad 100$ items produced

29. $y = 5x + 40, \quad 10$ items produced

30. $y = 8x + 10, \quad 40$ items produced

31. $y = 8x + 10, \quad 80$ items produced

32. $y = 50x + 1000, \quad 90$ items produced

33. $y = 50x + 1000, \quad 120$ items produced

34. Give four possible interpretations of average rate of change of a function over increments of its domain.

2.4 THE DERIVATIVE OF A FUNCTION

The derivative is the limit of the average rate of change as the increment of the independent variable approaches zero. The derivative will have similar interpretations to those of the average rate of change, except they will be instantaneous. That is, the derivative will evaluate instantaneous velocity, instantaneous growth rate, and instantaneous marginal cost, rather than averages. This list of possible interpretations is by no means exhaustive. We will show you many more uses for the derivative, but before that is possible, convenient methods for its evaluation are needed.

To find the derivative of a function at a point in its domain, it is necessary that the function be continuous at the point. Intuitively, a continuous function is one whose graph has no breaks. In Figure 2.5, $y = f(x)$ is discontinuous at $x = a$ and at $x = b$, where a single point is missing in its graph.

Definition

A function f is *continuous* at a if $\lim_{x \to a} f(x) = f(a)$. If a function is continuous for all values of x in an interval from a to b, then it will be said to be *continuous on the interval*.

The definition requires three things: $f(a)$ must exist, $\lim_{x \to a} f(x)$ must exist, and they must be equal.

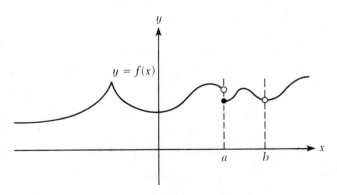

Figure 2.5 A discontinuous function

| **Example 1** | Is $f(x) = x^2 - 4$ continuous at $x = 3$? |
| **Solution** | Since $\lim_{x \to 3} f(x) = 9 - 4 = 5 = f(3)$, the function is continuous. |

| **Example 2** | Is $f(x) = 1/x$ continuous at $x = 0$? |
| **Solution** | Since $\lim_{x \to 0} (1/x)$ does not exist and $f(0)$ does not exist, $f(x) = 1/x$ is discontinuous at $x = 0$. The function $f(x)$ is continuous on the interval $1 \le x \le 17$, or the interval $-2 \le x < 0$, but not over an interval that contains 0. |

If the average rate of change of a function of x is taken over smaller and smaller intervals of its domain, it approaches "the instantaneous rate of change" of the function at x. Recall that the average rate of change had several interpretations; the instantaneous rate of change will also have many useful applications. In fact, because of the many potential applications, it is given a special name. Since its value is derived from the original function, it will be called the *derivative* of the function. Reference to the derivative as an instantaneous rate of change lies in the historic role the derivative has played in physical problems related to velocity and motion, where it refers to the instantaneous rate of change of position.

Definition

For $y = f(x)$, f a function, *the derivative of y with respect to x* is a new function with image found by

$$\lim_{\Delta x \to 0} \frac{\Delta y}{\Delta x} = \lim_{\Delta x \to 0} \frac{f(x + \Delta x) - f(x)}{\Delta x}$$

provided the limit exists. When $\lim_{\Delta x \to 0} (\Delta y / \Delta x)$ exists, y is said to be a *differentiable* function with respect to x. The derivative of f is a function of x.

Example 3   Find the derivative of y with respect to x for $y = x^2 + 2x$.

Solution
$$\lim_{\Delta x \to 0} \frac{\Delta y}{\Delta x} = \lim_{\Delta x \to 0} \frac{f(x + \Delta x) - f(x)}{\Delta x}$$
$$= \lim_{\Delta x \to 0} \frac{[(x + \Delta x)^2 + 2(x + \Delta x)] - (x^2 + 2x)}{\Delta x}$$
$$= \lim_{\Delta x \to 0} \frac{x^2 + 2x\Delta x + (\Delta x)^2 + 2x + 2\Delta x - x^2 - 2x}{\Delta x}$$
$$= \lim_{\Delta x \to 0} \frac{2x\Delta x + (\Delta x)^2 + 2\Delta x}{\Delta x}$$
$$= \lim_{\Delta x \to 0} (2x + \Delta x + 2) = 2x + 2$$

Therefore the derivative of y with respect to x is $2x + 2$.

The derivative has traditionally had several notations. They include y', f', dy/dx, df/dx, $D_x f$, and $D_x y$ for $y = f(x)$. Each of these notations reflects one or more special characteristics of the derivative. The choice of a particular form of notation depends on the meaning or characteristic most appropriate for a specific problem. The prime (') notations are compact and denote that the new function has been derived from the old. The dy/dx or df/dx notations remind one of the $\Delta y/\Delta x$ from the average rate of change, and later will remind the user of some of the fraction-like properties of the derivative. The $D_x f$ indicates the operation of differentiation working on a function. All of these notations will be used interchangeably.

In Example 3, the derivative of y with respect to x for $y = f(x) = x^2 + 2x$ was found to be $2x + 2$. Using some of the many notations for derivatives, this could be written

$$y' = 2x + 2 \qquad\qquad \frac{df}{dx} = 2x + 2$$
$$f'(x) = 2x + 2 \qquad\qquad D_x(x^2 + 2x) = 2x + 2$$
$$\frac{dy}{dx} = 2x + 2 \qquad\qquad D_x y = 2x + 2$$

Notice here that $f'(x)$ refers to the image of x with respect to the new function f'.

Example 4   Find $f'(x)$ for $f(x) = 1/x$.

Solution $f'(x) = \lim\limits_{\Delta x \to 0} \dfrac{f(x + \Delta x) - f(x)}{\Delta x}$

$$f'(x) = \lim_{\Delta x \to 0} \frac{\dfrac{1}{x + \Delta x} - \dfrac{1}{x}}{\Delta x}$$

$$= \lim_{\Delta x \to 0} \frac{\dfrac{x - (x + \Delta x)}{x(x + \Delta x)}}{\Delta x}$$

$$= \lim_{\Delta x \to 0} \frac{x - x - \Delta x}{(\Delta x \cdot x)(x + \Delta x)}$$

$$= \lim_{\Delta x \to 0} \frac{-\Delta x}{(\Delta x \cdot x)(x + \Delta x)}$$

$$= \lim_{\Delta x \to 0} \frac{-1}{x(x + \Delta x)} = \frac{-1}{x^2}$$

For $y = f(x)$, the derivative at x is the instantaneous rate of change of y with respect to x. That is, it is the limit of the average rate of change as $\Delta x \to 0$. For instance, if y gives the amount of bacteria in a culture at time x, then the derivative is the growth rate at a specific time. It is also the slope of the tangent to a curve for any given x. In Figure 2.6 points $P(x, y)$ and P_1 are located on the graph of $y = f(x)$. The slope of the line through P_1 and P is $\Delta y/\Delta x$. As Δx approaches zero, point P_1 moves closer to point P. The tangent line at P is the limiting position for the secant line.

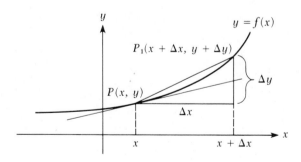

Figure 2.6 Geometrical interpretation of the derivative

Example 5   If $y = \sqrt{x}$, find the slope of the tangent line to the curve at $x = 4$ (see Figure 2.7).

Solution   The slope of the tangent is the derivative evaluated at $x = 4$.

$$y' = \lim_{\Delta x \to 0} \frac{\sqrt{x + \Delta x} - \sqrt{x}}{\Delta x}$$

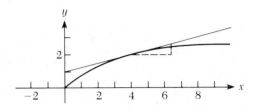

Figure 2.7 Graph of $y = \sqrt{x}$, $x \geq 0$, and $y \geq 0$

Rationalizing the numerator,

$$y' = \lim_{\Delta x \to 0} \left(\frac{\sqrt{x + \Delta x} - \sqrt{x}}{\Delta x} \right) \left(\frac{\sqrt{x + \Delta x} + \sqrt{x}}{\sqrt{x + \Delta x} + \sqrt{x}} \right)$$

$$= \lim_{\Delta x \to 0} \frac{x + \Delta x - x}{\Delta x (\sqrt{x + \Delta x} + \sqrt{x})}$$

$$= \lim_{\Delta x \to 0} \frac{1}{\sqrt{x + \Delta x} + \sqrt{x}}$$

$$= \frac{1}{\sqrt{x} + \sqrt{x}} = \frac{1}{2\sqrt{x}}$$

Since we know the slope of the tangent at $x = 4$ is y' at $x = 4$,

$$y'(4) = \frac{1}{2\sqrt{4}} = \frac{1}{4}$$

It is interesting to note the relationship between continuous functions and the derivative. If the derivative of a function exists for a specific argument of the function, then the function must be continuous at that value. If a function is continuous at a value, then the derivative may *or* may not exist. The proof of this statement can be found in any Engineering/Mathematics-oriented calculus text.

2.4 EXERCISES

In Exercises 1–4, find dy/dx.

1. $y = x^2$

2. $y = 7x + 4$

3. $y = -6x + 2$

4. $y = ax^2 + bx + c$, a, b, and c constants

In Exercises 5–8, find $D_x y$.

5. $y = 7 - x^2$

6. $y = x^3 - 2x$

7. $y = \dfrac{1}{x^2}$

8. $y = \dfrac{1}{\sqrt{x}}$

In Exercises 9–14, find the value of the derivative at the given point.

9. $f'(3)$ for $f(x) = \dfrac{x+1}{x}$

10. $f'(2)$ for $f(t) = 16t - t^2$

11. $h'(-1)$ for $h(z) = z^2 - 3z + 4$

12. $g'(4)$ for $g(\theta) = \theta^2 - 8\theta$

13. $f'(0)$ for $f(t) = 3t^3$

14. $f'(1)$ for $f(x) = \dfrac{1}{x+3}$

In Exercises 15–18, find the slope of the tangent line at the value indicated.

15. $f(x) = \dfrac{1}{x+1}$ at $x = 5$

16. $f(x) = \dfrac{x-1}{x+1}$ at $x = 0$

17. $f(x) = \dfrac{x}{x+1}$ at $x = 0$

18. $f(x) = x^4$ at $x = 2$

⑤ 19. The cost of producing x items is given in dollars by $c = 500 + x + (1/x)$. If the marginal cost is redefined as dc/dx, find the marginal cost of producing the tenth item.

20. By setting $dy/dx = 0$, find the point on the graph of $4y = x^2 - 8x + 12$ where the tangent is parallel to the x axis. Sketch a figure to illustrate.

⑤ 21. If $R = f(I)$, where R is the return on an investment of I amount, then dR/dI is called the marginal efficiency of investment. Find the marginal efficiency of investment if $R = 3I^2 - 2I$.

⑤ 22. If P is the measure of production of x machines or workers (that is, $P = f(x)$), then dP/dx is called the marginal physical productivity. Find the marginal physical productivity of a group of x lumberjacks, where P is given in thousands of board feet of lumber per day, and $P = -4x^2 + 10,000x$. According to this formula, if there are 2500 lumberjacks then $P = 0$; conjecture how such a large number of lumberjacks can result in no production.

In Exercises 23 and 24, are the functions continuous at the values indicated? State your reasons.

23. $f(x) = \dfrac{|x|}{x}$ at $x = 1$ **24.** $f(x) = \dfrac{|x|}{x}$ at $x = 0$

In Exercises 25 and 26, if the function f is defined by

$$f(x) = \begin{cases} x^2 & \text{for } x \in (-\infty, -2) \\ x + 1 & \text{for } x \in [-2, 1] \\ x^3 & \text{for } x \in (1, \infty) \end{cases}$$

is f continuous:

25.   at $x = -2$?   **26.**   at $x = 1$?

27.   Consider a function f such that $f'(x) = \dfrac{1}{x^2 + 1}$ for all real values of x. Is this function continuous at $x = -1$?

2.5   THE DERIVATIVE OF POLYNOMIALS

The definition of the derivative from the last section can be used to find the derivative of any function. However, this process often involves considerably more labor than necessary. The objective of this and the next few sections is to develop simple formulas to facilitate finding derivatives.

Theorem 2.14   If f is a function such that $y = f(x) = C$, C constant, then $y' = 0$.

Proof $f(x) = C$ for any x; therefore, $f(x + \Delta x) = C$. Then,

$$y' = \lim_{\Delta x \to 0} \frac{f(x + \Delta x) - f(x)}{\Delta x}$$

$$= \lim_{\Delta x \to 0} \frac{C - C}{\Delta x} = 0$$

This proves the theorem.

If we think of the derivative as the slope of the tangent line, the theorem becomes obvious. The graph of the equation $y = C$ for C constant is parallel to the x axis (see Figure 2.8). Hence, its slope for all x is the same as the slope of its tangent, which is 0.

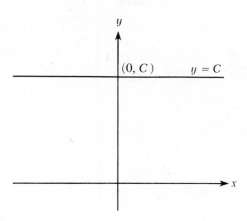

Figure 2.8 Graph of $y = C$

Example 1   If $f(x) = 7$, find $f'(x)$.

   Solution   From Theorem 2.14 we know

$$f'(x) = 0 \quad \text{for all } x$$

The next theorem is proved by use of the binomial theorem. The notation $n!$ is read "n factorial" and means

$$n(n - 1)(n - 2) \cdots 1$$

Thus, $3! = 3(2)(1) = 6$, and $5! = 5(4)(3)(2)(1) = 120$. $1!$ is defined equal to 1.

   The binomial theorem states that if a and b are real numbers, and n is a positive integer, then

$$(a + b)^n = a^n + \frac{na^{n-1}b^1}{1!} + \frac{n(n - 1)a^{n-2}b^2}{2!}$$
$$+ \frac{n(n - 1)(n - 2)a^{n-3}b^3}{3!} + \cdots + b^n$$

Example 2   Use the binomial theorem to find $(a + b)^3$.

   Solution $(a + b)^3 = a^3 + \dfrac{3a^2b}{1!} + \dfrac{(3)2ab^2}{2!} + b^3$

            $= a^3 + 3a^2b + 3ab^2 + b^3$

Theorem 2.15   If $y = x^n$ for n a positive integer, then $y' = nx^{n-1}$.

Proof   By definition of the derivative from the last section,

$$y' = \lim_{\Delta x \to 0} \frac{(x + \Delta x)^n - x^n}{\Delta x}$$

Using the binomial theorem to expand $(x + \Delta x)^n$ gives

$$y' = \lim_{\Delta x \to 0} \frac{\left(x^n + \dfrac{n\Delta x x^{n-1}}{1!} + \dfrac{n(n-1)(\Delta x)^2 x^{n-2}}{2!} + \cdots + (\Delta x)^n\right) - x^n}{\Delta x}$$

$$= \lim_{\Delta x \to 0} \frac{\dfrac{n\Delta x x^{n-1}}{1!} + \dfrac{n(n-1)(\Delta x)^2 x^{n-2}}{2!} + \cdots + (\Delta x)^n}{\Delta x}$$

$$= \lim_{\Delta x \to 0} \left[\frac{nx^{n-1}}{1!} + \frac{n(n-1)\Delta x x^{n-2}}{2!} + \cdots + (\Delta x)^{n-1}\right]$$

In the last line each term except the first contains at least one Δx as a factor. Therefore as $\Delta x \to 0$ each of these will vanish and the derivative is the first term.

$$y' = \frac{nx^{n-1}}{1!} = nx^{n-1}$$

Example 3   Find the derivatives of $y = x^3$ and $y = x^5$.

Solution   For $y = x^3$, $y' = 3x^2$. For $y = x^5$, $y' = 5x^4$.

Theorem 2.15 was proved for n a positive integer. It is also true for all rational numbers. Proofs for negative integers will be given in the section on the derivative of the quotient of two functions and for rational numbers in the section on the chain rule.

Example 4   Find y' for $y = 1/x^2$.

Solution   Write with negative exponent and apply Theorem 2.15. Then $y = 1/x^2$ is equivalent to $y = x^{-2}$. Therefore,

$$y' = -2x^{-2-1} = -2x^{-3} = \frac{-2}{x^3}$$

Example 5   Find $f'(x)$ for $f(x) = \sqrt{x}$.

Solution   Write $\sqrt{x}$ as $x^{1/2}$ and apply Theorem 2.15.

$$y = \sqrt{x} = x^{1/2}$$

Therefore,

$$y' = \frac{1}{2} x^{1/2-1} = \frac{1}{2} x^{-1/2} = \frac{1}{2x^{1/2}} = \frac{1}{2\sqrt{x}}$$

Example 6   Find the derivative of $y = x^{2/3}$.

Solution   For $y = x^{2/3}$, $y' = \frac{2}{3} x^{-1/3} = \frac{2}{3x^{1/3}}$.

Theorem 2.16   If $y = cf(x)$ for some function f, then $y' = cf'(x)$, which is to say the derivative of a constant times a function is the constant times the derivative of the function.

Proof   We must prove that if f is a differentiable function, c is a constant, and $y = cf(x)$, then $y' = cf'(x)$.

$$y' = \lim_{\Delta x \to 0} \frac{cf(x + \Delta x) - cf(x)}{\Delta x}$$

Applying limit theorems this becomes

$$y' = c \lim_{\Delta x \to 0} \frac{f(x + \Delta x) - f(x)}{\Delta x}$$
$$= cf'(x)$$

Example 7   Find $D_x y$ for $y = 3x^2$.

Solution   By Theorem 2.16 $D_x y$ is 3 times the derivative of x^2, which is $2x$. Therefore,

$$D_x y = 6x$$

Example 8   Find the derivative of $y = cx^n$, where c is a constant and n is a rational number.

    Solution $y = cx^n$, then $y' = cnx^{n-1}$.

Theorem 2.17   The derivative of the sum of differentiable functions is the sum of the derivatives. If f and g are differentiable functions, and $y = f(x) + g(x)$, then $y' = f'(x) + g'(x)$.

    Proof   From the definition of y',

$$\begin{aligned}
y' &= \lim_{\Delta x \to 0} \frac{f(x + \Delta x) + g(x + \Delta x) - [f(x) + g(x)]}{\Delta x} \\
&= \lim_{\Delta x \to 0} \frac{f(x + \Delta x) - f(x) + g(x + \Delta x) - g(x)}{\Delta x} \\
&= \lim_{\Delta x \to 0} \left[\frac{f(x + \Delta x) - f(x)}{\Delta x} + \frac{g(x + \Delta x) - g(x)}{\Delta x} \right] \\
&= \lim_{\Delta x \to 0} \frac{f(x + \Delta x) - f(x)}{\Delta x} + \lim_{\Delta x \to 0} \frac{g(x + \Delta x) - g(x)}{\Delta x} \\
&= f'(x) + g'(x)
\end{aligned}$$

Example 9   Find y' when $y = x^2 + x^3$.

    Solution   Since y' is the sum of the derivatives of x^2 and x^3,

$$y' = 2x + 3x^2$$

    Theorem 2.17 can be extended to the sum of any number of terms. Using this theorem and the others, the derivative of a polynomial is easily found.

Example 10   Find the derivative of $y = 7x^2 - 8x + 3$.

    Solution   The derivative of y is the sum of the derivatives of the three terms of the polynomial.

$$\begin{aligned}
D_x(7x^2) &= 2(7)x^1 = 14x \\
D_x(-8x) &= D_x(-8x^1) = (-8)(1)x^0 = -8 \\
D_x(3) &= 0
\end{aligned}$$

since the derivative of a constant is 0. Therefore,

$$y' = 14x - 8$$

Example 11   Find the derivative of $y = 7x^5 + 3x^3 + 7x^2 - x + x^{-2}$.

Solution   Take the derivative of each term. The fourth term, x, has derivative 1, since $x = x^1$; and $D_x(x^1) = 1x^0 = 1$, since $x^0 = 1$. Therefore,

$$y' = 35x^4 + 9x^2 + 14x - 1 - 2x^{-3}$$

Example 12   Find y' for $y = (x + 3)^2$.

Solution   First expand $(x + 3)^2$; then take the derivative term by term.

$$y = (x + 3)^2 = x^2 + 6x + 9$$
$$y' = 2x + 6$$

Example 13   Find y' if $y = \dfrac{3x^2 + 2}{\sqrt{x}}$.

Solution
$$y = \frac{3x^2 + 2}{\sqrt{x}}$$
$$= \frac{3x^2}{\sqrt{x}} + \frac{2}{\sqrt{x}}$$
$$= 3x^{3/2} + 2x^{-1/2}$$

Thus,

$$y' = 3\left(\frac{3}{2}\right)x^{1/2} + 2\left(-\frac{1}{2}\right)x^{-3/2}$$
$$= \frac{9}{2}x^{1/2} - x^{-3/2}$$

Once the form of the derivative has been found, the next natural step is the evaluation of the derivative for specific arguments.

Example 14   If $f(x) = \dfrac{3x^2 + 2}{\sqrt{x}}$, find $f'(81)$.

Solution   From the preceding example we know, if $f(x) = \dfrac{3x^2 + 2}{\sqrt{x}}$, then

$$f'(x) = \frac{9}{2}\,x^{1/2} - x^{-3/2}$$

Thus,

$$f'(81) = \frac{9}{2}\,(81)^{1/2} - (81)^{-3/2}$$

$$= \frac{9}{2}\,(9) - \frac{1}{(81)^{3/2}}$$

$$= \frac{81}{2} - \frac{1}{9^3}$$

$$= \left(\frac{81}{2}\right)\left(\frac{9^3}{9^3}\right) - \left(\frac{2}{2}\right)\left(\frac{1}{9^3}\right) = \frac{59{,}049 - 2}{1458} = \frac{59{,}047}{1458}$$

Example 15   If $f(x) = 3 - 4x + x^2$, find $f'(-1)$.

Solution $f'(x) = -4 + 2x$

Therefore,

$$f'(-1) = -4 + 2(-1)$$
$$= -4 - 2 = -6$$

2.5 EXERCISES

In Exercises 1–10, find dy/dx.

1. $y = x^3$ **2.** $y = x^2$

3. $y = 7x^5$ **4.** $y = 4x^4$

5. $y = 12x^{-3}$ **6.** $y = -x^{-5}$

7. $y = \sqrt{x}$ **8.** $y = x^{4/5}$

9. $y = 9x^{5/3}$ **10.** $y = -3x^{-4/3}$

In Exercises 11–16, find $D_x f$.

11. $f(x) = 3x^2 - 6x + 2$ **12.** $f(x) = 4 - 7x - 8x^3$

13. $f(x) = 3x^2 - 4x$ **14.** $f(x) = 4x^{12} + 5x^{11} - 6x^9 + 7000$

15. $f(x) = ax^2 + bx + c$,   **16.** $f(x) = 3ax^4 - \dfrac{bx^6}{6}$,
 $a, b,$ and c
 constants a and b constants

In Exercises 17–20, evaluate the derivative for the indicated value of the function's domain.

17. $f'(\frac{1}{2})$ for $f(x) = x^{-1} + x^{-2}$ **18.** $f'(0)$ for $f(x) = \sqrt[3]{x^2}$

19. $f'(1)$ for $f(x) = \dfrac{1}{x} + \dfrac{1}{x^3}$ **20.** $f'(16)$ for $f(x) = \sqrt{x} + \dfrac{4}{\sqrt{x}}$

In Exercises 21–26, find $D_x y$.

21. $y = \dfrac{x^5 + 3x^2 + 1}{x^2}$ **22.** $y = \dfrac{x + 1}{\sqrt{x}}$

23. $y = \dfrac{x^3 + 3x^2 + x}{\sqrt{x}}$ **24.** $y = (x + 4)^2$

25. $y = (x^2 + 3x)(x + 1)$ **26.** $y = x(x + 1)(x - 1)$

In Exercises 27–30, solve for y and find y'.

27. $xy + x = 2$ **28.** $x^2 y + 3x = 4x^2$

29. $x\sqrt{y} + 3x = 5$ **30.** $xy^2 = 16$

2.6 THE DERIVATIVE OF PRODUCTS AND QUOTIENTS

In the last section it was possible to change the products or the quotients of functions to polynomials. It is not always possible to do so.

Theorem 2.18 The derivative of the product of two functions is the first function times the derivative of the second plus the second function times the derivative of the first function. That is, if $v = v(x)$ and $u = u(x)$ represent differentiable functions, then

$$\frac{d(uv)}{dx} = u\frac{dv}{dx} + v\frac{du}{dx}$$

Proof We apply the definition of the derivative to $uv = u(x)v(x)$.

$$\frac{d(uv)}{dx} = \lim_{\Delta x \to 0} \frac{u(x + \Delta x)v(x + \Delta x) - u(x)v(x)}{\Delta x}$$

To the numerator, subtract and add a term $u(x + \Delta x)v(x)$.

$$\frac{d(uv)}{dx} = \lim_{\Delta x \to 0} \frac{u(x + \Delta x)v(x + \Delta x) - u(x + \Delta x)v(x) + u(x + \Delta x)v(x) - u(x)v(x)}{\Delta x}$$

$$= \lim_{\Delta x \to 0} \frac{u(x + \Delta x)[v(x + \Delta x) - v(x)] + v(x)[u(x + \Delta x) - u(x)]}{\Delta x}$$

$$= \lim_{\Delta x \to 0} \left[u(x + \Delta x) \frac{v(x + \Delta x) - v(x)}{\Delta x} + v(x) \frac{u(x + \Delta x) - u(x)}{\Delta x} \right]$$

$$= u(x)v'(x) + v(x)u'(x)$$

$$= u \frac{dv}{dx} + v \frac{du}{dx}$$

Example 1   Find y' for $y = (x + 4)(x^2 + 2)$.

Solution   Applying Theorem 2.18 for the derivative of the product of two functions, $u(x)$ corresponds to $x + 4$ and $v(x)$ corresponds to $x^2 + 2$. Thus,

$$y' = u \frac{dv}{dx} + v \frac{du}{dx} = (x + 4) \frac{d(x^2 + 2)}{dx} + (x^2 + 2) \frac{d(x + 4)}{dx}$$

$$= (x + 4)2x + (x^2 + 2)(1)$$
$$= 2x^2 + 8x + x^2 + 2$$
$$= 3x^2 + 8x + 2$$

In this example, it is possible to multiply the two functions and find the derivative term by term. This will not always be practical, but we do so now in case you are not convinced.

$$y = (x + 4)(x^2 + 2)$$
$$= x^3 + 4x^2 + 2x + 8$$

and therefore,

$$y' = 3x^2 + 8x + 2$$

Note that this checks with the result obtained using the first method.

Theorem 2.19   The derivative of the quotient of two differentiable functions is the denominator times the derivative of the numerator minus the numerator times the derivative of the denominator, all divided by the square of the denominator. That is, if $v = v(x)$ and $u = u(x)$ represent two differentiable functions, then

$$D_x\left(\frac{u}{v}\right) = \frac{vD_xu - uD_xv}{v^2}, \qquad v \neq 0$$

Proof

$$D_x\left(\frac{u}{v}\right) = \lim_{\Delta x \to 0} \frac{\dfrac{u(x + \Delta x)}{v(x + \Delta x)} - \dfrac{u(x)}{v(x)}}{\Delta x}$$

$$= \lim_{\Delta x \to 0} \frac{\dfrac{u(x + \Delta x)v(x) - u(x)v(x + \Delta x)}{v(x + \Delta x)v(x)}}{\Delta x}$$

Subtract and add $u(x)v(x)$ to the numerator,

$$D_x\left(\frac{u}{v}\right) = \lim_{\Delta x \to 0} \frac{u(x + \Delta x)v(x) - u(x)v(x) - u(x)v(x + \Delta x) + u(x)v(x)}{\Delta x v(x + \Delta x)v(x)}$$

$$= \lim_{\Delta x \to 0} \frac{v(x)\dfrac{u(x + \Delta x) - u(x)}{\Delta x} - u(x)\dfrac{v(x + \Delta x) - v(x)}{\Delta x}}{v(x + \Delta x)v(x)}$$

$$= \frac{v(x)u'(x) - u(x)v'(x)}{[v(x)]^2}$$

$$= \frac{vD_xu - uD_xv}{v^2}, \qquad v \neq 0$$

Example 2   Find y' for $y = \dfrac{x^2 + 1}{x^4 + 3x}$.

Solution $y' = \dfrac{(x^4 + 3x)D_x(x^2 + 1) - (x^2 + 1)D_x(x^4 + 3x)}{(x^4 + 3x)^2}$

$$= \frac{(x^4 + 3x)2x - (x^2 + 1)(4x^3 + 3)}{(x^4 + 3x)^2}$$

$$= \frac{2x^5 + 6x^2 - (4x^5 + 4x^3 + 3x^2 + 3)}{(x^4 + 3x)^2}$$

$$= \frac{2x^5 + 6x^2 - 4x^5 - 4x^3 - 3x^2 - 3}{(x^4 + 3x)^2}$$

$$= \frac{-2x^5 - 4x^3 + 3x^2 - 3}{(x^4 + 3x)^2}$$

Because of the simplicity of the formula, we have used the fact that if $y = x^n$ then $y' = nx^{n-1}$ for n a rational number, but we have only proved it for positive integers. Using the formula for the derivative of the quotient of two functions it is easy to show that if $y = x^n$ then $y' = nx^{n-1}$ for n a negative integer. Suppose that n is a negative integer. Then $-n$ is positive. The expression $y = x^n$ is equivalent to $y = 1/x^{-n}$. Now we apply the formula for the derivative of the quotient of two functions to $y = 1/x^{-n}$ with $v = x^{-n}$ and $u = 1$.

$$y' = \frac{x^{-n}D_x(1) - 1D_x(x^{-n})}{(x^{-n})^2}$$

The term $D_x(1) = 0$, because the derivative of a constant is zero. The term $D_x(x^{-n}) = -nx^{-n-1}$, because $-n$ is a positive integer, and the formula for the derivative of x to a positive integral power applies. Making these substitutions above:

$$y' = \frac{x^{-n}(0) - (-nx^{-n-1})}{x^{-2n}}$$

$$= \frac{nx^{-n-1}}{x^{-2n}} = nx^{n-1}$$

2.6 EXERCISES

In Exercises 1–14, find the derivative with respect to the independent variable, and simplify your answer using the formula for the derivative of the product of two functions.

1. $y = (3x + 1)(5x + 2)$

2. $y = (7x - 3)(1 - 3x)$

3. $y = (t^2 + 4t)(t + 1)$

4. $y = (5t^2 - 3t)(2 - t)$

5. $f(x) = (3x^2 + 5x)(2x^2 + 4)$

6. $f(x) = (4x^2 - 6x)(3x^2 - 2)$

7. $f(t) = (3t^2 + 5t + 6)(4t^2 - 2t)$

8. $f(t) = (5t^2 - 6t + 4)(t^2 - 4t - 2)$

9. $y = (3x^3 + 7x^2 + x)(6x^3 - 3x^2 + 5x + 2)$

10. $y = (x^4 - x^3 + 3x + 2)(5x^4 + 3x^3 - x)$

11. $y = (t^3 + 3t)\left(1 + \dfrac{5}{t}\right)$

12. $y = \left(5t^2 + \dfrac{1}{t}\right)\left(1 - \dfrac{1}{t^2}\right)$

13. $y = x(x + 1)(x - 1)$

14. $y = 2x(2x + 1)(x - 3)$

In Exercises 15–28, find the derivative using the quotient formula. Simplify your answer.

15. $y = \dfrac{x + 1}{x - 1}$

16. $y = \dfrac{2x + 3}{x + 2}$

17. $y = \dfrac{t}{t^2 + 1}$

18. $y = \dfrac{2t + 3}{t^2 - 1}$

19. $y = \dfrac{x^2 + 5}{x + 1}$

20. $y = \dfrac{3x^2 + 2x}{2x + 5}$

21. $y = \dfrac{t^2 + 2t + 1}{t^2 - 2t + 1}$

22. $y = \dfrac{3t^2 - 2t + 4}{2t^2 - 3t + 1}$

23. $y = \dfrac{x^3}{x^3 - 1}$

24. $y = \dfrac{x^4}{x^4 - 1}$

25. $y = \dfrac{t^{1/3}}{t^{1/3} - 1}$

26. $y = \dfrac{t^{1/4}}{t^{1/4} - 1}$

27. $y = \dfrac{1 - \sqrt{x}}{1 + \sqrt{x}}$

28. $y = \dfrac{1 - \sqrt[3]{x}}{1 + \sqrt[3]{x}}$

29. Find dy/dx at $x = 3$ for $x + 2y - 3x^2y = 0$.

30. Find dy/dx at $x = -7$ for $x^2 + yx + 7y = 4$.

2.7 THE DERIVATIVE OF COMPOSITE FUNCTIONS

A very important technique for finding derivatives is called the *chain rule*. It is used to find dy/dx when y is a function of u and u is a function of x; that is, when y is the composite function $y = f(u(x))$. Such a function is $y = \sqrt{u}$ and $u = x^2 + 5x$, which can be written $y = \sqrt{x^2 + 5x}$.

Theorem 2.20   (Chain rule)   If y is a function of u and u is a function of x, and if y is differentiable with respect to u and u is differentiable with respect to x, then

$$\frac{dy}{dx} = \frac{dy}{du} \cdot \frac{du}{dx}$$

Proof   Since $y = f(u(x))$, a change of Δx in x causes a change of Δu in $u(x)$, which in turn causes a change of Δy in $y = f(u(x))$. If a change in x of Δx causes no change in u, that is $\Delta u = 0$, then there can be no change in y, that is $\Delta y = 0$, and $dy/dx = 0$. If Δx and Δu are both not zero, then we may write

$$\frac{\Delta y}{\Delta x} = \frac{\Delta y}{\Delta x} \cdot \frac{\Delta u}{\Delta u} = \frac{\Delta y}{\Delta u} \cdot \frac{\Delta u}{\Delta x}$$

Therefore,

$$\frac{dy}{dx} = \lim_{\Delta x \to 0} \frac{\Delta y}{\Delta x}$$

$$= \lim_{\Delta x \to 0} \left(\frac{\Delta y}{\Delta u} \cdot \frac{\Delta u}{\Delta x} \right)$$

$$= \left(\lim_{\Delta x \to 0} \frac{\Delta y}{\Delta u} \right) \cdot \left(\lim_{\Delta x \to 0} \frac{\Delta u}{\Delta x} \right)$$

Recall that

$$\frac{dy}{du} = \lim_{\Delta u \to 0} \frac{\Delta y}{\Delta u} \quad \text{and} \quad \frac{du}{dx} = \lim_{\Delta x \to 0} \frac{\Delta u}{\Delta x}$$

Since u is a differentiable function, $\Delta u \to 0$ as $\Delta x \to 0$. Applying this information to the above,

$$\frac{dy}{dx} = \left(\lim_{\Delta x \to 0} \frac{\Delta y}{\Delta u} \right) \left(\lim_{\Delta x \to 0} \frac{\Delta u}{\Delta x} \right)$$

$$= \left(\lim_{\Delta u \to 0} \frac{\Delta y}{\Delta u} \right) \left(\lim_{\Delta x \to 0} \frac{\Delta u}{\Delta x} \right)$$

$$= \frac{dy}{du} \cdot \frac{du}{dx}$$

Example 1   Use the chain rule to find the derivative of $y = (2x + 3)^2$.

Solution   The function is a composite function of the form $y = f(u(x))$ where $u = u(x) = 2x + 3$ and $y = u^2$. According to the chain rule, y' is the product of the derivative of y with respect to u times the derivative of u with respect to x. Since $y = u^2$, then $dy/du = 2u$; and since $u = 2x + 3$, $du/dx = 2$; therefore,

$$y' = \frac{dy}{du} \cdot \frac{du}{dx} = (2u)(2)$$

Substituting $u = 2x + 3$,

$$y' = 2(2x + 3)(2) = 4(2x + 3) = 8x + 12$$

In Example 1, the derivative could have been found without resorting to the chain rule by first multiplying. That is, if $y = (2x + 3)^2$, then

$$y = 4x^2 + 12x + 9$$

Therefore,

$$y' = 8x + 12$$

This is the same answer found using the chain rule.

The chain rule can also be given in composite function notation as follows: If $y = (f \circ g)(x) = f(g(x))$ where f and g are functions, then $y' = D_x(f \circ g) = f'(g(x)) \cdot g'(x)$. To see this, let $u = g(x)$; then $y = f(u)$ and so $dy/du = f'(u) = f'(g(x))$ and $du/dx = g'(x)$. Thus

$$y' = \frac{dy}{dx} = \frac{dy}{du} \cdot \frac{du}{dx} = f'(g(x)) \cdot g'(x)$$

Example 2   Let f and g be functions such that $f(x) = x^2$ and $g(x) = 2x + 3$. Then $f'(x) = 2x$ and $g'(x) = 2$. Find $D_x(f \circ g)$.

Solution $D_x(f \circ g) = f'(g(x)) \cdot g'(x) = [2(2x + 3)] \cdot 2 = 8x + 12$.

Sometimes this process can be made more efficient by only identifying $f(u)$ mentally, as demonstrated in the following example. This is the same example we have already discussed in detail.

Example 3   Find y' for $y = (2x + 3)^2$.

Solution   Think of $y = (2x + 3)^2$ as $y = u^2$, where $u = 2x + 3$. Then $y' = 2uD_xu$. In operation it looks like this: If $y = (2x + 3)^2$, then

$$y' = \underbrace{2(2x + 3)}_{2u}\ \underbrace{2}_{D_xu}\ = 8x + 12$$

In Example 3, it was not necessary to use the chain rule to find the derivative. The chain rule is necessary when functions cannot be reduced to polynomials.

Example 4   Find $f'(x)$ for $f(x) = \sqrt[3]{x^4 + 5x^2 + 7}$.

Solution $f(x) = (x^4 + 5x^2 + 7)^{1/3}$.   Let $u = x^4 + 5x^2 + 7$,   and $f(u) = u^{1/3}$.
Therefore,

$$f'(x) = \frac{1}{3} (x^4 + 5x^2 + 7)^{-2/3} D_x(x^4 + 5x^2 + 7)$$

$$= \frac{1}{3} (x^4 + 5x^2 + 7)^{-2/3}(4x^3 + 10x)$$

$$= \frac{4x^3 + 10x}{3(x^4 + 5x^2 + 7)^{2/3}}$$

Example 5   Find y' for $y = x^2\sqrt{x^2 + 1}$.

Solution   The function y is the product of two functions x^2 and $\sqrt{x^2 + 1}$.
Therefore, the product rule must be applied and

$$y' = x^2 \cdot D_x\sqrt{x^2 + 1} + \sqrt{x^2 + 1} \cdot D_x x^2$$

The derivative of x^2, $D_x x^2$, is $2x$ but the chain rule must be used to
find $D_x\sqrt{x^2 + 1}$. Thus

$$\begin{aligned}
y' &= x^2 D_x\sqrt{x^2 + 1} + \sqrt{x^2 + 1} \cdot 2x \\
&= x^2 D_x(x^2 + 1)^{1/2} + 2x\sqrt{x^2 + 1} \\
&= x^2 \cdot \frac{1}{2} (x^2 + 1)^{-1/2} \cdot 2x + 2x\sqrt{x^2 + 1} \\
&= \frac{x^3}{\sqrt{x^2 + 1}} + 2x\sqrt{x^2 + 1} \\
&= \frac{x^3 + 2x(x^2 + 1)}{\sqrt{x^2 + 1}} \\
&= \frac{x^3 + 2x^3 + 2x}{\sqrt{x^2 + 1}} \\
&= \frac{3x^3 + 2x}{\sqrt{x^2 + 1}}
\end{aligned}$$

We have proved that if $y = x^n$ then $y' = nx^{n-1}$ for any *integer n*,
when $n \neq 0$. Using the chain rule, it is easy to prove this relation
is true for any *rational number n*.

Theorem 2.21   If $y = x^n$, $n = p/q$, then $y' = nx^{n-1}$, when p and q are integers,
$q \neq 0$.

Proof   If $y = x^{p/q}$, then $y^q = x^p$. Differentiating both sides with respect to x,

$$qy^{q-1}y' = px^{p-1}$$
$$y' = \frac{p}{q} \cdot \frac{x^{p-1}}{y^{q-1}}$$

But

$$y^{q-1} = (x^{p/q})^{q-1} = x^{p-p/q}$$

Thus,

$$y' = \frac{p}{q} \cdot \frac{x^{p-1}}{x^{p-p/q}}$$
$$= \frac{p}{q} x^{p/q-1}$$
$$= nx^{n-1}$$

2.7   EXERCISES

In Exercises 1–22, use the chain rule to find the derivative with respect to x. Simplify your answer.

1. $y = (x + 5)^6$

2. $y = 3(x^2 + 4)^5$

3. $y = (2x + 1)^{1/2}$

4. $y = (x^2 + x)^{1/2}$

5. $f(x) = \sqrt{x^2 + 1}$

6. $f(x) = \sqrt[3]{x^3 - 3x^2}$

7. $f(x) = \dfrac{1}{(x + 4)^{2/3}}$

8. $f(x) = \dfrac{1}{(x^2 + 2x + 2)^{1/2}}$

9. $y = \dfrac{1}{\sqrt[3]{x^3 + 2x}}$

10. $y = \dfrac{1}{\sqrt[4]{x^3 + 5x}}$

11. $y = x\sqrt{3x + 2}$

12. $y = x^2\sqrt{3x + 2}$

13. $y = \dfrac{5x^2}{\sqrt{x^2 + 9}}$

14. $y = \dfrac{3x^2 + 2x}{\sqrt{x^3 + 2x}}$

15. $y = (x + \sqrt{x})^{1/3}$

16. $y = (x + \sqrt[3]{x})^{1/6}$

17. $f(x) = \dfrac{\sqrt{x^2 + 5}}{x + 2}$

18. $f(x) = \dfrac{\sqrt{x^3 + 3x}}{x^2 + 1}$

19. $f(x) = \dfrac{\sqrt[3]{x^3 + 1}}{\sqrt{x^2 - 1}}$

20. $f(x) = \dfrac{\sqrt[3]{x^2 + 5x}}{\sqrt{4x^2 - 2}}$

21. $y = \sqrt{2x + \sqrt{x^2 + 1}}$

22. $y = \sqrt{3x^3 + x\sqrt{x^2 + 1}}$

23. Find the slope of the tangent to the curve $9x^2 - 4y^2 = 36$ at the point $(2, 0)$.

24. Find the slope of the tangent to the curve $16x^2 + 25y^2 = 400$ at the point $(-3, -\frac{16}{5})$.

25. Find $\dfrac{d}{dx}(x^{\sqrt{2}})$.   26. Find $\dfrac{d}{dx}(x^{\pi})$.

2.8 . HIGHER-ORDER DERIVATIVES

The derivative of a function is a function and is itself often differentiable. If $y = f(x)$ represents a differentiable function and f' is also differentiable, then the derivative of f' is called the *second derivative* of f, and is denoted as f''. The third derivative of f is the derivative of f'' and is written f'''. f'' is read "f double prime." f''' is read as "f triple prime." The prime notation is not used after the third derivative. Symbols for the fourth derivative are

$$y^{(4)}, \quad f^{(4)}, \quad D_x^4 y, \quad \frac{d^4 y}{dx^4}$$

The nth derivative is denoted

$$y^{(n)}, \quad f^{(n)}, \quad D_x^n y, \quad \frac{d^n y}{dx^n}$$

Higher-order derivatives are used in curve sketching and in curve approximations. In certain circumstances, they can also be interpreted as change in the growth rate, change in the marginal cost per item, or instantaneous acceleration.

Example 1   If $y = x^3 - 7x^2 - 2x$, find the successive derivatives of y.

Solution
$$y' = 3x^2 - 14x - 2$$
$$y'' = 6x - 14$$
$$y''' = 6$$
$$y^{(4)} = 0$$
$$y^{(5)} = 0$$
$$y^{(6)} = 0$$

All the derivatives after the third derivative are zero, since each is the derivative of a constant.

Example 2   Find d^2y/dx^2 for $y = \sqrt{x^2 + 1}$.

Solution $y = \sqrt{x^2 + 1} = (x^2 + 1)^{1/2}$

Therefore, using the chain rule,

$$\frac{dy}{dx} = \frac{1}{2}(x^2 + 1)^{-1/2}(2x)$$

$$= \frac{x}{\sqrt{x^2 + 1}}$$

$\dfrac{d^2y}{dx^2}$ is the derivative of $\dfrac{dy}{dx}$. Using the quotient formula

$$\frac{d^2y}{dx^2} = \frac{(\sqrt{x^2 + 1})(1) - x(1/2)(x^2 + 1)^{-1/2}(2x)}{[(x^2 + 1)^{1/2}]^2}$$

$$= \frac{\sqrt{x^2 + 1} - x^2(x^2 + 1)^{-1/2}}{(x^2 + 1)}$$

Multiplying the numerator and denominator by $(x^2 + 1)^{1/2}$ gives

$$\frac{d^2y}{dx^2} = \frac{x^2 + 1 - x^2}{(x^2 + 1)^{1/2}(x^2 + 1)}$$

$$= \frac{1}{(x^2 + 1)^{3/2}}$$

Example 3   Find $D_x^n x^{1/2}$ where n is any positive integer.

Solution   If $y = x^{1/2}$, then

$$D_x y = \frac{1}{2} x^{-1/2}$$

$$D_x^2 y = \frac{1}{2}\left(-\frac{1}{2}\right) x^{-3/2}$$

$$D_x^3 y = \frac{1}{2}\left(-\frac{1}{2}\right)\left(-\frac{3}{2}\right) x^{-5/2}$$

$$D_x^4 y = \frac{1}{2}\left(-\frac{1}{2}\right)\left(-\frac{3}{2}\right)\left(-\frac{5}{2}\right) x^{-7/2}$$

$$\cdot$$
$$\cdot$$
$$\cdot$$

$$D_x^n y = D_x^n x^{1/2} = \frac{1}{2}\left(-\frac{1}{2}\right)\left(-\frac{3}{2}\right) \cdots \left(-\frac{2n - 3}{2}\right) x^{-(2n-1)/2}$$

2.8 EXERCISES

In Exercises 1–20, find $D_x^2 y$ for the given functions.

1. $y = 3x$
2. $y = 28x$
3. $y = 2x^2 + 5x + 3$
4. $y = 5x^3 - 2x^2 + 7x$
5. $y = x(x + 4)(x - 4)$
6. $y = 5x^{-3} - 2x^{-1}$
7. $y = x^2 - 2x + \dfrac{1}{x^2}$
8. $y = \dfrac{x}{x + 1}$
9. $y = \dfrac{x + 1}{x - 1}$
10. $y = \sqrt{x^2 + 1}$
11. $y = \sqrt[3]{x^3 + 2}$
12. $y = \dfrac{x + 4}{x - 2}$
13. $y = (x^2 - 1)^2$
14. $y = x\sqrt{4 - x}$
15. $y = \sqrt{\dfrac{x - 1}{x + 1}}$ for $x > 1$
16. $\sqrt{x} + \sqrt{y} = 4$
17. $y = 3x + x^{7/2}$
18. $y = x^2 \sqrt{x^2 + 1}$
19. $y = (x^2 + 1)^{1/2} \cdot x^{1/2}$
20. $y - \sqrt{x} - 2x = 0$ for $y \geq 0, x \geq 0$

In Exercises 21–28, find $f'''(x)$ for each function.

21. $f(x) = 4x^3 + 2x^2 - x + 3$
22. $f(x) = 7x^2 + 8x + 2$
23. $f(x) = \dfrac{1}{x - 1}$
24. $f(x) = \sqrt{x + 2}$
25. $f(x) = \dfrac{x}{x + 1}$
26. $f(x) = \sqrt[3]{x}$
27. $f(x) = x^2$
28. $f(x) = \dfrac{1}{x^2}$

29. Find $D_x^{(100)} y$ for $y = \dfrac{1}{x^2}$.

30. Find $D_x^{(99)} y$ for $y = x^{-3/2}$.

31. Find the nth derivative of $y = x^a$ when $n < a$, and when $n > a$, n and a positive integers.

32. Find the nth derivative of $y = x^n$.

33. Graph $y = x^3 - 3x^2$ and draw tangents to the graph at the points where $f'(x) = 0$ and where $f''(x) = 0$.

34. Graph $y = x^2 - (x^3/6)$ and draw tangents to the curve at the points where $f'(x) = 0$ and where $f''(x) = 0$.

Here is a list of derivative formulas from this chapter. Let $u = u(x)$ and $v = v(x)$ be differentiable functions of x. Then

$$D_x C = 0, \qquad C \text{ constant}$$

$$D_x(u^n) = nu^{n-1}\frac{du}{dx}$$

$$D_x(u + v) = \frac{du}{dx} + \frac{dv}{dx}$$

$$D_x(u \cdot v) = u\frac{dv}{dx} + v\frac{du}{dx}$$

$$D_x\left(\frac{u}{v}\right) = \frac{v\dfrac{du}{dx} - u\dfrac{dv}{dx}}{v^2}, \qquad v \neq 0$$

NEW TERMS

Chain rule

Composite function

Continuous

Derivative

Discontinuous

First derivative

Fixed cost

Left-hand limit

Limit

Marginal cost

Point–slope form

Right-hand limit

Second derivative

Slope

Slope–intercept form

Variable cost

REVIEW

In Exercises 1–5, consider the total cost of manufacturing x items to be $c = 3x + 20$. If 70 items are produced, find:

1. Fixed cost

2. Variable cost

3. Marginal cost

4. Average cost per item

5. Total cost

In Exercises 6–15, find the first derivative of y with respect to the independent variable. Simplify your answer.

6. $y = 3x^2 + 5x + \dfrac{1}{x^2} + \dfrac{3}{\sqrt{x}}$

7. $xy^2 = 4, \quad y > 0$

8. $y = \dfrac{x + 3}{x + 2}$

9. $y = (t^3 + 3t)\sqrt{t^2 + 1}$

10. $y = (1 - \sqrt{x})(1 + \sqrt{x})$ **11.** $y = \dfrac{\sqrt{x^2 + 1}}{x^2}$

12. $y = (x + \sqrt{x^2 + 1})^{1/2}$ **13.** $y = x^{5/2}$

14. $y = x \sqrt{x(x + 1)}$ **15.** $y = \dfrac{1}{(x + \sqrt{x^2 + 1})^2}$

In Exercises 16 and 17, find the second derivative of y with respect to x.

16. $y = \sqrt{x^2 + 4}$ **17.** $y = (x^2 + 1)^{3/2}$

In Exercises 18 and 19, find the third derivative of y with respect to x.

18. $y = x^{-5/2}$ **19.** $y = \sqrt{x^2 + 1}$

20. The cost of producing x items is given in dollars by

$$c = 500 + \sqrt{x} + \frac{1}{x}$$

Find the marginal cost, dc/dx, of producing the 100th item.

21. If the number of grams of bacteria in a population after t hours is given by $y = t^2 \sqrt{4t + 1}$, find the growth rate of the population after 45 minutes.

3

More about Derivatives and Their Applications

This chapter will consider some of the applications of the derivative to economics, business, social science, and biological science as well as some additional methods for finding derivatives. Some of the examples are straightforward; other examples will require the development of background information. The first application, Newton's method of approximation, is mathematical. It shows how calculus concepts can be combined with other mathematical ideas to develop new solutions to old problems. The same kind of "new" combinations will be used in the other applications in this chapter.

3.1 NEWTON'S METHOD OF ROOT APPROXIMATION

Suppose in the process of determining the price a pro football team should charge for a season ticket, a consulting mathematical management specialist discovers he must find a solution to the equation $\sqrt[5]{x} - 5x + 6 = 0$. Unfortunately, this equation is not a type we know how to solve. In this section, we will examine how the derivative can be used to assist us to find a numerical solution. These problems require that we find the roots of nonlinear, nonquadratic equations. As the pro football example illustrates, it is not always possible to solve equations analytically.

Fortunately there are a number of methods for approximating roots of these difficult equations. The method that will be presented here was developed by Sir Isaac Newton about 1670. It can involve much tedious arithmetic when used for manual calculation of roots. However, it is extremely well suited for use by computers.

The roots of a function $y = f(x)$ are the values of x that, when substituted into the function, make its value zero. In Figure 3.1, the function crosses the x axis at a, b, and c. Since the range values of a function are zero for every point on the x axis, $x = a$, b, and c are roots of the function and $f(a) = f(b) = f(c) = 0$. Therefore the problem of solving $f(x) = 0$ for x can be viewed as finding the points at which $y = f(x)$ crosses the x axis, which is exactly what Newton's method does.

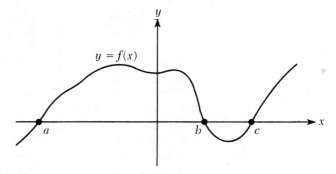

Figure 3.1   The roots of $f(x)$

Newton's method of root approximation involves guessing a root and then improving the accuracy of the guess. In Figure 3.2, f is

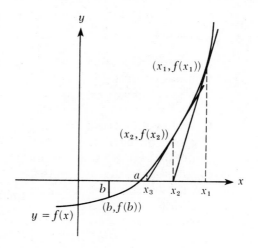

Figure 3.2   Newton's method of approximation

a continuous differentiable function. When b is substituted for x in $f(x), f(b)$ is negative; but $f(x_1)$ is positive. Therefore, since the functional value changes from positive to negative, there must be a real root somewhere between b and x_1. This root is denoted as a in Figure 3.2. We will arbitrarily take x_1 as the first estimate of a. An improved estimate is x_2 which is the x intercept of the tangent line at $(x_1, f(x_1))$. A third estimate is x_3, which is the x intercept of the tangent line at $(x_2, f(x_2))$. By repeated applications of this process, the root can be estimated to any desired degree of accuracy.

We use the point–slope form of a linear equation to find the equation of the line tangent to $y = f(x)$ at $(x_1, f(x_1))$. You should remember that the point–slope form of a linear equation is

$$y - y_1 = m(x - x_1)$$

where m is the slope of the line; x_1 is our first estimate for the root; $y_1 = f(x_1)$, and the slope, m, is $f'(x_1)$. Thus, the equation of the tangent line is

$$y - f(x_1) = f'(x_1)(x - x_1)$$

To find x_2, note that the ordinate of any point on the x axis is 0. Therefore, we substitute 0 for y and solve for $x = x_2$.

$$0 - f(x_1) = f'(x_1)(x_2 - x_1)$$

which implies

$$\frac{-f(x_1)}{f'(x_1)} = x_2 - x_1$$

and, finally

$$x_2 = x_1 - \frac{f(x_1)}{f'(x_1)}$$

After x_2 is computed, it can be used to find x_3 by replacing x_2 by x_3 and x_1 by x_2. That is,

$$x_3 = x_2 - \frac{f(x_2)}{f'(x_2)}$$

In general the nth approximation can be found by using the $(n - 1)$th approximation.

$$x_n = x_{n-1} - \frac{f(x_{n-1})}{f'(x_{n-1})}$$

Example 1   Use Newton's method to find a real root of $x^2 - 5x - 2 = 0$.

Solution $f(x) = x^2 - 5x - 2$ and $f'(x) = 2x - 5$

By trial and error we find that $f(5) = -2$ and $f(6) = 4$. Therefore, there is a root between 5 and 6. We could choose either 5 or 6 or some number between them for x_1. We choose $x_1 = 5$.

$$x_2 = x_1 - \frac{f(x_1)}{f'(x_1)}$$

$$= 5 - \frac{f(5)}{f'(5)}$$

$$= 5 - \left(\frac{-2}{5}\right)$$

$$= 5 + 0.4 = 5.4$$

Applying Newton's method a second time to find x_3,

$$x_3 = 5.4 - \frac{f(5.4)}{f'(5.4)} = 5.4 - \frac{0.16}{5.8}$$
$$= 5.4 - 0.0276 = 5.3724$$

Using the quadratic formula to solve the equation shows that, to three decimal places, $x = 5.3723$. Thus after just two applications of Newton's method, x_3 differs from the root by about 0.0001.

Newton's method fails if $f'(x_1) = 0$ where x_1 is an estimate of the root. Geometrically, this means that the slope of the tangent line is zero (see Figure 3.3). A line with a slope of zero is parallel

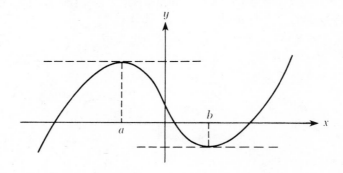

Figure 3.3 $f'(x) = 0$

to the x axis. Hence, it cannot intersect the x axis to yield a new and better estimate of the root. Certain specific curves (see Figure 3.4) are such that Newton's method does not improve an estimate of the root.

Newton's method can be used to find decimal estimates of irrational numbers.

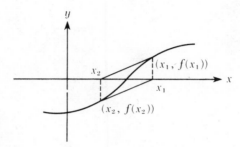

Figure 3.4 Newton's method fails

Example 2   Find a decimal approximation for $\sqrt{2}$ that is accurate to three decimal places using Newton's method three times.

Solution   Let $x = \sqrt{2}$, then $x^2 = 2$ and $x^2 - 2 = 0$. Use Newton's method to estimate a root of $f(x) = x^2 - 2$. The arithmetic is often simplified by applying the general formula before substituting constants.

$$x_2 = x_1 - \frac{f(x_1)}{f'(x_1)} = x_1 - \frac{x_1^2 - 2}{2x_1}$$
$$= \frac{2x_1^2 - x_1^2 + 2}{2x_1} = \frac{x_1^2 + 2}{2x_1}$$

Since $f(1) = -1$ and $f(2) = 2$ there is a root between 1 and 2. Take $x_1 = 1$; then

$$x_2 = \frac{1 + 2}{1(2)} = \frac{3}{2}$$

Find x_3 by substituting $\frac{3}{2}$ in

$$x_3 = \frac{x_2^2 + 2}{2x_2}$$

This gives

$$x_3 = \frac{(9/4) + 2}{2(3/2)} = \frac{9 + 8}{12} = \frac{17}{12}$$

Find x_4 by substituting $\frac{17}{12}$ in

$$x_4 = \frac{x_3^2 + 2}{2x_3}$$

This gives

$$x_4 = \frac{(17/12)^2 + 2}{2(17/12)} = \frac{(289/144) + 2}{17/6}$$

$$= \frac{289 + 288}{24(17)} = \frac{577}{408} = 1.414$$

3.1 EXERCISES

If a calculator is available, use it to help solve these problems.

In Exercises 1–4, use two applications of Newton's method to estimate the following irrational numbers.

1. $\sqrt{3}$ **2.** $\sqrt[3]{2}$

3. $\sqrt[4]{10}$ **4.** $\dfrac{1}{\sqrt{2}}$

In Exercises 5–12, find all real roots accurate to two decimal places.

5. $6x^2 + 7x - 24 = 0$ **6.** $x^4 - 5 = 0$

7. $x^2 + x + 1 = 0$ **8.** $5x^2 + 2x + 5 = 0$

9. $x^3 - 8x^2 + 4x + 7 = 0$ **10.** $x^3 + x^2 + x + 1 = 0$

11. $\sqrt[5]{x} - 5x + 6 = 0$ **12.** $x^2 - \sqrt{x} + 3 = 0$

In Exercises 13–16, estimate the root between the indicated values of x using two applications of Newton's method.

13. $x^2 - 8 = 0$, between $x = 2$ and $x = 3$

14. $10 - x^2 = 0$, between $x = 3$ and $x = 4$

15. $x^3 - 4x^2 + 3x + 2 = 0$, between $x = -1$ and $x = 0$

16. $x^3 - x^2 - x - 4 = 0$, between $x = 2$ and $x = 3$

A root for $f(x) = 0$ can be estimated by using the x intercept of the secant line formed between two points on $y = f(x)$, one above and the other below the x axis. The equation of such a secant is

$$y - y_2 = \frac{y_2 - y_1}{x_2 - x_1}(x - x_2)$$

where (x_1, y_1) and (x_2, y_2) are on opposite sides of the x axis. To find an estimate for a root let $y = 0$ and solve the above equation for x. The first estimate of the root, r_1, can then be used to find a second estimate by finding the x intercept of the secant line between $(r_1, f(r_1))$ and either (x_1, y_1) or (x_2, y_2), whichever is on the opposite side of the x axis from $(r_1, f(r_1))$.

As Exercises 17–20, use the equations given in Exercises 13–16 and estimate the root between the indicated values of x using two applications of the secant method.

21. Newton's method can be generalized to a formula for finding roots of expressions of the type

$$f(x) = x^k - a, \qquad k > 0$$

Show that Newton's method yields the following expression for the nth approximation in terms of the $n - 1$ approximation:

$$x_n = \left(\frac{k - 1}{k}\right) x_{n-1} + \frac{a}{k} \cdot \frac{1}{x_{n-1}^{k-1}}$$

In Exercises 22–25, use the formula given in Exercise 21 to find the following to 4 decimal places.

22. x such that $x^5 - 5 = 0$

23. x such that $x^{-3} = 2$ [*Hint:* Consider $x^3 - \frac{1}{2} = 0$.]

24. $\sqrt[4]{3}$ [*Hint:* Consider $x^4 - 3 = 0$.]

25. $\sqrt[10]{10}$

3.2 CURVE ANALYSIS

Functions can, of course, be used as mathematical models of many things. They may be used to relate supply to demand, sales to profit, or amounts of fertilizer to crop yield. In each case, the user of the model is primarily interested in how the dependent quantity relates to a set of values of the independent variable. Often, the simplest way to analyze relations is with a graph of the function involved. The derivative gives us a fundamental tool for the production of such sketches.

First and second derivatives can be used to sketch curves. In most applications an exact graph of a curve is not necessary, as long as the general shape is preserved. One criterion used to establish the "general shape" is simply whether the function is decreasing or increasing for some interval of x.

Definition

A function is *increasing* over an interval of its domain if, for every x_1 and x_2 in the interval, $f(x_2) > f(x_1)$ whenever $x_2 > x_1$.

Definition

A function is *decreasing* over an interval if, for every x_1 and x_2 in the interval, $f(x_2) < f(x_1)$ whenever $x_2 > x_1$.

Note that the definitions conform to an intuitive idea of increasing and decreasing functional values. An increasing function is such that the functional value increases as x increases (Figure 3.5a), while a decreasing function is one where the functional images decrease when x increases (Figure 3.5b).

It is not usually convenient to apply the definitions to determine the intervals of x for which a given function increases or decreases. A better method uses the first derivative. Recall that $f'(x) = \lim_{\Delta x \to 0} (\Delta y / \Delta x)$. Therefore, if a function is increasing, as in Figure 3.6a (page 120), Δx and Δy will both be positive or both negative. Thus their ratio is positive, and

$$\lim_{\Delta x \to 0} \frac{\Delta y}{\Delta x} = f'(x) > 0$$

If a function is decreasing, as in Figure 3.6b, Δx and Δy are of different signs so their ratio is negative. Therefore,

$$\lim_{\Delta x \to 0} \frac{\Delta y}{\Delta x} = f'(x) < 0$$

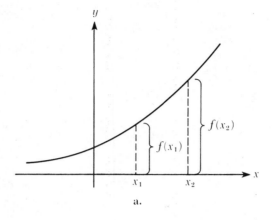

a.

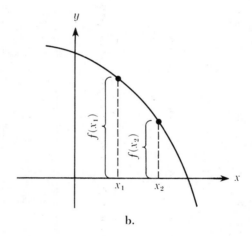

b.

Figure 3.5 **a.** An increasing function and
b. a decreasing function

This leads to the following theorem, which will be stated without proof.

Theorem 3.1 If f is a differentiable function and $f'(x) > 0$ for all x such that $x \in (a, b)$, then f is increasing on the interval from a to b. If $f'(x) < 0$ for all x such that $x \in (a, b)$, then f is decreasing on the interval from a to b.

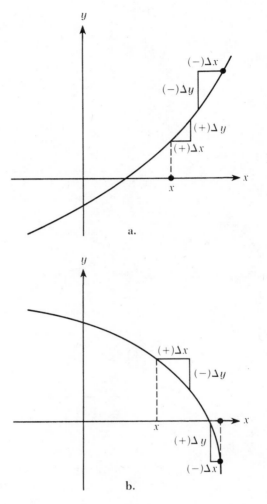

Figure 3.6 **a.** An increasing function, $\dfrac{\Delta y}{\Delta x} > 0$, and

   b. a decreasing function, $\dfrac{\Delta y}{\Delta x} < 0$.

Example 1   Find the values of x for which f given by $f(x) = x^2$ is increasing and those for which it is decreasing.

Solution   The derivative $f'(x) = 2x$; so $f'(x) > 0$ for all positive numbers and $f'(x) < 0$ for all negative numbers. Thus, by the above theorem, the function increases for x positive and decreases for x negative. This is clearly shown by the graph of the function in Figure 3.7.

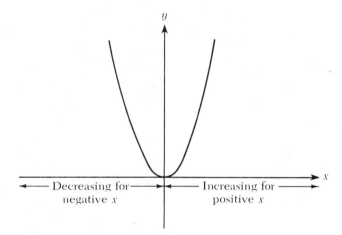

Figure 3.7 Graph of $y = x^2$

When $f'(x) = 0$ the slope of the tangent line is 0 and the tangent is parallel to the x axis. In this case the function is neither increasing nor decreasing. In the last example, $f'(x) = 0$ when $x = 0$.

The intervals of x for which a function is increasing or decreasing are slightly more difficult to find when the equation is cubic or higher in degree.

Example 2   Find the values of x for which f given by $f(x) = 2x^3 + 3x^2 - 72x$ is increasing and decreasing.

Solution   The derivative $f'(x) = 6x^2 + 6x - 72$ which, in factored form, is $f'(x) = 6(x + 4)(x - 3)$. By Theorem 3.1, the function is increasing when $f'(x) > 0$ and decreasing when $f'(x) < 0$. The sign of the derivative can be determined by examining the signs of its factors: if the factors are both positive or both negative, the derivative is positive; if one factor is positive and the other is negative, the derivative is negative. The graph of this function is shown in Figure 3.8.

The simplest way to analyze the sign of this expression is with a graph. Consider Figure 3.9. The first number line in Figure 3.9 shows the sign of $(x + 4)$. If x is less than -4, $(x + 4)$ is negative, if x is greater than -4, $(x + 4)$ is positive. The second number line analyzes $(x - 3)$ in a similar way. Below the two number lines the sign of the product is shown. The product is zero at $x = -4$ and at

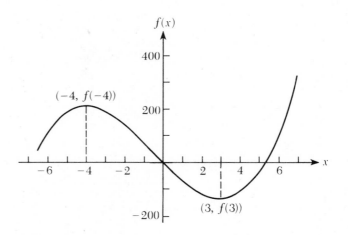

Figure 3.8 Graph of $f(x) = 2x^3 + 3x^2 - 72x$

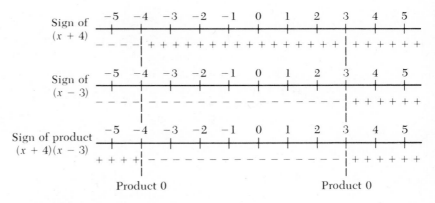

Figure 3.9 Sign of $(x + 4)(x - 3)$

$x = 3$. If x is less than -4 or greater than 3 the product is positive. Between -4 and 3 the product is negative. Therefore f is decreasing if $x \in (-4, 3)$, f is increasing if $x \in (3, \infty)$ or $x \in (-\infty, -4)$.

A second characteristic of curves, which is easily examined by use of derivatives (second derivatives to be exact), is called *concavity*. The first derivative, f', of a function is also a function. For certain values of x, f' will be an increasing function, for others decreasing. To discover the values of x for which f' is increasing (or decreasing), examine its derivative. This is the second derivative of f. If $f''(x) > 0$, then f' is increasing; if $f''(x) < 0$, then f' is decreasing. The derivative $f'(x)$ is the slope of the line tangent to the curve $y = f(x)$ at the point $(x, f(x))$. If f' is increasing, the slope of the tangent line is increasing; if f' is decreasing, the slope of the tangent line is decreasing. Figure 3.10 shows a curve for which the slope of the tangent lines increases from negative values to zero to positive values as x increases. Such a curve is said to be *concave upward*.

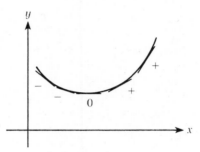

Figure 3.10   Concavity upward

Figure 3.11 illustrates a curve that is concave downward. Note that the slope of successive tangent lines decreases from positive to zero to negative values.

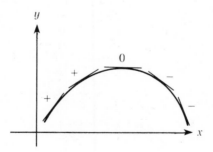

Figure 3.11   Concavity downward

Therefore, a curve is concave upward for values of x for which its first derivative is an increasing function. This happens when $f''(x) > 0$. By similar reasoning, a curve is concave downward for values of x for which $f''(x) < 0$. The point where a curve changes from concave downward to concave upward is called a *point of inflection*. Similarly, a point where a curve changes from concave upward to concave downward is also a point of inflection. Theorem 3.2 summarizes the above.

Theorem 3.2   If f is a differentiable function and $f''(x) > 0$ for all x such that $x \in (a, b)$, then f is *concave upward* on the interval from a to b.

If f is a differentiable function and $f''(x) < 0$ for all x such that $x \in (a, b)$, then f is *concave downward* on the interval from a to b.

Given a function f with $y = f(x)$, if the second derivative exists on an interval that contains a point of inflection, $(a, f(a))$, then $f''(a) = 0$. This follows from the fact that the curve is concave downward on one side of a point of inflection and concave upward on the other. Therefore the second derivative is negative on one side of a point of inflection and positive on the other. Then the second derivative must equal zero *at* the point of inflection, since by hypothesis it exists there. In Figure 3.12 a tangent line is drawn at the point of inflection.

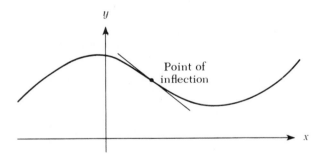

Figure 3.12   Point of inflection

Example 3   Find the points of inflection and values of x for which f given by $f(x) = x^3 + 3x^2$ is concave upward and those for which it is concave downward.

Solution $f'(x) = 3x^2 + 6x$ and $f''(x) = 6x + 6$

To find the points of inflection, set $f''(x)$ equal to zero, and solve for x.

$$6x + 6 = 0$$
$$6x = -6$$
$$x = -1$$

The second derivative, $6x + 6$, is negative for $x < -1$ and positive for $x > -1$. Therefore, $f(x)$ is concave downward for $x < -1$ and concave upward for $x > -1$; it has a point of inflection at $x = -1$, $y = 2$. Figure 3.13 is the graph of the curve.

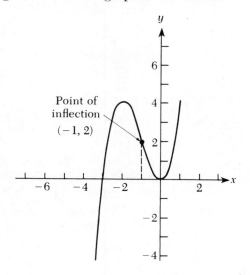

Figure 3.13   Graph of $f(x) = x^3 + 3x^2$

3.2 EXERCISES

In Exercises 1–12, is the given function increasing or decreasing, and is it concave upward or concave downward for the indicated value of x?

1. $f(x) = 3x^2 + 15x$, $x = -3$

2. $f(x) = 4x^2 - 28x + 7$, $x = -10$

3. $f(x) = 8x^3 + 9x^2 - 18x + 15$, $x = 0$

4. $f(x) = x^5 + x^3 - x^2 + x$, $x = -3$

5. $y = x^5 + 5x^3 + 18x^2$, $x = 0$

6. $y = x^3 + 3x^2 + 14$, $x = -2$

7. $y = 4x - 14$, $x = 7$ 　　　　**8.** $y = -3x + 9$, $x = -2$

9. $y = x^2 + \dfrac{1}{x^2}$, $x = 0$ 　　**10.** $y = \sqrt{x^2 + 25}$, $x = -1$

11. $y = \dfrac{1}{x^2 + 1}$, $x = -1$ 　　**12.** $y = \dfrac{x}{x^4 + 1}$, $x = -2$

In Exercises 13–18, find the values of the independent variable for which the given function is increasing or decreasing and for which it is concave upward or downward. Sketch the function.

13. $f(x) = x^2 + 2x + 1$

14. $f(x) = 3x^2 - 8x + 14$

15. $f(t) = 4t^3 - 45t^2 + 150t$

16. $f(t) = \frac{4}{3}t^3 + 4t^2 - 45t + 100$

17. $xy = 6$

18. $xy = 1$

In Exercises 19–22, find the points of inflection.

19. $y = x^3$

20. $y = x^{10} + x^8$

21. $y = \sqrt[3]{x}$

22. $y = x^{2/3}$

23. Sketch a continuous function with the following properties: $f(0) = 0; f(5) = -3; f'(5) = 0; f''(-1) = 0; f''(11) = 0; f''(x) < 0$ for $x < -1$ or $x > 11$, and $f''(x) > 0$ for $-1 < x < 11$.

24. Sketch a continuous function with the following properties: $f(-2) = f'(-2) = f''(-2) = 0; f'(x) < 0$ for $x \neq -2; f''(x) > 0$ for $x < -2; f''(x) < 0$ for $x > -2$.

3.3 MAXIMA—MINIMA

One of the most useful applications of the derivative is to find the largest and smallest values a function can take on over its domain. Consider the problem of the manufacture of a container designed to hold a fixed volume: by using the minimum amount of material necessary, the cost of manufacture can be held to a minimum, and hopefully the profit can be maximized. In a like manner, a farmer's profit depends on costs, and by minimizing costs profits can be maximized. If we assume that the quantities to be dealt with can be placed in some kind of functional relation, then the derivative can supply us with tools to find the maximum or minimum. This section will examine the mathematical theory necessary to solve these types of problems; the next section will apply the theory to specific problems.

In Figure 3.14, points P_1, P_3, and P_6 are relative maximum points. Point P_6 is also the absolute maximum of the function. Points P_2 and P_4 are relative minima. There is no absolute minimum. Point P_5 is neither a relative maximum nor a relative minimum point.

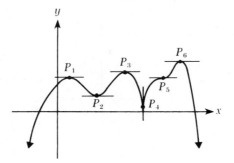

Figure 3.14 Maxima and minima

Definitions

If $f(a) > f(x)$ for all x near a, then $f(a)$ is called a *relative maximum* value of f. If $f(a) < f(x)$ for all x near a, then $f(a)$ is called a *relative minimum* value of f. If $f(a)$ is greater than $f(x)$ for all other values of x for which the function is defined, then $f(a)$ is the *absolute maximum* value of f. In the same manner, if $f(a)$ is less than all other images of f, then $f(a)$ is the *absolute minimum*. The plural of maximum is *maxima*; the plural of minimum is *minima*.

The words "relative" and "absolute" are usually dropped. Thus a minimum or maximum value of a function will mean either a relative or an absolute minimum or maximum value. An *extremum* of a function is a relative maximum or a relative minimum value. *Extrema* is the plural of extremum.

If $f'(a) = 0$ or $f'(a)$ is undefined, then a is called a *critical value*. Maxima and minima are found by examining critical values. If all critical values yielded either a maximum or a minimum value of the function, our problem would be greatly simplified. However, they do not. In Figure 3.14, the domain element of P_5 is a critical value since the slope of the tangent at P_5 is zero, but P_5 is neither a maximum nor a minimum value of the function. It is a point of inflection. If the behavior of a curve is known through studying its derivatives, its maxima and minima can easily be found. We can use a simple example to illustrate this.

Example 1   Find the extrema for y given by $y = x^2 - 2x$.

Solution   Since $y = x^2 - 2x$ is the equation of a parabola, it will have a single maximum or minimum point. In fact, since it is concave upward, the point we are looking for will be a minimum point. To find it we examine critical values, values of x that cause $y' = 2x - 2$ to equal zero or be undefined. Since y' is defined for all x, we need only set it equal to zero and solve for x.

$$2x - 2 = 0$$
$$2x = 2$$
$$x = 1$$

Substituting 1 for x in the original equation,

$$y = 1^2 - 2(1) = -1$$

Thus the minimum value for y is -1, and the extremum is at the point $(1, -1)$. Figure 3.15 shows the graph of the function and its minimum point. The tangent line at the minimum point has slope zero.

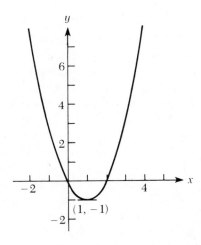

Figure 3.15   Graph of $y = x^2 - 2x$

The problem of finding maximum or minimum points is more difficult if the general shape of the curve of the function is not well known. In this case it is difficult to tell if a certain critical value is at a maximum point, minimum point, or neither. Of course, it is usually possible to test a critical value by substituting it into the equation of the function, then comparing the functional value with two others found by substituting values slightly greater and slightly less than the critical value into the function. In this manner we can determine whether the critical value yields a functional value greater than or less than those produced by values of x close to the critical value. An alternative and better method is to examine the first or second derivatives near the critical value.

In the graph of a relative maximum point, Figure 3.16a, note that the slope of the tangent line changes from a positive slope to zero slope and finally to a negative slope as x increases in value. In

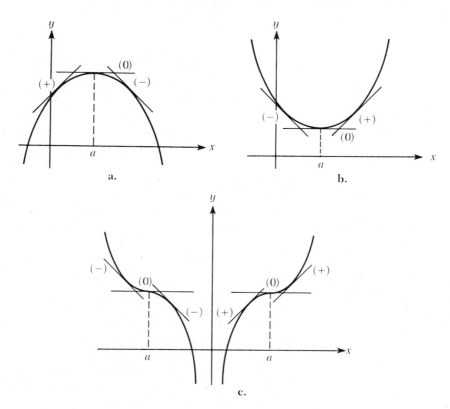

Figure 3.16 **a.** Relative maximum; $[+, 0, -]$; **b.** relative minimum; $[-, 0, +]$; **c.** neither maximum nor minimum; $[-, 0, -]$ or $[+, 0, +]$

the graph of a relative minimum point, Figure 3.16b, the slope of the tangent line changes from negative to zero and then to positive. For points that are neither relative maxima nor relative minima, as shown in Figure 3.16c, the tangent lines on either side of critical values both have positive slopes or both have negative slopes. These properties are summarized in the following test:

The First-Derivative Test for Extrema

If $x = a$ is a critical number and if f is continuous on an interval about a, then:

1. If $f'(x)$ changes sign from positive to negative as x increases in value over an interval containing a, $f(a)$ is a maximum value of f.

2. If $f'(x)$ changes sign from negative to positive as x increases in value over an interval containing a, $f(a)$ is a minimum value of f.

3. If $f'(x)$ does not change sign as x increases in value over an interval that contains a, $f(a)$ is neither a maximum nor a minimum value of f.

Example 2   Examine $y = 2x^3 + 3x^2 - 72x$ for maximum and minimum values.

Solution   We find critical values by taking the first derivative, setting it equal to zero, and solving for x.

$$y' = 6x^2 + 6x - 72 = 0$$
$$6(x^2 + x - 12) = 0$$
$$6(x + 4)(x - 3) = 0$$
$$x = -4 \qquad \text{or} \qquad x = 3$$

The question is: What kinds of points are produced by the critical values, -4 and 3?

We apply the first-derivative test to y' for an interval about -4. If x is slightly less than -4, then the factor $(x + 4)$, from $y' = 6(x + 4)(x - 3)$, is negative and the factor $(x - 3)$ is also negative. The derivative is the product of two negatives and hence is positive. If x is slightly greater than -4, then $(x + 4)$ is positive while $(x - 3)$ is negative. The derivative is negative. Therefore, as x increases in value over an interval that contains -4, the slope of the tangent line changes from positive to negative. By the first step of the first-derivative test, $(-4, f(-4))$ is a maximum point.

Applying the first-derivative test to $x = 3$, if x is slightly less than 3, then $(x + 4)$ is positive while $(x - 3)$ is negative. In this case the derivative is the product of a positive number and a negative number and is negative. If x is slightly greater than 3, then both $(x + 4)$ and $(x - 3)$ are positive and the derivative is positive. Thus as x increases in value over an interval that contains 3, the first derivative changes sign from negative to positive. By the second step of the first-derivative test, $(3, f(3))$ is a minimum point on the graph of f. The function $y = 2x^3 + 3x^2 - 72x$ is graphed in Figure 3.8.

The second derivative is often easier to use to test for maxima and minima since the concavity at any point is indicated by the second derivative. Recall that if the second derivative is negative at some point, the curve is concave downward there, and if the second derivative is positive at a point, then the curve is concave upward at that point. One type of maximum point, where the derivative exists, can be described as a point where the slope of the tangent line is zero and the curve is concave downward while some minima, again where the derivative exists, are points where the tangent line has zero slope and the curve is concave upward.

The Second-Derivative Test for Extrema

If $f'(x)$ and $f''(x)$ exist for an interval that contains a and if $f'(a) = 0$, then:

1. If $f''(a) < 0$, $f(a)$ is a maximum value of f.
2. If $f''(a) > 0$, $f(a)$ is a minimum value of f.

The second-derivative test cannot be used to describe the curve when $f''(a) = 0$. If $f''(a) = 0$, the point $(a, f(a))$ could be a point of inflection. However, for certain special curves it could also be a maximum or minimum value. When the second derivative is zero, the second-derivative test fails and we must resort to the first-derivative test.

Example 3

Examine $y = 2x^3 + 3x^2 - 72x$ for maximum and minimum values of y.

Solution

This is the same problem posed in the last example. The curve is drawn in Figure 3.8. The first and second derivatives are:

$$y' = 6x^2 + 6x - 72 \qquad \text{and} \qquad y'' = 12x + 6$$

The critical values were found to be $x = -4$ and $x = 3$ in the last example. By inspecting the second derivative, it is easy to see that $y'' < 0$ when $x = -4$, and $y'' > 0$ when $x = 3$. Therefore, by the second-derivative test, $x = -4$ yields a maximum point, and $x = 3$ yields a minimum.

Example 4   Examine $y = x^{2/3}$ for maximum and minimum points.

Solution $y' = \dfrac{2}{3} x^{-1/3} = \dfrac{2}{3 \sqrt[3]{x}}$

The derivative y' cannot equal zero and is undefined when $x = 0$. Therefore $x = 0$ is a critical value. To test it we apply the first-derivative test: If x is negative, $y' = 2/3 \sqrt[3]{x}$ is negative. If x is positive, y' is positive. Therefore y' changes from negative to positive as x increases. The critical value, $x = 0$, is at a minimum point. Thus $y = 0$ is the minimum value for y. There are no maximum values for y. This is clearly seen in the graph of the curve, Figure 3.17. Since y' is undefined at $x = 0$, the slope of the tangent at that point is undefined, and the tangent is vertical at the origin. Also, the second-derivative test shows the curve to be concave downward for all values of x except $x = 0$.

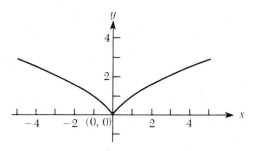

Figure 3.17   Graph of $f(x) = x^{2/3}$

3.3 EXERCISES

In Exercises 1–21, find the maximum and minimum points.

1. $y = 2x^2 + 8x + 3$ **2.** $y = 16 - 3x - 9x^2$

3. $y = 4x + 9$ **4.** $y = -8x + 2$

5. $y = x^3$ **6.** $y = 2x^3 - 6x$

7. $y = 4x^3 - 12x$ **8.** $y = x^3 + 3x^2 - 1$

9. $y = 2x^3 - 3x^2 - 36x + 4$ **10.** $y = x^4 - 2x^2$

11. $f(x) = x^4 - 2x^2 + 12$ **12.** $f(x) = x^5$

13. $f(x) = x^7$ **14.** $f(x) = x^4$

15. $f(x) = x + \dfrac{1}{x}$ **16.** $f(x) = 4x + \dfrac{16}{x}$

17. $f(x) = 3(x^{2/3} - x^{1/3})$ **18.** $f(x) = \dfrac{x - 1}{x + 1}$

19. $f(x) = x\sqrt{x - 1}$ **20.** $f(x) = \dfrac{x^2}{x - 1}$

21. $f(x) = x^{2/3} + x$

3.4 APPLICATIONS OF MAXIMA—MINIMA

The difficulty with the application of any mathematical theory, including the one presented here, to the solution of problems is that questions are usually stated in words rather than in equations. The problem of creating an equation is especially difficult in maximum–minimum problems, because any single exercise may use a number of different formulas. The key point is this: *The quantity to be maximized or minimized must be written as a function of some independent variable.* For instance, if income is to be maximized, then income must be written as a function of some independent variable, such as the number of items produced. If surface area is to be minimized, then it must be written as a function of an independent variable, such as length or width.

Example 1   Divide 36 into two parts whose product is maximum.

Solution   The product must be written as a function of one of the parts of 36. Let x be one of the parts of 36. Then $(36 - x)$ is the other part of 36. Let p represent the product.

$$p = x(36 - x)$$
$$= 36x - x^2$$

To maximize p, find the critical values of x by taking the first derivative of p with respect to x, setting it equal to zero, and solving for x.

$$p' = 36 - 2x = 0$$
$$-2x = -36$$
$$x = 18 \qquad \text{and} \qquad 36 - x = 18$$

Since $p = 36x - x^2$ is a parabola opening downward (when plotted on an x–p axis system), we know that the critical value is at a maximum point. Therefore, the two numbers 18 and 18 are the parts of 36 that give the maximum product.

Example 2

Suppose that a rancher has 100 feet of chicken wire to make a chicken yard. One side of the yard will be formed by the side of a barn so no wire will be needed there. What are the dimensions of the yard if it is to be rectangular in shape and contain maximum area?

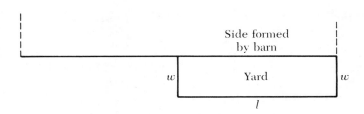

Figure 3.18 Area of yard

Solution

Since area is to be maximized, we need an equation for it. Let w be the width of the yard and l the length, as in Figure 3.18. Then, if A represents area,

$$A = wl \qquad \text{and} \qquad 2w + l = 100$$

In order to maximize area, we need to describe it as a function of a single variable. Therefore, we solve $2w + l = 100$ for l and substitute into the formula for area.

$$l = 100 - 2w$$

and

$$A = wl = w(100 - 2w)$$

so

$$A = 100w - 2w^2$$

To maximize A, we take its derivative, set the derivative equal to zero, and solve for w.

$$A' = 100 - 4w = 0$$
$$-4w = -100$$
$$w = 25$$

Substituting $w = 25$ into $l = 100 - 2w$, we solve for l.

$$l = 100 - 2(25)$$
$$= 100 - 50 = 50$$

The yard should be 25 feet wide and 50 feet long in order to have maximum area.

Example 3

A certain television set manufacturer will, on the average, sell 1000 television sets per month at $500 per set, and he can sell an additional 100 sets per month for each $20 decrease in price. What price per set will bring the greatest income?

Solution Let the price of the television set be x dollars. Since the income is to be maximized, we need an expression for income in terms of the independent variable, price. It is clear that the income will be equal to the price per television set times the number of sets sold. If I stands for the income, then

$$I = (\text{Price per set})(\text{Number of sets sold})$$

But x stands for the price per set, so income is

$$I = x(\text{Number of sets sold})$$

The number of sets sold will be 1000 sets plus 100 more sets for each $20 price reduction from $500. The price reduction is the original price of $500 minus the new price of x dollars, or $(500 - x)$. For each $20 in the reduction 100 more sets will be sold. The question is, how many units of $20 are in the reduction $(500 - x)$? The answer is $(500 - x)/20$. Thus the number of sets sold in excess of 1000 will be

$$100 \left(\frac{500 - x}{20} \right)$$

Therefore, the total number of sets sold will be 1000 plus the extra sets sold because of the price reduction, or

$$1000 + 100 \left(\frac{500 - x}{20} \right)$$

Since the income I is the price per set times the number of sets sold,

$$\begin{aligned}
I &= x \left[1000 + 100 \left(\frac{500 - x}{20} \right) \right] \\
&= x[1000 + 5(500 - x)] \\
&= x[1000 + 2500 - 5x] \\
&= x[3500 - 5x] \\
&= 3500x - 5x^2
\end{aligned}$$

Taking the derivative, setting it equal to zero, and solving for x gives:

$$\begin{aligned}
I' = 3500 - 10x &= 0 \\
-10x &= -3,500 \\
x &= 350
\end{aligned}$$

The price that will yield maximum income is $350.

The following example and some of the exercises require a knowledge of certain area and volume formulas. The list below is offered for easy reference.

Let r represent radius, l length, w width, h height, and b length of base.

1. Area of a triangle: $A = \frac{1}{2}bh$
2. Area of a trapezoid: $A = \frac{1}{2}h(b_1 + b_2)$
3. Surface area of a sphere: $A = 4\pi r^2$
4. Lateral surface area of a right circular cylinder: $A = 2\pi rh$
5. Total surface area of a right circular cylinder: $A = 2\pi rh + 2\pi r^2$
6. Volume of a right circular cylinder: $V = \pi r^2 h$
7. Volume of sphere: $V = \frac{4}{3}\pi r^3$
8. Volume of a rectangular solid: $V = lwh$
9. Volume of a right circular cone: $V = \frac{1}{3}\pi r^2 h$

Example 4   Find the dimensions of the cylindrical can that would hold 250 cubic centimeters of tuna and would require the least amount of material to construct.

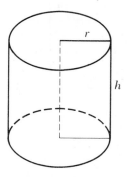

Figure 3.19   A cylindrical can with a volume of 250 cubic centimeters

Solution   A tuna can is a right circular cylinder as in Figure 3.19. It must have a volume of 250 cubic centimeters. The material in the can will be minimum when the surface area is minimum. The equation for total surface area, Equation 5 above, is:

$$A = 2\pi rh + 2\pi r^2$$

We need to express the area formula as a function of a single variable, either r or h, in order to minimize area. The fact that the volume must be 250 cubic centimeters can be used to express h in terms of r. From Equation 6 above, the volume of a right circular cylinder is

$$V = \pi r^2 h = 250$$

Solving for h,

$$h = \frac{250}{\pi r^2}$$

Substituting for h in Equation 5 and simplifying,

$$A = 2\pi r \frac{250}{\pi r^2} + 2\pi r^2 = \frac{500}{r} + 2\pi r^2 = 500r^{-1} + 2\pi r^2$$

To minimize area, we need A'.

$$A' = -500r^{-2} + 4\pi r = \frac{-500}{r^2} + 4\pi r$$

where A' will be undefined if $r = 0$. However, $r = 0$ is not the value we seek, since it is impossible to have a right circular cylinder of radius 0. To find the value of r that will minimize surface area, set A' equal to zero and solve for r.

$$-\frac{500}{r^2} + 4\pi r = 0$$

Multiply by r^2.

$$-500 + 4\pi r^3 = 0$$
$$4\pi r^3 = 500$$
$$r^3 = \frac{500}{4\pi} = \frac{125}{\pi}$$
$$r = \sqrt[3]{\frac{125}{\pi}} = \frac{5}{\sqrt[3]{\pi}} = \frac{5}{\pi^{1/3}}$$

To find h, substitute $r = 5/\pi^{1/3}$ in $h = 250/\pi r^2$.

$$h = \frac{250}{\pi(5/\pi^{1/3})^2} = \frac{250}{\pi(25/\pi^{2/3})} = \frac{10}{\pi^{1/3}}$$

Therefore, the dimensions of the tuna can that will contain 250 cubic centimeters of tuna and require the minimum amount of material are

$$r = \frac{5}{\pi^{1/3}} \text{ centimeters} \qquad \text{and} \qquad h = \frac{10}{\pi^{1/3}} \text{ centimeters}$$

3.4   EXERCISES

1. Find two numbers whose sum is 30 and whose product is maximum.

2. Find two numbers whose sum is 30 such that the sum of their squares is minimum.

3. Find two numbers whose sum is 30 such that the sum of the square of one plus ten times the other is minimum.

4. Find the dimensions of the open box with the largest volume that can be made from a 12-inch square piece of cardboard by cutting equal squares from the corners of the cardboard and turning up the sides.

5. Find the volume, to the nearest one-tenth of a cubic inch, of the open box with the greatest volume that can be made from an 8-inch by 15-inch rectangular piece of sheet metal by cutting equal squares from the corners and folding up the sides.

6. What are the dimensions of the rectangular field with maximum area that has a perimeter of 1000 feet?

7. A rectangular garden is to be created using 600 feet of fencing. At one corner of the garden and at right angles to each other there will be a 20-foot shed and a 20-foot gate. Find the dimensions of the garden with a maximum area if no fencing is needed for the spaces occupied by the shed and the gate.

8. Find the volume of the box with the greatest volume that can be made from 1200 square inches of material if the box is to be rectangular with a square base and no top.

9. A rectangular box with a square base is to contain 540 cubic inches. If the top costs $0.30 per square inch of material, the bottom $0.20 and the sides $0.10, find the dimensions of the box so that the cost is minimum.

10. A poster is to contain 200 square inches of illustrations and instructions. The top and bottom margins are to be 4 inches wide while the side margins are 2 inches wide. Find the length and width of the poster that would use the least amount of paper.

11. Answer Exercise 10 for a poster that is to contain 100 square inches of printed material with the same margins.

12. What dimensions of a cylindrical coffee can would require the least amount of metal if the can is to contain 64 cubic inches of coffee?

13. In making a cylindrical tin can to contain 64 cubic inches of coffee, there is no waste involved in making the vertical side of the can. However, the circular top and bottom are cut from squares of material with the corner parts of the square wasted. What are the height and radius of the most economical can?

14. An ammonia tank is to hold c cubic feet. It is constructed of a cylinder capped on each end by a hemisphere. Find the dimensions of the most economical tank if the cylindrical part costs $10.00 per square foot to build while the hemispheres cost $15.00 per square foot.

15. The cross-section of a culvert is in the shape of an isosceles triangle with vertex downward. The equal sides of the triangle are 10 meters long. Such a culvert will handle maximum water flow when it has maximum cross-sectional area. At what angle should the equal sides be to one another in order for the culvert to handle maximum water flow?

16. The culvert of Exercise 15 is to be made in the shape of an isosceles trapezoid with shorter base downward. The equal sides and the short base are all 10 meters long. How long should the other base be so that the culvert will carry maximum water flow?

$ 17. A golf pro-shop operator has found that 500 sets a year of top grade clubs at $300 per set can be sold. For each $5.00 the price drops, 10 more sets of clubs can be sold. What price would give the largest gross income?

$ 18. A resort hotel will provide a dinner, dance and drinks for a local organization at $50 a couple, with the agreement that there be at least 100 couples attending. However, it is also agreed that the price will be reduced $2.00 per couple for each 10 couples in excess of 100. How many couples will maximize the hotel's total revenue?

$ 19. The board of directors of the organization described in Exercise 18 agreed to the $50 but thought that the price should be reduced $0.20 per couple for each couple in excess of 100. Now how many couples will it take to maximize the hotel's gross revenue?

$ 20. When traveling at s miles per hour, the cost per hour of operating a truck is $12 + (s^2/48)$, which includes the driver's wages, depreciation, cost of fuel, maintenance, insurance, and the like (in dollars). At what speed is the cost minimum during a 96-mile trip?

$ 21. At what speed is the cost minimum for the truck in Exercise 20 if the cost per hour is $20 + (s^2/100)$ and the trip is 500 miles?

22. A power line runs due north. Town A is 2 miles due east from point c on the power line. Town B is 10 miles due east from point d on the power line. Points c and d are 5 miles apart. At what point between c and d on the power line should a transformer be located so that when power lines are run from it to towns A and B, they will be of minimum length?

23. Three towns are located at the vertices of an isosceles triangle. Towns B and C are at the base angles while A is at the vertex of the triangle. The distance from B to C is 16 miles and the altitude through A is 10 miles. How far from A, along the altitude through A, should a well be located so that it will use minimum pipe when supplying water to all three towns?

§ 24. A printer has contracted to print 500,000 campaign posters. A large number of metal copies of the poster could be set in type so that each impression of the printing press would create a number of posters, or a single metal copy of the poster could be set and 500,000 impressions made on the printing press. A single metal copy of the poster costs $0.80 to set in type. It costs $5.00 per hour to run the printing press. How many metal copies of the poster should be set to minimize the cost of printing if the press runs 1000 impressions per hour?

§ 25. For a certain commodity, x items will be sold per week when the price is $P = 100 - 0.10x$ dollars. The cost of x items is $C = 50x + 1000$ dollars. How many should be produced each week in order to maximize profits?

§ 26. Answer Exercise 25 for a price $P = 8000 - 7x$ dollars and a cost $C = 25x + 950$ dollars.

3.5 THE DIFFERENTIAL

Up to this point, the symbol dy/dx (or df/dx) has simply denoted the first derivative of a function with respect to x. However, it is useful to be able to think of dy/dx as the quotient of the two symbols dy and dx. The following definitions will allow us to think of dy/dx as either the derivative of a function with respect to x or the ratio of dy and dx. This will, in turn, expand possible interpretations of the derivative. In fact, dy and dx must be separate entities in the study of integral calculus.

Definition

If f is a differentiable function of x expressed by $y = f(x)$, then the *differential* of x, dx, is an increment of x. That is, $dx = \Delta x$. The *differential* of y is $dy = f'(x)\,dx$.

Example 1   If $y = 2x^3 + 7x$, find the differential of y.

Solution $dy = f'(x)\,dx$, but $f'(x) = 6x^2 + 7$; so,

$$dy = (6x^2 + 7)\,dx$$

The differential dy is also denoted by df. Therefore, $df = dy = f'(x)\,dx$. Note that if both sides of

$$dy = f'(x)\,dx$$

are divided by dx, we have

$$\frac{dy}{dx} = f'(x), \qquad dx \neq 0$$

Thus, dy/dx can be thought of as the symbol for the first derivative of f, or as the quotient of differentials.

Although it is true that $dx = \Delta x$, in general, $dy \neq \Delta y$. This can be seen in Figure 3.20.

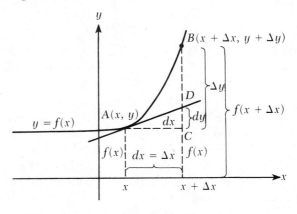

Figure 3.20   The differential

The line through points A and D is tangent to the curve at point A. The derivative evaluated at point A gives the slope of the tangent line. But the ratio of the lengths of line segments $\overline{DC}$ to $\overline{AC}$ is also the slope of the tangent line. Therefore,

$$f'(x) = \frac{\overline{DC}}{\overline{AC}}$$

But the length of $\overline{AC}$ is dx; therefore dy must equal the measure of $\overline{DC}$ since

$$f'(x) = \frac{\overline{DC}}{\overline{AC}} = \frac{\overline{DC}}{dx}$$

Thus,

$$\overline{DC} = f'(x)\, dx = dy$$

Now for any change, Δx, in x, $\Delta y = \overline{BC} = f(x + \Delta x) - f(x)$. This is also shown in Figure 3.20, where the difference between Δy and dy is given by the line segment $\overline{BD}$. Although dy is not equal to Δy, it is a good estimate of Δy when dx is small. To see this, note that in Figure 3.20, if dx were made very small, then the difference between dy and Δy would also become very small.

Example 2

Find Δy and dy for $y = f(x) = x^2 + 4x$ when $x = 3$ and $\Delta x = 0.2$.

Solution

$$\begin{aligned}
\Delta y &= f(x + \Delta x) - f(x) \\
&= f(3 + 0.2) - f(3) \\
&= f(3.2) - f(3) \\
&= (3.2)^2 + 4(3.2) - [3^2 + 4(3)] \\
&= 10.24 + 12.8 - (9 + 12) \\
&= 23.04 - 21 = 2.04
\end{aligned}$$

To compute $dy = f'(x)\, dx$, we note that $dx = \Delta x = 0.2$, and $f'(x) = 2x + 4$; so,

$$\begin{aligned}
dy &= f'(3)\, dx \\
&= [2(3) + 4](0.2) \\
&= (10)(0.2) = 2
\end{aligned}$$

Thus $\Delta y = 2.04$ and $dy = 2$. They differ by 0.04.

Example 3   Given a function f such that

$$f(x) = \frac{x^2}{\sqrt{x^2 + 21}}$$

estimate, using the differential, the change in the function value if x is changed from 10 to 9.7.

Solution   If $x = 10$, then $dx = \Delta x = -0.3$. We can estimate Δy with dy.

$$
\begin{aligned}
dy = f'(x)\, dx &= \frac{\sqrt{x^2 + 21}\,(2x) - x^2 \cdot \dfrac{2x}{2\sqrt{x^2 + 21}}}{x^2 + 21}\, dx \\[2mm]
&= \frac{\sqrt{x^2 + 21}\,(2x) - \dfrac{x^3}{\sqrt{x^2 + 21}}}{x^2 + 21}\, dx \\[2mm]
&= \frac{(x^2 + 21)(2x) - x^3}{(x^2 + 21)^{3/2}}\, dx \\[2mm]
&= \frac{2x^3 + 42x - x^3}{(x^2 + 21)^{3/2}}\, dx \\[2mm]
&= \frac{x^3 + 42x}{(x^2 + 21)^{3/2}}\, dx
\end{aligned}
$$

With $x = 10$ and $dx = -0.3$,

$$
\begin{aligned}
dy &= \frac{10^3 + 42(10)}{(10^2 + 21)^{3/2}}(-0.3) \\[2mm]
&= \frac{1000 + 420}{(121)^{3/2}}(-0.3) \\[2mm]
&= \frac{1420}{11^3}(-0.3) \\[2mm]
&= -\frac{426}{1331} = -0.32
\end{aligned}
$$

Example 4   Find an approximation for the maximum possible error in calculating the volume of a sphere when its radius is 5 inches with a possible error of ±0.01 inch.

Solution   We are being asked what effect an error in the radius of a sphere will have on its volume. The volume of a sphere is $V = \frac{4}{3}\pi r^3$. The change in V is approximately the differential of the volume where r is the independent variable, V is the dependent variable, $r = 5$, and $dr = \pm0.01$.

$$dV = V' \, dr, \qquad \text{but} \quad V' = 4\pi r^2$$
$$dV = 4\pi r^2 \, dr$$
$$= 4\pi (5)^2 (\pm 0.01)$$
$$= 100\pi (\pm 0.01)$$
$$= \pm \pi \approx \pm 3.14 \text{ cubic inches}$$

Thus an error of ± 0.01 inch in the radius produces an error of π cubic inches in the volume.

Formulas for differentials can be obtained by multiplying derivative formulas by dx. Recall that if u and v are differentiable functions of x, and $y = u/v$, then

$$\frac{dy}{dx} = \frac{v \dfrac{du}{dx} - u \dfrac{dv}{dx}}{v^2}$$

To find the differential of y, dy, multiply both sides by dx.

$$dy = \frac{v \, du - u \, dv}{v^2}$$

Other formulas can be found in a similar manner. For instance, if u and v are differentiable functions of x and c is a constant, then

$$dc = 0$$
$$d(u^n) = nu^{n-1} \, du$$
$$d(u + v) = du + dv$$
$$d(u \cdot v) = u \, dv + v \, du$$

and so forth. These formulas are important in integration, the second major concept of calculus, which will be introduced later. For the time being the differential of the dependent variable is easily found using the definition

$$dy = f'(x) \, dx$$

3.5 EXERCISES

In Exercises 1–8, find dy.

1. $y = 8x^2$

2. $y = 5x^3 + 7x^2 + 8$

3. $y = \sqrt{x^2 + 25}$

4. $y = -\sqrt{16 - x^2}$

5. $y = (x^2 + 1)(x^2 + 2)$

6. $y = \dfrac{x + 1}{x}$

7. $y = \dfrac{x}{x + 1}$

8. $y = \dfrac{5x^2}{3x^3 + 4}$

In Exercises 9–12, evaluate df for the given functions and the given values of x and dx.

9. $f(x) = x^3 + 2x$, $\quad x = 2, dx = 0.01$

10. $f(x) = \sqrt[3]{x}$, $\quad x = -32, dx = 3$

11. $f(x) = \dfrac{1}{x}$, $\quad x = 10, dx = -0.003$

12. $f(x) = x$; $\quad x = \dfrac{\pi}{2}, dx = -0.01$

In Exercises 13–16, find dy and Δy for the given functions and the given values of x and dx.

13. $y = x^2$, $\quad x = 5, dx = 0.2$

14. $y = x^2 + 7x + 2$, $\quad x = -3, dx = 0.01$

15. $y = x^3$, $\quad x = 2, dx = 0.1$

16. $y = x^3$, $\quad x = 2, dx = -0.1$

In Exercises 17–20, use dy to estimate Δy.

17. $y = \dfrac{x^2}{\sqrt{x^2 + 21}}$, $\quad x = 10, dx = 0.1$

18. $y = \dfrac{x}{\sqrt{x + 1}}$, $\quad x = 24, dx = 1$

19. $y = x^3 - 4x^2 + 2x + 1$, $\quad x = 8, \Delta x = -0.3$

20. $y = \dfrac{1}{x^3}$, $\quad x = 10, \Delta x = -1$

21. Use differentials to estimate the volume of a cube that is 6.1 inches on a side.

22. A small adding machine is sold with the following instruction for finding the square root of a number near 16.

$$\sqrt{16 + dx} = 4 + \frac{dx}{8}$$

What is the corresponding instruction for finding the square root of a number near 25?

23. Answer Exercise 22 for a number near 100.

24. A cubical metal box has an interior measurement of exactly 5 inches along each edge. If the metal sides and the top and bottom of the box are all 0.5 inch thick use differentials to approximate the volume of metal used in constructing the box.

25. The radius of a sphere is to be measured and that measurement is to be used to compute the volume of the sphere. However, the radius can only be measured to within $\pm\frac{1}{4}$ inch of its actual length but the volume can be in error by at most 12 cubic inches. What is, approximately, the radius of the largest sphere whose volume will be in error by at most 12 cubic inches when its radius is in error $\pm\frac{1}{4}$ inch?

3.6 IMPLICIT DIFFERENTIATION

Functions are regarded as a pairing of variables by some rule. Consider the equations $2x + 2y - 3 = 0$ and $y = (3 - 2x)/2$. If we want to use these equations to define y as a function of x, both define identical functions. Any pair of values that satisfies one of the equations will satisfy the other. To differentiate between the two types of definitions, we say that $y = (3 - 2x)/2$ defines y as an *explicit function* of x, whereas the equation $2x + 2y - 3 = 0$ defines y as an *implicit function* of x.

Expressions such as $y = x^2$ or $y = \sqrt{x^2 + 5x} + x$ define y as explicit functions of x, whereas expressions such as $x^2 + y^2 = 1$ and $x^4y^5 + 3xy + y^3 + 5 = 0$ define y as implicit functions of x. This means that the expressions can either be solved for y in terms of x or, if this is impractical or impossible, the set of ordered pairs that satisfy the expression form a function of x. In practice it is often necessary to restrict our attention to certain intervals of x or y in order to insure a functional relationship. The expression $x^2 + y^2 = 1$ defines many functions of x, two of which are

$$f_1: \quad f_1(x) = y = \sqrt{1 - x^2} \qquad \text{and} \qquad f_2: \quad f_2(x) = y = -\sqrt{1 - x^2}$$

Many equations that define implicit functions (y of x) cannot readily be solved for y in terms of x. However, it may not be difficult to find dy/dx for such functions. The method is called *implicit differentiation*. We will assume that y is implicitly a function of x in any expression containing x and y. Thus if the term of y^2 is encountered, we can take its derivative with respect to x by assuming y is an implicit function of x and applying the chain rule.

$$D_x(y^2) = 2y \frac{dy}{dx}$$

Example 1   Use implicit differentiation to find dy/dx for $x^2 + y^2 = 1$.

Solution   Take the derivative of each side with respect to x and solve for dy/dx.

$$D_x(x^2) + D_x(y^2) = D_x(1)$$
$$2x + 2y \frac{dy}{dx} = 0$$
$$2y \frac{dy}{dx} = -2x$$
$$\frac{dy}{dx} = -\frac{x}{y}$$

Example 2   Find y' for $xy = 1$.

Solution   The derivative of 1, the right side of the equation, is zero. The left side of the equation, xy, is the product of x and a function of x, namely, y. Therefore, we take x times the derivative of y plus y times the derivative of x, the rule for finding the derivative of the product of two functions of x.

$$D_x(xy) = D_x(1)$$
$$x \frac{dy}{dx} + y(1) = 0$$
$$x \frac{dy}{dx} = -y$$
$$\frac{dy}{dx} = -\frac{y}{x}$$

or

$$y' = -\frac{y}{x}$$

Solving $xy = 1$ for y gives $y = 1/x$. If we substitute $y = 1/x$ in $y' = -y/x$, we get $y' = -1/x^2$, which is the same result that comes from finding y' explicitly for $y = 1/x$.

Example 3   Find dy/dx for $2x^4 - 3x^2y^2 + y^4 = 0$.

Solution   Take the derivative with respect to x term by term. Note that the middle term will be treated as the product of two functions of x.

$$2D_x(x^4) - 3D_x(x^2y^2) + D_x(y^4) = 0$$

$$8x^3 - 3\left(2x^2y\frac{dy}{dx} + 2xy^2\right) + 4y^3\frac{dy}{dx} = 0$$

$$8x^3 - 6x^2y\frac{dy}{dx} - 6xy^2 + 4y^3\frac{dy}{dx} = 0$$

Solving for dy/dx,

$$4y^3\frac{dy}{dx} - 6x^2y\frac{dy}{dx} = 6xy^2 - 8x^3$$

$$(4y^3 - 6x^2y)\frac{dy}{dx} = 6xy^2 - 8x^3$$

$$\frac{dy}{dx} = \frac{6xy^2 - 8x^3}{4y^3 - 6x^2y}$$

$$= \frac{3xy^2 - 4x^3}{2y^3 - 3x^2y}$$

Example 4   Find the slope of the tangent line to the ellipse $4x^2 + 9y^2 = 40$ at the point $(1, 2)$.

Solution   To find the slope we need only evaluate y' at the point $(1, 2)$. We could solve the equation for y and then find y'. However, it is easier to find y' by implicit differentiation. If

$$4x^2 + 9y^2 = 40$$

then

$$8x + 18y(y') = 0$$

$$18y(y') = -8x$$

$$y' = \frac{-8x}{18y} = \frac{-4x}{9y}$$

We now substitute the values 1 and 2 for x and y.

$$y' = \frac{-4x}{9y}$$

$$= \frac{-4(1)}{9(2)} = -\frac{2}{9}$$

The slope is $-\frac{2}{9}$.

It is also possible to find second derivatives by implicit differentiation.

Example 5   Find y'' for $xy^3 = 1$.

Solution   First we find y'.

$$xy^3 = 1$$
$$3xy^2 \cdot y' + y^3 = 0$$
$$3xy^2 \cdot y' = -y^3$$
$$y' = \frac{-y^3}{3xy^2} = -\frac{y}{3x}$$

Now we take the derivative of y' to get y''.

$$y'' = -\frac{1}{3}\left[\frac{x \cdot y' - y}{x^2}\right]$$

Substituting $y' = -y/3x$ in y'' and simplifying,

$$y'' = -\frac{1}{3}\left[\frac{x(-y/3x) - y}{x^2}\right]$$
$$= -\frac{1}{3}\left[\frac{(-y/3) - y}{x^2}\right]$$
$$= -\frac{1}{3}\left[\frac{(-4/3)y}{x^2}\right]$$
$$= \frac{4y}{9x^2}$$

3.6   EXERCISES

In Exercises 1–12, find dy/dx by implicit differentiation.

1. $x^2 - y^2 = 1$
2. $4x^2 + 25y^2 = 100$
3. $xy + x = 2$
4. $3xy^2 + xy = 1$
5. $x^3 - xy + y^2 = 0$
6. $x^{1/3} + y^{1/3} = 4$
7. $y\sqrt{x+1} = 4$
8. $x\sqrt{y} = y\sqrt{x}$
9. $xy + x + y = x^2y^2$
10. $\dfrac{x}{y} + \dfrac{y}{x} = 2$
11. $x^{2/3} + y^{2/3} = 3^{2/3}$
12. $x^2y^2 = 1$

In Exercises 13–18, find the slope of the tangent line to the given function at the given point.

13. $x^2 + y^2 = 25$, $(-4, 3)$ **14.** $x^2 + 4x - y^2 = 3$, $(2, 3)$

15. $xy^2 = 12$, $(3, -2)$ **16.** $x^2y = -18$, $(3, -2)$

17. $2xy - 2x + y + 14 = 0$, $(2, -2)$

18. $x^3 - 2x^2 + 3y^2 - 4xy + 2x + y - 58 = 0$, $(4, -1)$

In Exercises 19–26, find y''. Simplify your answer.

19. $x^2 - y^2 = 1$ **20.** $x^2y^2 = 1$

21. $4x^2 + 9y^2 = 36$ **22.** $x^{1/2} + y^{1/2} = 1$

23. $y^2 = x^2 \cdot$ **24.** $y^2 - xy + 2 = 0$

25. $3x^2 + 4xy + y^2 = 0$ **26.** $x^{1/3} + y^{1/3} = 1$

3.7   RELATED RATES, IMPLICIT FUNCTIONS OF TIME

In many physical and economic applications of mathematics variables are implicit functions of time. For instance, if $C = f(x)$ defines a function that relates cost, C, to the number of items sold, x, it may also be true that both x and C are functions of time. That is, we might expect to sell 25 items per hour, which in turn might give a cost of $100 per hour, which means that over a definite number N of hours we will sell $25N$ items and incur $100N$ dollars of cost.

The instantaneous rate of change in x per unit of time is noted by dx/dt, and the instantaneous rate of change in cost per unit of time is denoted dC/dt. If some value for x and dx/dt at a specific time can be computed, then we can find dC/dt at that time.

Example 1

The cost in dollars of selling x items is given by $C = 500 + x + (1/x)$. When the 50th item is sold, it is noted that the rate of sales is 20 items per hour. What is the rate of change of the cost with respect to time at that moment?

Solution

We are given $dx/dt = 20$ and $x = 50$ and are asked to find dC/dt. We know C and x are both implicit functions of time. Therefore, we use the chain rule to take the derivative with respect to time of both sides of $C = 500 + x + (1/x)$.

$$\frac{dC}{dt} = \frac{d(500)}{dt} + \frac{dx}{dt} + \frac{d(1/x)}{dt}$$

$$= 0 + \frac{dx}{dt} + \left(-\frac{1}{x^2}\right)\frac{dx}{dt}$$

$$= \frac{dx}{dt} - \frac{1}{x^2}\frac{dx}{dt}$$

$$= \left(1 - \frac{1}{x^2}\right)\frac{dx}{dt}$$

Substituting 50 for x and 20 for dx/dt gives

$$\frac{dC}{dt} = \left(1 - \frac{1}{50^2}\right)20$$

$$= \left(1 - \frac{1}{2500}\right)20$$

$$= \frac{2499}{2500}(20)$$

$$= \frac{2499}{125} \approx \$20\,\text{per hour}$$

This means that the cost at the time the 50th item is sold is increasing at the rate of about \$20 per hour.

In this example, the cost per hour is a function of both the number of items sold and the rate of sales. If the rate of sales is 5 items per hour at the time the second item is sold, then the cost per hour is only increasing at \$3.75, which is computed as follows.

$$x = 2, \qquad \frac{dx}{dt} = 5, \qquad \frac{dC}{dt} = \left(1 - \frac{1}{x^2}\right)\frac{dx}{dt}$$

so

$$\frac{dC}{dt} = \left(1 - \frac{1}{4}\right)5$$

$$= \frac{3}{4}(5) = \$3.75$$

The cost per hour is a function of the number of items sold, as well as the rate of sales, because an increase in the number of items sold causes greater production costs, and an increase in the rate of sales causes greater sales cost due to a need for more salespeople, delivery personnel, and so forth.

In physical problems, the most common example of an instantaneous rate of change of a variable with respect to time is velocity, because velocity is the rate of change of distance with respect to time, and physical problems often involve movement over distances.

Example 2   Suppose that a 13-foot ladder, shown in Figure 3.21, is leaning against a vertical wall with the foot of the ladder on a horizontal surface. The foot of the ladder is drawn away from the wall at the rate of 3 feet per second. How fast is the top of the ladder moving down the wall when the foot of the ladder is 5 feet from the wall?

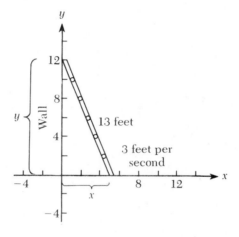

Figure 3.21   The base of the ladder is pulled away from the wall at the rate of 3 feet per second

Solution   Let the height of the ladder as measured along the wall be y and the distance of the foot of the ladder from the wall be x. Since the ladder is 13 feet long, the Pythagorean theorem gives

$$x^2 + y^2 = 13^2$$

Our problem is to find dy/dt when $x = 5$ and $dx/dt = 3$. Implicit differentiation with respect to time gives

$$2x \frac{dx}{dt} + 2y \frac{dy}{dt} = 0$$

$$\frac{dy}{dt} = - \frac{x}{y} \frac{dx}{dt}$$

We need the value of y when $x = 5$.

$$x^2 + y^2 = 169$$
$$25 + y^2 = 169$$
$$y^2 = 144$$
$$y = 12$$

Therefore,

$$\frac{dy}{dt} = -\frac{x}{y}\frac{dx}{dt}$$
$$= -\frac{5}{12}(3)$$
$$= -\frac{5}{4} = -1.25 \text{ feet per second}$$

The top of the ladder is moving down the wall at -1.25 feet per second. The negative sign in -1.25 indicates that the y value is decreasing.

In Example 2, it was necessary to find dy/dt when $x = 5$. Many students make the error of substituting 5 for x before taking the implicit derivative of the function. Since 5 is a constant and the derivative of constants is zero, this causes the derivative of the x terms to be zero which, in turn, leads to some nonsense conclusion, such as $dy/dt = 0$. Therefore, *substitute values after implicit differentiation.*

3.7 EXERCISES

1. The cost in dollars of selling x items is given by $C = 0.75x + 1000$. What is the rate of change in the cost with respect to time at the instant that the 100th item is sold if the items are selling at the rate of 4 per hour?

2. Find the rate of change in cost, when the cost in dollars is given by $C = 700 + 8\sqrt{x}$ and the 36th item is sold, if the items are selling at the rate of 50 per hour.

3. Find the rate of change per week in cost, when the cost in dollars is given by $C = 100 + 2x + (1/x^2)$ when the first item is sold, if the items are selling at 1 a week.

[$] **4.** Find the rate of change of cost, if the cost in dollars is given by $C = 800 + 9x^{2/3}$, when the 8th item is sold, and the items are selling at the rate of 100 per day.

[$] **5.** Find the rate of change in cost, where cost is given by $C = 0.0002x^3 - 0.05x^2 + 20x + 20,000$, when the 1000th item is sold, if the rate of sales is 50 per day.

[$] **6.** What is the rate of change in profit, given by $P = 4x^2 + 5x - 1000$, at the time the 100th item is sold, given that the items are selling at the rate of 70 a day?

[$] *In Exercises 7–9, assume that the profit, P, resulting from the sales of x data-wacks is given by $P = x^2 + 500x - 100$, and exactly two data-wacks are sold every day.*

7. How is the profit changing when the 100th data-wack is sold?

8. How is the profit changing when the 125th data-wack is sold?

9. How is the profit changing when the 150th data-wack is sold?

10. A 26-foot ladder is leaning against a vertical wall with its foot on a horizontal floor. The foot of the ladder is being moved away from the wall at the rate of 5 feet per second. Find the rate at which the top of the ladder is moving down the wall when the foot of the ladder is 24 feet from the wall.

11. If the radius of a circle is decreasing at the rate of $\frac{1}{2}$ inch per minute, at what rate is its area changing when the radius is 12 inches? When the radius is 2 inches?

12. The volume of a sphere is $V = \frac{4}{3}\pi r^3$. Suppose that a balloon retains a spherical shape as it is being blown up. How fast is the radius of the balloon increasing at the instant it is 3 inches, if air is being pumped into the balloon at 5 cubic inches per second?

13. Rice is being poured onto the ground. The pile formed is always conical in shape with its radius twice its height. Find the rate that the height of the pile is increasing when it is 10 feet if rice is being poured at the rate of 20 cubic feet per second.

14. The volume of a sphere is increasing at the rate of 25 cubic inches per second. How fast is the surface increasing when the radius is 10 inches?

15. The base of a triangle is increasing at the rate of 2 feet per second while the height is decreasing at 4 feet per second. How is the area of the triangle changing when its base is 10 feet and its height is 6 feet?

16. The radius of a cone is increasing at the rate of 3 inches per minute and the altitude is increasing at the rate of 4 inches per minute. Find the rate at which the volume is changing when the radius is 12 inches and the altitude is 20 inches.

17. A boat is being pulled to a dock by a rope that is tied to the deck of the boat. The point from which the boat is being pulled is 5 feet above its deck. How fast is the boat approaching the dock when it is 12 feet from the dock if the rope is being pulled in at the rate of 25 feet per minute?

18. Two ships leave the same port at the same time. One ship steams directly north at 15 knots. The other steams due west at 20 knots. How fast is the distance between them increasing 3 hours after they leave port?

19. One ship leaves port and steams due north at 10 knots. Three hours later another ship leaves the same port and steams due west at 30 knots. How fast is the distance between them increasing when the first ship has been out of port for 5 hours?

20. An airplane is flying horizontally at the rate of 1000 feet per second at an elevation of 20,000 feet. At what rate is the distance to an observer on the ground decreasing when the plane is 40,000 feet from the observer and flying toward this person?

$ 21. The demand equation for a certain item is $p = 3x^2 + 2x + 5$. Find the rate of change of p when $x = 4$ and x is increasing at the rate of 0.5 item per week.

22. A point is moving on the graph of $x^2 - 4y^2 - 36 = 0$. How fast is y changing at the point $(10, -4)$ if x is increasing at a rate of 3 units per second?

3.8   THE DERIVATIVES OF INVERSE FUNCTIONS

In Chapter 1 we examined how a function may be related to a second function called its *inverse*. In this section we will look at the manner in which the derivatives of a function and its inverse are related.

Consider a function f with inverse f^{-1}. We know that $y = f^{-1}(x)$ if and only if $f(y) = x$. If we differentiate $f(y) = x$ implicitly,

$$D_x f(y) = D_x x \qquad (1)$$

However, using the chain rule,

$$D_x f(y) = D_y f(y) \cdot D_x y$$

Therefore, taking the derivative of both sides of (1),

$$D_y f(y) D_x y = 1$$

or, provided $D_y f(y) \neq 0$,

$$D_x y = \frac{1}{D_y f(y)}$$

or, since $y = f^{-1}(x)$,

$$D_x f^{-1}(x) = \frac{1}{D_y f(y)}$$

Here we have to be very careful; $D_x f^{-1}(x)$ means the derivative of $f^{-1}(x)$, with respect to x. However, $D_y f(y)$ means considering the function f, with y as its argument and then differentiating with respect to y.

Example 1   Given a function f, such that $f(x) = (x + 3)/(x - 2)$, find the derivative of f^{-1}.

Solution $D_x f(x) = \dfrac{(x - 2)(1) - (x + 3)(1)}{(x - 2)^2} = \dfrac{-5}{(x - 2)^2}$

Then

$$D_y f(y) = \frac{-5}{(y - 2)^2}$$

Therefore, if $y = f^{-1}(x)$,

$$D_x y = D_x f^{-1}(x)$$

$$= \frac{1}{D_y f(y)}$$

$$= \frac{1}{\dfrac{-5}{(y - 2)^2}}$$

$$= \frac{-(y - 2)^2}{5}$$

There is a second way to approach this problem.

Example 2   Given a function f such that $f(x) = (x + 3)/(x - 2)$, find $f^{-1}(x)$ and $D_x f^{-1}(x)$.

Solution   The inverse $f^{-1}(x)$ can be found by interchanging x and y in the equation $y = f(x)$ and solving for y in terms of x.

$$y = \frac{x + 3}{x - 2}$$

becomes

$$x = \frac{y + 3}{y - 2}$$
$$x(y - 2) = y + 3$$
$$xy - 2x = y + 3$$
$$xy - y = 2x + 3$$
$$y(x - 1) = 2x + 3$$
$$y = \frac{2x + 3}{x - 1}$$

Therefore,

$$f^{-1}(x) = \frac{2x + 3}{x - 1}$$

and

$$D_x f^{-1}(x) = \frac{(x - 1)(2) - (2x + 3)(1)}{(x - 1)^2}$$
$$= \frac{2x - 2 - 2x - 3}{(x - 1)^2} = \frac{-5}{(x - 1)^2}$$

Compare Examples 1 and 2. Which is correct? Does $D_x f^{-1}(x) = -5/(x - 1)^2$ or does $D_x f^{-1}(x) = -(y - 2)^2/5$, where $y = f^{-1}(x)$?

The answer is that both are correct. If $y = f^{-1}(x)$, then in Example 2 we found that $y = (2x + 3)/(x - 1)$. If we substitute this expression for y in the result of Example 1, we get

$$D_x y = D_x f^{-1}(x) = \frac{-(y - 2)^2}{5} = \frac{-\left[\left(\dfrac{2x + 3}{x - 1}\right) - 2\right]^2}{5}$$

$$= \frac{-\left[\dfrac{2x + 3 - 2(x - 1)}{(x - 1)}\right]^2}{5}$$

$$= \frac{- \left[\dfrac{2x + 3 - 2x + 2}{(x - 1)} \right]^2}{5}$$

$$= \frac{- \left[\dfrac{5}{(x - 1)} \right]^2}{5}$$

$$= \frac{-5^2}{(x - 1)^2 \cdot 5} = \frac{-5}{(x - 1)^2}$$

This is the same expression we found for $D_x f^{-1}(x)$ in the second example. Thus there are two ways to find the derivative of a function defined as the inverse of a given function. We can try to find an explicit expression for the inverse and differentiate it in the usual manner, or we can use the formula

$$D_x f^{-1}(x) = \frac{1}{D_y f(y)}$$

and then make some kind of replacement for y in the right-hand expression. This latter method will be the key to developing the calculus of the log/exponential functional pairs, and the calculus of the inverse trigonometric functions in chapters to follow.

Example 3   Given a function f defined by $f(x) = x^3 + x + 1$, find $D_x f^{-1}(x)$.

Solution   If $y = f^{-1}(x)$, then

$$D_x f^{-1}(x) = \frac{1}{D_y f(y)} = \frac{1}{D_y(y^3 + y + 1)}$$

$$= \frac{1}{3y^2 + 1}$$

There is no easy way to find y as a function of x explicitly in this problem.

Example 4   Given a function g defined by $g(x) = 2/x^3$, find $D_x g^{-1}(x)$.

Solution   If $y = g^{-1}(x)$, then

$$D_x y = D_x g^{-1}(x) = \frac{1}{D_y g(y)}$$

$$= \frac{1}{D_y(2/y^3)}$$

$$= \frac{1}{-6/y^4} = \frac{-y^4}{6}$$

Note that if we find g^{-1} by switching x and y and solving for y, $x = 2/y^3$, then $y^3 = 2/x$.

$$g^{-1}(x) = y = \frac{\sqrt[3]{2}}{x^{1/3}} = \sqrt[3]{2} \cdot x^{-1/3}$$

Thus,

$$y^4 = \frac{2^{4/3}}{x^{4/3}} = \frac{2\sqrt[3]{2}}{x^{4/3}}$$

By substitution,

$$D_x y = \frac{-y^4}{6} = -\frac{2\sqrt[3]{2}}{x^{4/3}} \cdot \frac{1}{6} = \frac{-\sqrt[3]{2}}{3x^{4/3}} = \frac{-\sqrt[3]{2}}{3} x^{-4/3}$$

By direct calculation, with $y = \sqrt[3]{2}\, x^{-1/3}$,

$$D_x y = D_x(\sqrt[3]{2}\, x^{-1/3}) = -\tfrac{1}{3} \sqrt[3]{2} \cdot x^{-4/3}$$

Note that the procedure considered above is valid only if $D_y f(y) \neq 0$ for the set of values of y we are interested in.

3.8   EXERCISES

In Exercises 1–10, find explicit forms for f^{-1}, if it is a function, for each f given and use direct differentiation to find $D_x f^{-1}(x)$.

1. $f(x) = 3x + 2$ **2.** $f(x) = 4 - 7x$

3. $f(x) = x^5 + 2$ **4.** $f(x) = 3 - x^3$

5. $f(x) = \dfrac{x + 4}{x - 2}$ **6.** $f(x) = \dfrac{3 - x}{x + 3}$

7. $f(x) = \dfrac{2x + 5}{x - 1}$ **8.** $f(x) = \dfrac{3x + 4}{2 - x}$

9. $f(x) = \dfrac{2x + 3}{5 - 4x}$ **10.** $f(x) = \dfrac{(\frac{1}{2})x + \frac{1}{3}}{(\frac{2}{3})x + \frac{1}{4}}$

As Exercises 11–20, use the $D_x f^{-1}(x) = \dfrac{1}{D_y f(y)}$ formula to find $D_x f^{-1}(x)$ for Exercises 1–10.

In Exercises 21–30, for each function f find $D_x f^{-1}(x)$ in terms of y, where $y = f^{-1}(x)$.

21. $f(x) = (x - 3)^3$ **22.** $f(x) = (2 - x)^3$

23. $f(x) = \dfrac{1}{x^3}$ **24.** $f(x) = \dfrac{1}{x^5}$

25. $f(x) = x^3 - 2x + 3$ **26.** $f(x) = x - x^3$

27. $f(x) = \dfrac{1}{x^2}, \quad x > 0$ **28.** $f(x) = \dfrac{1}{x^2 + 1}, \quad x > 0$

29. $f(x) = \dfrac{x}{x^2 + 1}, \quad x > 0$ **30.** $f(x) = \dfrac{x}{1 - x^2}, \quad x > 0$

3.9 ROLLE'S THEOREM AND THE MEAN VALUE THEOREM

Most of the topics we have considered have had some direct application. The subjects of this section do not. They are necessary to prove certain calculus theorems that will be useful later.

Consider the behavior of the function $f(x) = x^2 - 5x + 4$ between $x = 1$ and $x = 4$. The graph of this function is shown in Figure 3.22.

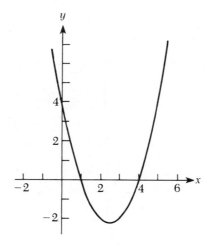

Figure 3.22 Graph of $y = f(x) = x^2 - 5x + 4$

The interval of special interest lies between the points where the functional value is zero. That is, $f(1) = f(4) = 0$. Notice from the graph that the function appears to have a minimum value at

$x = \frac{5}{2}$. This can be verified by using calculus. Since

$$f(x) = x^2 - 5x + 4$$

then

$$f'(x) = 2x - 5$$

Setting $f'(x)$ equal to zero,

$$2x - 5 = 0$$
$$x = \frac{5}{2}$$

Checking $f''(x)$, $f''(\frac{5}{2}) = 2$. The conclusion is that $f(\frac{5}{2}) = (\frac{5}{2})^2 - 5(\frac{5}{2}) + 4 = -\frac{9}{4}$ is a minimum value of the function. Rolle's theorem is concerned with a similar type of situation, namely a function defined and continuous over an interval with a zero functional value at each endpoint. Rolle's theorem draws an important conclusion about the derivative of such a function.

Theorem 3.3   (Rolle's theorem) Let f be a function such that:

1. f is continuous for all x such that $a \leqslant x \leqslant b$
2. f' exists for all x such that $a < x < b$
3. $f(a) = f(b) = 0$

Then there exists a number c such that $a < c < b$ and $f'(c) = 0$.

In words, Rolle's theorem states that if a function is continuous and differentiable over an interval and is zero at both endpoints of the interval, then at least once within the interval the derivative of the function must be zero.

Consider an unknown function which meets the conditions of Rolle's theorem. What can we deduce about the function? Suppose $f(x)$ is continuous on the interval $a \leqslant x \leqslant b$, $f'(x)$ exists on this interval, and $f(a) = f(b) = 0$. This case is illustrated in Figure 3.23. All the curves in Figure 3.23 meet the conditions of the theorem. It seems reasonable to conclude that each function illustrated must have a maximum or minimum. Since the derivative of the function must exist at maximum or minimum values—in fact, by the assumption it exists everywhere in the interval—we then know that its value must be zero at that point. Hence, Rolle's theorem seems a very reasonable one. A formal proof can be found in any standard Engineering/Mathematics-oriented calculus text.

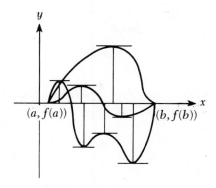

Figure 3.23 An illustration of Rolle's theorem

Example 1

Does $f(x) = x^{4/3} - 3x^{1/3}$ for $0 \leqslant x \leqslant 3$ meet the requirements of Rolle's theorem?

Solution

$f(0) = 0^{4/3} - 3(0)^{1/3} = 0 - 0 = 0$

$f(3) = 3^{4/3} - 3(3)^{1/3} = 3^{4/3} - 3^{4/3} = 0$

$f'(x) = \dfrac{4}{3} x^{1/3} - x^{-2/3}$

The derivative $f'(x)$ exists for $0 < x < 3$, and $f(x)$ exists and is continuous on $0 \leqslant x \leqslant 3$. Thus, the requirements of Rolle's theorem have been met. The conclusion is there must exist a number c such that $0 < c < 3$ and $f'(c) = 0$. Let's attempt to find c. We know that

$$f'(c) = \frac{4}{3} c^{1/3} - c^{-2/3} = 0$$

$$\frac{4}{3} c^{1/3} - \frac{1}{c^{2/3}} = 0$$

$$\frac{4c^{1/3}c^{2/3}}{3c^{2/3}} - \frac{3}{3c^{2/3}} = 0$$

$$\frac{4c - 3}{3c^{2/3}} = 0$$

This is only true if

$$4c - 3 = 0$$

$$c = \frac{3}{4}$$

Notice this value of c does meet the required condition, because

$$0 < \frac{3}{4} < 3$$

This curve is shown in Figure 3.24, which demonstrates that $x = \frac{3}{4}$ corresponds to a minimum value of the function.

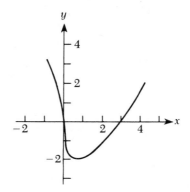

Figure 3.24   Graph of $y = f(x) = x^{4/3} - 3x^{1/3}$

Example 2   Consider $f(x) = x^3 - 16x$ for $-4 \leqslant x \leqslant 4$. Show that the conditions of Rolle's theorem are met, and find the value c guaranteed by the theorem.

Solution   The function $f(x) = x^3 - 16x$ is continuous everywhere and thus on the interval $-4 \leqslant x \leqslant 4$; $f'(x) = 3x^2 - 16$ also exists for $-4 < x < 4$.

$$f(4) = 4^3 - 16(4) = 64 - 64 = 0$$
$$f(-4) = (-4)^3 - 16(-4) = -64 + 64 = 0$$

Therefore the conditions of Rolle's theorem are met. The curve of this function is shown in Figure 3.25. Now we must find the c value that Rolle's theorem tells us must exist. Setting $f'(c)$ equal to zero,

$$f'(c) = 3c^2 - 16$$
$$3c^2 - 16 = 0$$
$$c = \pm \frac{4}{\sqrt{3}} \approx \pm 2.31$$

There appear to be two values, but which one is correct? Both lie in

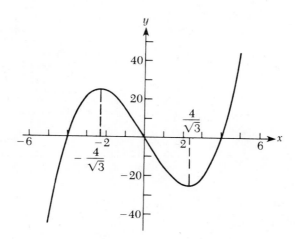

Figure 3.25 Graph of $y = f(x) = x^3 - 16x$

the interval: that is, $-4 < -4/\sqrt{3} < 4$ and $-4 < 4/\sqrt{3} < 4$. Thus either value will do. This is fine, since Rolle's theorem assures at least one value, but does not rule out more than one value.

Rolle's theorem can be applied to prove one of the fundamental theorems of elementary calculus. This is the Mean Value Theorem.

Theorem 3.4 (Mean Value Theorem) Let f be a function such that f is continuous for $a \leq x \leq b$, and $f'(x)$ exists for $a < x < b$. Then there exists a number d such that $a < d < b$, and

$$f'(d) = \frac{f(b) - f(a)}{b - a}$$

The theorem is illustrated in Figure 3.26. In effect, the Mean Value Theorem states, among other things, that the slope of the tangent to the curve (as measured by $f'(d)$) will, somewhere in the interval, equal the slope of the line connecting the endpoints of the interval.

Rolle's theorem provides an easy proof of the theorem.

Proof Consider a function f meeting the conditions of the Mean Value Theorem: that is, $f(x)$ is continuous for $a \leq x \leq b$, and $f'(x)$ exists

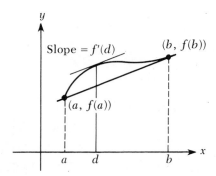

Figure 3.26 The Mean Value Theorem

for $a < x < b$. Consider a second function F, such that

$$F(x) = f(x) - \frac{f(b) - f(a)}{b - a}(x - a) - f(a)$$

where $F(x)$ measures the vertical distance from $f(x)$ to its secant line. The function F is also continuous for $a \leqslant x \leqslant b$.

$$
\begin{aligned}
F(a) &= f(a) - \frac{f(b) - f(a)}{b - a}(a - a) - f(a) \\
&= f(a) - f(a) = 0 \\
F(b) &= f(b) - \frac{f(b) - f(a)}{b - a}(b - a) - f(a) \\
&= f(b) - [f(b) - f(a)] - f(a) \\
&= f(b) - f(b) + f(a) - f(a) = 0 \\
F'(x) &= f'(x) - \frac{f(b) - f(a)}{b - a}
\end{aligned}
$$

and $F'(x)$ exists for $a < x < b$, as $f'(x)$ exists. The function F meets the conditions of Rolle's theorem. Therefore, there exists a number d such that $a < d < b$ and $F'(d) = 0$. However,

$$F'(d) = f'(d) - \frac{f(b) - f(a)}{b - a}$$

Therefore,

$$f'(d) - \frac{f(b) - f(a)}{b - a} = 0$$

and

$$f'(d) = \frac{f(b) - f(a)}{b - a}$$

Example 3   Find the value of d for $f(x) = x^2 - 5x + 6$ in the interval $1 \le x \le 5$, which is predicted by the Mean Value Theorem.

Solution   The graph of this function is shown in Figure 3.27. The function f meets the requirements of the Mean Value Theorem. The problem is to find the d value whose existence is assured by the theorem.

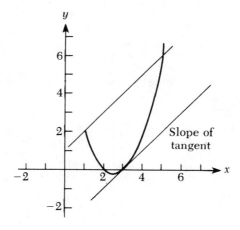

Figure 3.27   Graph of $y = f(x) = x^2 - 5x + 6$, $1 \le x \le 5$

$$f(1) = 1 - 5 + 6 = 2$$
$$f(5) = 25 - 25 + 6 = 6$$

Therefore,

$$\frac{f(b) - f(a)}{b - a} = \frac{f(5) - f(1)}{5 - 1} = \frac{6 - 2}{5 - 1} = \frac{4}{4} = 1$$

and

$$f'(d) = 2d - 5$$

If we set $f'(d) = 1$,

$$2d - 5 = 1$$
$$2d = 6$$
$$d = 3$$

The required d is 3:

$$f'(3) = \frac{f(5) - f(1)}{5 - 1}$$

and

$$1 < 3 < 5$$

Example 4   Consider $f(x) = x^3 - 2x^2$, $1 \leqslant x \leqslant 3$. Show that this function satisfies the conditions of the Mean Value Theorem and find the value of d.

Solution   The function is continuous on $1 \leqslant x \leqslant 3$ and $f'(x) = 3x^2 - 4x$ exists for $1 < x < 3$.

$$f(3) = 27 - 18 = 9$$
$$f(1) = 1 - 2 = -1$$

Therefore,

$$\frac{f(b) - f(a)}{b - a} = \frac{f(3) - f(1)}{3 - 1} = \frac{9 - (-1)}{3 - 1} = \frac{10}{2} = 5$$

Setting

$$f'(x) = \frac{f(b) - f(a)}{b - a}$$

we get

$$3x^2 - 4x = 5$$
$$3x^2 - 4x - 5 = 0$$
$$x = \frac{4 \pm \sqrt{16 - 4(-5)(3)}}{6} = \frac{4 \pm \sqrt{16 + 60}}{6}$$
$$= \frac{4 \pm \sqrt{76}}{6} \approx \frac{4 \pm 8.7}{6} \approx \frac{12.7}{6} \quad \text{or} \quad \frac{-4.7}{6}$$
$$\approx 2.12 \quad \text{or} \quad -0.78$$

Since -0.78 is not in the interval, and 2.12 is, $(4 + \sqrt{76})/6$ is the desired value, and $d = (4 + \sqrt{76})/6 \approx 2.12$.

Suppose that $D_x f(x) = 2x$. What is $f(x)$? A little trial and error would lead to $f(x) = x^2$. Now suppose that $D_x g(x) = 2x$. What is $g(x)$? Must $f(x) = g(x) = x^2$? The answer is that $f(x)$ and $g(x)$ must differ, at most, by a constant. That is, $f(x) = g(x) + C$, for some constant C. For instance, if $g(x) = x^2 + 91$, then $D_x g(x) = 2x$. An important idea that will be used often in later chapters is that if the derivatives of two functions are equal, the functions can differ only by a constant amount. The Mean Value Theorem is used to prove this.

Theorem 3.5 If $f(x)$ and $g(x)$ are two functions defined for all x, such that $f'(x) = g'(x)$, then there is a constant C such that $f(x) = g(x) + C$.

Proof Let $h(x)$ be a function such that

$$h(x) = f(x) - g(x) \qquad (1)$$

Then

$$h'(x) = f'(x) - g'(x)$$

By hypothesis, $f'(x) = g'(x)$. Therefore, $f'(x) - g'(x) = 0$ and

$$h'(x) = 0$$

Now the derivative of any constant function is zero, but if a derivative of a function such as $h'(x)$ is zero, can we conclude that the function is a constant function? Assume that $h(x)$ is *not* a constant function. Then there would exist two distinct numbers a and b such that $a < b$ and $h(a) \neq h(b)$. Also, by the Mean Value Theorem, we know that there is a number d such that $a < d < b$ and

$$\frac{h(b) - h(a)}{b - a} = h'(d)$$

Since, by hypothesis, $h'(x) = 0$ for all x, it must be that $h'(d) = 0$. Therefore,

$$\frac{h(b) - h(a)}{b - a} = 0$$

which implies that $h(a) = h(b)$. This is contrary to our assumption that $h(x)$ was not a constant function.

Therefore $h(x)$ is constant. Let $h(x) = C$. Then from (1) we have that

$$h(x) = f(x) - g(x) = C$$

or $f(x) = g(x) + C$, and the proof is complete.

The Mean Value Theorem is used in the proof of a number of other important theorems, notably in a rigorous proof of the Fundamental Theorem of Calculus, a topic which appears in a later chapter.

3.9 EXERCISES

In Exercises 1–5, verify that the premises of Rolle's theorem have been met and find an appropriate value of c.

1. $f(x) = x^2 - 7x + 10$,   for $2 \leqslant x \leqslant 5$
2. $f(x) = x^3 - 5x^2 - 17x + 21$,   for $-3 \leqslant x \leqslant 7$
3. $f(x) = x^3 - 4x^2 + x + 6$,   for $x \in [-1, 2]$
4. $f(x) = x^3 - 4x^2 + x + 6$,   for $x \in [2, 3]$
5. $f(x) = x^4 - 28x^2 + 75$,   for $x \in [\sqrt{3}, 5]$

In Exercises 6–10, verify that the conditions of the Mean Value Theorem have been met and find an appropriate value of d.

6. $f(x) = x^2 - 4x + 10$,   for $0 \leqslant x \leqslant 3$
7. $f(x) = x^3 - 3x^2 + 3$,   for $0 \leqslant x \leqslant 3$
8. $f(x) = \dfrac{x}{x + 1}$,   for $x \in [1, 2]$
9. $f(x) = \dfrac{1}{x - 3}$,   for $3.1 \leqslant x \leqslant 3.5$
10. $f(x) = (x + 2)(x + 3)$,   for $x \in [2, 3]$
11. The function $f(x) = x^3 - 2x^2$, $1 \leqslant x \leqslant 3$, was considered as an example related to the Mean Value Theorem in this section. Sketch the graph of the function and plot the tangent at d.

In Exercises 12–15, determine in what way the functions fail to meet the conditions of Rolle's theorem on the indicated interval.

12. $f(x) = \dfrac{x-1}{x-3}$, for $x \in [1, 3]$

13. $f(x) = 4x^2 - 8x$, for $x \in [0, 3]$

14. $f(x) = \dfrac{3x^2 - 2x + 4}{x - 2}$, for $1 \leqslant x \leqslant 5$

15. $f(x) = \dfrac{3x + 2}{(x - 1)^2}$, for $\dfrac{1}{2} \leqslant x \leqslant 2$

16. Show by supplying an example that the converse of Rolle's theorem does not hold; that is, if $f'(c) = 0$ for some c such that $a < c < b$ and $f(x)$ is continuous for $a \leqslant x \leqslant b$ and $f'(x)$ exists for $a < x < b$, then $f(b)$ or $f(a)$ or both are not necessarily equal to zero.

In Exercises 17–20, determine in what way each function fails to meet the conditions of the Mean Value Theorem on the indicated interval, if it fails.

17. $f(x) = \dfrac{x^2 - 2x}{x}$, for $x \in [-1, 2]$

18. $f(x) = \sqrt{4 - x^2}$, for $x \in [-2, 5]$

19. $f(x) = 3x + 2$, for $x \in [-1, 2]$

20. $f(x) = \dfrac{3}{x^2 - 4}$, for $x \in [0, 5]$

NEW TERMS

Absolute maximum

Absolute minimum

Concave downward

Concave upward

Critical value

Decreasing function

Differential

Explicit function

Extremum, extrema

First-derivative test

Implicit differentiation

Implicit function

Increasing function

Maximum, maxima

Mean Value Theorem

Minimum, minima

Point of inflection

Relative maximum

Relative minimum

Rolle's theorem

Second-derivative test

REVIEW

1. Use Newton's method to find the real roots of $x^2 + 4x - 10 = 0$, accurate to two decimal places.

2. Find the real root of $x^3 - 6x - 2 = 0$ that is between -1 and 0, accurate to two decimal places.

3. Find the values of x for which $y = x^2 + 4x + 1$ is increasing, decreasing, concave upward, and concave downward. Also give values of x for which the function is maximum or minimum, and give points of inflection.

4. Repeat Exercise 3 for $y = x^3 + 4x$.

In Exercises 5–10, find dy/dx. Simplify your answer.

5. $3x^2 - 4y^2 = 10$ 6. $x^{1/2} + y^{1/2} = 4$

7. $x^3 - 4xy + x^2 = 0$ 8. $y = x^{2/3} + y^{2/3} - 8^{2/3}$

9. $y = xy + y^2 + x^2$ 10. $y = 2x + 2y + 2xy$

In Exercises 11–13, find y″. Simplify your answer.

11. $x^2 - y^2 = 4$ 12. $y^2 = x^2$

13. $y = x^2 + y^2$

14. If $y = f(x) = 2x + 5$, find $f[f^{-1}(x)]$.

15. The cost of selling x items is given by $C = 500 + 10\sqrt{x}$. What is the rate of change in the cost with respect to time at the instant that the 100th item is sold if the items are selling at the rate of 40 per hour?

16. How much metal, approximately, is used in the construction of a cubic container if the outer edges of the container are 6 inches and the sides are 0.1 inch thick?

17. Find dy and Δy for $y = 3x^2 + 8x - 4$.

18. The base of a triangle is increasing at the rate of 4 feet per minute, while the height is increasing at 2 feet per minute. How fast is the area increasing when the base and height are both 6 feet?

19. A rectangular corral is to be made inside a large barn. The corral is to be constructed in the corner of the barn, so the walls of the barn will form two of its sides. What should be the dimensions of the corral if it is to have maximum area when 250 feet of fencing is available?

$ **20.** An excursion-boat company will make its services available to any organization that will guarantee 50 passengers at $12 per passenger. For each passenger over 50 the cost per passenger is reduced $0.25. How many passengers would give the company the maximum gross income?

21. Town A is located on one side of a 100-foot wide canal and town B on the other. Town B is 1000 feet downstream from town A. A pipeline is to be laid from town A to town B. The pipeline costs $20 per foot when laid in the soil along the canal bank and $30 per foot when laid in the canal itself. How far from town A should the pipeline enter the canal so that its cost will be minimum?

In Exercises 22 and 23, verify that the conditions of Rolle's theorem apply to the indicated functions and find an appropriate value of c.

22. $f(x) = x^2 - 6x - 7$ on the interval $[-1, 7]$

23. $f(x) = x^3 + x^2 - 6x$ on the interval $[0, 2]$

4

Introduction to the Integral

In the preceding chapters we have been concerned with functions and their derivatives, and we have seen how the derivatives of a given function can aid in the analysis of the function. We have also been introduced to various applications of derivatives in the world of business, economics, and other areas where maximizing or minimizing of quantities is applicable. In this chapter we are going to examine functions that are possible derivatives of other functions, and for the given derivative, find the function from which it was derived. This process is called *finding an antiderivative,* or *integration*.

4.1 ANTIDERIVATIVES

> **Definition**
>
> Let f be a function. Then any function F for which $F'(x) = f(x)$ is an *antiderivative* of f.

Notation   For a given function f,

$$\int f(x)\, dx$$

will denote an antiderivative of f with respect to x.

Example 1   Find $\int 3x^2\,dx$.

Solution $\int 3x^2\,dx = x^3$

because if $F(x) = x^3$, then

$$F'(x) = 3x^2 = f(x)$$

Is the solution for the above example unique? The answer must be no, because the function F, where $F(x) = x^3 + 7$, yields the same derivative, $F'(x) = 3x^2 = f(x)$. For that matter, $F(x) = x^3 + C$, where C is any constant, satisfies the condition, since $D_x(C) = 0$.

In general, any function which has an antiderivative has *infinitely many* antiderivatives. But, as suggested in the above discussion, all of the derivatives of a function differ from each other only by a constant amount. For this reason we will refer to the *general antiderivative* of a function.

Definition

For any function f, if one can find a second function F such that $F'(x) = f(x)$, then F is an antiderivative of f and the expression $F(x) + C$, where C is an arbitrary constant, is called the *general antiderivative* of $f(x)$:

$$\int f(x)\,dx = F(x) + C$$

The symbol $\int f(x)\,dx$ will denote either the general antiderivative or a specific antiderivative as required by the context of the discussion. Recall that the process of finding an antiderivative is also called integration; another name for the antiderivative is *indefinite integral*.

Example 2   Find $\int 3x^2\,dx$.

Solution $\int 3x^2\,dx = x^3 + C$

It should be evident from the definition and the examples that finding antiderivatives is the inverse process of finding derivatives;

or alternately, that integration is the inverse process of differentiation. That is to say, it poses the problem of finding a function whose derivative (or differential) is given.

The following integration formulas are given for the solution of simple problems. Their proofs are based on the theorems on the derivative and the differential. Further techniques of integration are given in subsequent chapters, and a Table of Integrals has been compiled and appears inside the cover to aid in the integration of many problems.

$$\int dx = x + C$$

If u is a function of some other variable, such as x, and du stands for $u'(x)\,dx$, then

$$\int a\,du = a\int du \qquad (a \text{ is a constant})$$
$$\int [f(x) \pm g(x)]\,dx = \int f(x)\,dx \pm \int g(x)\,dx$$
$$\int u^n\,du = \frac{u^{n+1}}{n+1} + C \qquad (\text{if } n \text{ is rational and } n \neq -1)$$

Example 3   Find $\int x^3\,dx$; $\int \sqrt{x}\,dx$; and $\int \frac{1}{x^3}\,dx$.

Solution
$$\int x^3\,dx = \frac{x^{3+1}}{3+1} + C = \frac{x^4}{4} + C$$
$$\int \sqrt{x}\,dx = \int x^{1/2}\,dx = \frac{x^{(\frac{1}{2})+1}}{(\frac{1}{2})+1} + C = \frac{2x^{3/2}}{3} + C$$
$$\int \frac{1}{x^3}\,dx = \int x^{-3}\,dx = \frac{x^{-3+1}}{-3+1} + C = \frac{-1}{2x^2} + C$$

The last formula can easily be verified. Let u be a function of x such that

$$F(x) = \frac{u^{n+1}}{n+1} + C$$

Then

$$F'(x)\,dx = (n+1)\frac{u^{n+1-1}}{n+1}\,du = u^n\,du$$

This formula is called the *power formula*, since it involves a function raised to a power. Notice the restriction, $n \neq -1$. If $n = -1$ the denominator would be zero and division by zero has no meaning. The integral $\int u^{-1}\,du$ involves a logarithmic function and will be discussed later.

Example 4   Find $\int 3\,dx$.

Solution $\int 3\,dx = 3\int dx = 3x + C$

Example 5   Find $\int (x^2 + 5x + 1)\,dx$.

Solution $\displaystyle \int (x^2 + 5x + 1)\,dx = \int x^2\,dx + 5\int x\,dx + \int dx = \frac{x^3}{3} + \frac{5x^2}{2} + x + C$

Example 6   Find $\int (x^2 + 2)^3(2x)\,dx$.

Solution   A substitution, $u = x^2 + 2$, makes this seemingly complicated problem quite simple. If $u = x^2 + 2$, then $du = 2x\,dx$, and the expression $u^3\,du$ has an antiderivative, $u^{3+1}/(3 + 1) = u^4/4$.

$$\int u^3\,du = \frac{1}{4}u^4 + C$$

Since $u = x^2 + 2$,

$$\int \underbrace{(x^2 + 2)^3}_{u^3}\underbrace{(2x)\,dx}_{du} = \frac{1}{4}(x^2 + 2)^4 + C$$

To check the solution, recall that the integral is an antiderivative, and if $F = \int f\,dx$, then $F' = f$.

$$F(x) = \frac{1}{4}(x^2 + 2)^4 + C \qquad \text{implies} \qquad F'(x) = (x^2 + 2)^3(2x) = f(x)$$

Example 7   Find $\int x^3\sqrt{x^4 + 2}\,dx$.

Solution   Again a substitution simplifies the problem. Let

$$u = x^4 + 2; \quad \text{then} \quad du = 4x^3\,dx$$

We can rewrite $x^3\sqrt{x^4 + 2}\,dx$ as $\frac{1}{4}\sqrt{x^4 + 2}\,(4x^3)\,dx$ (multiply and divide by 4). Thus,

$$\int x^3 \sqrt{x^4 + 2}\, dx = \int \frac{1}{4} u^{1/2}\, du = \frac{1}{4} \int u^{1/2}\, du$$

$$= \frac{1}{4} \frac{u^{3/2}}{3/2} + C = \frac{1}{6} u^{3/2} + C$$

$$= \frac{1}{6} (x^4 + 2)^{3/2} + C$$

To check this result, take the derivative of $\frac{1}{6}(x^4 + 2)^{3/2} + C$ and obtain

$$\frac{1}{4} (x^4 + 2)^{1/2}(4x^3)$$

which simplifies to $x^3 \sqrt{x^4 + 2}$.

Example 8   Find $\int \dfrac{3}{(2x + 5)^3}\, dx$.

Solution $\int \dfrac{3}{(2x + 5)^3}\, dx = 3 \int (2x + 5)^{-3}\, dx$

Let $u = 2x + 5$;   then

$$du = 2\, dx \qquad \text{and} \qquad dx = \frac{1}{2}\, du$$

By making the appropriate substitutions,

$$3 \int (2x + 5)^{-3}\, dx = 3 \int u^{-3} \left(\frac{1}{2}\, du \right) = \frac{3}{2} \int u^{-3}\, du$$

$$= \frac{3}{2} \left(\frac{u^{-3+1}}{-3 + 1} \right) + C = \frac{3}{2} \left(\frac{u^{-2}}{-2} \right) + C$$

$$= - \frac{3}{4(2x + 5)^2} + C$$

Example 9   Find $\int \dfrac{x\, dx}{\sqrt{a^2 + x^2}}$ (a is a constant).

Solution   The substitution $u = a^2 + x^2$ is advisable; then $du = 2x\, dx$.

$$\int \frac{x\, dx}{\sqrt{a^2 + x^2}} = \frac{1}{2} \int \frac{2x\, dx}{\sqrt{a^2 + x^2}} = \frac{1}{2} \int \frac{du}{\sqrt{u}}$$

$$= \frac{1}{2} \int u^{-1/2}\, du = \frac{1}{2} \frac{u^{1/2}}{1/2} + C = u^{1/2} + C$$

$$= \sqrt{a^2 + x^2} + C$$

Therefore,

$$\int \frac{x\,dx}{\sqrt{a^2 + x^2}} = \sqrt{a^2 + x^2} + C$$

Check: If

$$F(x) = \sqrt{a^2 + x^2} + C$$

then

$$F'(x) = \frac{1}{2}(a^2 + x^2)^{-1/2}(2x) = \frac{x}{\sqrt{a^2 + x^2}} = f(x)$$

In summary,

$$\int dx = x + C$$

$$\int du = u + C \quad \text{(where } u \text{ is a function of some}$$
$$\text{other variable)}$$

$$\int a\,du = a \int du \qquad (a \text{ is a constant})$$

$$\int [f(x) \pm g(x)]\,dx = \int f(x)\,dx \pm \int g(x)\,dx$$

$$\int u^n\,du = \frac{u^{n+1}}{n+1} + C \quad \text{(if } n \text{ is rational, and } n \neq -1)$$

4.1 EXERCISES

In Exercises 1–31, integrate and check your answers by differentiation.

1. $\int 3x\,dx$

2. $\int (5x^2 + 3x + 2)\,dx$

3. $\int (x^3 + x^2 - 1)\,dx$

4. $\int \sqrt{x}\,dx$

5. $\int \sqrt[3]{x}\,dx$

6. $\int (x^{10} - 2x^6 + x^3)\,dx$

7. $\int \frac{dx}{x^2}$

8. $\int \frac{dx}{x^3}$

9. $\int (x + 2)^2\,dx$

10. $\int (2x + 3)^2\,dx$

11. $\int \left(\frac{2}{x^2} + \frac{3}{x^3} \right) dx$

12. $\int x^{2a}\,dx$ (a is a positive constant)

13. $\int x\,\sqrt{x^2 + 5}\,dx$

14. $\int x\,\sqrt[3]{x^2 + 5}\,dx$

15. $\displaystyle\int \frac{x}{\sqrt{1-x^2}}\,dx$

16. $\displaystyle\int \frac{x}{\sqrt[3]{1-x^2}}\,dx$

17. $\displaystyle\int 3(6x + 9)^4\,dx$

18. $\displaystyle\int \frac{2\,dx}{(x+5)^5}$

19. $\displaystyle\int (x^2 - 1)(4 - x^2)\,dx$

20. $\displaystyle\int (2x + 5)(x^2 + 5x + 3)^{10}\,dx$

21. $\displaystyle\int (4x^2 + 2x + 1)^3(8x + 2)\,dx$

22. $\displaystyle\int \sqrt{3-x}\,dx$

23. $\displaystyle\int \frac{4x-3}{(2x^2 - 3x + 1)^3}\,dx$

24. $\displaystyle\int \frac{dx}{\sqrt[3]{(2x+1)^2}}$

25. $\displaystyle\int \frac{1 + \sqrt{x}}{x^2}\,dx$

26. $\displaystyle\int (x^2 - 2x + 1)^{5/3}\,dx$

27. $\displaystyle\int \frac{x^3}{\sqrt{x^4 - 1}}\,dx$

28. $\displaystyle\int (x - 3)^{3/2}\,dx$

29. $\displaystyle\int \frac{3x^4 + x^3 + x - 3}{x^2 + 1}\,dx$
[*Hint:* Use long division to simplify the expression first.]

30. $\displaystyle\int \frac{3x^3 + 6x^2 + 3x + 1}{x^2 + 2x + 1}\,dx$

31. $\displaystyle\int \frac{12 + \sqrt{x}}{3x^2}\,dx$

Often, one can find an appropriate value to give to the arbitrary constant in a general antiderivative by the use of outside or boundary conditions. For example, suppose you are told that

$$F(x) = \int f(x)\,dx = \int x^3\,dx$$

and that

$$F(1) = 1$$

Then since

$$F(x) = \int x^3\,dx = \frac{x^4}{4} + C$$

we have

$$F(1) = \frac{1^4}{4} + C = 1$$

thus

$$C = \frac{3}{4} \qquad and \qquad F(x) = \frac{x^4}{4} + \frac{3}{4}$$

In Exercises 32–36, find specific values for the constant of integration, that is, C.

32. $F(x) = \int 3x^2 \, dx, \quad F(0) = 1$

33. $F(x) = \int (3x^2 - 4) \, dx, \quad F(2) = 0$

34. $F(x) = \int x^{-1/2} \, dx, \quad F(1) = 0$

35. $F(x) = \int x^{1/3} \, dx, \quad F(8) = 10$

36. $F(x) = \int \dfrac{x^{-2/3}}{3} \, dx, \quad F(1) = -6$

37. Find $D_x \left(\int \dfrac{1}{x^2 - 4} \, dx \right)$.

38. Find $D_x(\int x^{5/3} \, dx)$ and $\int (D_x x^{5/3}) \, dx$.

39. Find $D_x[\int (x^2 - x) \, dx]$ and $\int [D_x(x^2 - x)] \, dx$.

40. In general, does $\int [D_x(f(x))] \, dx = D_x[\int f(x) \, dx]$ for any function? Explain the difference.

4.2   AREA BOUNDED BY A CURVE
—APPROXIMATION BY RECTANGLES

The area under a curve is as basic to integral calculus as the slope of the tangent line is to differential calculus. The slope of the tangent line had numerous interpretations, which led to a number of applications for the derivative of a function. In a similar manner, the area under a curve will point to the *integral*, which has many applications in business, economics, social and other behavioral sciences, statistics, biology, physics, and many others too numerous to mention. Selected applications will be presented throughout the rest of this text.

The topics of this chapter, antiderivatives, areas, and summation might, at first glance, seem totally unrelated, but they are not. Areas can be computed by using summations; the definite integral is defined in terms of the limit of an area; and the antiderivative is a key to evaluating the definite integral.

Finding the area of a rectangle or circle presents no problem. Formulas for these areas are readily available. However, the area bounded by the curve $y = x^2 + 1$ and the lines $y = 0$, $x = 0$, and $x = 2$ is a little more difficult to find (see Figure 4.1).

We can approximate the area by dividing it into two rectangles with equal widths, as shown in Figures 4.2 and 4.3. The width of

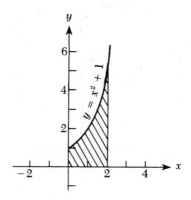

Figure 4.1   The shaded region
is the required area

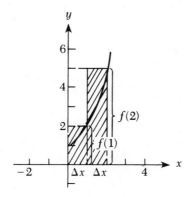

Figure 4.2 Approximation
of an irregular
area by rectangles

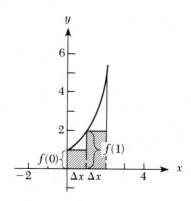

Figure 4.3 Approximation
of an irregular
area by rectangles

each rectangle is Δx, where $\Delta x = (2 - 0)/n$, and n is the number of
rectangles used. Since $n = 2$, $\Delta x = 1$. Figure 4.2 shows the heights
of the rectangles as $f(1)$ and $f(2)$, respectively, so that the area may
be approximated as the sum of the areas of the rectangles: $1(f(1)) +$
$1(f(2)) = 2 + 5 = 7$ square units. Figure 4.3 used $f(0)$ and $f(1)$ as
the heights of the rectangles, so that the approximation of the area
is $1(f(0)) + 1(f(1)) = 1 + 2 = 3$ square units. From the diagrams it
is evident that the first approximation overestimates the area and
the second approximation underestimates it, so that the area A is
such that $3 < A < 7$. A better approximation would be obtained by
partitioning the region into four rectangles ($n = 4$), as in Figure 4.4.
The area will again be overestimated or underestimated, depending on the values selected for $f(x)$. It should be clear that, as n

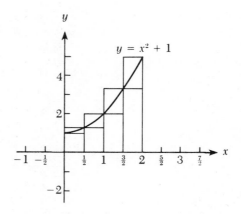

Figure 4.4 A better approximation of an irregular area by rectangles

increases, the approximation gets closer and closer to the actual area to be found.

If the region is partitioned into n rectangles, each with base $\Delta x = x_i - x_{i-1} = (2 - 0)/n$, where $i = 1, 2, 3, \ldots, n$, and with height $f(x_i)$, so that the area of each rectangle is $f(x_i)\,\Delta x$, then

$$\lim_{n \to \infty} [f(x_1)\,\Delta x + f(x_2)\,\Delta x + \cdots + f(x_i)\,\Delta x + \cdots + f(x_n)\,\Delta x]$$

is the exact area bounded by the curve $y = x^2 + 1$, $y = 0$, $x = 0$, and $x = 2$ (see Figure 4.5).

An illustration of a typical partition of the interval $[a, b]$ on the x axis is shown in Figure 4.6.

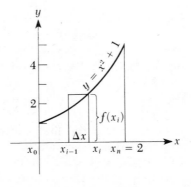

Figure 4.5 The rectangle $f(x_i)\,\Delta x$

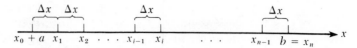

Figure 4.6 A typical partition on the x axis

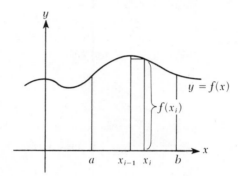

Figure 4.7 Area approximation

In general, consider a continuous function, f, which is on or above the x axis ($f(x) \geq 0$), and the region bounded by the curve, the x axis, and the lines $x = a$ and $x = b$, where $a < b$ (Figure 4.7).

Partition the region into n rectangles with equal width, $\Delta x = (b - a)/n$. As $n \to \infty$, $\Delta x \to 0$; therefore, for the ith given rectangle, $f(x_{i-1}) \to f(x_i)$; that is to say, the difference in the two function values becomes negligibly small. The required area, A, is given by the following definition.

Definition

$$A = \lim_{n \to \infty} [f(x_1) \, \Delta x + f(x_2) \, \Delta x + \cdots + \\ + f(x_i) \, \Delta x + \cdots + f(x_n) \, \Delta x]$$

This exact area is also known as the *definite integral* of the function from $x = a$ to $x = b$, and is written

$$A = \int_a^b f(x) \, dx$$

The function f is called the *integrand*, a and b are the *limits of integration*, and x is the *variable of integration*.

Since the exact area, or the actual value of the definite integral, involves a limit as n tends to infinity, techniques for evaluating this limit will have to be devised. The next section on summation will present some of these techniques, but the Fundamental Theorem of Calculus, which relies on antidifferentiation for evaluating the definite integral, will provide an elegant and relatively simple way in most instances.

In the meantime, we will rely on approximations using a finite number of rectangles to evaluate areas.

Example 1   Find an approximation of $\int_2^4 (2x + 1)\, dx$ by partitioning the required area into four rectangles of width $\Delta x = x_i - x_{i-1} = (b - a)/n$ and altitude $f(x_i)$.

Solution $\Delta x = \dfrac{b - a}{n} = \dfrac{4 - 2}{4} = \dfrac{1}{2}$

$x_0 = 2$

$x_1 = \dfrac{5}{2}, \quad f(x_1) = 6 \qquad A_1 = f(x_1)\,\Delta x = 6\left(\dfrac{1}{2}\right) = 3$

$x_2 = 3, \quad f(x_2) = 7 \qquad A_2 = f(x_2)\,\Delta x = 7\left(\dfrac{1}{2}\right) = \dfrac{7}{2}$

$x_3 = \dfrac{7}{2}, \quad f(x_3) = 8 \qquad A_3 = f(x_3)\,\Delta x = 8\left(\dfrac{1}{2}\right) = 4$

$x_4 = 4, \quad f(x_4) = 9 \qquad A_4 = f(x_4)\,\Delta x = 9\left(\dfrac{1}{2}\right) = \dfrac{9}{2}$

$$A \approx 3 + \frac{7}{2} + 4 + \frac{9}{2} = 15 \text{ square units}$$

Since the required region (Figure 4.8) is a trapezoid, we can compute the actual area by the formula

$$A = \frac{1}{2}(b_1 + b_2)h = \frac{1}{2}[f(2) + f(4)]2 = 14 \text{ square units}$$

Thus,

$$\int_2^4 (2x + 1)\, dx = 14 \text{ square units}$$

Now we use $n = 8$ to see how much closer the approximation becomes. When $n = 8$,

$$\Delta x = \frac{4 - 2}{8} = \frac{1}{4}$$

$x_0 = 2$

$x_1 = \dfrac{9}{4}, \quad f(x_1) = \dfrac{11}{2} \qquad A_1 = \dfrac{11}{8}$

$x_2 = \dfrac{5}{2}, \quad f(x_2) = 6 \qquad A_2 = \dfrac{3}{2}$

$x_3 = \dfrac{11}{4}, \quad f(x_3) = \dfrac{13}{2} \qquad A_3 = \dfrac{13}{8}$

$x_4 = 3, \quad f(x_4) = 7 \qquad A_4 = \dfrac{7}{4}$

$x_5 = \dfrac{13}{4}, \quad f(x_5) = \dfrac{15}{2} \qquad A_5 = \dfrac{15}{8}$

$x_6 = \dfrac{7}{2}, \quad f(x_6) = 8 \qquad A_6 = 2$

$x_7 = \dfrac{15}{4}, \quad f(x_7) = \dfrac{17}{2} \qquad A_7 = \dfrac{17}{8}$

$x_8 = 4, \quad f(x_8) = 9 \qquad A_8 = \dfrac{9}{4}$

$$A \approx \frac{11}{8} + \frac{3}{2} + \frac{13}{8} + \frac{7}{4} + \frac{15}{8} + 2 + \frac{17}{8} + \frac{9}{4} = 14\tfrac{1}{2} \text{ square units}$$

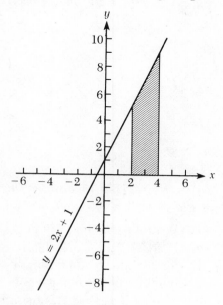

Figure 4.8   Area under $y = 2x + 1$,
$2 \leqslant x \leqslant 4$

To summarize what we have done, we look at the process used for finding the value of the definite integral. Each evaluation is an estimate or approximation of the actual value. If, as we improve the estimates by taking more and more subintervals (letting n equal larger and larger numbers) these estimates can be shown to have a limit (the limit as n tends to infinity exists), then we call that limit the definite integral.

It should be pointed out that, although we have taken only *regular* partitions, $\Delta x = (b - a)/n$, these subintervals do not have to be equal in length. However, unequal subintervals will require that the definition of the definite integral be changed to include the limit as the *largest* of the unequal subintervals in the partition tends to zero.

Secondly, the function need not be positive over the interval, but the limit process considered above, even though it may lead to a definite integral, does not lead to an area.

4.2 EXERCISES

In Exercises 1–10, approximate each integral as the sum of the areas of n rectangles, using the given information and by evaluating $f(x_i)$ at the right end of each subinterval. Sketch each area.

1. $\int_0^4 x^2 \, dx, \quad n = 4$

2. $\int_0^4 x^2 \, dx, \quad n = 8$

3. $\int_1^5 (x^2 - 3x + 4) \, dx, \quad n = 4$

4. $\int_1^3 (x^3 + 1) \, dx, \quad n = 4$

5. $\int_{-2}^1 (2x^2 + x + 3) \, dx, \quad n = 3$

6. $\int_0^2 (x^2 + 1) \, dx, \quad n = 8$

7. $\int_{-2}^1 (2x^2 + x + 3) \, dx, \quad n = 6$

8. $\int_3^4 \frac{x^3 + 2}{3} \, dx, \quad n = 4$

9. $\int_1^4 (x^2 + 2) \, dx, \quad n = 6$

10. $\int_3^4 \frac{x^3 + 2}{3} \, dx, \quad n = 8$

In Exercises 11–15, consider the area bounded by the graphs of $y = x^2$, $y = 0$, $x = 1$, and $x = 3$.

11. Write this area as a definite integral and sketch the area.

12. Construct an *overestimate* of this area by partitioning the interval into 4 equal-length subintervals and using the right-hand endpoint of each subinterval.

13. Repeat Exercise 12 using left-hand endpoints to find an *underestimate* of the area.

14. Repeat Exercise 12 using 8 subintervals.

15. Repeat Exercise 13 using 8 subintervals.

In Exercises 16–20, consider the area bounded by the graphs of $y = 1/x$, $x = 1$, $x = 4$, *and* $y = 0$.

16. Write this area as a definite integral and sketch the area.

17. Construct an underestimate of this area by using the right-hand endpoint of each of 6 equal subintervals.

18. Repeat Exercise 17, but find an overestimate.

19. Repeat Exercise 17 using a partition with 12 equal subintervals.

20. Repeat Exercise 18 using a partition with 12 equal subintervals.

4.3 SUMMATION NOTATION

In the preceding section many sums were written in long and tedious ways. The *summation* notation is a very efficient method for writing sums, and it will be used throughout this text. It should be pointed out that this is the principal notation used in statistics, that powerful tool of the social scientist.

Before we can use summation notation, it is necessary to define a *sequence*.

Definition

A *sequence* is a function k whose domain is the set of natural numbers, 1,2,3, . . . ,n, and whose range is $k(1), k(2), . . . , k(n)$. An alternate notation for $k(n)$ is k_n.

From the above definition, we see that we can write the terms of a sequence as $k_1, k_2, k_3,$ . . . ,k_n.

Example 1 Write the first five terms of the sequence whose nth term is $k_n = 3n + 2$.

Solution If $k_n = 3n + 2$, then

$$k_1 = 3(1) + 2 = 5$$
$$k_2 = 3(2) + 2 = 8$$
$$k_3 = 3(3) + 2 = 11$$
$$k_4 = 3(4) + 2 = 14$$
$$k_5 = 3(5) + 2 = 17$$

The first five terms of the sequence are 5, 8, 11, 14, and 17.

A useful, compact notation for the *sum* of n terms of the sequence $k_1, k_2, k_3, \ldots, k_n$ is

$$\sum_{i=1}^{n} k_i \qquad \text{(read: "The sum of } k_i \text{ as } i \text{ goes from 1 to } n \text{.")}$$

Since $\sum$ is the Greek capital letter *sigma*, the notation is sometimes called the *sigma notation*. Using this notation, we can write the following sums.

$$1 + 2 + 3 + 4 + 5 = \sum_{i=1}^{5} i$$

$$2 + 4 + 6 + 8 + 10 + 12 = \sum_{i=1}^{6} 2i$$

$$1 + 4 + 9 + 16 + \cdots + n^2 = \sum_{i=1}^{n} i^2$$

$$\frac{1}{2} + \frac{2}{5} + \frac{3}{10} + \frac{4}{17} = \sum_{i=1}^{4} \frac{i}{i^2 + 1}$$

The following theorems on summation are useful and easy to justify.

Theorem 4.1 $\displaystyle\sum_{i=1}^{n} a_i = na,$ if $a_i = a$, a constant for every i

Theorem 4.2 $\displaystyle\sum_{i=1}^{n} (k_i \pm g_i) = \sum_{i=1}^{n} k_i \pm \sum_{i=1}^{n} g_i$

Theorem 4.3 $\displaystyle\sum_{i=1}^{n} ak_i = a \sum_{i=1}^{n} k_i,$ if a is constant

Proof of Theorem 4.1 $\displaystyle\sum_{i=1}^{n} a_i = a + a + a + \cdots + a = na$

Proof of Theorem 4.2

$$\sum_{i=1}^{n} (k_i \pm g_i) = (k_1 \pm g_1) + (k_2 \pm g_2) + (k_3 \pm g_3) + \cdots + (k_n \pm g_n)$$

$$= (k_1 + k_2 + k_3 + \cdots + k_n) \pm (g_1 + g_2 + g_3 + \cdots + g_n)$$

$$= \sum_{i=1}^{n} k_i \pm \sum_{i=1}^{n} g_i$$

Proof of Theorem 4.3

$$\sum_{i=1}^{n} ak_i = ak_1 + ak_2 + ak_3 + \cdots + ak_n$$

$$= a(k_1 + k_2 + k_3 + \cdots + k_n)$$

$$= a \sum_{i=1}^{n} k_i$$

The above theorems, in addition to the following formulas (which can be proved by mathematical induction), enable us to compute many integrals.

Theorem 4.4

$$\sum_{i=1}^{n} i = \frac{n(n + 1)}{2}$$

Theorem 4.5

$$\sum_{i=1}^{n} i^2 = \frac{n(n + 1)(2n + 1)}{6}$$

Theorem 4.6

$$\sum_{i=1}^{n} i^3 = \left[\frac{n(n + 1)}{2}\right]^2$$

Example 2

Evaluate each of the following:

   a. $1 + 2 + 3 + \cdots + 20$

   b. $1 + 4 + 9 + \cdots + 64$

   c. $1 + 8 + 27 + 64 + 125$

Solution

   a. $1 + 2 + 3 + \cdots + 20 = \sum_{i=1}^{20} i = \dfrac{20(20 + 1)}{2}$

$$= 210 \quad \text{by Theorem 4.4}$$

   b. $1 + 4 + 9 + \cdots + 64 = \sum_{i=1}^{8} i^2 = \dfrac{8(8 + 1)(16 + 1)}{6}$

$$= 204 \quad \text{by Theorem 4.5}$$

   c. $1 + 8 + 27 + 64 + 125 = \sum_{i=1}^{5} i^3 = \left[\dfrac{5(5 + 1)}{2}\right]^2$

$$= 225 \quad \text{by Theorem 4.6}$$

We can also express the integral $\int_a^b f(x)\, dx$ as the limit of an expression in sigma notation.

Definition

Let f be a continuous function on the closed interval $[a, b]$. Then

$$\int_a^b f(x)\, dx = \lim_{n \to \infty} \sum_{i=1}^n f(x_i)\, \Delta x$$

where $\Delta x = (b - a)/n$, provided this limit exists.

This sum is usually called a *Riemann sum* in honor of the German mathematician G. F. B. Riemann (1826–1866). In the sum as defined above, Δx is a constant width. This is not a requirement for a Riemann sum, but for the purposes of our discussion we will limit the partitions to equal width subintervals.

Example 3   Evaluate $\int_2^4 (x^2 + x + 1)\, dx$.

Solution   By definition,

$$\int_2^4 (x^2 + x + 1)\, dx = \lim_{n \to \infty} \sum_{i=1}^n f(x_i)\, \Delta x$$

where $\Delta x = (b - a)/n = (4 - 2)/n = 2/n$, x_i is the right endpoint of the width of the ith rectangle, and $f(x_i)$ is the height of the ith rectangle (Figure 4.9).

$$x_0 = 2, \quad x_1 = 2 + \Delta x, \quad x_2 = 2 + 2\,\Delta x,$$
$$x_3 = 2 + 3\,\Delta x, \quad x_i = 2 + i\,\Delta x$$

$$\begin{aligned}
f(x_i) = f(2 + i\,\Delta x) &= (2 + i\,\Delta x)^2 + (2 + i\,\Delta x) + 1 \\
&= 4 + 4i\,\Delta x + i^2(\Delta x)^2 + 2 + i\,\Delta x + 1 \\
&= i^2(\Delta x)^2 + 5i\,\Delta x + 7
\end{aligned}$$

$$\sum_{i=1}^n f(x_i)\, \Delta x = \sum_{i=1}^n [i^2(\Delta x)^2 + 5i(\Delta x) + 7]\, \Delta x$$

Applying the theorems for sigma notation, and remembering that Δx is constant, we can write

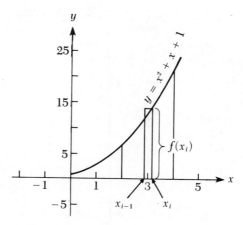

Figure 4.9 $\displaystyle\int_2^4 (x^2 + x + 1)\, dx$

$$\sum_{i=1}^n \left[i^2(\Delta x)^2 + 5i(\Delta x) + 7 \right] \Delta x = (\Delta x)^3 \sum_{i=1}^n i^2 + 5(\Delta x)^2 \sum_{i=1}^n i + 7n\, \Delta x$$

By Theorem 4.5,

$$(\Delta x)^3 \sum_{i=1}^n i^2 = (\Delta x)^3\, \frac{n(n+1)(2n+1)}{6}$$

By Theorem 4.4,

$$5(\Delta x)^2 \sum_{i=1}^n i = 5(\Delta x)^2\, \frac{n(n+1)}{2}$$

$$\sum_{i=1}^n f(x_i)\, \Delta x = (\Delta x)^3\, \frac{n(n+1)(2n+1)}{6} + 5(\Delta x)^2\, \frac{n(n+1)}{2} + 7n\, \Delta x$$

But $\Delta x = \dfrac{b-a}{n} = \dfrac{2}{n}$; therefore,

$$\lim_{n \to \infty} \sum_{i=1}^n f(x_i)\, \Delta x = \lim_{n \to \infty} \left[\frac{8}{n^3}\, \frac{n(n+1)(2n+1)}{6} + \frac{20}{n^2}\, \frac{n(n+1)}{2} + 7n \cdot \frac{2}{n} \right]$$

$$= \lim_{n \to \infty} \left(\frac{80n^2 + 42n + 4}{3n^2} \right) = \frac{80}{3}$$

Thus,

$$\int_2^4 (x^2 + x + 1)\, dx = \frac{80}{3}$$

It should be pointed out that the letter i in the symbol $\displaystyle\sum_{i=1}^{n}$ stands for the *index* of the summation but the letters j or k are also frequently used. For example,

$$\sum_{j=1}^{4} (2j + j^2) \qquad \text{or} \qquad \sum_{k=1}^{52} \frac{k-1}{k^2}$$

In the preceding section, the definite integral, $\displaystyle\int_{a}^{b} f(x)\,dx$, was defined as an area, when $a < b$ and the function was nonnegative in the closed interval $[a, b]$. These restrictions are not necessary for our new definition and, as a result, it is not always meaningful to regard the integral as an area. In order to allow an integral to be a negative number, or even zero, the following definitions are given.

Definitions

If $a < b$,

$$\int_{a}^{b} f(x)\,dx = -\int_{b}^{a} f(x)\,dx$$

If $b = a$,

$$\int_{a}^{b} f(x)\,dx = \int_{a}^{a} f(x)\,dx = 0$$

4.3 EXERCISES

In Exercises 1–5, write each sum in expanded notation, and compute the sums.

1. $\displaystyle\sum_{i=1}^{5} (2i - 3)$

2. $\displaystyle\sum_{j=1}^{4} (j^2 + 3j + 1)$

3. $\displaystyle\sum_{k=1}^{6} \frac{1}{k}$

4. $\displaystyle\sum_{i=1}^{3} \frac{i^2 - 2}{i}$

5. $\displaystyle\sum_{k=1}^{4} \left[\frac{1}{k} - \frac{1}{k+1} \right]$

In Exercises 6–10, use the theorems and formulas given in this section to compute the sums.

6. $\displaystyle\sum_{i=1}^{n} (2i - 3)$

7. $\displaystyle\sum_{i=1}^{n} (i^2 + 3i + 1)$

8. $\displaystyle\sum_{i=1}^{n} i(2i + 1)$

9. $\displaystyle\sum_{i=1}^{n} (i + 2)(2i + 3)$

10. $\displaystyle\sum_{i=1}^{n} (i - 2)(3i^2)$

In Exercises 11–16, find the exact value of each of the following integrals by the method outlined in this section. Compare these values with the approximations obtained in the exercises for Section 4.2.

11. $\displaystyle\int_{0}^{4} x^2 \, dx$

12. $\displaystyle\int_{1}^{4} (x^2 + 2) \, dx$

13. $\displaystyle\int_{0}^{2} (x^2 + 1) \, dx$

14. $\displaystyle\int_{1}^{3} (x^3 + 1) \, dx$

15. $\displaystyle\int_{-2}^{1} (2x^2 + x + 3) \, dx$

16. $\displaystyle\int_{3}^{4} \frac{x^3 + 2}{3} \, dx$

17. Evaluate $\displaystyle\int_{3}^{3} (x^3 - 3x^2 + 5x - 2) \, dx$.

18. Evaluate $\displaystyle\int_{-2}^{-2} (3x^5 - 4x + 9) \, dx$.

19. If $\displaystyle\int_{-3}^{2} (x^2 + 3x - 1) \, dx = -5/6$, find $\displaystyle\int_{2}^{-3} (x^2 + 3x - 1) \, dx$.

20. If $\displaystyle\int_{0}^{3} (6x^5 - 2) \, dx = 723$, find $\displaystyle\int_{3}^{0} (6x^5 - 2) \, dx$.

In Exercises 21–26, consider the function f such that $f(x) = x^3$ on the interval from $x = -2$ to $x = 2$.

21. Sketch the graph of this function.

22. Partition the interval into 4 equal length subintervals and form a Riemann sum using the right-hand endpoint of each subinterval. Evaluate this sum.

23. Repeat Exercise 22 using 8 equal subintervals.

24. Repeat Exercise 22 using 16 equal subintervals.

25. Use Theorem 4.6 to evaluate the definite integral $\displaystyle\int_{-2}^{2} x^3 \, dx$.

26. Does the integral in Exercise 25 correspond to the area under the curve as in Exercises 21–24? Discuss.

27. The Little Profit Manufacturing Company has a policy of increasing each employee's salary by $200 per year. If Mr. Rapidtype's starting salary is $5000, how much money will he have earned at the end of 12 years?

§ **28.**   If $600 is invested at 5% simple interest, what will the investment amount to after 10 years?

4.4 PROPERTIES OF THE INTEGRAL

In Section 4.3 the integral of a continuous function over the closed interval $[a, b]$ was defined as

$$\int_a^b f(x)\, dx = \lim_{n \to \infty} \sum_{i=1}^n f(x_i)\, \Delta x, \qquad \text{where} \qquad \Delta x = \frac{b - a}{n}$$

provided this limit exists.

If f is a continuous function for all x in the interval $[a, b]$, then $\int_a^b f(x)\, dx$ exists, and the function is said to be *integrable* on $[a, b]$.

The definitions $\int_a^b f(x)\, dx = -\int_b^a f(x)\, dx$, when $a < b$, and $\int_a^a f(x)\, dx = 0$, together with the following four theorems, form the key properties of the integral.

Theorem 4.7 $\displaystyle \int_a^b k \cdot f(x)\, dx = k \int_a^b f(x)\, dx,$ where k is any constant

Theorem 4.8   Let c be in $[a, b]$; then

$$\int_a^c f(x)\, dx + \int_c^b f(x)\, dx = \int_a^b f(x)\, dx$$

The student can easily verify these theorems.

Theorem 4.9   If f and g are continuous functions on the interval from $x = a$ to $x = b$, then

$$\int_a^b (f(x) + g(x))\, dx = \int_a^b f(x)\, dx + \int_a^b g(x)\, dx$$

This theorem is derived from the fact that the limit of a sum is the sum of the limits, provided these limits exist.

Another theorem which shall be stated without proof is the Mean Value Theorem for Integrals.

Theorem 4.10 (Mean Value Theorem for Integrals) Let f be a continuous function on the closed interval $[a, b]$. There exists a number c in $[a, b]$ such that

$$\int_a^b f(x)\, dx = f(c)(b - a)$$

The value $f(c)$ is said to be the *mean value* of the function, or its *average* value.

If the function f is nonnegative over the entire interval, then for every area bounded by the graph of f, the x axis, and the lines $x = a$ and $x = b$ there exists a rectangle of equal area, with length $(b - a)$ and height $f(c)$ for some value c in $[a, b]$ (see Figure 4.10).

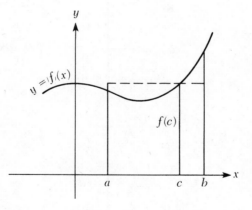

Figure 4.10 $\int_a^b f(x)\, dx = f(c)(b - a)$

The Mean Value Theorem for Integrals guarantees the existence of the number c such that the required area $A = f(c)(b - a)$, but this number is usually not easy to find, unless the value of the integral is known. Thus, if

$$\int_3^6 x^2\, dx = 63$$

then $A = f(c)(b - a) = 63$, $b - a = 6 - 3 = 3$; $f(c) = A/(b - a) = 63/3 = 21$; $c^2 = 21$, $c = \sqrt{21} \approx 4.58$, and 4.58 is between 3 and 6 as required by the theorem.

4.4 EXERCISES

In Exercises 1–5, for each integral, find the number c guaranteed by the Mean Value Theorem for Integrals. Sketch each curve and the corresponding rectangle with dimensions $f(c)$ and $(b - a)$.

1. $\int_0^4 x^2 \, dx = \dfrac{64}{3}$

2. $\int_1^4 (x^2 + 2) \, dx = 27$

3. $\int_0^2 (x^2 + 1) \, dx = \dfrac{14}{3}$

4. $\int_1^3 (x^3 + 1) \, dx = 22$

5. $\int_{-2}^1 (3x^2 + x + 3) \, dx = 16\frac{1}{2}$

6. If $f(x) = x^2 - 2x + 1$ and $\int_{-1}^3 f(x) \, dx = \dfrac{16}{3}$, find two values of c such that

$$\int_{-1}^3 f(x) \, dx = f(c)(3 + 1)$$

In Exercises 7–16, if $\int_0^1 x^3 \, dx = \dfrac{1}{4}$, $\int_0^1 x^2 \, dx = \dfrac{1}{3}$, and $\int_1^2 x^2 \, dx = \dfrac{7}{3}$, evaluate:

7. $\int_0^2 x^2 \, dx$

8. $\int_0^1 (x^3 + x^2) \, dx$

9. $\int_2^1 x^2 \, dx$

10. $\int_0^1 4x^3 \, dx$

11. $\int_0^1 (3x^3 + 2x^2) \, dx$

12. $\int_0^2 6x^2 \, dx$

13. $\int_2^2 x^2 \, dx$

14. $\int_1^0 x^3 \, dx$

15. $\int_0^1 (x^3 - x^2) \, dx$

16. $\int_0^1 (x^2 - x^3) \, dx$

In Exercises 17–20, if $\int_1^3 x^2 \, dx = \dfrac{26}{3}$, $\int_3^5 x^2 \, dx = \dfrac{98}{3}$, $\int_1^3 x^4 \, dx = \dfrac{242}{5}$, and $\int_3^5 x^4 \, dx = \dfrac{2882}{5}$, evaluate:

17. $\int_1^3 (2x^2 + 3x^4) \, dx$

18. $\int_1^5 10x^2 \, dx$

19. $\int_3^5 (x^4 - 2x^2) \, dx$

20. $\int_3^1 (x^2 + 5x^4) \, dx$

In Exercises 21–24, state whether the Mean Value Theorem for Integrals applies. If it does not apply, say why not.

21. $\displaystyle\int_0^3 \frac{x}{x-1}\,dx$ 　　　　　　　**22.** $\displaystyle\int_1^5 \frac{x-1}{x+1}\,dx$

23. $\displaystyle\int_{-10}^{20}(x^9 + x^2 + 3)\,dx$ 　　　**24.** $\displaystyle\int_2^5 \sqrt{16-x^2}\,dx$

25. If

$$f(x) = \begin{cases} x & \text{for } -1 \leqslant x < 1 \\ 2-x & \text{for } 1 \leqslant x < 10 \end{cases}$$

does the Mean Value Theorem for Integrals apply to $\displaystyle\int_0^4 f(x)\,dx$? Justify your answer by sketching $f(x)$ and shading the desired area.

In Exercises 26–30, if $f(x) = x^2$, find the average value of the function on the indicated interval $[a, b]$, using the given value of the integral $\displaystyle\int_a^b x^2\,dx$.

26. $\displaystyle\int_0^2 x^2\,dx = \frac{8}{3}$ 　　　　　　**27.** $\displaystyle\int_1^3 x^2\,dx = 8\tfrac{2}{3}$

28. $\displaystyle\int_{-3}^{-1} x^2\,dx = 8\tfrac{2}{3}$ 　　　　**29.** $\displaystyle\int_4^8 x^2\,dx = 149\tfrac{1}{3}$

30. $\displaystyle\int_{-2}^2 x^2\,dx = \frac{16}{3}$

4.5   THE FUNDAMENTAL THEOREM OF CALCULUS

In the preceding sections of this chapter we studied antidifferentiation as the inverse process of differentiation; we defined the definite integral for a continuous function on a closed interval as the limit of a Riemann sum; and then we did some very cumbersome computations to evaluate some simple integrals. At that time the thought must have occurred that there has to be an easier way! There is. The Fundamental Theorem of Calculus, which links antidifferentiation and the definite integral, will provide the way for finding the value of a definite integral in a relatively quick and simple manner.

Theorem 4.11   (Fundamental Theorem of Calculus) If f is a function defined and continuous on the closed interval $[a, b]$, then

  (i)   the function A with the rule $A(x) = \int_a^x f(t)\, dt$ is an antiderivative of f on $[a, b]$, and

  (ii) if F is any antiderivative of f on $[a, b]$, then

$$\int_a^b f(x)\, dx = F(b) - F(a)$$

Example 1   Use the Fundamental Theorem of Calculus to evaluate

$$\int_{-1}^2 (3x^2 + 2)\, dx$$

Solution   The function $f(x) = 3x^2 + 2$ is continuous and defined for all values of x between -1 and 2. An antiderivative of $3x^2 + 2$ is $x^3 + 2x$. Thus $F(x) = x^3 + 2x$; $F(-1) = (-1)^3 + 2(-1) = -3$; $F(2) = (2)^3 + 2(2) = 12$.

$$\int_{-1}^2 (3x^2 + 2)\, dx = F(2) - F(-1) = 12 - (-3) = 15$$

The following is a proof of the Fundamental Theorem of Calculus.

Proof   The general proof of this theorem may be found in the more rigorous calculus texts used in engineering courses. It relies on the Mean Value Theorem for integrals and the properties of the integral stated in Section 4.4. We shall prove the fundamental theorem under the special conditions that $\int_a^b f(x)\, dx$ represents the area above the x axis and under the graph of $y = f(x)$ between $x = a$ and $x = b$, and $y = f(x)$ is a positive, continuous, increasing function over the interval $[a, b]$, as shown in Figure 4.11.

Let $F(x)$ be any antiderivative of $f(x)$. That is, $F'(x) = f(x)$. Let A be a new function that represents the area above the x axis and below the graph of $y = f(x)$ between the vertical lines at a and x. Thus $A(a) = 0$, since the area below $f(x)$ between a and a is zero. $A(b)$ is the area under $f(x)$ between a and b, which means $A(b) = \int_a^b f(x)\, dx$. A positive change in x of Δx would cause a change in A of ΔA as shown in Figure 4.11. The area ΔA is greater than that of the rectangle whose base is Δx and whose height is $f(x)$, and the

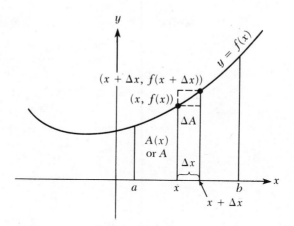

Figure 4.11 $\displaystyle\int_a^b f(x)\,dx = F(b) - F(a)$

area ΔA is less than that of the rectangle with base Δx and height $f(x + \Delta x)$. That is,

$$f(x)\,\Delta x \leqslant \Delta A \leqslant f(x + \Delta x)\,\Delta x$$

Dividing by Δx gives

$$f(x) \leqslant \frac{\Delta A}{\Delta x} \leqslant f(x + \Delta x)$$

Since $\displaystyle\lim_{\Delta x \to 0} f(x + \Delta x) = f(x)$ (because f is continuous), taking the limit of the above inequality as Δx approaches 0 implies that

$$\lim_{\Delta x \to 0} \frac{\Delta A}{\Delta x} = f(x)$$

But

$$\lim_{\Delta x \to 0} \frac{\Delta A}{\Delta x} = A'(x)$$

So,

$$A'(x) = f(x)$$

By hypothesis, $F'(x) = f(x)$. Therefore, $F(x)$ and $A(x)$ are both anti-derivatives of $f(x)$, which means they can differ, at most, by a constant. Let C be the constant such that

$$A(x) = F(x) + C$$

Since $A(a) = 0$,

$$A(a) = F(a) + C$$
$$0 = F(a) + C$$
$$C = -F(a)$$

Therefore by replacing C with $-F(a)$,

$$A(x) = F(x) - F(a)$$

Replacing x with b gives

$$A(b) = F(b) - F(a)$$

As stated earlier,

$$A(b) = \int_a^b f(x)\, dx$$

So

$$\int_a^b f(x)\, dx = F(b) - F(a)$$

This proves the Fundamental Theorem of Calculus for the conditions stated; that is, $\int_a^b f(x)\, dx$ represents an area above the x axis, under the graph of $y = f(x)$, between $x = a$ and $x = b$, and f is a positive, continuous, increasing function over the interval $[a, b]$.

In order to appreciate this theorem, refer to the method of Sections 4.2 and 4.3, using $\lim\limits_{n \to \infty} \sum\limits_{i=1}^{n} f(x_i)\, \Delta x$ to evaluate

$$\int_2^4 (x^2 + x + 1)\, dx$$

Now note that the Fundamental Theorem of Calculus applies, since $x^2 + x + 1$ is defined and continuous on the closed interval $[2, 4]$. To find $F(x)$, we take the antiderivative

$$\int (x^2 + x + 1)\, dx = \frac{x^3}{3} + \frac{x^2}{2} + x + C$$

Thus,

$$F(x) = \frac{x^3}{3} + \frac{x^2}{2} + x + C$$

$$F(4) = \frac{4^3}{3} + \frac{4^2}{2} + 4 + C = \frac{64}{3} + 8 + 4 + C = \frac{100}{3} + C$$

$$F(2) = \frac{2^3}{3} + \frac{2^2}{2} + 2 + C = \frac{8}{3} + 2 + 2 + C = \frac{20}{3} + C$$

$$F(4) - F(2) = \left(\frac{100}{3} + C\right) - \left(\frac{20}{3} + C\right) = \frac{80}{3}.$$

Note: The constant C will always cancel when computing the definite integral $\int_a^b f(x)\,dx$, so it may be omitted.

Having discussed the method of applying the Fundamental Theorem in this example, let us write the problem in a more compact form.

$$\int_2^4 (x^2 + x + 1)\,dx = \frac{x^3}{3} + \frac{x^2}{2} + x \,\Big]_2^4$$

The right-hand side of the statement is the indefinite integral of $(x^2 + x + 1)$, and the symbol $\Big]_2^4$ indicates that we are to evaluate the integral $F(x)$ for $F(4) - F(2)$.

Example 2   Evaluate $\int_{-1}^3 (x^3 + 1)\,dx$.

Solution   An antiderivative of $x^3 + 1$ is $(x^4/4) + x$.

$$\int_{-1}^3 (x^3 + 1)\,dx = \frac{x^4}{4} + x \,\Big]_{-1}^3 = \left(\frac{3^4}{4} + 3\right) - \left[\frac{(-1)^4}{4} + (-1)\right]$$

$$= \frac{93}{4} - \left(-\frac{3}{4}\right) = 24$$

Example 3   Evaluate $\int_0^3 \sqrt{x - 2}\,dx$.

Solution   An antiderivative, $\int \sqrt{x - 2}\,dx$, is readily found by letting $u = x - 2$; then $du = dx$. Thus,

$$\int \sqrt{x - 2}\,dx = \int u^{1/2}\,du = \frac{u^{3/2}}{3/2} + C$$

$$= \frac{2}{3}(x - 2)^{3/2} + C$$

An antiderivative of $\sqrt{x-2}$ is, therefore, $\frac{2}{3}(x-2)^{3/2}$. Before we fall into the trap of evaluating the expression $\frac{2}{3}(x-2)^{3/2}\Big]_0^3$, we should realize that $\sqrt{x-2}$ is *not* defined for all x in $[0, 3]$, and the Fundamental Theorem does not apply.

Example 4   Evaluate $\displaystyle\int_0^5 \frac{dx}{(x+2)^2}$.

Solution   Here the Fundamental Theorem of Calculus applies, since $1/(x+2)^2$ is continuous and defined for all x in the interval $[0, 5]$. An antiderivative is found by substituting $u = x + 2$, whence $du = dx$ and

$$\int u^{-2}\, du = -u^{-1} + C = \frac{-1}{x+2} + C$$

Then

$$\int_0^5 \frac{dx}{(x+2)^2} = \frac{-1}{x+2}\Big]_0^5 = \frac{-1}{7} - \left(-\frac{1}{2}\right) = \frac{5}{14}$$

4.5 EXERCISES

In Exercises 1–5, for each integral $\int_a^b f(x)\, dx$, find values for a and b for which the integral may be evaluated by the Fundamental Theorem of Calculus.

1. $\displaystyle\int_a^b \frac{dx}{x^2}$

2. $\displaystyle\int_a^b \sqrt{3x-2}\, dx$

3. $\displaystyle\int_a^b (x^2 + 3)^3\, dx$

4. $\displaystyle\int_a^b \frac{x\, dx}{(4 - x^2)^{3/2}}$

5. $\displaystyle\int_a^b \frac{dx}{x^2 + x - 2}$

In Exercises 6–26, evaluate each integral.

6. $\displaystyle\int_1^5 (x^2 + x)\, dx$

7. $\displaystyle\int_{-3}^{-1} (x^2 + x + 3)\, dx$

8. $\displaystyle\int_0^3 x(x^2 + 1)^4\, dx$

9. $\displaystyle\int_1^3 \left(2x^3 - \frac{4}{x^2}\right) dx$

10. $\int_0^1 (x^9 + 1)\, dx$ **11.** $\int_0^1 x^8(x^9 + 1)^2\, dx$

12. $\int_0^1 x(x^2 + 3)^3\, dx$ **13.** $\int_1^4 \sqrt{x}\, dx$

14. $\int_{-1}^1 \dfrac{dx}{x^2 + 4x + 4}$ **15.** $\int_1^4 \dfrac{1}{\sqrt{x}}\, dx$

16. $\int_1^2 x^2 \sqrt{1 + x^3}\, dx$

17. $\int_0^a x \sqrt{a - x}\, dx$ (a is a positive constant)

18. $\int_0^a \dfrac{dx}{\sqrt{x + a}}$ (a is a positive constant)

19. $\int_{-1}^0 (1 - x)(2 + x^2)\, dx$ **20.** $\int_0^{1/2} \dfrac{x\, dx}{\sqrt{1 - x^2}}$

21. $\int_{-1}^1 \dfrac{x\, dx}{(x^2 + 1)^2}$ **22.** $\int_0^5 \dfrac{dx}{(x + 2)^2}$

23. $\int_{-1}^4 (y + 2)^2\, dy$ **24.** $\int_1^4 (w^{3/2} + w^{1/2} - w^{-1/2})\, dw$

25. $\int_1^4 (z^{3/2} + z^{1/2} - z^{-1/2})\, dz$

26. $\int_1^4 ax^{a-1}\, dx$ (a is a positive constant)

In Exercises 27–30, use the Fundamental Theorem of Calculus to find the areas between the curves and lines. Sketch the graphs.

27. $y = 2x + 7, \quad y = 0, x = 1, x = 5$

28. $y = 4 - x^2, \quad y = 0, x = 1, x = 2$

29. $y = x^2 - 6x + 9, \quad y = 0, x = 2, x = 4$

30. $y = 1 - x^2, \quad y = 0, x = 0, x = \frac{1}{4}$

4.6 INTEGRATION BY USE OF TABLES

In spite of the fact that we have now learned how to integrate $\int dx$, $\int du$, $\int u\, du$, and $\int u^n\, du$, where n is a rational number other than -1, we are not ready to integrate many functions. What happens when $n = -1$ in the form $\int u^n\, du$? Even a deceptively simple looking integral such as

$$\int \frac{dx}{x^2 + 1}$$

is beyond the methods we have learned so far for finding antiderivatives.

In subsequent chapters we will learn how to handle many different techniques of integration, but at this time we introduce the student to the Table of Integrals, which is found inside the cover of this text. The table lists 57 indefinite integral forms. Many of these involve logarithmic and exponential functions, and trigonometric and inverse trigonometric functions. Most of these will be considered at a later time. We will restrict our attention to the simpler types found in the table, but hope that the student will at least be aware that the other types are there to be used when necessary.

In order to use the table, the user must first be able to identify the proper form. This usually involves some algebraic manipulation of the integrand, as the following examples will demonstrate.

Example 1   Find $\displaystyle\int \frac{dx}{x^2 - 6x + 9}$.

Solution   In searching through the table, one would conclude that either form 22, 23, or 24 would apply. All consider

$$\int \frac{dx}{ax^2 + bx + c}$$

In this case, $a = 1$, $b = -6$, and $c = 9$. Therefore, $b^2 = 36$, $4ac = 36$, and $b^2 = 4ac$. Thus, form 24 applies. Here

$$\int \frac{dx}{ax^2 + bx + c} = \frac{-2}{2ax + b}$$

or

$$\int \frac{dx}{x^2 - 6x + 9} = \frac{-2}{2(1)x - 6} = \frac{-2}{2x - 6} = \frac{-2}{2(x - 3)}$$

$$= -\frac{1}{x - 3} \quad \text{or} \quad -\frac{1}{x - 3} + C$$

Note that even though forms 22 and 23 look identical to form 24, the different values of a, b, and c result in an inverse tangent antiderivative in form 22, and a logarithmic antiderivative in form 23.

We might also recognize that the problem in the above example could quickly be worked with the antidifferentiation methods we already know:

$$\int \frac{dx}{x^2 - 6x + 9} = \int \frac{dx}{(x - 3)^2} = \int (x - 3)^{-2} \, dx$$

$$= \frac{(x - 3)^{-1}}{-1} + C$$

$$= -\frac{1}{x - 3} + C$$

The major problem in using the table is that the form of the integral is not always obvious. For example

$$\int 2x \sqrt{x^2 + 3} \, dx$$

$$\int \frac{3 \sqrt{x}}{2} (x^{3/2} + 2)^{-4} \, dx$$

and

$$\int (x^4 - 2x^2 + 1)^7 (4x^3 - 4x) \, dx$$

are all examples of form 2 in the table, that is,

$$\int u^n \, du = \frac{u^{n+1}}{n + 1}$$

In $\int 2x \sqrt{x^2 + 3} \, dx$, if we let $u = x^2 + 3$, then $du = 2x \, dx$ and

$$\int 2x \sqrt{x^2 + 3} \, dx = \int u^{1/2} \, du = \frac{2u^{3/2}}{3} + C$$

$$= \frac{2}{3} (x^2 + 3)^{3/2} + C$$

In $\int (3 \sqrt{x}/2)(x^{3/2} + 2)^{-4} \, dx$, if we let $u = x^{3/2} + 2$, then $du = \frac{3}{2} x^{1/2} \, dx$ and

$$\int \frac{3 \sqrt{x}}{2} (x^{3/2} + 2)^{-4} \, dx = \int u^{-4} \, du$$

$$= \frac{u^{-3}}{-3} + C = \frac{-1}{3} (x^{3/2} + 2)^{-3} + C$$

In $\int (x^4 - 2x^2 + 1)^7 (4x^3 - 4x)\, dx$, we let $u = x^4 - 2x^2 + 1$, and then $du = (4x^3 - 4x)\, dx$ and

$$\int (x^4 - 2x^2 + 1)^7 (4x^3 - 4x)\, dx = \int u^7\, du$$

$$= \frac{u^8}{8} + C = \frac{1}{8}(x^4 - 2x^2 + 1)^8 + C$$

The actual use of the table then requires the user to identify the proper form.

As stated above, logarithmic and exponential functions, as well as trigonometric and inverse trigonometric functions will be studied in subsequent chapters. However, a quick introduction to the significance of some of these functions, and the application of the table to integrate expressions involving the number e and the logarithm to the base e is appropriate.

The constant e, approximately equal to 2.71828, is an irrational number. It is the key constant in describing curves formed by suspended cables, problems involving population growth, and radioactive decay, to name just a few applications. The function f such that $f(x) = e^x$ is called an exponential function, and its inverse function, f^{-1}, is the logarithmic function to the base e. Thus $f(x) = e^x$ and $f^{-1}(x) = \log_e x$. The logarithm to the base e is called the *natural* logarithm, and will be written ln. (A detailed discussion of these functions, their derivatives, and their antiderivatives is presented in Chapter 6.) Numerical tables are also provided for evaluating these functions, but for the purpose of this section we shall not use these tables, but concentrate on the use of the Table of Integrals for integrating expressions involving these functions.

From form 42,

$$\int e^u\, du = e^u + C$$

Using this form, we find

$$\int e^x\, dx = e^x + C$$

so

$$\int e^{2x}\, dx = \frac{1}{2}e^{2x} + C$$

by letting $u = 2x$, $du = 2\, dx$, $dx = \frac{1}{2}\, du$, and making the appropriate substitution. Also,

$$\int \frac{dx}{e^x} = \int e^{-x}\, dx = -e^{-x} + C = \frac{-1}{e^x} + C$$

By finding the appropriate form in the table we can now integrate $\int u^n\, du$ with $n = -1$, or $\int du/u$. Form 3 tells us that

$$\int \frac{du}{u} = \ln|u| + C$$

that is, the general antiderivative of $1/u$ is the natural logarithm of $|u|$ plus some constant. The use of absolute value is necessary to insure that the expression is defined. Recall that we can only find logarithms of *positive* numbers. Since u originally appeared in the denominator of the integrand, we can be assured that it does not equal zero, but it could be negative; therefore, the absolute value of u is needed.

Example 2   Use the Table of Integrals to find $\int \frac{dx}{3x + 1}$.

Solution   This integral matches form 5, with $a = 3$ and $b = 1$. Therefore

$$\int \frac{dx}{3x + 1} = \frac{1}{3} \ln|3x + 1| + C$$

Often, a form requires more than one application of integration, as Example 3 demonstrates.

Example 3   Use the Table of Integrals to find $\int x^2 e^{3x}\, dx$.

Solution   Since $n > 0$, form 43 applies; $n = 2$ and $a = 3$.

$$\int x^2 e^{3x}\, dx = \frac{1}{3} x^2 e^{3x} - \frac{2}{3} \int x e^{3x}\, dx$$

The new integral again fits form 43, but $n = 1$, while a is still equal to 3.

$$\int x e^{3x}\, dx = \frac{1}{3} x e^{3x} - \frac{1}{3} \int e^{3x}\, dx \qquad (x^{1-1} = x^0 = 1 \text{ if } x \neq 0)$$

Now $\int e^{3x}\, dx = \frac{1}{3} e^{3x}$ by form 42 and an appropriate substitution.

Combining these steps and simplifying,

$$\frac{1}{3} x^2 e^{3x} - \frac{2}{3} \left[\frac{1}{3} x e^{3x} - \frac{1}{3} \int e^{3x} \, dx \right]$$

$$= \frac{1}{3} x^2 e^{3x} - \frac{2}{3} \left[\frac{1}{3} x e^{3x} - \frac{1}{3} \left(\frac{1}{3} e^{3x} \right) \right]$$

$$= \frac{1}{3} x^2 e^{3x} - \frac{2}{9} x e^{3x} + \frac{2}{27} e^{3x}$$

and

$$\int x^2 e^{3x} \, dx = \frac{1}{3} e^{3x} \left(x^2 - \frac{2}{3} x + \frac{2}{9} \right) + C$$

The process required two applications of form 43 and one application of form 42.

Example 4 Use the Table of Integrals to find $\displaystyle\int \frac{dx}{3x^2 + 2x - 4}$.

Solution This calls for form 22, 23, or 24, depending on the relationship between b^2 and $4ac$. We have $b = 2$, $b^2 = 4$, $a = 3$, $c = -4$, and $4ac = -48$. Since $4 > -48$, form 23 applies.

$$\int \frac{dx}{3x^2 + 2x - 4} = \frac{1}{\sqrt{52}} \ln \left| \frac{6x + 2 - \sqrt{52}}{6x + 2 + \sqrt{52}} \right| + C$$

$$= \frac{1}{2\sqrt{13}} \ln \left| \frac{3x + 1 - \sqrt{13}}{3x + 1 + \sqrt{13}} \right| + C$$

4.6 EXERCISES

In Exercises 1–20, use the Table of Integrals to find each of the following. (In Exercises 19 and 20, use form 52 or 53.)

1. $\displaystyle\int (2x + 3)^4 (2x) \, dx$
2. $\displaystyle\int x \sqrt{3x + 2} \, dx$

3. $\displaystyle\int \frac{dx}{4x + 5}$
4. $\displaystyle\int \frac{dx}{9 - x^2}$

5. $\displaystyle\int e^{4x} \, dx$
6. $\displaystyle\int e^{-4x} \, dx$

7. $\displaystyle\int \frac{dx}{x^2 + 10x + 25}$
8. $\displaystyle\int \frac{dx}{x(2x + 3)}$

9. $\displaystyle\int \frac{dx}{3x \sqrt{2x + 5}}$

10. $\displaystyle\int \frac{dx}{x \sqrt{9 - 4x^2}}$

11. $\displaystyle\int x(2x + 5)^5 \, dx$

12. $\displaystyle\int xe^{5x} \, dx$

13. $\displaystyle\int x^2 e^{4x} \, dx$

14. $\displaystyle\int \ln x \, dx$

15. $\displaystyle\int \frac{dx}{x \ln x}$

16. $\displaystyle\int \frac{e^x}{x^2} \, dx$

17. $\displaystyle\int \frac{dx}{x^2 + 5x + 5}$

18. $\displaystyle\int \frac{dx}{x^2 + 4x + 4}$

19. $\displaystyle\int_0^\infty x^3 e^{-x} \, dx$

20. $\displaystyle\int_0^\infty e^{-2x^2} \, dx$

In Exercises 21–25, use the Table of Integrals and the Fundamental Theorem of Calculus to evaluate each integral.

21. $\displaystyle\int_1^3 x(2x + 3)^2 \, dx$

22. $\displaystyle\int_0^7 3x \sqrt{x + 2} \, dx$

23. $\displaystyle\int_2^4 \frac{dx}{2x + 3}$ (Leave the answer in logarithmic form.)

24. $\displaystyle\int_0^3 xe^{2x} \, dx$ (Leave the answer in exponential form.)

25. $\displaystyle\int_1^2 2x \sqrt{x^2 + 3} \, dx$

4.7   AREA BOUNDED BY A CURVE
—APPROXIMATION BY TRAPEZOIDS

In Section 4.2 we learned to approximate areas by using rectangles and summing the areas of the rectangles. We defined the definite integral as an area, provided the function to be integrated is nonnegative for the entire interval $[a, b]$, where a and b are the limits of integration. The Fundamental Theorem of Calculus enabled us to evaluate certain integrals very readily, and the tedious task of approximation of an area by using successively more rectangles was eliminated. However, it is often necessary to evaluate integrals that cannot be evaluated by using the Fundamental Theorem, or even the techniques presented in later chapters. In these cases, the integral can often be approximated to any degree of accuracy by using the *trapezoidal rule*. This is a method which uses a partitioning of the desired area into trapezoids rather than rectangles, and it reaches a desired degree of accuracy much faster than the rectangular method.

The trapezoidal rule has another advantage: it is very well suited to computer use, and when computer time is available, even integrals that can be analytically evaluated are often estimated by the trapezoidal rule.

Refer to Section 4.2, and look at Figures 4.2, 4.3, and 4.4 to see how a partitioning of an area into rectangles overestimates or underestimates the desired area. Then look at Figure 4.12 to

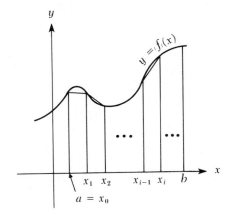

Figure 4.12   Approximation by trapezoids

see how a better approximation can be achieved more rapidly by partitioning an area into trapezoids. The method used is described below.

Partition the interval $[a, b]$ into n equal intervals, and construct trapezoids with parallel sides $f(x_{i-1})$ and $f(x_i)$ (where $i = 1,2,3, \ldots ,n$), and altitude $\Delta x = x_i - x_{i-1}$. Since the area of a trapezoid, A_T, is equal to half the product of the altitude and the sum of the lengths of the parallel sides, the area of the ith trapezoid is

$$A_{T_i} = \frac{1}{2} \Delta x \left[f(x_{i-1}) + f(x_i) \right]$$

The total area of all trapezoids in the region is

$$A_{T_1} + A_{T_2} + A_{T_3} + \cdots + A_{T_n}$$

where

$$A_{T_1} = \frac{1}{2} \Delta x \left[f(x_0) + f(x_1) \right]$$

$$A_{T_2} = \frac{1}{2} \Delta x \, [f(x_1) + f(x_2)]$$

$$A_{T_3} = \frac{1}{2} \Delta x \, [f(x_2) + f(x_3)]$$

$$\cdot$$
$$\cdot$$
$$\cdot$$

$$A_{T_{n-1}} = \frac{1}{2} \Delta x \, [f(x_{n-2}) + f(x_{n-1})]$$

$$A_{T_n} = \frac{1}{2} \Delta x \, [f(x_{n-1}) + f(x_n)]$$

Using the associative, commutative, and distributive properties of the real numbers, we combine these areas to obtain the sum of the areas of these trapezoids,

$$S_T = \frac{1}{2} \Delta x \, [f(x_0) + 2f(x_1) + 2f(x_2) + \cdots$$
$$+ 2f(x_i) + \cdots + 2f(x_{n-1}) + f(x_n)]$$

Since $\Delta x = (b - a)/n$, this yields the formula

$$\text{Total area} \approx S_T = \frac{b - a}{2n} \, [f(x_0) + 2f(x_1) + \cdots$$
$$+ 2f(x_i) + \cdots + 2f(x_{n-1}) + f(x_n)]$$

It is again obvious that as $n \to \infty$ and $\Delta x \to 0$, $S_T \to \int_a^b f(x) \, dx$.

The trapezoidal rule for approximating areas is stated below for easy reference.

Trapezoidal Rule

If f is a nonnegative continuous function on $[a, b]$, then

$$\int_a^b f(x) \, dx \approx \frac{b - a}{2n} \, [f(x_0) + 2f(x_1) + \cdots$$
$$+ 2f(x_i) + \cdots + 2f(x_{n-1}) + f(x_n)]$$
$$= \frac{b - a}{2n} \, [f(x_0) + f(x_n) + 2 \sum_{i=1}^{n-1} f(x_i)]$$

Example 1   Use the approximation by trapezoids to find $\int_0^2 (x^2 + 1)\, dx$, when $n = 4$.

Solution $\Delta x = \dfrac{b - a}{n} = \dfrac{2 - 0}{4} = \dfrac{1}{2}$ $\dfrac{b - a}{2n} = \dfrac{1}{4}$

$$x_0 = 0 \quad f(x_0) = 1$$
$$x_1 = \frac{1}{2} \quad f(x_1) = \frac{5}{4} \quad \text{and} \quad 2f(x_1) = \frac{5}{2}$$
$$x_2 = 1 \quad f(x_2) = 2 \quad \text{and} \quad 2f(x_2) = 4$$
$$x_3 = \frac{3}{2} \quad f(x_3) = \frac{13}{4} \quad \text{and} \quad 2f(x_3) = \frac{13}{2}$$
$$x_4 = 2 \quad f(x_4) = 5$$

$$A \approx S_T = \frac{b - a}{2n} \left[f(x_0) + 2f(x_1) + 2f(x_2) + 2f(x_3) + f(x_4) \right]$$
$$= \frac{1}{4} \left[1 + \frac{5}{2} + 4 + \frac{13}{2} + 5 \right]$$
$$= \frac{19}{4} \text{ square units} = 4.75 \text{ square units}$$

We can easily compute the exact area by using the Fundamental Theorem of Calculus.

$$\int_0^2 (x^2 + 1)\, dx = \frac{x^3}{3} + x \Big]_0^2 = \frac{14}{3} \approx 4.67$$

The approximation by use of four trapezoids is not far off. Figure 4.13 illustrates the example.

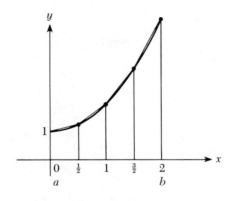

Figure 4.13 $f(x) = x^2 + 1$; four trapezoids

3/26/80

4.7 Exercises

Approximate each of the following integrals by the trapezoidal rule for the given number n. Whenever possible, compare the results by using the Fundamental Theorem of Calculus. (Use the Table of Integrals when necessary.) Be sure to check that the function is nonnegative over the desired interval.

1. $\int_0^4 x^2\, dx, \quad n = 4$ **2.** $\int_2^5 (x^3 - 3x + 4)\, dx, \quad n = 3$

3. $\int_{-2}^1 (2x^2 + x + 3)\, dx, \quad n = 3$ **4.** $\int_1^4 (x^2 + 2)\, dx, \quad n = 6$

5. $\int_3^4 \frac{x^3 + 2}{3}\, dx, \quad n = 4$ **6.** $\int_1^3 (x^3 + 1)\, dx, \quad n = 4$

7. $\int_1^4 x(x^2 - 1)\, dx, \quad n = 3$ **8.** $\int_2^4 \frac{dx}{x}, \quad n = 4$

9. $\int_0^3 \frac{dx}{2x + 1}, \quad n = 6$ **10.** $\int_0^1 \frac{dx}{x^2 + 1}, \quad n = 4$

11. $\int_0^1 (1 - x^2)^{1/2}\, dx, \quad n = 4$ **12.** $\int_0^1 (1 + x^2)^{1/2}\, dx, \quad n = 4$

NEW TERMS

Antiderivative

Definite integral

Fundamental Theorem of
    Calculus

Indefinite integral

Integral

Integrand

Integration

Limits of integration

Mean Value Theorem for
    Integrals

Riemann sum

Sequence

Summation (or sigma) notation

Trapezoidal rule

Variable of integration

REVIEW

In Exercises 1–5, for each integral

a. approximate as a sum of the areas of n rectangles;

b. approximate as a sum of the areas of n trapezoids;

c. use the Fundamental Theorem of Calculus to find the exact numerical answer.

1. $\int_0^2 x^2 \, dx, \quad n = 4$
2. $\int_1^4 (2x + 1) \, dx, \quad n = 6$
3. $\int_0^2 (2x^3 + 3x + 4) \, dx, \quad n = 4$
4. $\int_2^4 (x^2 + x - 3) \, dx, \quad n = 4$
5. $\int_1^3 (x + 1) \, dx, \quad n = 4$

6. Use the definition of $\int_a^b f(x) \, dx$ to find the area between the curve $y = 4 - x^2$ and the x axis.

7. Use the method of Exercise 6 to find the area bounded by the curve $y = x^2 + 1$, the x axis, and the lines $x = 0$ and $x = 3$.

8. For the integral $\int_0^4 (x^2 + 1) \, dx$, find the number c guaranteed by the Mean Value Theorem for integrals such that $\int_0^4 (x^2 + 1) \, dx = f(c) \cdot (b - a)$.

In Exercises 9–23, find the antiderivatives and check your answers by differentiation.

9. $\int (x^2 + 2) \, dx$
10. $\int (3x^5 + 2x^3 - x^2 + 1) \, dx$

11. $\int \dfrac{dx}{x^2}$
12. $\int \dfrac{dx}{\sqrt{x}}$

13. $\int (x + 3)^{3/2} \, dx$
14. $\int (x \sqrt{x} + \sqrt{x} - 3) \, dx$

15. $\int (2x - 1)^{1/3} \, dx$
16. $\int (1 - 2x)^{1/3} \, dx$

17. $\int \dfrac{dx}{(2x - 1)^{1/3}}$
18. $\int \dfrac{dx}{\sqrt{2x - 1}}$

19. $\int \dfrac{x}{\sqrt{x^2 - 1}} \, dx$
20. $\int \dfrac{x}{\sqrt{1 - x^2}} \, dx$

21. $\int (x^3 + 3x + 4)^4 (x^2 + 1) \, dx$
22. $\int \dfrac{3x^2 + 2}{(x^3 + 2x - 4)^5} \, dx$

23. $\int (\sqrt{x} + 1)^2 \, dx$

In Exercises 24–35, use the Fundamental Theorem of Calculus to evaluate each integral.

24. $\int_1^2 (x^2 - 2x) \, dx$
25. $\int_1^3 (x^2 + 2x + 1) \, dx$

26. $\int_0^4 (2x + 3) \, dx$
27. $\int_{-1}^3 (4 - x) \, dx$

28. $\int_{-1}^4 \left(x^2 + 2x - \dfrac{1}{x^2} \right) dx$
29. $\int_{10}^{13} (2x - 7) \, dx$

30. $\displaystyle\int_{-10}^{13} (2 - x + x^2)\, dx$

31. $\displaystyle\int_{-1/2}^{1/2} (x - 2)^2\, dx$

32. $\displaystyle\int_{0}^{5} \frac{dx}{(x + 2)^2}$

33. $\displaystyle\int_{0}^{1} (x^9 + 1)\, dx$

34. $\displaystyle\int_{1}^{2} \frac{1}{\sqrt{x + 3}}\, dx$

35. $\displaystyle\int_{2}^{3} \sqrt{x - 2}\, dx$

In Exercises 36–40, use the Table of Integrals to find the antiderivatives.

36. $\displaystyle\int x \sqrt{2x + 3}\, dx$

37. $\displaystyle\int x^2 e^x\, dx$

38. $\displaystyle\int 3x(4x^2 + 1)^4\, dx$

39. $\displaystyle\int \frac{dx}{x^2 - 6x + 6}$

40. $\displaystyle\int \frac{dx}{3x + 7}$

The Trigonometric Functions

The trigonometric functions have important uses in calculus. In fact, certain parts of integral calculus are dependent upon trigonometry. Take another look at the Table of Integrals and notice how many integrals involve trigonometric functions. In preparation for dealing with derivatives and integrals of functions which are either trigonometric themselves, or result in trigonometric functions, we present a short review of trigonometry. Following the review Sections 5.1 and 5.2, we will discuss limits involving trigonometric functions, derivatives of trigonometric and inverse trigonometric functions, and integrations of trigonometric functions and functions whose integrals are trigonometric.

5.1 TRIGONOMETRIC FUNCTIONS

Consider a circle with radius r and center at the origin of a set of coordinate axes. Let P be any point on the circle. Figure 5.1 shows an angle of measure α in standard position with vertex at the center of a circle of radius r and initial side along the positive x axis. Point P has coordinates (x, y). The six trigonometric functions of α are defined below.

Definition

$$\sin \alpha = \frac{y}{r} \qquad \cos \alpha = \frac{x}{r} \qquad \tan \alpha = \frac{y}{x}$$

$$\csc \alpha = \frac{r}{y} \qquad \sec \alpha = \frac{r}{x} \qquad \cot \alpha = \frac{x}{y}$$

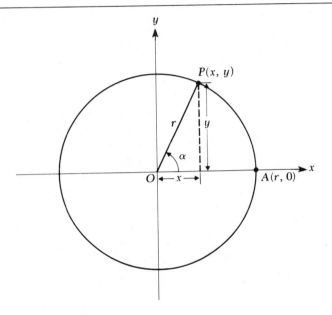

Figure 5.1   An angle in standard position

Example 1   List the values of the six trigonometric functions for $\alpha = 0$.

Solution   An angle in standard position and of measure 0 would have both initial and terminal sides on the x axis. In Figure 5.1 the terminal side is at point $A(r, 0)$. Therefore, we substitute $x = r$ and $y = 0$ in the definitions of the trigonometric functions:

$$\sin \alpha = \frac{y}{r} = \frac{0}{r} = 0 \qquad \cos \alpha = \frac{x}{r} = \frac{r}{r} = 1$$

$$\tan \alpha = \frac{y}{x} = \frac{0}{r} = 0 \qquad \csc \alpha = \frac{r}{y} = \frac{r}{0} \quad \text{(undefined)}$$

$$\sec \alpha = \frac{r}{x} = \frac{r}{r} = 1 \qquad \cot \alpha = \frac{x}{y} = \frac{r}{0} \quad \text{(undefined)}$$

Both radians and degrees will be used to measure angles.

Definition

An angle has a measure of one *radian* if and only if, when its vertex is placed at the center of a circle, the angle subtends an arc whose length is equal to the radius of the circle.

The measure of angle AOB in Figure 5.2 is 1 radian because it subtends an arc of length r, the radius of the circle. The circumference of the circle is $C = 2\pi r$. Thus 2π or about 6.28

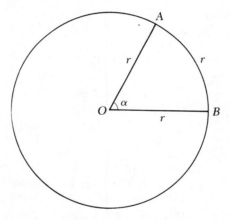

Figure 5.2 Angle α measures 1 radian

radii can be placed on the circumference of a circle. This means that an angle that is a complete rotation measures 2π radians. The degree measure of a complete rotation is 360°. Therefore,

$$2\pi \text{ radians} = 360° \qquad \text{or} \qquad \pi \text{ radians} = 180°$$

which implies that

$$1 \text{ radian} = \frac{180°}{\pi} \approx 57.2958°$$

and

$$1° = \frac{\pi}{180} \text{ radian} \approx 0.01745 \text{ radian}$$

When angles are measured in radians, the word "radians" will not be written. An angle of measure $\pi/3$ will be understood to be measured in radians.

Example 2 What is the degree measure of an angle of measure $\pi/3$?

Solution Since $\pi = 180°$, $\pi/3 = 180°/3 = 60°$.

Example 3 Change 90° to an equivalent number of radians.

Solution Since $1° = \pi/180$ then 90° will equal 90 times $\pi/180$. That is,

$$90° = 90 \cdot \frac{\pi}{180} = \frac{\pi}{2}$$

Some of the most often used degree-radian equivalences are:

$$30° = \frac{\pi}{6}, \quad 45° = \frac{\pi}{4}, \quad 60° = \frac{\pi}{3}, \quad 90° = \frac{\pi}{2},$$

$$180° = \pi, \quad 270° = \frac{3\pi}{2}, \quad 360° = 2\pi$$

The trigonometric functions for $\pi/6$, $\pi/4$, and $\pi/3$ occur often enough that it is necessary to be able to recall them. Figure 5.3 shows a 30°—60°—90° triangle with two different orientations, and a 45°—45°—90° triangle. From plane geometry it can be shown that no matter what the size of triangles similar to these, the sides will be in the ratios shown in the figure. Using these triangles and the definitions of the trigonometric functions, any function of $\pi/6$ or $\pi/4$ or $\pi/3$ is easily found.

Example 4 Find $\tan(\pi/6)$, $\cos(\pi/3)$, and $\sec(\pi/4)$.

Solution Think of the triangle in Figure 5.3a as having the vertex of the $\pi/6$ radian angle at the origin of a rectangular coordinate system with $x = \sqrt{3}$, $y = 1$, and $r = 2$. Then

$$\tan \frac{\pi}{6} = \tan 30° = \frac{y}{x} = \frac{1}{\sqrt{3}}$$

Similarly for the triangles 5.3b and 5.3c

$$\cos \frac{\pi}{3} = \cos 60° = \frac{x}{r} = \frac{1}{2}$$

$$\sec \frac{\pi}{4} = \sec 45° = \frac{r}{x} = \frac{\sqrt{2}}{1} = \sqrt{2}$$

The triangles of Figure 5.3 can also be used to find the values of trigonometric functions of angles greater than 90°.

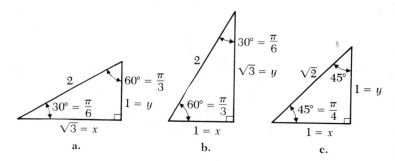

Figure 5.3

Example 5   Find $\cos (3\pi/4)$.

Solution   Figure 5.4a shows an angle of measure $3\pi/4$ in standard position. Angle BOP is 45°. We let the radius of the circle be $\sqrt{2}$. Then

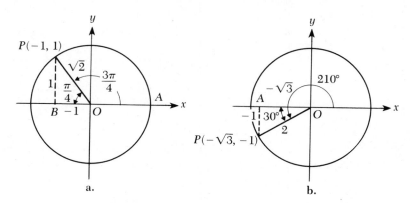

Figure 5.4

the other two sides of triangle BOP are both 1 unit in length and point P has coordinates $(-1, 1)$. Thus

$$\cos \frac{3\pi}{4} = \frac{x}{r} = \frac{-1}{\sqrt{2}}$$

Example 6   Find sin 210°.

Solution   In Figure 5.4b note that the terminal side of a 210° angle makes a 30° angle with the negative portion of the x axis. Choose 2 for the radius of the circle and assign the other two sides the ratios for a 30°—60°—90° triangle, with appropriate signs. Point P has coordinates $(-\sqrt{3}, -1)$. Therefore.

$$\sin 210° = \frac{y}{r} = \frac{-1}{2} = -\frac{1}{2}$$

The trigonometric functions for other numbers are found in Table V of the Appendix. The table contains angles in degrees from 1° to 90°. The radian equivalent is given next to each degree. Use the trigonometric functions at the top of the table for angles from 0° to 45°. The functions along the bottom of the table are for angles from 45° to 90°.

Example 7   Find cot 40° and sec 64°.

Solution   The value of cot 40° is found opposite 40° in the table and under the column headed by cot: cot 40° = 1.192. The value of sec 64° is found opposite 64° in the table in the column that has sec at its base: sec 64° = 2.281.

The trigonometric functions of angles greater than 90° can be found by using Table V, the concept of *reference angle*, and Figure 5.5. The figure shows which functions are positive in each quadrant. For instance in the second quadrant, the sine and cosecant functions are positive, while the cosine, tangent, secant, and cotangent are all negative.

Definition

If angle α is in standard position, then the least angle that the terminal side of α makes with the x axis is called the *reference angle* of α.

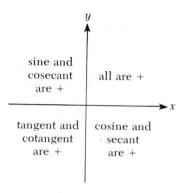

Figure 5.5   The algebraic sign
of the trigonometric
functions

Example 8   Find sin 345°.

Solution   In Figure 5.6a, the reference angle of 345° is shown to be 15°. The
numerical value of the sine of any angle will equal that of its
reference angle. The algebraic sign of the sine function will

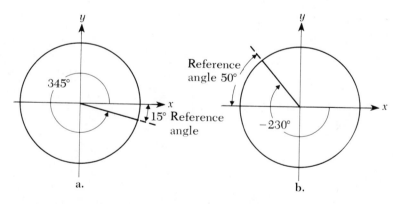

a.   b.

Figure 5.6   Reference angles

depend upon which quadrant contains the terminal side of the
angle, as shown in Figure 5.5. The terminal side of 345° is in
quadrant IV, where the sine function is negative. Therefore,

$$\sin 345° = -\sin 15° = -0.259$$

Negative angles are created when the terminal side of an
angle in standard position is rotated in a clockwise direction.

Example 9   Find $\sin(-230°)$.

Solution   An angle of $-230°$ is shown in Figure 5.6b, along with its reference angle. In the second quadrant the sine function is positive. Therefore,

$$\sin(-230°) = +\sin 50° = 0.766$$

The table below is created by choosing values for x and then solving for the corresponding value for y in $y = \sin x$. The decimal values for the sines of the special numbers were found using Table V.

x	0	$\frac{\pi}{6}$	$\frac{\pi}{4}$	$\frac{\pi}{3}$	$\frac{\pi}{2}$	$\frac{2\pi}{3}$	$\frac{3\pi}{4}$	$\frac{5\pi}{6}$	π
y	0	0.5	0.707	0.866	1	0.866	0.707	0.5	0

x	$\frac{7\pi}{6}$	$\frac{5\pi}{4}$	$\frac{4\pi}{3}$	$\frac{3\pi}{2}$	$\frac{8\pi}{3}$	$\frac{7\pi}{4}$	$\frac{11\pi}{6}$	2π
y	-0.5	-0.707	-0.866	-1	-0.866	-0.707	-0.5	0

When the points are plotted and connected with a smooth curve the result is the graph of $y = \sin x$ in Figure 5.7. The figure

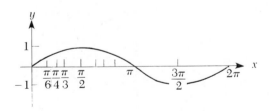

Figure 5.7   Graph of $y = \sin x$

represents a single period of the graph of the sine function. The entire graph of the sine function is made up of endless repetitions—in both directions—of the period shown.

The following table is offered as a handy reference for the values of the trigonometric functions of commonly used angle measurements.

| x | | | | | | | |
Degrees	Radians	sin x	cos x	tan x	sec x	csc x	cot x
0	0	0	1	0	1	—	—
30	$\pi/6$	1/2	$\sqrt{3}/2$	$1/\sqrt{3}$	$2/\sqrt{3}$	2	$\sqrt{3}$
45	$\pi/4$	$1/\sqrt{2}$	$1/\sqrt{2}$	1	$\sqrt{2}$	$\sqrt{2}$	1
60	$\pi/3$	$\sqrt{3}/2$	1/2	$\sqrt{3}$	2	$2/\sqrt{3}$	$1/\sqrt{3}$
90	$\pi/2$	1	0	—	—	1	0
180	π	0	−1	0	−1	—	—
270	$3\pi/2$	−1	0	—	—	−1	0

The graphs of cos x, tan x, sec x, csc x, and cot x are shown in Figures 5.8–5.12.

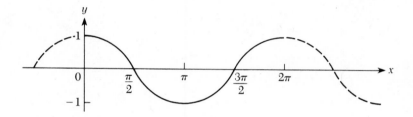

Figure 5.8 Graph of $y = \cos x$

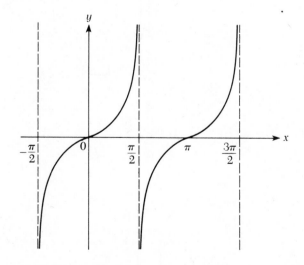

Figure 5.9 Graph of $y = \tan x$

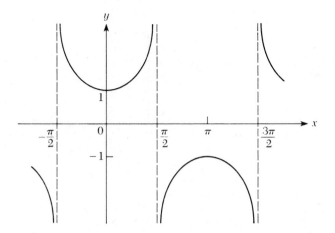

Figure 5.10   Graph of $y = \sec x$

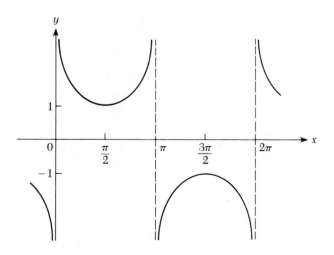

Figure 5.11   Graph of $y = \csc x$

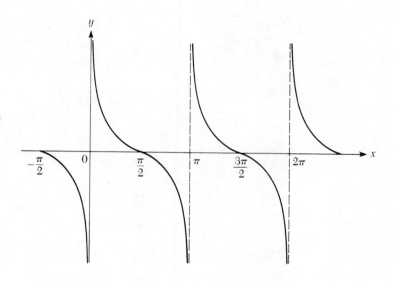

Figure 5.12 Graph of $y = \cot x$

5.1 EXERCISES

In Exercises 1–10, change to an equivalent degree measure.

1. $\dfrac{3\pi}{2}$ 2. $\dfrac{5\pi}{4}$ 3. $\dfrac{2\pi}{3}$

4. $\dfrac{3\pi}{4}$ 5. $\dfrac{7\pi}{6}$ 6. 3π

7. $-\dfrac{\pi}{2}$ 8. $-\dfrac{3\pi}{2}$ 9. $-\dfrac{7\pi}{3}$

10. $-\dfrac{9\pi}{8}$

In Exercises 11–18, change to an equivalent number of radians.

11. $270°$ 12. $210°$ 13. $315°$

14. $300°$ 15. $-120°$ 16. $-135°$

17. $-750°$ 18. $-3000°$

In Exercises 19–46, evaluate without using a table.

19. $\sin \dfrac{\pi}{3}$ **20.** $\cos \dfrac{\pi}{3}$ **21.** $\cot \dfrac{\pi}{3}$

22. $\sec \dfrac{\pi}{4}$ **23.** $\tan \dfrac{\pi}{4}$ **24.** $\csc \dfrac{\pi}{6}$

25. $\cot \dfrac{\pi}{6}$ **26.** $\sin \dfrac{\pi}{6}$ **27.** $\sec \dfrac{2\pi}{3}$

28. $\tan \left(-\dfrac{3\pi}{4} \right)$ **29.** $\cot \left(-\dfrac{3\pi}{4} \right)$ **30.** $\sin \left(-\dfrac{2\pi}{3} \right)$

31. $\sin 0$ **32.** $\sec 0$ **33.** $\tan \dfrac{\pi}{2}$

34. $\cot 0$ **35.** $\csc 0$ **36.** $\cot \pi$

37. $\csc \pi$ **38.** $\cot \left(-\dfrac{\pi}{2} \right)$ **39.** $\tan \left(-\dfrac{3\pi}{2} \right)$

40. $\cot 6\pi$ **41.** $\sec (-8\pi)$ **42.** $\tan (-10\pi)$

43. $\sin \left(\dfrac{-7\pi}{6} \right)$ **44.** $\sin \left(\dfrac{-11\pi}{6} \right)$ **45.** $\cos \left(-\dfrac{23\pi}{6} \right)$

46. $\cos \left(-\dfrac{35\pi}{6} \right)$

In Exercises 47–62, evaluate either by using Table V in the Appendix, or by using a calculator.

47. $\sec 35°$ **48.** $\tan 18°$ **49.** $\sec 80°$

50. $\sin 64°$ **51.** $\cot 61°$ **52.** $\csc 47°$

53. $\sin 92°$ **54.** $\cos 131°$ **55.** $\tan 175°$

56. $\csc 227°$ **57.** $\csc 264°$ **58.** $\tan 290°$

59. $\sin 305°$ **60.** $\sin 352°$ **61.** $\tan 400°$

62. $\csc 539°$

5.2 TRIGONOMETRIC IDENTITIES

The equations encountered in algebra are, for the most part, conditional. Some values of the variable satisfy the equation, others do not. For instance, $3x + 6 = 0$ is true only for $x = -2$. The trigonometric equations of this section are identities; that is, equations that are true for *all* values of the variable for which the trigonometric functions involved in the equation are defined. These identities are used to simplify or change the form of trigonometric expressions. In integral calculus we will see that certain trigonometric forms are more easily manipulated than others.

The following are called *fundamental identities*:

1. $\sin \alpha = \dfrac{1}{\csc \alpha}$ **5.** $\csc \alpha = \dfrac{1}{\sin \alpha}$

2. $\cos \alpha = \dfrac{1}{\sec \alpha}$ **6.** $\sec \alpha = \dfrac{1}{\cos \alpha}$

3. $\tan \alpha = \dfrac{1}{\cot \alpha}$ **7.** $\cot \alpha = \dfrac{1}{\tan \alpha}$

4. $\tan \alpha = \dfrac{\sin \alpha}{\cos \alpha}$ **8.** $\cot \alpha = \dfrac{\cos \alpha}{\sin \alpha}$

The fundamental identities are proved using the definitions given in the last section for the six trigonometric functions. To prove identity 4, $\tan \alpha = \sin \alpha / \cos \alpha$, we refer to Figure 5.1.

$$\frac{\sin \alpha}{\cos \alpha} = \frac{y/r}{x/r} = \frac{y}{r} \cdot \frac{r}{x} = \frac{y}{x} = \tan \alpha$$

The fundamental identities can be used to graph $y = \csc x$ by relating $\csc x$ to $\sin x$. From identity 5 we have $y = \csc x = 1/\sin x$. That is, the cosecant is the reciprocal of the sine. The dotted line in Figure 5.13 represents the graph of $y = \sin x$. The cosecant is

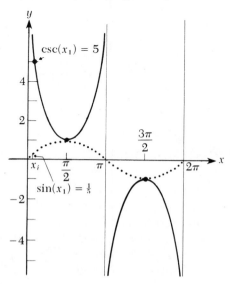

Figure 5.13   Graph of $y = \csc x$

graphed by estimating reciprocal values from the graph of sine. In the figure note the position of x_1. The corresponding y value for sine is shown in the figure as sin (x_1) and is equal to $\frac{1}{5}$. The cosecant value corresponding to x_1 is the reciprocal of the sine value or csc $(x_1) = 1/\frac{1}{5} = 5$, as shown in the figure. Other points on the graph of $y = $ csc x can be found by estimating the reciprocals of the sine values. The lines $x = \pi$ and $x = 2\pi$ are asymptotes of the graph. At $x = \pi$ and $x = 2\pi$ the sine is zero and the reciprocal of zero is undefined.

The following identities are sometimes called the *Pythagorean identities*. They are used to simplify or change given identities to more manageable forms.

9. $\sin^2 \alpha + \cos^2 \alpha = 1$

10. $1 + \tan^2 \alpha = \sec^2 \alpha$

11. $\cot^2 \alpha + 1 = \csc^2 \alpha$

To prove 9, note in Figure 5.1 that a circle of radius r has equation $x^2 + y^2 = r^2$. Dividing by r^2 gives

$$\frac{x^2}{r^2} + \frac{y^2}{r^2} = 1$$

which is equivalent to

$$\left(\frac{x}{r}\right)^2 + \left(\frac{y}{r}\right)^2 = 1$$

but cos $\alpha = x/r$ and sin $\alpha = y/r$, so

$$(\cos \alpha)^2 + (\sin \alpha)^2 = 1$$
$$\sin^2 \alpha + \cos^2 \alpha = 1$$

Identities 10 and 11 are similarly proved.

Example 1 will show the proof of an identity that is not a special case, as are the identities so far mentioned. The technique of proof will be to alter one side of the supposed identity by substituting from the fundamental and Pythagorean identities and simplifying algebraically, until the altered side is shown to be equivalent to the unaltered side. There are other logically valid techniques for proving identities, but the one described uses the same techniques required to apply calculus to trigonometry.

Example 1   Prove that $\sin^4 x - \cos^4 x = 1 - 2 \cos^2 x$ is an identity.

Solution   One side of the equation will be shown equivalent to the other. We choose to work with the left side. A good general practice when proving identities is to work with the more algebraically complicated side—in this case, the left side. The left side of the identity, $\sin^4 x - \cos^4 x$, can be factored:

$$\sin^4 x - \cos^4 x = (\sin^2 x - \cos^2 x)(\sin^2 x + \cos^2 x)$$

but $\sin^2 x + \cos^2 x = 1$, so

$$\sin^4 x - \cos^4 x = (\sin^2 x - \cos^2 x)1$$

Solving identity 9 for $\sin^2 x$ gives $\sin^2 x = 1 - \cos^2 x$. Substituting above:

$$\begin{aligned} &= (1 - \cos^2 x) - \cos^2 x \\ &= 1 - \cos^2 x - \cos^2 x \\ &= 1 - 2 \cos^2 x \end{aligned}$$

Therefore, the left side of the given equation is shown to be equivalent to the right side, and the identity is proved.

Example 2   Prove that $\sin A + \sin B = \sin A \sin B(\csc A + \csc B)$ is an identity.

Solution   We will alter the right side of the equation:

$$\sin A \sin B(\csc A + \csc B) = \sin A \sin B \csc A + \sin A \sin B \csc B$$
$$= \sin A \sin B \cdot \frac{1}{\sin A} + \sin A \sin B \cdot \frac{1}{\sin B}$$
$$= \sin B + \sin A = \sin A + \sin B$$

The next two identities to be reviewed are:

12. $\sin (A + B) = \sin A \cos B + \cos A \sin B$

13. $\cos (A + B) = \cos A \cos B - \sin A \sin B$

Example 3   Use identity 12 to evaluate $\sin\left(\dfrac{\pi}{6} + \dfrac{\pi}{3}\right)$.

Solution
$$\sin\left(\frac{\pi}{6} + \frac{\pi}{3}\right) = \sin\frac{\pi}{6}\cos\frac{\pi}{3} + \cos\frac{\pi}{6}\sin\frac{\pi}{3}$$
$$= \frac{1}{2}\cdot\frac{1}{2} + \frac{\sqrt{3}}{2}\cdot\frac{\sqrt{3}}{2}$$
$$= \frac{1}{4} + \frac{3}{4} = 1$$

The answer 1 is correct since $\sin\left(\dfrac{\pi}{6} + \dfrac{\pi}{3}\right) = \sin\dfrac{\pi}{2} = 1.$

Example 4   Evaluate $\cos 20° \cos 25° - \sin 20° \sin 25°.$

Solution   From identity 13 we have

$$\cos 20° \cos 25° - \sin 20° \sin 25° = \cos(20° + 25°)$$
$$= \cos 45°$$
$$= \frac{1}{\sqrt{2}}$$

It is important to notice that identities 12 and 13 lead to two useful identities when $A = B$.

$$\sin(A + A) = \sin 2A = \sin A \cos A + \sin A \cos A = 2\sin A \cos A$$

or simply

$$\sin 2A = 2\sin A \cos A$$

Similarly, when $A = B$,

$$\cos(A + A) = \cos 2A = \cos A \cos A - \sin A \sin A$$
$$= \cos^2 A - \sin^2 A$$

Restating these two results,

14. $\sin 2A = 2\sin A \cos A$
15. $\cos 2A = \cos^2 A - \sin^2 A$

5.2 EXERCISES

In Exercises 1–8, use a circle and the definitions of the trigonometric functions to prove each of the following, and state which values of α must be excluded $(0 \leqslant \alpha \leqslant 2\pi)$.

1. $\tan \alpha \cot \alpha = 1$ **2.** $\sin \alpha \csc \alpha = 1$

3. $\sin \alpha = \dfrac{1}{\csc \alpha}$ **4.** $\cos \alpha = \dfrac{1}{\sec \alpha}$

5. $\cot \alpha = \dfrac{\cos \alpha}{\sin \alpha}$ **6.** $\sec \alpha = \dfrac{1}{\cos \alpha}$

7. $1 + \tan^2 \alpha = \sec^2 \alpha$ **8.** $\cot^2 \alpha + 1 = \csc^2 \alpha$

In Exercises 9–21, use identity 12 or 13 to evaluate:

9. $\cos \dfrac{7\pi}{12}$ **10.** $\cos \dfrac{5\pi}{12}$

11. $\sin \dfrac{7\pi}{12}$ **12.** $\sin \dfrac{5\pi}{12}$

13. $\cos \left(\theta + \dfrac{\pi}{2}\right)$ **14.** $\sin \left(\theta + \dfrac{\pi}{2}\right)$

15. $\sin (\theta + \pi)$ **16.** $\cos (\theta + \pi)$

17. $\tan (\theta + \pi)$

18. $\sin 20° \cos 10° + \cos 20° \sin 10°$

19. $\sin 20° \cos 40° + \cos 20° \sin 40°$

20. $\cos 50° \cos 40° - \sin 50° \sin 40°$

21. $\cos 22° \cos 23° - \sin 22° \sin 23°$

In Exercises 22–32, use identities 1–15 given in this section to prove:

22. $\sin x \cdot \sec x = \tan x$

23. $\cos^2 x - \sin^2 x = 1 - 2 \sin^2 x$

24. $\sec^2 x + \csc^2 x = \sec^2 x \cdot \csc^2 x$

25. $(\sin x + \cos x)^2 = 1 + 2 \sin x \cos x$

26. $\dfrac{\sin x + \cos x}{\cos x} = 1 + \tan x$

27. $1 - 2 \sin^2 x = \dfrac{1 - \tan^2 x}{1 + \tan^2 x}$

28. $\tan (A + B) = \dfrac{\tan A + \tan B}{1 - \tan A \tan B}$

29. $\cot (A + B) = \dfrac{\cot A \cot B - 1}{\cot A + \cot B}$

30. $\dfrac{\sin 2x + \sin x}{\cos 2x + \cos x + 1} = \tan x$

31. $\cos^2 x = \dfrac{\cos 2x + 1}{2}$ **32.** $\sin^2 x = \dfrac{1 - \cos 2x}{2}$

5.3 LIMITS OF TRIGONOMETRIC FUNCTIONS

Since trigonometric functions are important to the solution of many problems in calculus, it is essential to consider their limits and whether the limits exist.

If a trigonometric function f is defined at a, then

$$\lim_{x \to a} f(x) = f(a)$$

Example 1

$$\lim_{x \to \pi} \sin x = \sin \pi = 0$$

$$\lim_{x \to \pi/4} \tan x = \tan \frac{\pi}{4} = 1$$

But what happens to the limits of trigonometric functions when the functions are not defined at a? Recall that $\tan x$ does not exist for $x = \pi/2$; neither does $\lim_{x \to \pi/2} \tan x$ exist. Figure 5.14 shows the graph of $f(x) = \tan x$ over the interval $[-\pi, \pi]$. Notice that as values of x are taken closer and closer to $\pi/2$ but *less than*

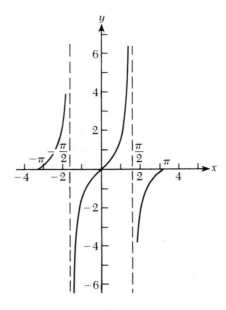

Figure 5.14 Graph of $y = \tan x$

$\pi/2$, $f(x)$ takes on successively larger positive values. That is to say, as x approaches $\pi/2$ from the *left*, tan x becomes unboundedly large. Similarly, as x approaches $\pi/2$ from the *right*, tan x takes on negative values which become unboundedly large in absolute value. Symbolically,

$$\lim_{x \to \pi/2^-} \tan x = \infty$$

$$\lim_{x \to \pi/2^+} \tan x = -\infty$$

By referring to the graphs of sec x, csc x, and cot x, in Figures 5.10, 5.11, and 5.12, respectively, the limits of these functions may be determined for specific values in their respective domains.

It is significant to note that sin x and cos x are continuous for all values of x, and thus

$$\lim_{x \to a} \sin x = \sin a$$

$$\lim_{x \to a} \cos x = \cos a$$

A very important limit is

$$\lim_{\theta \to 0} \frac{\sin \theta}{\theta}$$

It is easily seen that substituting 0 for θ in the expression $(\sin \theta)/\theta$ results in 0/0, an undefined expression. But it turns out that this *limit* actually exists and equals 1. To prove

$$\lim_{\theta \to 0} \frac{\sin \theta}{\theta} = 1$$

we will first show that

$$\lim_{\theta \to 0^+} \frac{\sin \theta}{\theta} = 1$$

and then that

$$\lim_{\theta \to 0^-} \frac{\sin \theta}{\theta} = 1$$

Figure 5.15 is a circle with center at O and radius r. The central angle θ is such that $0 < \theta < \pi/2$. Line segment $\overline{BC}$ is tangent

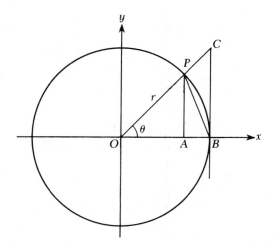

Figure 5.15

to the circle at B; $\overline{PA}$ is perpendicular to $\overline{OB}$. The area of $\triangle OBC$ is greater than the area of sector OPB, which is greater than the area of $\triangle OBP$. Since $\tan\theta = \overline{BC}/\overline{OB} = \overline{BC}/r$, the area of

$$\triangle OBC = \frac{1}{2}\,\overline{OB}\cdot\overline{BC} = \frac{1}{2}\,r\cdot r\tan\theta = \frac{1}{2}r^2\tan\theta$$

The area of sector

$$OPB = \frac{\theta}{2\pi}\cdot(\text{Area of circle}) = \frac{\theta}{2\pi}\cdot\pi r^2 = \frac{\theta r^2}{2}$$

The area of

$$\triangle OBP = \frac{1}{2}\,\overline{OB}\cdot\overline{PA} = \frac{1}{2}\,r\cdot r\sin\theta = \frac{1}{2}r^2\sin\theta$$

Therefore,

$$\frac{1}{2}r^2\tan\theta > \frac{\theta r^2}{2} > \frac{1}{2}r^2\sin\theta$$

Multiplying all three terms by $2/r^2$ gives

$$\tan\theta > \theta > \sin\theta$$

Since $\tan \theta = (\sin \theta)/(\cos \theta)$,

$$\frac{\sin \theta}{\cos \theta} > \theta > \sin \theta$$

Multiplying by $1/(\sin \theta)$ yields

$$\frac{1}{\cos \theta} > \frac{\theta}{\sin \theta} > 1$$

From algebra we know that if a and b are positive numbers and $a > b$, then $1/a < 1/b$. Since $1/(\cos \theta)$, $\theta/(\sin \theta)$, and 1 are positive numbers for $0 < \theta < \pi/2$, taking their reciprocals results in

$$\cos \theta < \frac{\sin \theta}{\theta} < 1$$

The $\lim_{\theta \to 0^+} [(\sin \theta)/\theta]$ is between $\lim_{\theta \to 0^+} \cos \theta$ and $\lim_{x \to 0^+} 1$. However both the latter limits are 1. So:

$$\lim_{\theta \to 0^+} \frac{\sin \theta}{\theta} = 1$$

By taking θ as a negative angle between 0 and $-\pi/2$, the same method as above can be used to show

$$\lim_{\theta \to 0^-} \frac{\sin \theta}{\theta} = 1$$

Since the right- and left-hand limits are equal,

$$\lim_{\theta \to 0} \frac{\sin \theta}{\theta} = 1$$

The name of the variable does not affect the limit so $\lim_{x \to 0} [(\sin x)/x] = 1$, and $\lim_{y \to 0} [(\sin 3y)/3y] = 1$.

We know that $\lim_{y \to 0} [(\sin 3y)/3y] = 1$, because as y approaches 0 so does $3y$.

Recalling the theorem stating that $\lim_{x \to a} k \cdot f(x) = k \lim_{x \to a} f(x)$ for k constant, and applying this theorem, the following limits can be found.

Example 2   Find $\lim\limits_{x \to 0} \dfrac{\sin 2x}{x}$.

Solution $\lim\limits_{x \to 0} \dfrac{\sin 2x}{x} = \lim\limits_{x \to 0} \dfrac{2 \sin 2x}{2x} = 2 \lim\limits_{x \to 0} \dfrac{\sin 2x}{2x} = 2(1) = 2$

A second important limit is

$$\lim_{\theta \to 0} \frac{1 - \cos \theta}{\theta} = 0$$

Proof

$$\lim_{\theta \to 0} \frac{1 - \cos \theta}{\theta} = \lim_{\theta \to 0} \frac{1 - \cos \theta}{\theta} \cdot \frac{1 + \cos \theta}{1 + \cos \theta}$$

$$= \lim_{\theta \to 0} \frac{1 - \cos^2 \theta}{\theta(1 + \cos \theta)}$$

$$= \lim_{\theta \to 0} \frac{\sin^2 \theta}{\theta(1 + \cos \theta)}$$

$$= \lim_{\theta \to 0} \frac{\sin \theta}{\theta} \cdot \frac{\sin \theta}{1 + \cos \theta}$$

$$= \lim_{\theta \to 0} \frac{\sin \theta}{\theta} \cdot \lim_{\theta \to 0} \frac{\sin \theta}{1 + \cos \theta}$$

$$= (1) \frac{0}{1 + 1} = (1)0 = 0$$

Example 3   Find $\lim\limits_{\theta \to 0} \dfrac{\sin \theta(1 - \cos \theta)}{\theta^2}$.

Solution $\lim\limits_{\theta \to 0} \dfrac{\sin \theta(1 - \cos \theta)}{\theta^2} = \lim\limits_{\theta \to 0} \dfrac{\sin \theta}{\theta} \lim\limits_{\theta \to 0} \dfrac{1 - \cos \theta}{\theta}$

$$= 1 \cdot 0 = 0$$

Example 4   Find $\lim\limits_{x \to 0} \dfrac{2x^2 \cot^2 x}{x \csc x}$.

Solution

$$\lim_{x \to 0} \frac{2x^2 \cot^2 x}{x \csc x} = 2 \lim_{x \to 0} x \cdot \frac{1}{\csc x} \cdot \cot^2 x$$

$$= 2 \lim_{x \to 0} x \cdot \sin x \cdot \frac{\cos^2 x}{\sin^2 x}$$

$$= 2 \lim_{x \to 0} x \frac{\cos^2 x}{\sin x}$$

$$= 2 \lim_{x \to 0} \frac{x}{\sin x} \cdot \cos^2 x$$

$$= 2 \lim_{x \to 0} \frac{1}{(\sin x)/x} \cdot \cos^2 x$$

$$= 2 \left(\frac{1}{1}\right) 1^2 = 2$$

The $\lim_{x \to \infty} \sin x$ does not exist because the sine function will take on the value of each real number between -1 and 1 for each 2π interval of its domain and approaches none of them. However, $\lim_{x \to \infty} [(\sin x)/x] = 0$, because for any number C in the range of sine, $C \lim_{x \to \infty} (1/x) = 0$.

Example 5   Find $\lim_{x \to \infty} (3x + \cos x)$.

Solution   As x increases in value so does $3x$; however, $\cos x$ is bounded between -1 and 1. Thus the sum must increase. Therefore,

$$\lim_{x \to \infty} (3x + \cos x) = \infty$$

To summarize some of the important limits of this section:

$$\lim_{x \to a} \sin x = \sin a$$

$$\lim_{x \to a} \cos x = \cos a$$

$$\lim_{x \to \pi/2^-} \tan x = \infty$$

$$\lim_{x \to \pi/2^+} \tan x = -\infty$$

$$\lim_{x \to 0} \frac{\sin x}{x} = 1$$

$$\lim_{x \to 0} \frac{1 - \cos x}{x} = 0$$

5.3 EXERCISES

In Exercises 1–30, find the indicated limits.

1. $\lim_{x \to \pi/4} 2 \tan x$

2. $\lim_{\theta \to \pi/3} \frac{\sin \theta}{\theta}$

3. $\displaystyle\lim_{y\to\pi} \frac{1-\cos y}{y}$

4. $\displaystyle\lim_{x\to\pi/6} \tan x \cot x$

5. $\displaystyle\lim_{x\to\pi/2^+} \sec x$

6. $\displaystyle\lim_{x\to2\pi^-} \csc x$

7. $\displaystyle\lim_{x\to0^+} \cot x$

8. $\displaystyle\lim_{x\to3\pi/4^+} \sin x$

9. $\displaystyle\lim_{x\to\infty} x\left(1-\cos\frac{1}{x}\right)$

10. $\displaystyle\lim_{x\to\infty} (x+\sin x)$

11. $\displaystyle\lim_{x\to\infty} x\sin\left(\frac{1}{x}\right)$

12. $\displaystyle\lim_{x\to\infty} (3x+\sin x)$

13. $\displaystyle\lim_{x\to0} \frac{1+\cos x}{x}$

14. $\displaystyle\lim_{x\to0} \frac{2x^2\tan^2 x}{x\sec x}$

15. $\displaystyle\lim_{x\to0} \frac{\sin 3x}{x}$

16. $\displaystyle\lim_{x\to0} \frac{\sin 5x}{8x}$

17. $\displaystyle\lim_{x\to0} \frac{\sin nx}{x}$, $\ n$ constant $n \neq 0$

18. $\displaystyle\lim_{x\to0} \frac{x}{\sin x}$

19. $\displaystyle\lim_{x\to0} \frac{x}{1-\cos x}$

20. $\displaystyle\lim_{x\to0} \frac{\sin^2 x}{x}$

21. $\displaystyle\lim_{x\to0} \frac{\sin^2 x}{x^2}$

22. $\displaystyle\lim_{x\to0} \frac{\sin x}{\tan x}$

23. $\displaystyle\lim_{x\to0} \frac{\tan x}{x}$

24. $\displaystyle\lim_{x\to0} \frac{-2x}{\tan 3x}$

25. $\displaystyle\lim_{x\to0} \sin 2x \cot x$

26. $\displaystyle\lim_{x\to0} \frac{\cot ax}{\cot bx}$, $\ a$ and b constants, $ab \neq 0$

27. $\displaystyle\lim_{x\to0} \frac{1-\cos x}{\sin x}$

28. $\displaystyle\lim_{x\to0} \frac{1-\cos x}{\tan x}$

29. $\displaystyle\lim_{x\to0} \frac{\sec x - \tan x}{\sin x}$

30. $\displaystyle\lim_{x\to\pi/2} \frac{\cos x}{(\pi/2) - x}$

5.4   THE DERIVATIVES OF sin u AND cos u

To find the derivative of $\sin x$, we can make use of the definition of the derivative of $f(x)$ and two facts about limits that were established earlier:

$$\lim_{\Delta x\to0} \frac{\sin \Delta x}{\Delta x} = 1 \qquad \text{and} \qquad \lim_{\Delta x\to0} \frac{\cos \Delta x - 1}{\Delta x} = 0$$

If $y = \sin x$, then by definition of $f'(x)$,

$$y' = \lim_{\Delta x \to 0} \frac{\sin (x + \Delta x) - \sin x}{\Delta x}$$

Applying the trigonometric identity for $\sin (A + B)$,

$$\sin (A + B) = \sin A \cos B + \cos A \sin B$$

$$y' = \lim_{\Delta x \to 0} \frac{\sin x \cos \Delta x + \sin \Delta x \cos x - \sin x}{\Delta x}$$

$$= \lim_{\Delta x \to 0} \left[\frac{\sin x \cos \Delta x - \sin x}{\Delta x} + \frac{\sin \Delta x \cos x}{\Delta x} \right]$$

$$= \lim_{\Delta x \to 0} \left[\sin x \left(\frac{\cos \Delta x - 1}{\Delta x} \right) + \left(\frac{\sin \Delta x}{\Delta x} \right) \cos x \right]$$

Since

$$\lim_{\Delta x \to 0} \frac{\cos \Delta x - 1}{\Delta x} = 0 \qquad \text{and} \qquad \lim_{\Delta x \to 0} \frac{\sin \Delta x}{\Delta x} = 1$$

$$y' = (\sin x)0 + 1(\cos x) = \cos x$$

Therefore if $y = \sin x$, $y' = \cos x$. If u is a differentiable function of x, then, by the chain rule,

$$D_x \sin u = \cos u \, \frac{du}{dx}$$

We could also find the derivative of $\cos x$ by application of the definition, but instead we will make use of the chain rule. From trigonometry we have the identity

$$\cos x = \sin \left(\frac{\pi}{2} - x \right)$$

Therefore to find $D_x \cos x$, we take the derivative of $\sin u$, where $u = (\pi/2) - x$.

$$D_x \cos x = D_x \sin \left(\frac{\pi}{2} - x \right)$$

$$= \cos \left(\frac{\pi}{2} - x \right) D_x \left(\frac{\pi}{2} - x \right)$$

$$= \cos \left(\frac{\pi}{2} - x \right)(-1)$$

Since $\cos\left(\dfrac{\pi}{2} - x\right) = \sin x$,

$$D_x \cos x = (\sin x)(-1) = -\sin x$$

If u is a differentiable function of x, then the chain rule applies, and

$$D_x \cos u = -\sin u \, \frac{du}{dx}$$

Example 1   If $y = \cos 2x$, then

$$y' = D_x \cos 2x = -\sin 2x \cdot D_x(2x) = -2 \sin 2x$$

Example 2   Find y' for $y = \cos 2x \sin x$.

Solution   The function y is the product of two functions: $\cos 2x$ and $\sin x$. Therefore, the formula for the derivative of the product of two functions applies. The derivative y' will be the first function, $\cos 2x$, times the derivative of $\sin x$ plus the second function, $\sin x$, times the derivative of $\cos 2x$.

$$\begin{aligned} y' &= \cos 2x \cdot D_x \sin x + \sin x \cdot D_x \cos 2x \\ &= \cos 2x \cdot \cos x + \sin x(-2 \sin 2x) \\ &= \cos 2x \cos x - 2 \sin x \sin 2x \end{aligned}$$

Example 3   Find $f'(x)$ when $f(x) = \sin^2 x$.

Solution   The function $f(x)$ is of the form $f(x) = u^2$, where $u = \sin x$. By the chain rule,

$$\begin{aligned} f'(x) &= 2uD_x u = 2 \sin x \cdot D_x \sin x \\ &= 2 \sin x \cos x = \sin 2x \end{aligned}$$

Higher-order derivatives of the sine and cosine functions have an interesting cyclic property.

Let

$$y \quad = \sin x$$

then

$$\begin{aligned} y' \quad &= \cos x \\ y'' \quad &= -\sin x \end{aligned}$$

$$y^{[3]} = -\cos x$$
$$y^{[4]} = \sin x$$

From this pattern it can be seen that the fifth derivative of sin x is the same as the first derivative, the sixth derivative is the same as the second derivative, and so on. A similar cycle is true for successive derivatives of $y = \cos x$.

Let

$$y \quad = \cos x$$

then

$$y' \quad = -\sin x$$
$$y'' \quad = -\cos x$$
$$y^{[3]} = \sin x$$
$$y^{[4]} = \cos x$$

Higher derivatives of the remaining four trigonometric functions unfortunately do not have this cyclic property.

5.4   EXERCISES

In Exercises 1–24, find the derivative of y with respect to x or t.

1. $y = \sin 5x$
2. $y = \sin(-3x)$
3. $y = \cos 4x$
4. $y = \cos(-2x)$
5. $y = 3 \sin t$
6. $y = -2 \sin 2t$
7. $y = -4 \cos 6t$
8. $y = -2 \cos(-t)$
9. $y = \sin x - \cos x$
10. $y = 4 \cos 3x + 5 \sin 2x$
11. $y = \cos^2 x$
12. $y = \cos x^2$
13. $y = x \cos x^2$
14. $y = x^2 \cos^2 x$
15. $y = \cos x \sin x$
16. $y = \cos^2 x \sin x$
17. $y = \sqrt{\sin x}, \quad 0 < x \leq \dfrac{\pi}{2}$
18. $y = \sqrt[3]{\cos x}$
19. $y = \dfrac{\sin x}{x}$
20. $y = \dfrac{1 + \cos x}{\sin x}$
21. $y = \dfrac{1 - \sin t}{1 + \sin t}$
22. $y = \dfrac{\sin x + \cos x}{\sin x - \cos x}$
23. $y = \cos^2 \sqrt{x}$
24. $y = \sin^2 \sqrt{x^2 + 1}$

25. If the number of grams of bacteria in a population after t hours of growth is given by $y = t^2 \sin t$, find the rate of growth of the population at $t = \pi$ hours.

26. Find the growth rate of the population in Exercise 25 at 6.28 hours.

27. If the income of a company is $(x^3 \cos 2x)10^3$ dollars in x years, find the rate of change of income in 3.14 years.

28. Find the rate of change of income of a company in 2 years, if income in x years is $(x^3 \sin 3x)10^3$ dollars.

29. If $y = \sin 2x$, find y''.

30. If $y = \cos 3x$, find y''.

31. If $y = x \cos x$, find the third derivative of y with respect to x.

32. If $y = \sin (1/x)$, find y''.

33. Find the slope of the tangent line to the graph of $y = \sin x$ at $x = \pi/2$, and write an equation of the tangent line.

34. Find the slope of the tangent line to the graph of $y = \cos 2x$ when $x = \pi/3$. Write an equation of the tangent line.

35. For what value(s) of x does the function $y = \sin 3x$ have maximum or minimum values in the interval $[0, 2\pi]$?

36. For what value(s) of x does the function $y = \cos 4x$ have maximum or minimum values in the interval $[-\pi/2, \pi/2]$?

5.5 DERIVATIVES OF tan u, cot u, sec u, AND csc u

The derivatives of $\tan x$, $\cot x$, $\sec x$, and $\csc x$ are found by writing each in terms of $\sin x$ and $\cos x$, and then applying the formula for the derivative of the quotient of two functions or the chain rule.

If $y = \tan x$, then

$$y' = D_x \tan x = D_x \left(\frac{\sin x}{\cos x} \right)$$

Applying the quotient formula,

$$y' = \frac{\cos x(D_x \sin x) - \sin x(D_x \cos x)}{\cos^2 x}$$
$$= \frac{\cos x(\cos x) - \sin x(-\sin x)}{\cos^2 x}$$
$$= \frac{\cos^2 x + \sin^2 x}{\cos^2 x}$$

But $\cos^2 x + \sin^2 x = 1$; hence

$$y' = \frac{1}{\cos^2 x} = \sec^2 x$$

Therefore,

$$D_x \tan x = \sec^2 x$$

If $y = \cot x$, then y' can be found as follows:

$$
\begin{aligned}
D_x \cot x &= D_x \left(\frac{\cos x}{\sin x} \right) \\
&= \frac{\sin x D_x \cos x - \cos x D_x \sin x}{\sin^2 x} \\
&= \frac{\sin x(-\sin x) - \cos x(\cos x)}{\sin^2 x} \\
&= \frac{-\sin^2 x - \cos^2 x}{\sin^2 x} \\
&= \frac{-(\sin^2 x + \cos^2 x)}{\sin^2 x} \\
&= \frac{-1}{\sin^2 x} \\
&= -\csc^2 x
\end{aligned}
$$

To find the derivative of sec x, write it as $1/(\cos x)$ and use the quotient formula.

$$
\begin{aligned}
D_x \sec x &= D_x \left(\frac{1}{\cos x} \right) = \frac{(\cos x)(0) - 1(-\sin x)}{\cos^2 x} \\
&= \frac{\sin x}{\cos^2 x} = \frac{1}{\cos x} \cdot \frac{\sin x}{\cos x} \\
&= \sec x \tan x
\end{aligned}
$$

The derivative for csc x is established in the same way as that for sec x.

$$
\begin{aligned}
D_x \csc x &= D_x \left(\frac{1}{\sin x} \right) = \frac{(\sin x)(0) - 1 \cos x}{\sin^2 x} \\
&= \frac{-\cos x}{\sin^2 x} = \frac{-1}{\sin x} \cdot \frac{\cos x}{\sin x} \\
&= -\csc x \cot x
\end{aligned}
$$

If u is a differentiable function of x, these four derivatives can be stated:

$$D_x \tan u = \sec^2 u \cdot D_x u$$
$$D_x \cot u = -\csc^2 u \cdot D_x u$$
$$D_x \sec u = \sec u \tan u \cdot D_x u$$
$$D_x \csc u = -\csc u \cot u \cdot D_x u$$

Example 1   Find y' for $y = \tan 3x$.

Solution   We use the formula for the derivative of $\tan u$, where $u = 3x$.

$$y' = \sec^2 3x \cdot D_x 3x$$
$$= (\sec^2 3x)3 = 3 \sec^2 3x$$

Example 2   Find y' for $y = \sec^2 x$.

Solution   The function y is in the form u^2 where $u = \sec x$

$$y' = D_x \sec^2 x = 2 \sec x \cdot D_x \sec x$$
$$= 2 \sec x \cdot \sec x \tan x = 2 \sec^2 x \tan x$$

Example 3   Find y' for $y = x^2 \cot 2x$.

Solution   We use the formula for the derivative of the product of two functions.

$$y' = x^2 \cdot D_x(\cot 2x) + \cot 2x \, D_x(x^2)$$
$$= x^2(-2 \csc^2 2x) + 2x \cot 2x$$
$$= -2x^2 \csc^2 2x + 2x \cot 2x$$

In Section 5.4 we discussed the cyclic nature of successive derivatives of the sine and cosine functions. The following example illustrates that this cyclic property does not hold for the secant function, and the exercises will provide an opportunity for finding out that second and third derivatives of the remaining trigonometric functions can be quite messy.

Example 4   Find $D_x^2 y$ when $y = \sec x$.

Solution
$$D_x y = \sec x \tan x$$
$$D_x^2 y = \sec x \cdot D_x(\tan x) + \tan x \cdot D_x(\sec x)$$
$$= \sec x \cdot \sec^2 x + \tan x \cdot \sec x \tan x$$
$$= \sec^3 x + \sec x \tan^2 x$$

5.5 EXERCISES

In Exercises 1–20, differentiate each function.

1. $y = \sec 3x$

2. $y = 4 \sec 2x$

3. $y = 3 \tan 2x$

4. $y = \tan 9x$

5. $y = \cot x^2$

6. $y = \cot \sqrt{x}$

7. $y = -4 \csc x^2$

8. $y = x \csc x$

9. $f(t) = \tan^2 t$

10. $f(t) = \sqrt{\tan t}$

11. $f(t) = 3t^2 + 7t + \sec t^2$

12. $f(t) = \sec (t^3 + 5t)$

13. $f(t) = t^2 \csc \sqrt{t}$

14. $f(x) = x \cot^3 (x^2 + 5)$

15. $f(t) = \tan (\sin t)$

16. $f(t) = \sin (\sec t)$

17. $f(x) = \dfrac{\sin 2x}{\tan x}$

18. $f(x) = \dfrac{\cot 2x}{\sec 3x}$

19. $f(x) = \dfrac{\tan x - 1}{\tan x + 1}$

20. $f(x) = \dfrac{\sec x - \tan x}{\sec x + \tan x}$

In Exercises 21–30, find $D_x^2 y$ for each function.

21. $y = \tan x$

22. $y = \cot x$

23. $y = \sec x$

24. $y = \csc x$

25. $y = x^2 \tan x$

26. $y = \tan^2 3x$

27. $y = \dfrac{1 + \tan^2 x}{2}$

28. $y = \sec^2 x$

29. $y = x \cot 3x$

30. $y = 1 + \tan x$

31. Find the slope of the tangent line to the graph of $y = \tan x$ at $x = \pi/4$.

32. Find the slope of the tangent line to the graph of $y = \sec x$ at $x = \pi/3$.

33. Find the slope of the tangent line to the graph of $y = x \tan 2x$ at $x = \pi/2$.

34. Find the slope of the tangent line to the graph of $y = 1/(1 + \csc x)$ at $x = \pi/3$.

35. Find the third derivative of $y = \tan x$ with respect to x.

36. If $f(x) = \cot x$, find $f^{[3]}(x)$.

5.6   INVERSE TRIGONOMETRIC FUNCTIONS

In Chapter 1 the concept of the inverse of a function was introduced. We shall now apply this concept to finding the inverses of the trigonometric functions.

The inverse of the sine function $y = \sin x$ is

$$x = \sin y$$

We have formed this inverse by interchanging x and y in the equation $y = \sin x$. Up to this point we have, after interchanging x and y, solved for y. We cannot algebraically solve $x = \sin y$ for y. However, we will adopt a notation that allows the defining equation of the inverse to have y as a function of x, its domain element. It is important to note that $x = \sin y$ is not a function of x unless we restrict the domain, for if $x = 1$ in $x = \sin y$, y could be $\pi/2$, $5\pi/2$, or $-3\pi/2$. Any value of x can be paired with any number of values for y. To form the inverse function we must restrict the domain of sine. The restricted domain must be such that each value of x is paired with one and only one value of y, so that when we interchange x and y, the inverse will be a function, and there must be a domain element for each y such that $-1 \leqslant y \leqslant 1$. Figure 5.16 shows the graph of $y = \sin x$ and a restricted interval of x, where the interval

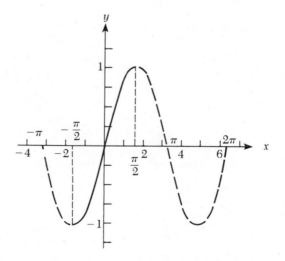

Figure 5.16   Graph of $y = \sin x$ with the principal value interval shown between $-\pi/2$ and $\pi/2$

$[-\pi/2, \pi/2]$ is called the *principal value domain* for the sine function.

We can now define Sine (spelled with a capital S) so that it will have an inverse function sine^{-1}. If

$$y = \text{Sin } x \qquad \text{and} \qquad x \in \left[-\frac{\pi}{2}, \frac{\pi}{2}\right]$$

then the inverse of the Sine function is

$$x = \text{Sin } y \qquad \text{and} \qquad y \in \left[-\frac{\pi}{2}, \frac{\pi}{2}\right]$$

The graph of sine^{-1} is shown in Figure 5.17. Note that it is a function. It was drawn by using the fact that the graph of a function and its inverse is symmetric about the line $y = x$.

We now introduce a notation that allows the defining equation of the inverse of the Sine function to have y as a function of x.

Definition

$$y = \sin^{-1} x \text{ if and only if } x = \text{Sin } y.$$

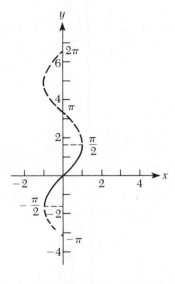

Figure 5.17 The graph of sine^{-1} is shown as a solid line

The expression $y = \sin^{-1} x$ can be read "y equals the inverse sine of x" or "y equals the arc sine of x" or "y is the angle whose Sine is x." The last phrase, "y is the angle whose Sine is x," gives the intuitive meaning to $y = \sin^{-1} x$. When asked to evaluate $\sin^{-1}$ for certain values of x, it is sometimes useful to recall that y is now taking the place of the "angle."

Example 1   Evaluate $\sin^{-1}(1/\sqrt{2})$.

Solution   By definition $y = \sin^{-1} x$ if and only if $x = \text{Sin } y$. Then, if we let $y = \sin^{-1}(1/\sqrt{2})$, we have $1/\sqrt{2} = \text{Sin } y$. This means, "$y$ is the number whose Sine is $1/\sqrt{2}$." Since we are restricted to $y \in [-\pi/2, \pi/2]$, $y = \pi/4$. Therefore,

$$\sin^{-1} \frac{1}{\sqrt{2}} = \frac{\pi}{4}$$

Example 2   Evaluate $\sin^{-1}(-1)$.

Solution   Many students answer incorrectly, $3\pi/2$. Remember, the $\sin^{-1}$ is defined only for $-\pi/2 \leqslant y \leqslant \pi/2$. Therefore,

$$\sin^{-1}(-1) = -\frac{\pi}{2}$$

Example 3   Evaluate $\sin^{-1} \frac{1}{2}$ and $\sin^{-1}(-\frac{1}{2})$.

Solution $\sin^{-1}\left(-\frac{1}{2}\right) = -\frac{\pi}{6}$ and $\sin^{-1} \frac{1}{2} = \frac{\pi}{6}$

The same method as that used in developing $\sin^{-1}$ is used with the other trigonometric functions. Figures 5.18 and 5.19 show the graphs of the inverse cosine and tangent functions. These were graphed by graphing the standard cosine and tangent functions, then "reflecting" these graphs across the line $y = x$.

Using capital letters to indicate functions with restricted domains, we define

$$y = \text{Cos } x \qquad \text{and} \qquad x \in [0, \pi]$$

$$y = \text{Tan } x \qquad \text{and} \qquad x \in \left(-\frac{\pi}{2}, \frac{\pi}{2}\right)$$

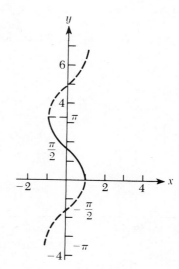

Figure 5.18 The graph of cosine⁻¹ is
shown as a solid line

Now the inverse functions can be defined.

$$\text{cosine}^{-1}: \quad x = \text{Cos } y \qquad \text{and} \qquad y \in [0, \pi]$$

$$\text{tangent}^{-1}: \quad x = \text{Tan } y \qquad \text{and} \qquad y \in \left(-\frac{\pi}{2}, \frac{\pi}{2}\right)$$

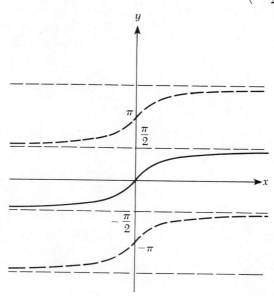

Figure 5.19 The graph of tangent⁻¹ is
shown as a solid line

Definition

$$y = \cos^{-1} x \quad \text{if and only if} \quad x = \text{Cos } y$$
$$y = \tan^{-1} x \quad \text{if and only if} \quad x = \text{Tan } y$$

The capital "C" in $x = \text{Cos } y$ and the capital "T" in $x = \text{Tan } y$ indicate the cosine and tangent functions have restricted domains. Thus, $y = \cos^{-1} x$ is a function with domain $[-1, 1]$ and with range $[0, \pi]$, whereas $y = \tan^{-1} x$ has domain $(-\infty, \infty)$ and range $(-\pi/2, \pi/2)$.

Example 4   Evaluate $\cos^{-1} 1$, $\cos^{-1} (-1)$, $\cos^{-1} (\sqrt{3}/2)$, $\cos^{-1} (-\sqrt{3}/2)$, $\tan^{-1} 1$, and $\tan^{-1} (-1/\sqrt{3})$.

Solution

$$\cos^{-1} 1 = 0 \qquad\qquad \cos^{-1} (-1) = \pi$$
$$\cos^{-1} \frac{\sqrt{3}}{2} = \frac{\pi}{6} \qquad \cos^{-1} \left(-\frac{\sqrt{3}}{2}\right) = \frac{5\pi}{6}$$
$$\tan^{-1} 1 = \frac{\pi}{4} \qquad \tan^{-1} \left(-\frac{1}{\sqrt{3}}\right) = -\frac{\pi}{6}$$

Following the Established Pattern

1. $y = \cot^{-1} x$ Domain   all real numbers
    Range $y \in (0, \pi)$

2. $y = \sec^{-1} x$ Domain $x \leqslant -1$ or $x \geqslant 1$
    Range $y \in [0, \pi]$ and $y \neq \frac{\pi}{2}$

3. $y = \csc^{-1} x$ Domain $x \leqslant -1$ or $x \geqslant 1$
    Range $y \in \left[-\frac{\pi}{2}, \frac{\pi}{2}\right]$ and $y \neq 0$

Example 5   Evaluate $\cot^{-1} 1$, $\cot^{-1} (-1)$, $\csc^{-1} \sqrt{2}$, $\csc^{-1} (-\sqrt{2})$, $\sec^{-1} 2$, and $\sec^{-1} (-2)$.

Solution

$$\cot^{-1} 1 = \frac{\pi}{4} \qquad\qquad \cot^{-1} (-1) = \frac{3\pi}{4}$$
$$\csc^{-1} \sqrt{2} = \frac{\pi}{4} \qquad \csc^{-1} (-\sqrt{2}) = -\frac{\pi}{4}$$
$$\sec^{-1} 2 = \frac{\pi}{3} \qquad \sec^{-1} (-2) = \frac{2\pi}{3}$$

Example 6	Find $\sin\left[\cos^{-1}\left(\sqrt{3}/2\right)\right]$.
Solution	This asks for the sine of the number whose Cosine is $\sqrt{3}/2$. Now, $\cos^{-1}\left(\sqrt{3}/2\right) = \pi/6$. Therefore,

$$\sin\left(\cos^{-1}\frac{\sqrt{3}}{2}\right) = \sin\frac{\pi}{6} = \frac{1}{2}$$

5.6 EXERCISES

In Exercises 1–20, evaluate without using a table.

1. $\cos^{-1}\dfrac{1}{2}$ **2.** $\sin^{-1}\left(-\dfrac{1}{2}\right)$

3. $\sin^{-1} 1$ **4.** $\sin^{-1} 0$

5. $\sin^{-1}\left(-\dfrac{\sqrt{3}}{2}\right)$ **6.** $\cos^{-1}\left(-\dfrac{\sqrt{3}}{2}\right)$

7. $\cos^{-1}\left(-\dfrac{1}{\sqrt{2}}\right)$ **8.** $\tan^{-1}(-1)$

9. $\tan^{-1} 0$ **10.** $\sec^{-1}\dfrac{2}{\sqrt{3}}$

11. $\csc^{-1}(-2)$ **12.** $\cot^{-1}(-\sqrt{3})$

13. $\sin\left(\cos^{-1}\dfrac{\sqrt{3}}{2}\right)$ **14.** $\cos\left(\sin^{-1}\dfrac{\sqrt{3}}{2}\right)$

15. $\tan(\sin^{-1} 1)$ **16.** $\cot(\cos^{-1} 1)$

17. $\sin^{-1}\left(\sin\dfrac{\pi}{2}\right)$ **18.** $\cos^{-1}\left(\cos\dfrac{\pi}{4}\right)$

19. $\tan(\tan^{-1}(-1))$ **20.** $\cot(\cot^{-1} 1)$

In Exercises 21–30, use Table V in the Appendix to evaluate in radian measure.

21. $\sin^{-1} 0.629$ **22.** $\cos^{-1} 0.799$

23. $\tan^{-1} 1.235$ **24.** $\cot^{-1} 0.231$

25. $\sec^{-1} 1.942$ **26.** $\csc^{-1} 1.040$

27. $\sin^{-1} 0.875$ **28.** $\tan^{-1} 19.08$

29. $\sec^{-1} 1.010$ **30.** $\csc^{-1} 1.942$

In Exercises 31–36, for what values of x is each true?

31. $\sin (\sin^{-1} x) = x$ **32.** $\sin^{-1} (\sin x) = x$

33. $\tan (\tan^{-1} x) = x$ **34.** $\cos^{-1} (\cos x) = x$

35. $\sec (\sec^{-1} x) = x$ **36.** $\cot^{-1} (\cot x) = x$

37. If $\tan^{-1} (3/4) = x$, find $\sin x$, $\cos x$, $\sec x$, $\tan x$, $\cot x$, and $\csc x$. A triangle correctly labeled will help in finding these values.

38. If $\sin^{-1} (2/3) = x$, find $\tan x$, $\cot x$, and $\sec x$.

5.7 DERIVATIVES OF THE INVERSE TRIGONOMETRIC FUNCTIONS

The derivatives of the inverse trigonometric functions are found by using the definition of these functions and implicit differentiation.

We first find $D_x(\sin^{-1} x)$. If $y = \sin^{-1} x$, then by definition $x = \text{Sin } y$. Taking the derivative of both sides of $x = \text{Sin } y$ with respect to x,

$$1 = \cos y \, \frac{dy}{dx}$$

$$\frac{dy}{dx} = \frac{1}{\cos y}$$

We want dy/dx as a function of x rather than y. From the identity $\cos^2 y + \sin^2 y = 1$, we have $\cos y = \pm\sqrt{1 - \sin^2 y}$. But $y = \sin^{-1} x$ requires $y \in [-\pi/2, \pi/2]$ and for these values of y, $\cos y$ is positive. Therefore, $\cos y = \sqrt{1 - \sin^2 y}$. Since $x = \sin y$, we substitute x^2 for $\sin^2 y$ and have $\cos y = \sqrt{1 - x^2}$. Thus,

$$\frac{dy}{dx} = \frac{1}{\cos y} = \frac{1}{\sqrt{1 - x^2}}$$

$$D_x(\sin^{-1} x) = \frac{1}{\sqrt{1 - x^2}}$$

We now find $D_x(\tan^{-1} x)$. If $y = \tan^{-1} x$, then by definition $x = \text{Tan } y$. Taking the derivative of both sides of $x = \text{Tan } y$ with respect to x gives

$$1 = \sec^2 y \, \frac{dy}{dx}$$

$$\frac{dy}{dx} = \frac{1}{\sec^2 y}$$

But $\sec^2 y = 1 + \tan^2 y$. Since $x = \tan y$, we can substitute x^2 for $\tan^2 y$ and $\sec^2 y = 1 + x^2$. Therefore,

$$\frac{dy}{dx} = \frac{1}{\sec^2 y} = \frac{1}{1 + x^2}$$

$$D_x(\tan^{-1} x) = \frac{1}{1 + x^2}$$

To find $D_x(\sec^{-1} x)$, we note that if $y = \sec^{-1} x$, then $x = \text{Sec } y$. Taking the derivative of both sides of $x = \text{Sec } y$ gives

$$1 = \sec y \tan y \, \frac{dy}{dx}$$

$$\frac{dy}{dx} = \frac{1}{\sec y \tan y}$$

In the above equation we wish to replace $\sec y$ and $\tan y$ with equivalent functions of x. $\text{Sec } y$ is easy, since it equals x. From the identity $1 + \tan^2 y = \sec^2 y$, we have $\tan y = \pm\sqrt{\sec^2 y - 1}$, and substituting x^2 for $\sec^2 y$ gives $\tan y = \pm\sqrt{x^2 - 1}$. Then

$$\frac{dy}{dx} = \frac{1}{\sec y \tan y} = \frac{1}{x(\pm\sqrt{x^2 - 1})}$$

In order to decide about the ambiguous $\pm$ sign, note that it depends upon $\tan y$. For $y = \sec^{-1} x$, x must be greater than 1 or less than -1. When x is positive $(x > 1)$, $0 < y < \pi/2$, and $\tan y$ is positive. In this case x is positive and the positive part of $\pm\sqrt{x^2 - 1}$ applies. When x is negative $(x < -1)$, $\pi/2 < y < \pi$, and $\tan y$ is negative. In this case, x is negative, and so is $\sqrt{x^2 - 1}$. Therefore, the product $x\sqrt{x^2 - 1}$ is always positive. To show this we use absolute value symbols.

$$D_x(\sec^{-1} x) = \frac{1}{|x|\sqrt{x^2 - 1}}$$

The derivatives of the three remaining inverse trigonometric functions are found in the same way. Their proofs are left as exercises. Below are the derivatives of the six inverse trigonometric functions of u, where u is a function of x.

$$D_x(\sin^{-1} u) = \frac{1}{\sqrt{1 - u^2}} \frac{du}{dx} \qquad D_x(\cos^{-1} u) = \frac{-1}{\sqrt{1 - u^2}} \frac{du}{dx}$$

$$D_x(\tan^{-1} u) = \frac{1}{1 + u^2} \frac{du}{dx} \qquad D_x(\cot^{-1} u) = \frac{-1}{1 + u^2} \frac{du}{dx}$$

$$D_x(\sec^{-1} u) = \frac{1}{|u| \sqrt{u^2 - 1}} \frac{du}{dx} \qquad D_x(\csc^{-1} u) = \frac{-1}{|u| \sqrt{u^2 - 1}} \frac{du}{dx}$$

Example 1   Find $D_x(\sin^{-1} 3x)$.

Solution

$$D_x(\sin^{-1} 3x) = \frac{1}{\sqrt{1 - (3x)^2}} D_x(3x)$$

$$= \frac{3}{\sqrt{1 - 9x^2}}$$

Example 2   Find y' for $y = x \tan^{-1} x$.

Solution   Treat $x \tan^{-1} x$ as any other product of two functions of x.

$$y' = x \left(\frac{1}{1 + x^2} \right) + (\tan^{-1} x)(1)$$

$$= \frac{x}{1 + x^2} + \tan^{-1} x$$

5.7   EXERCISES

1. Prove $D_x(\cos^{-1} x) = \frac{-1}{\sqrt{1 - x^2}}$.

2. Prove $D_x(\cot^{-1} x) = \frac{-1}{1 + x^2}$.

3. Prove $D_x(\csc^{-1} x) = \frac{-1}{|x| \sqrt{x^2 - 1}}$.

In Exercises 4–13, find y'.

4. $y = \sin^{-1} 5x$

5. $y = \sin^{-1} x^2$

6. $y = \tan^{-1} (3x^2 + 2x)$

7. $y = \tan^{-1} (3x - 4x^3)$

8. $y = x \cot^{-1} x$ **9.** $x^2 \cot^{-1} 2x$

10. $y = \sec^{-1} (\cos x)$ **11.** $y = \sec^{-1} (\sec x)$

12. $y = [\sin^{-1} x]^2$ **13.** $y = [4 \tan^{-1} x^2]^2$

In Exercises 14–19, find y″.

14. $y = \sin^{-1} x$ **15.** $y = \cos^{-1} x$

16. $y = \tan^{-1} x$ **17.** $y = \cot^{-1} x$

18. $y = \sec^{-1} x$ **19.** $y = \csc^{-1} x$

20. Find the slope of the tangent line to $y = \sin^{-1} x$ when $x = \frac{1}{2}$.

21. Find the slope of the tangent line to $y = \tan^{-1} x$ when $x = \sqrt{3}$.

22. A 6-foot high picture is placed on a wall with its lower edge 2 feet above the level of an observer's eye. How far from the wall should the observer stand so that the angle of vision will be maximum?

23. A lighthouse is located 1 mile from point P on the shore; P is the nearest point on the shore to the lighthouse and the shoreline is perfectly straight. The beacon in the lighthouse makes two revolutions per minute. It throws a beam of light that moves along the shoreline as the beacon revolves. How fast is the spot of light moving on the shoreline when it is 3 miles from point P?

5.8 INTEGRATION—TRIGONOMETRIC FUNCTIONS

Integrals involving trigonometric functions are important for the nonengineering student chiefly as a tool for integrating other functions. Some of these applications will be considered in Chapter 6. We start with the antiderivatives of the six trigonometric functions, which are listed below and can easily be verified by differentiation.

$$\int \sin x \, dx = -\cos x + C \qquad \int \cos x \, dx = \sin x + C$$
$$\int \sec^2 x \, dx = \tan x + C \qquad \int \sec x \tan x \, dx = \sec x + C$$
$$\int \csc^2 x \, dx = -\cot x + C \qquad \int \csc x \cot x \, dx = -\csc x + C$$

From previous experience, it is evident, for example, that if u is a function of x, then

$$\int \sin u \, du = -\cos u + C$$

Example 1   Find $\int \sin 3x \, dx$.

Solution $\int \sin 3x \, dx = \frac{1}{3} \int \sin 3x(3) \, dx = -\frac{1}{3} \cos 3x + C$

(Note that we let $u = 3x$; then $du = 3 \, dx$.)

Example 2   Find $\int \sec^2 5\theta \, d\theta$.

Solution $\int \sec^2 5\theta \, d\theta = \frac{1}{5} \int \sec^2 5\theta(5) \, d\theta = \frac{1}{5} \tan 5\theta + C$

The Fundamental Theorem of Calculus applies to evaluating definite integrals which satisfy the conditions of the theorem.

Example 3   Find $\int_0^{\pi/3} \cos x \, dx$.

Solution   The Fundamental Theorem of Calculus applies, because $\cos x$ is defined and continuous on the closed interval $[0, \pi/3]$. An antiderivative of $\cos x$ is $\sin x$.

$$\int_0^{\pi/3} \cos x \, dx = \sin x \Big]_0^{\pi/3} = \sin \frac{\pi}{3} - \sin 0 = \frac{\sqrt{3}}{2}$$

Example 4   Evaluate $\int_0^{\pi} \sin \theta \, d\theta$.

Solution   An antiderivative of $\sin \theta$ is $-\cos \theta$.

$$\int_0^{\pi} \sin \theta \, d\theta = -\cos \theta \Big]_0^{\pi} = (-\cos \pi) - (-\cos 0) = 2$$

The following integrals result in inverse trigonometric functions. Assume that u is a differentiable function of x. The integrals can easily be checked by differentiating the inverse trigonometric functions.

$$\int \frac{du}{\sqrt{1 - u^2}} = \sin^{-1} u + C \qquad \int \frac{-du}{\sqrt{1 - u^2}} = \cos^{-1} u + C$$

$$\int \frac{du}{1 + u^2} = \tan^{-1} u + C \qquad \int \frac{-du}{1 + u^2} = \cot^{-1} u + C$$

$$\int \frac{du}{|u| \sqrt{u^2 - 1}} = \sec^{-1} u + C \qquad \int \frac{-du}{|u| \sqrt{u^2 - 1}} = \csc^{-1} u + C$$

Example 5   Find $\displaystyle\int \frac{dx}{\sqrt{1 - 9x^2}}$.

Solution $\displaystyle\int \frac{dx}{\sqrt{1 - 9x^2}} = \frac{1}{3} \int \frac{du}{\sqrt{1 - u^2}}$

where $u = 3x$ and $du = 3\ dx$. We multiply and divide the first integral by 3, placing the $\frac{1}{3}$ outside the integral. Then

$$\frac{1}{3} \int \frac{du}{\sqrt{1 - u^2}} = \frac{1}{3} \sin^{-1} u + C = \frac{1}{3} \sin^{-1} 3x + C$$

Example 6   Find $\displaystyle\int \frac{dx}{9 + x^2}$.

Solution $\displaystyle\int \frac{dx}{9 + x^2} = \int \frac{dx}{9\left(1 + \dfrac{x^2}{9}\right)} = \frac{1}{9} \int \frac{dx}{1 + \left(\dfrac{x}{3}\right)^2}$

Let $u = x/3$; then $du = \frac{1}{3} dx$. Multiply and divide by $\frac{1}{3}$, and obtain

$$\frac{1}{3} \int \frac{du}{1 + u^2} = \frac{1}{3} \tan^{-1} u + C$$
$$= \frac{1}{3} \tan^{-1} \left(\frac{x}{3}\right) + C$$

Since the situation in Example 6 arises frequently, that is, instead of $1 + u^2$, or $1 - u^2$ in the integrand of a problem whose antiderivative is an inverse trigonometric function, we find $a^2 + u^2$, or $a^2 - u^2$, $a \neq 1$, the following formulas are useful and timesaving.

$$\int \frac{du}{\sqrt{a^2 - u^2}} = \sin^{-1}\left(\frac{u}{a}\right) + C \qquad \int \frac{-du}{\sqrt{a^2 - u^2}} = \cos^{-1}\left(\frac{u}{a}\right) + C$$

$$\int \frac{du}{a^2 + u^2} = \frac{1}{a} \tan^{-1}\left(\frac{u}{a}\right) + C \qquad \int \frac{du}{|u|\sqrt{u^2 - a^2}} = \frac{1}{a} \sec^{-1}\left(\frac{u}{a}\right) + C$$

Example 7   Find $\displaystyle\int \frac{dx}{\sqrt{9 - 4x^2}}$.

Solution   This form results in an inverse sine function. We have $a^2 = 9$ and

$u^2 = 4x^2$; thus $a = 3$, $u = 2x$, and $du = 2\,dx$, or $dx = \frac{1}{2}\,du$. By making the appropriate substitutions.

$$\int \frac{dx}{\sqrt{9 - 4x^2}} = \frac{1}{2}\int \frac{du}{\sqrt{a^2 - u^2}} = \frac{1}{2}\sin^{-1}\left(\frac{u}{a}\right) + C$$

$$= \frac{1}{2}\sin^{-1}\left(\frac{2x}{3}\right) + C$$

Example 8   Find $\displaystyle\int \frac{dx}{x\,\sqrt{4x^2 - 9}}$, $\quad x > \dfrac{3}{2}$.

Solution $\displaystyle\int \frac{dx}{x\,\sqrt{4x^2 - 9}} = \int \frac{dx}{x\,\sqrt{(2x)^2 - 9}}$

$$= \int \frac{2\,dx}{2x\,\sqrt{(2x)^2 - 9}}$$

$$= \int \frac{du}{u\,\sqrt{u^2 - a^2}}$$

$$= \frac{1}{a}\sec^{-1}\frac{u}{a} + C$$

$$= \frac{1}{3}\sec^{-1}\frac{2x}{3} + C$$

Example 9   Find $\displaystyle\int \frac{dx}{\sqrt{2 + x - x^2}}$.

Solution   This situation usually calls for completing the square under the radical.

$$2 + x - x^2 = 2 - \left(x^2 - x + \frac{1}{4} - \frac{1}{4}\right)$$

$$= \frac{9}{4} - \left(x - \frac{1}{2}\right)^2$$

Thus,

$$\int \frac{dx}{\sqrt{2 + x - x^2}} = \int \frac{dx}{\sqrt{9/4 - [x - (1/2)]^2}}$$

Now let $u = (x - \frac{1}{2})$; then $du = dx$, and the integral becomes

$$\int \frac{du}{\sqrt{a^2 - u^2}} \quad \left(\text{where } a = \frac{3}{2}\right)$$

$$\int \frac{du}{\sqrt{a^2 - u^2}} = \sin^{-1}\left(\frac{u}{a}\right) + C$$

$$= \sin^{-1}\left[\frac{x - (1/2)}{3/2}\right] + C$$

$$= \sin^{-1}\left[\frac{2[x - (1/2)]}{3}\right] + C$$

$$= \sin^{-1}\left[\frac{2x - 1}{3}\right] + C$$

Many times an integral requires much manipulation of the basic integration forms, and we once again resort to the Table of Integrals to minimize the work involved.

Example 10   Find $\sec^7 x \, dx$.

Solution   Form 38 of the Table of Integrals applies.

$$\int \sec^n u \, du = \frac{\tan u \, \sec^{n-2} u}{n - 1} + \frac{n - 2}{n - 1}\int \sec^{n-2} u \, du \quad (n \neq 1)$$

Applying this formula, where $u = x$ and $n = 7$,

$$\int \sec^7 x \, dx = \frac{\tan x \, \sec^5 x}{6} + \frac{5}{6}\int \sec^5 x \, dx$$

Applying form 38 again with $n = 5$,

$$\int \sec^7 x \, dx = \frac{\tan x \, \sec^5 x}{6} + \frac{5}{6}\left(\frac{\tan x \, \sec^3 x}{4} + \frac{3}{4}\int \sec^3 x \, dx\right)$$

Applying form 38 for the third time with $n = 3$ gives

$$\int \sec^7 x \, dx = \frac{\tan x \, \sec^5 x}{6} + \frac{5 \tan x \, \sec^3 x}{24}$$
$$+ \frac{15}{24}\left(\frac{\tan x \, \sec x}{2} + \frac{1}{2}\int \sec x \, dx\right)$$

Now a new problem arises. The apparently simple integral, $\int \sec x \, dx$ is not one we have encountered before. We have not come across a function whose derivative is the secant. Searching the table, we find form 30, which supplies $\int dx/(\cos ax)$, and we recall the identity

$$\sec x = \frac{1}{\cos x}$$

so form 30 applies. But note that this integral results in a natural logarithm. Applying form 30, with $a = 1$, and simplifying,

$$\int \sec^7 x \, dx = \frac{\tan x \sec^5 x}{6} + \frac{5 \tan x \sec^3 x}{24} + \frac{15 \tan x \sec x}{48}$$
$$+ \frac{15}{48} \ln |\sec x + \tan x| + C$$

5.8 EXERCISES

In Exercises 1–28, perform the indicated integrations.

1. $\int \cos 2x \, dx$ **2.** $\int \sin 3\theta \, d\theta$

3. $\int \sec^2 \frac{\theta}{2} \, d\theta$ **4.** $\int \sin x \cos x \, dx$

5. $\int \tan^3 x \sec^2 x \, dx$ **6.** $\int \tan \theta \sec \theta \, d\theta$

7. $\int \frac{1}{\sin^2 x} \, dx$ **8.** $\int \frac{\sin 3t}{\sec^3 3t} \, dt$

9. $\int \frac{\cos \theta}{\csc^2 \theta} \, d\theta$ **10.** $\int \frac{\sin 2x}{\sin x \cos x} \, dx$

11. $\int \frac{dx}{\sqrt{1 - 3x^2}}$ **12.** $\int \frac{dx}{2 + x^2}$

13. $\int \frac{dx}{-1 - x^2}$ **14.** $\int \frac{dx}{x \sqrt{x^2 - 4}}, \quad x > 2$

15. $\int \frac{dx}{4 + 9x^2}$ **16.** $\int \frac{dx}{x \sqrt{2x^2 - 1}}, \quad x > \frac{1}{\sqrt{2}}$

17. $\int \frac{dx}{\sqrt{1 + 2x - x^2}}$ **18.** $\int \frac{dx}{\sqrt{2 - 2x^2 + x}}$

19. $\int \frac{\cos x}{1 + \sin^2 x} \, dx$ **20.** $\int \frac{\sin x}{\sqrt{1 - \cos^2 x}} \, dx$

21. $\int \dfrac{1}{x^2} \sec \dfrac{1}{x} \tan \dfrac{1}{x} \, dx$

22. $\int \tan x \cot x \, dx$

23. $\int \dfrac{\sin x}{1 + \cos^2 x} \, dx$

24. $\int \dfrac{dx}{\sqrt{4 - 16x^2}}$

25. $\int \cos^3 \dfrac{x}{2} \sin \dfrac{x}{2} \, dx$

26. $\int \sec^2 x \tan x \, dx$

27. $\int \dfrac{\sin (1/x)}{x^2} \, dx$

28. $\int x \sin^7 (5x^2) \cos (5x^2) \, dx$

In Exercises 29–33, evaluate each integral.

29. $\displaystyle\int_0^{1/2} \dfrac{dx}{\sqrt{1 - x^2}}$

30. $\displaystyle\int_{-1}^{1} \dfrac{dx}{x^2 + 1}$

31. $\displaystyle\int_{-\pi/4}^{\pi/4} \sec^2 x \, dx$

32. $\displaystyle\int_0^{\pi/2} \sin x \, dx$

33. $\displaystyle\int_{\pi/2}^{\pi} \cos x \, dx$

In Exercises 34–45, use the Table of Integrals and any appropriate identities to find the antiderivatives.

34. $\int \sin^2 x \, dx$

35. $\int \cos^2 x \, dx$

36. $\int \cos^2 \dfrac{x}{2} \, dx$

37. $\int \sin^2 3x \, dx$

38. $\int \sin 3x \cos 2x \, dx$

39. $\int \sin 3x \sin 5x \, dx$

40. $\int x^3 \cos x \, dx$

41. $\int \sqrt{9 - x^2} \, dx$

42. $\int \dfrac{dx}{x \sqrt{3x - 1}}$

43. $\int \dfrac{\sqrt{x^2 - 4}}{x} \, dx$

44. $\int \sin^3 3x \, dx$

45. $\int \cos^3 2x \, dx$

46. To find $\int \dfrac{dx}{\sin x + \cos x}$, make the following substitution: Let $z = \tan (x/2)$. Use the identity

$$\tan \frac{x}{2} = \sqrt{\frac{1 - \cos x}{1 + \cos x}}$$

to find $\cos x$, $\sin x$, and dx in terms of z. Then integrate by use of the table.

47. Use the substitution and data from Exercise 46 to find $\int \sqrt{1 + \sin x} \, dx$.

NEW TERMS

Cosecant   Radian
Cosine   Reference angle
Cotangent   Secant
Fundamental identities   Sine
Inverse trigonometric functions   Tangent
Pythagorean identities   Trigonometric functions

REVIEW

In Exercises 1–6, evaluate.

1. $\sin 300°$ **2.** $\cos(-45°)$ **3.** $\tan \dfrac{5\pi}{4}$

4. $\sec \dfrac{3\pi}{2}$ **5.** $\csc 98°$ **6.** $\cot(-600°)$

In Exercises 7 and 8, evaluate using trigonometric identities.

7. $\sin \dfrac{7\pi}{12}$ **8.** $\cos 55° \cos 35° - \sin 55° \sin 35°$

In Exercises 9 and 10, prove the identities.

9. $\cos^2 x - \sin^2 x = 2\cos^2 x - 1$

10. $\tan^2 x + \csc^2 x + 1 = \sec^2 x \csc^2 x$

In Exercises 11–16, find the limits.

11. $\lim\limits_{x \to 0^-} \csc x$ **12.** $\lim\limits_{x \to 0} \dfrac{x^2}{1 - \cos x}$

13. $\lim\limits_{x \to 3} \sin x$ **14.** $\lim\limits_{x \to 0} \dfrac{\sin 5x}{2x}$

15. $\lim\limits_{x \to \infty} x \sin x \csc x$ **16.** $\lim\limits_{x \to \pi^+} \tan \dfrac{x}{2}$

In Exercises 17–26, find each derivative with respect to the independent variable.

17. $y = \dfrac{\sin x}{x^2}$ **18.** $y = \tan^2 x^2$

19. $y = \sec^2 3x$ **20.** $y = \csc x^2$

21. $y = \tan x \cot x$ **22.** $y = x \sin^{-1} x$

23. $y = \tan^{-1} \sqrt{x + 1}$ **24.** $y = \cos^{-1} (\cos x)$

25. $y = \sec^{-1} x^2$ **26.** $y = x \tan^{-1} \dfrac{1}{x}$

27. If $y = \tan x$, find y''.   **28.** If $f(x) = \csc^{-1} x$, find $f''(x)$.

29. Find $D_x^{(100)} y$ for $y = \sin x$.   **30.** Find $D_x^{(99)} y$ for $y = \cos x$.

 31. If the number of grams of bacteria in a population after t hours is given by $y = t^2 \cos t$, find the growth rate of the population after 45 minutes.

In Exercises 32–50, integrate using the Table of Integrals if necessary.

32. $\displaystyle\int \sin 3x \, dx$ **33.** $\displaystyle\int \cos \dfrac{x}{2} \, dx$

34. $\displaystyle\int \sec^2 (2x + 3) \, dx$ **35.** $\displaystyle\int \sin^2 x \cos x \, dx$

36. $\displaystyle\int \dfrac{\sin x}{\cos^3 x} \, dx$ **37.** $\displaystyle\int \dfrac{\tan^3 x}{\cos^2 x} \, dx$

38. $\displaystyle\int x \cos x^2 \, dx$ **39.** $\displaystyle\int 3x \sin^2 x^2 \cos x^2 \, dx$

40. $\displaystyle\int \dfrac{dx}{1 + 2x^2}$ **41.** $\displaystyle\int \dfrac{dx}{\sqrt{2 - x^2}}$

42. $\displaystyle\int \dfrac{dx}{\csc x \sqrt{1 - \cos^2 x}}$ **43.** $\displaystyle\int \sin \dfrac{x}{2} \cos \dfrac{x}{2} \, dx$

44. $\displaystyle\int \sin \dfrac{x}{2} \cos \dfrac{x}{4} \, dx$ **45.** $\displaystyle\int \sin^2 3x \, dx$

46. $\displaystyle\int \cos^2 (2x + 1) \, dx$ **47.** $\displaystyle\int \dfrac{dx}{x^2 + x + 1}$

48. $\displaystyle\int \sin^3 x \, dx$ **49.** $\displaystyle\int \dfrac{dx}{|x| \sqrt{x^2 - 9}}$

50. $\displaystyle\int x \sin 2x^2 \, dx$

In Exercises 51–55, evaluate each integral by the Fundamental Theorem of Calculus.

51. $\displaystyle\int_0^{\pi/2} \sin 2x \, dx$ **52.** $\displaystyle\int_0^1 \dfrac{dx}{1 + x^2}$

53. $\displaystyle\int_{-\pi/2}^0 (1 - \cos x) \, dx$ **54.** $\displaystyle\int_0^{\pi/2} \sin^2 x \, dx$

55. $\displaystyle\int_0^{\sqrt{2}/2} \dfrac{dx}{\sqrt{1 - x^2}}$

6

Logarithmic and Exponential Functions: Techniques of Integration

Chapters 4 and 5 have dealt with very mechanical mathematics: integration of elementary algebraic and trigonometric functions, as well as limits and derivatives of trigonometric functions and their inverses. The student is probably beginning to wonder when and how these concepts are going to be applied. Before we are ready for these applications, however, we must learn about two more functions, the logarithmic function and the exponential function. Some applications of these functions are in the areas of sociology for determining population growth, in the life sciences for calculating radioactive decay, and in business and economics for computing capital value of future income. Some of these applications will be presented in later chapters.

In this chapter we present the logarithmic and exponential functions, their derivatives, and their antiderivatives. We also demonstrate more integration techniques, and last, but by no means least, improper integrals will be examined.

6.1 THE NATURAL LOGARITHM

When $\int x^n \, dx$ was discussed, it was emphasized that

$$\int x^n \, dx = \frac{x^{n+1}}{n+1} + C$$

as long as $n \neq -1$, since $n = -1$ would cause division by 0. Thus $\int (1/x) \, dx$ has not been defined. However, Figure 6.1 is a graph of $f(t) = 1/t$, and, as this graph illustrates, $\int_1^x (1/t) \, dt$ does have a real measurable value.

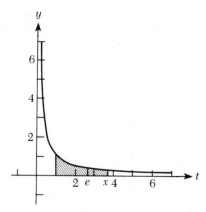

Figure 6.1   Graph of $f(t) = 1/t$

Definition

$$\ln x = \int_1^x \frac{1}{t} \, dt \quad \text{(where } x > 0)$$

To see if this definition is reasonable, let us consider the graph of $f(t) = 1/t$. The shaded region in Figure 6.1 represents $\ln x$. When $x = 1$,

$$\ln 1 = \int_1^1 \frac{1}{t} \, dt = 0$$

If $x < 1$, then

$$\ln x = \int_1^x \frac{1}{t}\, dt = -\int_x^1 \frac{1}{t}\, dt$$

It is obvious from the graph that t can never equal zero, since $1/t$ is not defined at $t = 0$. Furthermore, from the definition

$$\ln x = \int_1^x \frac{1}{t}\, dt \quad (x > 0)$$

it is clear that if $\ln x$ is to fulfill our definition of a function, t must be positive.

The expression "$\ln x$" is read "log x" and is defined as the *natural logarithm of x*. The properties of logarithms, which are reviewed in Chapter 1, also hold for natural logarithms.

To compare these properties, the following chart is helpful. Let a and x be any positive numbers, and b any positive number other than 1; then:

	Logarithm ($\log_b x$)	Natural logarithm ($\ln x$)
1.	$\log_b (ax) = \log_b a + \log_b x$	$\ln (ax) = \ln a + \ln x$
2.	$\log_b \dfrac{x}{a} = \log_b x - \log_b a$	$\ln \dfrac{x}{a} = \ln x - \ln a$
3.	$\log_b x^r = r \log_b x,$ if r is a rational number	$\ln x^r = r \ln x,$ if r is a rational number

Three other important properties of logarithms also hold for $\ln x$:

4. $\log_b b = 1$

5. $\log_b 1 = 0$

6. $\log_b b^r = r$

In order to show these properties, $\ln$ must have a base, b. Justification and discussion of all six properties follow.

To show $\ln (ax) = \ln a + \ln x$, let a be constant, and define $f(x) = \ln (ax)$. Then by the chain rule, $f'(x) = (1/ax) \cdot a = 1/x$, since $\int (1/x)\, dx = \ln x$ by definition.

Thus $\ln (ax)$ and $\ln x$ have the same derivative for all positive a and x, which means that $\ln (ax)$ and $\ln x$ must differ by, at most, a constant.

$$\ln (ax) - \ln x = C$$

When $x = 1$, $\ln x = \ln 1 = 0$, and $\ln a = C$. Thus, $\ln (ax) - \ln x = \ln a$, which implies $\ln (ax) = \ln a + \ln x$.

To show that $\ln (x/a) = \ln x - \ln a$, substitute $x = 1/a$ in the statement

$$\ln (ax) = \ln x + \ln a$$

Then,

$$\ln a \left(\frac{1}{a}\right) = \ln \frac{1}{a} + \ln a$$

But

$$\ln a \left(\frac{1}{a}\right) = \ln 1 = 0$$

thus

$$0 = \ln \frac{1}{a} + \ln a$$

and

$$\ln \frac{1}{a} = -\ln a$$

Thus,

$$\ln \frac{x}{a} = \ln \left(x \frac{1}{a}\right) = \ln x + \ln \frac{1}{a} = \ln x - \ln a$$

The proof for the third statement, $\ln x^r = r \ln x$, is similar to the proof for the first two statements, and should be attempted by the student.

Finally, if $\ln x$ is a logarithm like those studied in algebra, it must have a base. Recall that $\log_b b^r = r$, for any logarithm of base b. Let e be the number such that $\ln e^r = r$. By property 3, $\ln e^r = r \ln e$. Therefore,

$$\ln e^r = r \qquad \text{and} \qquad \ln e^r = r \ln e$$

which implies that

$$r \ln e = r$$

or

$$\ln e = 1$$

By the definition of $\ln x$,

$$\ln e = \int_1^e \frac{1}{t}\, dt = 1$$

This means that e is a number on the t axis of Figure 6.1 such that the area above the t axis and below the graph of $f(t) = 1/t$ between 1 and e is 1. It is easy to show that e must be between 2 and 3. Using the trapezoidal rule, the area under $f(t) = 1/t$ between 1 and 2 is about 0.69. The area under $f(t) = 1/t$ between 1 and 3 is about 1.1. Therefore e, the number such that the area under $f(t) = 1/t$ and between 1 and e is 1, is between 2 and 3. In fact, e is approximately 2.71828 and is irrational, as is π.

This discussion shows that $\ln x = \log_e x$. The appendix contains a table of logarithms for $\ln x$ and tables of powers of e (Tables II–IV).

From the discussion so far, you might wonder why we bother with the natural logarithm and its base, e. The reason is that they occur often in nature. The number e is the key constant in equations that describe the curves formed by suspended cables such as those used in bridge construction. Hence engineers use e and natural logarithms. We will use these concepts in problems of population growth, radioactive decay, and carbon-14 dating.

To summarize, for $x > 0$,

$$\ln x = \int_1^x \frac{1}{t}\, dt$$

$$D_x(\ln x) = D_x \int_1^x \frac{1}{t}\, dt = \frac{1}{x}$$

Thus, if $f(x) = \ln x$, then $f'(x) = 1/x$. Also, if u is a positive differentiable function of x, then if $f(x) = \ln u, f'(x) = (1/u)(du/dx)$.

Example 1   If $f(x) = \ln (2x + 1)$, find $f'(x)$.

Solution   Let $u = 2x + 1$; then $du/dx = 2$, and if $f(x) = \ln u$,

$$f'(x) = \frac{1}{u} \cdot \frac{du}{dx} = \frac{1}{2x + 1}\,(2) = \frac{2}{2x + 1}$$

Example 2   If $f(x) = \ln (3x^2 + 4)^{2/3}$, find $f'(x)$.

Solution   Recall that $\ln (3x^2 + 4)^{2/3} = \frac{2}{3} \ln (3x^2 + 4)$. Let $u = 3x^2 + 4$. Then

$$\frac{du}{dx} = 6x$$

$$f(x) = \frac{2}{3} \ln u$$

and

$$f'(x) = \frac{2}{3} \left(\frac{1}{u}\right) \frac{du}{dx}$$

$$= \frac{2}{3} \left(\frac{1}{3x^2 + 4}\right) (6x)$$

$$= \frac{4x}{3x^2 + 4}$$

Example 3   If $f(x) = \ln [(x^2 + 3)(x - 2)]$, find $f'(x)$.

Solution   Since $\ln (ax) = \ln a + \ln x$, $\ln [(x^2 + 3)(x - 2)] = \ln (x^2 + 3) + \ln (x - 2)$.

$$f(x) = \ln (x^2 + 3) + \ln (x - 2)$$
$$f'(x) = \frac{1}{x^2 + 3} (2x) + \frac{1}{x - 2}$$
$$= \frac{2x}{x^2 + 3} + \frac{1}{x - 2}$$
$$= \frac{3x^2 - 4x + 3}{(x^2 + 3)(x - 2)}$$

Example 4   If $f(x) = \ln [(x + 3)^2/x^3]$, find $f'(x)$.

Solution   Recall that $\ln (x/a) = \ln x - \ln a$, and that $\ln x^r = r \ln x$. Then

$$f(x) = \ln \left[\frac{(x + 3)^2}{x^3}\right]$$
$$= \ln (x + 3)^2 - \ln x^3$$
$$= 2 \ln (x + 3) - 3 \ln x$$

Thus

$$f'(x) = 2 \cdot \frac{1}{x + 3} - 3 \cdot \frac{1}{x}$$
$$= \frac{2x - 3(x + 3)}{x(x + 3)}$$
$$= -\frac{x + 9}{x(x + 3)}$$

Figure 6.1 shows ln x as an area. Specificially, it shows the shaded region as the area of $\int_{1}^{x} (1/t)\,dt$ in the closed interval $[1, x]$, where $1 < x$. Since the natural logarithm is merely another logarithmic function, the logarithm to the base e, its graph has all the properties of a logarithmic curve. Figure 6.2 is the graph of ln x.

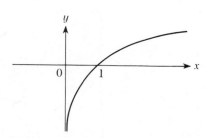

Figure 6.2   Graph of $y = \ln x$

Note that the function is increasing for all x in the domain of the function, that $\lim_{x \to \infty} \ln x = \infty$, that the line $x = 0$ is a vertical asymptote of this function and $\lim_{x \to 0^{+}} \ln x = -\infty$, that the curve is concave downward for all values of x, and that the graph crosses the x axis at $x = 1$ (ln $1 = 0$). The student will be asked to verify some of the above statements in the exercises by using calculus.

A number of formulas from economics can be stated in terms of natural logarithms. One of these is the coefficient of elasticity of demand. It measures the effect lowering or raising the price of an item has on the total revenue from the sales of the item. A commodity like salt is said to be price inelastic since any reasonable change in price has no effect on the amount of salt consumed, whereas a small decrease in the price of a certain brand of Scotch whiskey may greatly increase its total revenue (which is to say that the demand for Scotch is price elastic).

Let $y = f(x)$ be a function that relates the number of items sold, x, to the price, y. The function $f(x)$ is a demand function. One measure of the change in price with respect to change in demand is, of course, dy/dx. Economists do not use this measure because it involves absolute units of y and x that may not be comparable for different commodities. For instance, a 5 cents change in the price of one candy bar does not mean the same thing as a 5 cents change in the price of one refrigerator. Therefore the coefficient of elasticity compares a percentage change in y with a percentage change in x. If y changes by Δy, then Δy is the $\Delta y/y$ part of y. In the same manner, a change of Δx is the $\Delta x/x$ part of x. The coefficient of elasticity at a point of $y = f(x)$, denoted by the Greek letter eta, η, is:

$$\begin{aligned}\eta &= -\lim_{\Delta x \to 0} \frac{\Delta x}{x} \div \frac{\Delta y}{y} \\ &= -\lim_{\Delta x \to 0} \frac{y}{x} \cdot \frac{\Delta x}{\Delta y} = -\lim_{\Delta x \to 0} \frac{y}{x} \cdot \frac{1}{\Delta y/\Delta x} \\ &= -\frac{f(x)}{x f'(x)}\end{aligned}$$

When the price is reduced (Δy is negative), the demand increases (Δx is positive). Thus the ratio $\Delta y/\Delta x$ is negative. The minus sign is inserted in the formula to make the answer positive.

In other words η compares a percentage change in price with a percentage change in demand. The easiest way to remember η is not by the formula above, but by

$$\begin{aligned}\eta &= -\frac{f(x)}{x f'(x)} \\ &= -\frac{1}{x} \cdot \frac{y}{dy/dx} \\ &= -\frac{1}{x} \div \left(\frac{1}{y} \cdot \frac{dy}{dx}\right)\end{aligned}$$

But $D_x(\ln y) = \dfrac{1}{y}\dfrac{dy}{dx}$ and $D_x(\ln x) = 1/x$, so

$$\eta = -\frac{D_x \ln x}{D_x \ln y}$$

Example 6
§

If the number of items sold, x, is related to price, y, for a certain item by $y = -10x + 50$, find the coefficient of price elasticity with respect to demand when 3 units are sold.

Solution

$$\begin{aligned}\eta &= -\frac{D_x \ln x}{D_x \ln y} = -\frac{D_x \ln x}{D_x \ln(-10x + 50)} \\ &= -\frac{1/x}{\dfrac{-10}{-10x + 50}} \\ &= \frac{-10x + 50}{10x}\end{aligned}$$

When $x = 3$,

$$\eta = \frac{-10(3) + 50}{10(3)}$$
$$= \frac{-30 + 50}{30}$$
$$= \frac{20}{30} = 0.667$$

An η greater than 1 means that any percentage change in price will result in a greater percentage change in demand, and the demand is *elastic*. If $\eta = 1$, then any percentage change in price results in the same percentage change in demand, and the demand has *unit elasticity*. If $\eta < 1$, as in this example, then demand is said to be *inelastic*.

6.1 EXERCISES

In Exercises 1–20, differentiate each function.

1. $y = \ln(3x + 2)$ 2. $y = \ln(x^2 + 2x)$

3. $y = \ln(x^2 + 3)^2$ 4. $y = \ln\left(\frac{x + 2}{x - 3}\right)$

5. $y = \ln\left(\frac{x^2 + 3}{2x^2 - 1}\right)$ 6. $y = \ln\left(\frac{\sqrt{x + 3}}{x^2 + 1}\right)$

7. $y = x \ln x$ 8. $y = \ln(2x - x^3)$

9. $y = \ln\sqrt{x^2 + 3}$ 10. $y = \sqrt{\ln(x^2 + 3)}$

11. $y = \ln(x + \sqrt{x^2 + 1})$ 12. $y = \frac{x}{\ln x}$

13. $y = \ln(x^3 + 3x)^{1/3}$ 14. $y = [\ln(x^3 + 3x)]^{1/3}$

15. $y = \ln\sqrt{\frac{2x + 1}{2x - 1}}$ 16. $y = \ln^2(3x)$

17. $y = \frac{\ln x}{x^2}$ 18. $y = x \ln\left(\frac{1}{x}\right)$

19. $y = \ln(\ln x)$ 20. $y = \ln|x|$

21. The function $f(x) = \ln x$ is defined only for positive values of x. Use this fact, plus the fact that a function is increasing at all points for which the slope of the tangent to the curve at that point is positive, to prove that f is an increasing function for all x in the domain of f.

22. Use the properties of calculus to prove that the graph of ln x is concave downward for all values of x for which ln x is defined.

23. Find the slope of the tangent to the graph of ln x where the curve crosses the x axis.

24. For what value(s) of x does the graph of $y = \ln x$ have a horizontal tangent?

In Exercises 25–28, the given demand function relates the price, $y = f(x)$, to the quantity sold, x, of a commodity. Determine whether the demand is elastic, unit elastic, or inelastic at the given price.

25. $f(x) = -100x + 1000,$
$x = 4$

26. $f(x) = \dfrac{1}{x}, \quad x = 75$

27. $f(x) = \dfrac{1}{x^2}, \quad x = 8000$

28. $f(x) = -20x + 300, \quad x = 6$

6.2   THE NATURAL LOGARITHM—INTEGRATION

The natural logarithm was defined in the preceding section as

$$\ln x = \int_1^x \frac{1}{t}\, dt$$

and only positive values of t are considered. In order to insure the positive nature of x in the general antiderivative of

$$\int \frac{1}{x}\, dx$$

we write this antiderivative as $\ln |x| + C$. To verify the legitimacy of the use of absolute value, we differentiate:

$$y = \ln |x|$$

$$D_x(\ln |x|) = \frac{1}{|x|} \cdot \frac{|x|}{x} = \frac{1}{x}$$

and thus

$$\int \frac{1}{x}\, dx = \ln |x| + C$$

or, if u is a differentiable function in x, and $u \neq 0$, then

$$\int \frac{1}{u} \, du = \ln|u| + C$$

We are now ready to integrate.

Example 1   Find $\int \dfrac{dx}{3x + 1}$.

Solution   Let $u = 3x + 1$; then $du = 3 \, dx$.

$$\int \frac{dx}{3x + 1} = \frac{1}{3} \int \frac{3 \, dx}{3x + 1} = \frac{1}{3} \int \frac{du}{u}$$

$$= \frac{1}{3} \ln|u| + C = \frac{1}{3} \ln|3x + 1| + C$$

Example 2   Find $\int \dfrac{dx}{x \ln(x^2)}$.

Solution $\ln(x^2) = 2 \ln x$, so that

$$\int \frac{dx}{x \ln(x^2)} = \int \frac{dx}{2x \ln x} = \frac{1}{2} \int \frac{dx}{x \ln x}$$

Let $u = \ln x$. Then

$$du = \frac{1}{x} \, dx = \frac{dx}{x}$$

Therefore,

$$\frac{1}{2} \int \frac{dx}{x \ln x} \qquad \text{becomes} \qquad \frac{1}{2} \int \frac{du}{u}$$

and

$$\frac{1}{2} \int \frac{du}{u} = \frac{1}{2} \ln|u| + C$$

Replacing u with $\ln x$, we have

$$\frac{1}{2} \ln|u| + C = \frac{1}{2} \ln|\ln x| + C$$

or

$$\ln \sqrt{\ln x} + C$$

Example 3   Evaluate $\displaystyle\int_1^3 \frac{dx}{x}$.

Solution $\displaystyle\int_1^3 \frac{dx}{x} = \ln x \Big]_1^3 = \ln 3 - \ln 1 = \ln 3 - 0 = \ln 3$

Example 4   Find $\displaystyle\int \frac{6x^2 + 5}{x - 1} \, dx$.

Solution   Since the numerator is of higher degree than the denominator, simplify the integrand first by division.

$$\frac{6x^2 + 5}{x - 1} = 6x + 6 + \frac{11}{x - 1}$$

Therefore,

$$\int \frac{6x^2 + 5}{x - 1} \, dx = \int \left(6x + 6 + \frac{11}{x - 1} \right) dx$$
$$= \int 6x \, dx + \int 6 \, dx + 11 \int \frac{dx}{x - 1}$$
$$= 3x^2 + 6x + 11 \ln|x - 1| + C$$

6.2 EXERCISES

In Exercises 1–20, integrate:

1. $\displaystyle\int \frac{dx}{x + 2}$

2. $\displaystyle\int \frac{dx}{2x + 1}$

3. $\displaystyle\int \frac{6x}{3x^2 - 1} \, dx$

4. $\displaystyle\int \frac{dx}{1 - x}$

5. $\displaystyle\int \frac{x \, dx}{3 - x^2}$

6. $\displaystyle\int \frac{1}{x \ln x} \, dx$

7. $\displaystyle\int \frac{1}{x \ln \sqrt{x}} \, dx$

8. $\displaystyle\int \frac{4x^2 + 2}{2x + 1} \, dx$

9. $\displaystyle\int \frac{x^3 + 4x^2}{x + 3} \, dx$

10. $\displaystyle\int \frac{4x + 1}{2x^2 + x + 5} \, dx$

11. $\displaystyle\int \frac{x^2}{1 - 5x^3} \, dx$

12. $\displaystyle\int \frac{3x^2}{3x - 1} \, dx$

13. $\displaystyle\int \frac{x}{2x^2 + 3} \, dx$

14. $\displaystyle\int \frac{2x + 6}{(x + 3)^2} \, dx$

15. $\displaystyle\int \frac{x + 1}{x^2 + 2x + 5} \, dx$

16. $\displaystyle\int \frac{dx}{\sqrt{x}\,(1 + \sqrt{x})}$

17. $\displaystyle\int \frac{x^3}{x - 1} \, dx$

18. $\displaystyle\int \frac{10x^3 - 4x}{(5x^2 - 2)^2} \, dx$

19. $\displaystyle\int \frac{x^2 + 1}{x^3 + 3x} \, dx$

20. $\displaystyle\int \frac{x \, dx}{x^2 - a^2}$, $\quad a$ is constant, $a^2 \neq x^2$

In Exercises 21–26, evaluate:

21. $\displaystyle\int_{-1}^{3} \frac{dx}{x + 2}$

22. $\displaystyle\int_{0}^{4} \frac{3x}{x^2 + 2} \, dx$

23. $\displaystyle\int_{0}^{5} \frac{16x^2 - 3}{4x + 1} \, dx$

24. $\displaystyle\int_{2}^{5} \frac{x}{x^2 - 1} \, dx$

25. $\displaystyle\int_{1}^{3} \frac{3x}{3x - 1} \, dx$

26. $\displaystyle\int_{0}^{3} \frac{6x}{2x^2 + 5} \, dx$

27. Find the area enclosed by the lines $x = 1$, $x = 4$, $y = 0$, and the curve $y = 1/(x + 2)$.

28. Find the area enclosed by the graph of $y = 1/x$, the x axis, and the lines $x = 1$ and $x = e$.

6.3 THE EXPONENTIAL FUNCTION

We know the algebraic definition of a logarithm requires

$$x = \log_b y \quad \text{if and only if} \quad y = b^x$$

where b is a positive number not equal to 1, and $y > 0$. The function $y = b^x$, or $f(x) = b^x$, with the stated restrictions, is called an *exponential function*.

Definition

The exponential function with base b is defined by

$$f(x) = b^x \qquad (b > 0, b \neq 1)$$

for x any real number.

Figure 6.3 illustrates the graphs of the exponential function for $0 < b < 1$ and for $b > 1$. Since $b^0 = 1$ for all values of b for which the exponential function is defined, the graphs intersect at the point $(0, 1)$.

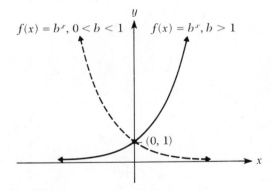

Figure 6.3 The exponential function

The exponential function is the *inverse* of the logarithmic function. If $f(x) = \log_b x$, then $f^{-1}(x) = b^x$, and

$$f(f^{-1}(x)) = \log_b b^x = x \log_b b = x$$

Also

$$f^{-1}(f(x)) = b^{\log_b x} = x$$

Figure 6.4 illustrates the graph of $f(x) = \log_b x$ and its inverse, $f(x) = b^x$ for $b > 1$. Figure 6.5 is the graph of the same for $0 < b < 1$. Compare these figures with Figure 6.3.

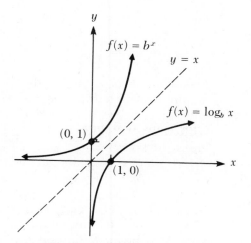

Figure 6.4 The logarithmic function and the exponential function for $b > 1$

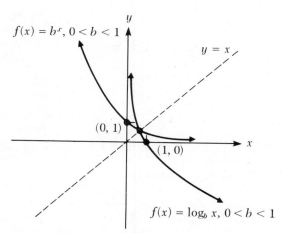

Figure 6.5 The logarithmic function and the exponential function for $0 < b < 1$

We now have learned the base of the natural logarithm, $\ln x$, is the number e (named for the Swiss mathematician Euler), so that

$$y = \ln x \quad \text{if and only if} \quad x = e^y$$

or, if

$$f(x) = \ln x$$

then

$$f^{-1}(x) = e^x$$

Also, since $\log_b b = 1$ for all $b > 0$ and $b \neq 1$, $\ln e = 1$. Since the natural logarithm function and its inverse, the exponential function with base e, are so important in calculus, this function is often written exp x, which means e^x. Let us examine exp x further, and find its derivative.

As the inverse of the function f such that $f(x) = \ln x$, the domain of $\ln x$ becomes the range of its inverse, e^x, and the range of $\ln x$ becomes the domain of e^x. The domain of $\ln x$ is the set of all positive real numbers. The range of $\ln x$ is all real numbers. Thus e^x has as its domain all real numbers, and as its range the positive real numbers. Also, by the definition of e^x as the inverse of $\ln x$, we now have included powers of e that are *irrational*.

To differentiate exp x, we go back to the definition

$$y = e^x \quad \text{if and only if} \quad \ln y = x$$

By implicit differentiation,

$$\frac{1}{y} \frac{dy}{dx} = 1$$

$$\frac{dy}{dx} = y = e^x$$

Thus, if $y = e^x$, $dy/dx = e^x$, and $d^2y/dx^2 = e^x$. In fact, $d^n(e^x)/dx^n = e^x$. If $u = f(x)$, then $d(e^u)/dx = e^u \, du/dx$.

Example 1   Find the derivative of e^{3x+2}.

Solution   Let $f(x) = e^{3x+2} = e^u$, where $u = 3x + 2$ and $du/dx = 3$.

$$f'(x) = e^{3x+2}(3) = 3e^{3x+2}$$

Example 2   Find $\int e^{3x} \, dx$.

Solution $\int e^{3x} \, dx = \frac{1}{3} \int e^{3x}(3) \, dx = \frac{1}{3} e^{3x} + C$

Similarly, since $d(e^u)/du = e^u$, $\int e^u \, du = e^u + C$.

Example 3   Find dy/dx if $y = e^{1/x}$.

Solution   Let $u = 1/x$; then

$$\frac{du}{dx} = -\frac{1}{x^2}$$

and

$$\frac{dy}{dx} = e^u \frac{du}{dx} = \frac{-e^{1/x}}{x^2}$$

It is often necessary to find the derivative of an expression of the form

$$y = a^x \qquad (a > 0, a \neq 1, a \neq e)$$

If $y = a^x$, then $\ln y = \ln a^x$, and $\ln y = x \ln a$ if x is a rational number.

Also, $a^x = e^{x \ln a}$, and therefore

$$D_x(a^x) = D_x(e^{x \ln a}) = e^{x \ln a}(\ln a) = a^x \ln a$$

It can be shown that this statement is true for all real numbers, x, but the proof for this statement is beyond the scope of this text. Thus, in general, for all x and for $a > 0, a \neq 1$,

$$D_x(a^x) = a^x \ln a$$

and if $u = f(x)$, then

$$D_x(a^u) = a^u \ln a \frac{du}{dx}$$

Clearly this formula is true if $a = e$, since $\ln e = 1$.

Example 4   If $y = 2^{3x^2+1}$, find dy/dx.

Solution
$$\frac{dy}{dx} = 2^{3x^2+1}(\ln 2)(6x)$$
$$= 6x \, 2^{3x^2+1} \ln 2$$

Example 5   Find $\int 2^{4x}\, dx$.

Solution   This integral involves a general solution for $\int a^u\, du$. Since

$$\frac{d(a^u)}{du} = a^u \ln a$$
$$d(a^u) = a^u \ln a\, du$$

If we multiply and divide $\int a^u\, du$ by $\ln a$, we obtain

$$\frac{1}{\ln a} \int a^u \ln a\, du = \frac{1}{\ln a}(a^u) + C$$

Therefore,

$$\int 2^{4x}\, dx = \frac{1}{\ln 2} \int 2^{4x} \ln 2\, dx$$

But

$$u = 4x \quad \text{implies} \quad du = 4\, dx$$

therefore,

$$\int 2^{4x}\, dx = \frac{1}{4 \ln 2} \int 2^{4x} 4 \ln 2\, dx$$
$$= \frac{1}{4 \ln 2}(2^{4x}) + C$$

Example 6   If $y = \log_b x$, find dy/dx.

Solution   This problem can be solved by several different methods. If $b = e$, then the solution is trivial, so assume $b \neq e$.

$$y = \log_b x \quad \text{if and only if} \quad x = b^y$$

Taking the natural logarithm of each side of the exponential equation, we obtain

$$\ln x = y \ln b$$

and

$$y = \frac{\ln x}{\ln b} = \log_b x$$

Recognizing that $\ln b$ is *constant*, we can now differentiate.

$$D_x(\log_b x) = D_x \left(\frac{\ln x}{\ln b} \right) = \frac{1}{\ln b} \cdot \frac{1}{x} = \frac{1}{x \ln b}$$

To summarize the forms used in this section,

$$D_x e^u = e^u \frac{du}{dx}$$

$$D_x a^u = a^u \ln a \frac{du}{dx}$$

$$D_x(\log_b u) = \frac{1}{u \ln b} \frac{du}{dx}$$

$$\int e^u \, du = e^u + C$$

$$\int a^u \, du = \frac{1}{\ln a} (a^u) + C$$

where u is a differentiable function of x.

6.3 EXERCISES

In Exercises 1–24, find dy/dx.

1. $y = e^{x^2}$

2. $y = 3xe^{2x}$

3. $y = \frac{1}{2}(e^x - e^{-x})$

4. $y = \frac{e^x - e^{-x}}{e^x + e^{-x}}$

5. $y = e^{-x^2}$

6. $y = \ln e^{ax+b}$, where a and b are constants

7. $y = x^e$

8. $y = 3^x$

9. $y = \ln e^x$

10. $\ln y + e^{\sqrt{x}} = 2$

11. $x = e^y$

12. $e^x - e^y = 1$

13. $y = 3^{x^2}$

14. $y = \log_3 4x$

15. $y = \log_{10} x^2$

16. $y = \ln 3^x$

17. $e^x + x^e = y$

18. $y = x^2 e^x$

19. $y = 5^{2x}e^{2x}$

20. $y = \frac{3}{e^{x^2}}$

21. $y = x \ln x$ **22.** $y = x \ln 2$

23. $y = e^x \ln x$ **24.** $e^{2x} - e^{x^2} + y = 0$

In Exercises 25–40, integrate:

25. $\int e^{3x}\, dx$ **26.** $\int x e^{x^2}\, dx$

27. $\int \dfrac{e^x}{1 + e^x}\, dx$ **28.** $\int \dfrac{e^x + 1}{e^x}\, dx$

29. $\int 2^{3x+2}\, dx$ **30.** $\int \dfrac{x}{2^{x^2}}\, dx$

31. $\int \dfrac{dx}{e^x}$ **32.** $\int x^e\, dx$

33. $\int \dfrac{e^x - e^{-x}}{2}\, dx$ **34.** $\int e^{\ln x}\, dx$

35. $\int x^4 e^{x^5}\, dx$ **36.** $\int \dfrac{1}{2^x}\, dx$

37. $\int \dfrac{e^{1/x}}{x^2}\, dx$ **38.** $\int \ln e^x\, dx$

39. $\int \dfrac{1}{\ln e^x}\, dx$ **40.** $\int \dfrac{1}{x \ln x}\, dx$

In Exercises 41–45, evaluate:

41. $\displaystyle\int_2^4 e^{3x}\, dx$ **42.** $\displaystyle\int_0^3 e^{-2x}\, dx$

43. $\displaystyle\int_0^2 x\, 3^{2x^2}\, dx$ **44.** $\displaystyle\int_0^2 3^{2x}\, dx$

45. $\displaystyle\int_0^1 \dfrac{x}{e^{x^2}}\, dx$

46. Let f be a function such that $f(x) = e^{x^2}$. Use the first and second derivatives of this function to analyze its graph in terms of any maximum and minimum values, concavity, and for what values of x the function is decreasing and increasing.

47. Use an analysis similar to the one in Exercise 46 to sketch the graph of $y = xe^x$ over the interval $[-2, 5]$.

48. Find the area enclosed by the y axis, the x axis, the line $x = 4$, and the graph of $y = xe^{x^2}$.

49. Find the area bounded by the y axis, the x axis, the line $x = \ln 4$, and the graph of $y = 10e^{-x}$.

50. What is the slope of the tangent line to the graph in Exercise 49 when $x = \ln 4$?

6.4   INTEGRATION BY PARTS

Frequently, a problem in integration does not lend itself to a substitution of the form $u\ du$. For example, $\int \ln x\ dx$. If we let $u = \ln x$, then $du = (1/x)\ dx$, and since there is no way of introducing a factor of $1/x$ to the integrand without further complicating matters, this substitution is not practical.

Another way of looking at the integral $\int \ln x\ dx$ is to recognize that there is no simple function F such that $F'(x) = \ln x$.

Before offering a solution to this problem, let us examine the following.

Let u and v be functions of x. Then

$$d(u \cdot v) = u\ dv + v\ du$$

and

$$\int d(u \cdot v) = u \cdot v = \int u\ dv + \int v\ du + C$$

So that

$$\int u\ dv = u \cdot v - \int v\ du + C \tag{1}$$

The integral on the left, $\int u\ dv$, offers a solution for an integrand which is the product of a function, u, and the differential of a function, v. For example, in the problem $\int \ln x\ dx$, let

$$u = \ln x \quad \text{and} \quad dv = dx$$

Then

$$du = \frac{1}{x}\ dx \quad \text{and} \quad v = x + C_1$$

By Equation 1,

$$
\begin{aligned}
\int \ln x\ dx &= uv - \int v\ du + C \\
&= \ln x(x + C_1) - \int (x + C_1)\frac{1}{x}\ dx + C \\
&= \ln x(x + C_1) - x - C_1 \ln x + C \\
&= x \ln x + C_1 \ln x - x - C_1 \ln x + C \\
&= x \ln x - x + C
\end{aligned}
$$

When using this method of integration, the constant, C_1, associated with v always cancels, so it is not necessary to include it when finding v.

Example 1   Find $\int xe^x \, dx$.

Solution   Again no immediate solution is available unless the integrand is expressed in the form $u \, dv$. Let

$$u = x \quad \text{and} \quad dv = e^x \, dx$$

Then

$$du = dx \quad \text{and} \quad v = e^x$$

$$\int xe^x \, dx = xe^x - \int e^x \, dx + C$$
$$= xe^x - e^x + C$$
$$= e^x(x - 1) + C$$

The student may ask how to select u and dv, and whether the choice makes any difference. The answers to these questions are not easy, since each problem must be analyzed individually. If an integrand is a product of an algebraic function (a polynomial, or x^p, where p is a real number) and a logarithmic function or an inverse trigonometric function, then the algebraic function is selected as the dv part, since these functions have simple antiderivatives, whereas this cannot be said for logarithmic or inverse trigonometric functions. For example, there was only one way to find $\int \ln x \, dx$. In the example $\int xe^x \, dx$, if we had let $u = e^x$ and $dv = x \, dx$, then $du = e^x \, dx$ and $v = x^2/2$, and the resulting expression would have been more complicated than the original, since

$$\int xe^x \, dx = \frac{x^2}{2} e^x - \int \frac{x^2}{2} e^x \, dx + C$$

There are times, however, when the selection is entirely arbitrary.

Example 2   Find $\int e^x \cos x \, dx$.

Solution   Let $u = \cos x$ and $dv = e^x \, dx$; then $du = -\sin x \, dx$ and $v = e^x$.

$$\int e^x \cos x \, dx = e^x \cos x - \int e^x(-\sin x) \, dx + C$$

The new integrand presents the same difficulties as the original integral, so parts are once again used. In the new integral, let

$$u = -\sin x \quad \text{and} \quad dv = e^x \, dx$$

Then

$$du = -\cos x \, dx \quad \text{and} \quad v = e^x$$

$$\int e^x \cos x \, dx = e^x \cos x - \left[e^x(-\sin x) - \int e^x(-\cos x) \, dx \right] + C$$
$$= e^x \cos x + e^x \sin x - \int e^x \cos x \, dx + C$$

Thus,

$$2 \int e^x \cos x \, dx = e^x \cos x + e^x \sin x + C$$
$$\int e^x \cos x \, dx = \frac{1}{2} (e^x \cos x + e^x \sin x + C)$$

Note that after integration by parts was used twice, the original integral appeared on the right side (preceded by a minus sign) and when it was added to both sides we had the solution. In this problem, the selection of u and dv was arbitrary and could have been reversed. However, when the second application of parts was necessitated, the pattern of the first selection had to be repeated. The student should check what would happen if the second application had been changed to let $u = e^x$ and $dv = -\sin x \, dx$.

Example 3   Find $\int x \tan^{-1} x \, dx$.

Solution   Integration by parts is indicated. We must select $u = \tan^{-1} x$, since we don't know an antiderivative for this function. Let

$$u = \tan^{-1} x \quad \text{and} \quad dv = x \, dx$$

Then

$$du = \frac{dx}{1 + x^2} \quad \text{and} \quad v = \frac{x^2}{2}$$

$$\int x \tan^{-1} x \, dx = \frac{x^2}{2} \tan^{-1} x - \frac{1}{2} \int \frac{x^2}{1 + x^2} \, dx + C$$

$$= \frac{x^2}{2} \tan^{-1} x - \frac{1}{2} \int \left(1 - \frac{1}{x^2 + 1} \right) dx + C$$

$$= \frac{x^2}{2} \tan^{-1} x - \frac{1}{2} (x - \tan^{-1} x) + C$$

$$= \frac{x^2}{2} \tan^{-1} x - \frac{x}{2} + \frac{\tan^{-1} x}{2} + C$$

For the *definite* integral,

$$\int_a^b u \, dv = uv \, \Big]_a^b - \int_a^b v \, du$$

Example 4   Evaluate $\int_1^3 x^2 \ln x \, dx$.

Solution   Let

$$u = \ln x \quad \text{and} \quad dv = x^2 \, dx$$

Then

$$du = \frac{1}{x} dx \quad \text{and} \quad v = \frac{x^3}{3}$$

$$\int_1^3 x^2 \ln x \, dx = \frac{x^3}{3} \ln x \, \Big]_1^3 - \frac{1}{3} \int_1^3 x^2 \, dx$$

$$= \frac{x^3}{3} \ln x \, \Big]_1^3 - \frac{1}{3} \left(\frac{x^3}{3} \right) \Big]_1^3$$

$$= 9 \ln 3 - \frac{1}{3} \ln 1 - \frac{1}{3} (9) + \frac{1}{3} \left(\frac{1}{3} \right)$$

$$= 9 \ln 3 - \frac{26}{9}$$

It is often necessary to integrate the expressions $\int \sec x \, dx$ and $\int \csc x \, dx$. If we write $\int \csc x \, dx = \int dx/\sin x$ and let $u = \sin x$, then we cannot find a ready expression for du, since $u = \sin x$ implies $du = \cos x \, dx$ and there is no $\cos x$ in the integrand. The following identities are useful for integrating $\int \sec x \, dx$ and $\int \csc x \, dx$.

$$\sec x = \sec x \, \frac{\sec x + \tan x}{\sec x + \tan x} \qquad \csc x = \csc x \, \frac{\csc x - \cot x}{\csc x - \cot x}$$

Using these identities, it is easy to perform the required integrations.

$$\int \sec x \, dx = \int \frac{\sec x (\sec x + \tan x)}{\sec x + \tan x} \, dx$$

Now let $u = \sec x + \tan x$. Then

$$du = (\sec x \tan x + \sec^2 x) \, dx = \sec x (\sec x + \tan x) \, dx$$

and

$$\int \sec x \, dx = \int \frac{du}{u} = \ln|u| + C$$
$$= \ln|\sec x + \tan x| + C$$

The second identity may be used to find $\int \csc x \, dx$. These integrals are given below as an aid.

$$\int \sec x \, dx = \ln|\sec x + \tan x| + C$$
$$\int \csc x \, dx = \ln|\csc x - \cot x| + C$$

6.4 EXERCISES

In Exercises 1–16, find the antiderivatives.

1. $\int e^x \sin x \, dx$ **2.** $\int x \cos x \, dx$

3. $\int x^2 \ln x \, dx$ **4.** $\int x^2 e^x \, dx$

5. $\int x^2 e^{-x} \, dx$ **6.** $\int x^3 e^{-x^2} \, dx$

7. $\int \ln(x+1) \, dx$ **8.** $\int x \sec^2 x \, dx$

9. $\int x^2 \sin x \, dx$ **10.** $\int \sin^{-1} 2x \, dx$

11. $\int 3x \tan^{-1} x \, dx$

12. $\int \sec^3 x \, dx$ [*Hint:* $\sec^3 x = \sec x \sec^2 x$]

13. $\int x \cos^2 x \, dx$ $\left[Hint: \cos^2 x = \dfrac{1 + \cos 2x}{2} \right]$

14. $\int e^{ax} \sin bx \, dx$ (*a* and *b* are constants)

15. $\int x e^{-ax} \, dx$ (*a* is a constant)

16. $\int x a^x \, dx$ (*a* is a constant, $a > 0$, and $a \neq 1$)

In Exercises 17–25, evaluate:

17. $\displaystyle\int_1^e \ln x \, dx$ **18.** $\displaystyle\int_1^e x^e \ln x \, dx$

19. $\displaystyle\int_1^4 x e^{-x} \, dx$ **20.** $\displaystyle\int_0^2 x^3 e^{x^2} \, dx$

21. $\displaystyle\int_1^3 x \ln x \, dx$ **22.** $\displaystyle\int_0^{\pi/2} x \cos x \, dx$

23. $\displaystyle\int_0^{\pi/2} x \sin^2 x \, dx$ **24.** $\displaystyle\int_{\pi/4}^{\pi/2} x^2 \cos x \, dx$

25. $\displaystyle\int_0^{\pi/2} x \sin x \, dx$

6.5 TRIGONOMETRIC SUBSTITUTIONS

Integrals containing the terms $\sqrt{a^2 - u^2}$, $\sqrt{a^2 + u^2}$, $\sqrt{u^2 - a^2}$, $a^2 + u^2$, $a^2 - u^2$ can often be simplified and interpreted by means of a trigonometric substitution. Some of these integrals were introduced earlier. In this section we will explore these problems in more depth.

Let us recall the following identities.

$$\sin^2 x + \cos^2 x = 1 \qquad \cos x = \sqrt{1 - \sin^2 x}$$

$$\tan^2 x + 1 = \sec^2 x \qquad \sec x = \sqrt{\tan^2 x + 1} \qquad \tan x = \sqrt{\sec^2 x - 1}$$

These identities, together with a little algebra, frequently make the following substitutions practical.

1. For $\sqrt{a^2 - u^2}$, let $u = a \sin \theta$.
2. For $\sqrt{a^2 + u^2}$, let $u = a \tan \theta$.
3. For $\sqrt{u^2 - a^2}$, let $u = a \sec \theta$.
4. For $a^2 + u^2$, let $u = a \tan \theta$.
5. For $a^2 - u^2$, let $u = a \sin \theta$.

Proper care must be taken to insure values of θ for which the expression is defined. The following examples show applications of the above substitutions.

Example 1   Find $\int \sqrt{4 - x^2}\, dx$.

Solution   If we let $u = 4 - x^2$, then $du = -2x\, dx$. Thus this approach does not work. However, using substitution 1 above, where $a^2 = 4$ and $u^2 = x^2$, let $x = 2 \sin \theta$. Then

$$4 - x^2 = 4 - 4 \sin^2 \theta = 4(1 - \sin^2 \theta)$$
$$= 4 \cos^2 \theta$$

and

$$\sqrt{4 - x^2} = \sqrt{4 \cos^2 \theta} = 2 \cos \theta$$

If $x = 2 \sin \theta$,

$$dx = 2 \cos \theta\, d\theta$$

Thus $\int \sqrt{4 - x^2}\, dx$ becomes $\int 2 \cos \theta\, (2 \cos \theta\, d\theta) = 4 \int \cos^2 \theta\, d\theta$. Using the identity $\cos^2 \theta = (1 + \cos 2\theta)/2$ to integrate the expression,

$$4 \int \cos^2 \theta \, d\theta = 2 \int (1 + \cos 2\theta) \, d\theta$$
$$= 2\theta + \sin 2\theta + C$$
$$= 2\theta + 2 \sin \theta \cos \theta + C$$

since $\sin 2\theta = 2 \sin \theta \cos \theta$.

To express this solution in terms of x, recall that we let

$$x = 2 \sin \theta$$

Therefore

$$\frac{x}{2} = \sin \theta$$

and

$$\sin^{-1} \frac{x}{2} = \theta$$

Also

$$\sqrt{4 - x^2} = 2 \cos \theta$$

So

$$2\theta + 2 \sin \theta \cos \theta + C = 2 \sin^{-1} \frac{x}{2} + \frac{x}{2} \sqrt{4 - x^2} + C$$

Example 2   Find $\int \sqrt{4 + x^2} \, dx$.

Solution   If we let $u = 4 + x^2$, then $du = 2x \, dx$, so this approach is impractical. Using substitution 2 from above, let

$$x = 2 \tan \theta$$

Then

$$x^2 = 4 \tan^2 \theta$$
$$4 + x^2 = 4 + 4 \tan^2 \theta = 4(1 + \tan^2 \theta) = 4 \sec^2 \theta$$
$$\sqrt{4 + x^2} = 2 \sqrt{\sec^2 \theta} = 2 \sec \theta$$

Also, if

$$x = 2 \tan \theta$$

then

$$dx = 2 \sec^2 \theta \, d\theta$$

$$\int \sqrt{4 + x^2} \, dx = \int 2 \sec \theta (2 \sec^2 \theta) \, d\theta$$
$$= 4 \int \sec^3 \theta \, d\theta$$

By noting that $\sec^3 \theta = \sec \theta \sec^2 \theta$, we can integrate by using parts.

$$4 \int \sec^3 \theta \, d\theta = 4 \int \sec \theta \sec^2 \theta \, d\theta$$

Let

$$u = \sec \theta \qquad \text{and} \qquad dv = \sec^2 \theta \, d\theta$$

Then

$$du = \sec \theta \tan \theta \, d\theta \qquad \text{and} \qquad v = \tan \theta$$

$$\int \sec^3 \theta \, d\theta = \sec \theta \tan \theta - \int \sec \theta \tan^2 \theta \, d\theta + C_1$$
$$= \sec \theta \tan \theta - \int \sec \theta (\sec^2 \theta - 1) \, d\theta + C_1$$
$$= \sec \theta \tan \theta - \int \sec^3 \theta \, d\theta + \int \sec \theta \, d\theta + C_1$$
$$= \sec \theta \tan \theta - \int \sec^3 \theta \, d\theta + \ln \left| \sec \theta + \tan \theta \right| + C_1 + C_2$$

$$2 \int \sec^3 \theta \, d\theta = \sec \theta \tan \theta + \ln \left| \sec \theta + \tan \theta \right| + C_1 + C_2$$
$$4 \int \sec^3 \theta \, d\theta = 2 \sec \theta \tan \theta + 2 \ln \left| \sec \theta + \tan \theta \right| + C$$

[Note that $2(C_1 + C_2)$ is still a constant, so it may be written as C.]

Figure 6.6 illustrates the assumption $x = 2 \tan \theta$, or $\tan \theta = x/2$. Using the theorem of Pythagoras and the basic definitions of

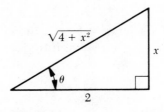

Figure 6.6

the trigonometric functions, we can label the sides of the right triangle. Then, from the diagram we read

$$\sec \theta = \frac{\sqrt{4 + x^2}}{2} \qquad \tan \theta = \frac{x}{2}$$

Therefore,

$$2 \sec \theta \tan \theta + 2 \ln \left| \sec \theta + \tan \theta \right|$$
$$= 2 \frac{\sqrt{4 + x^2}}{2} \left(\frac{x}{2} \right) + 2 \ln \left| \frac{\sqrt{4 + x^2}}{2} + \frac{x}{2} \right|$$
$$= \frac{x \sqrt{4 + x^2}}{2} + 2 \ln \left| \frac{x + \sqrt{4 + x^2}}{2} \right|$$

So that

$$\int \sqrt{4 + x^2}\, dx = \frac{x\sqrt{4 + x^2}}{2} + 2\ln \left| \frac{x + \sqrt{4 + x^2}}{2} \right| + C$$

Example 3   Determine $\int \dfrac{dx}{9 - 4x^2}$.

Solution   The denominator $9 - 4x^2$ is of the form $a^2 - u^2$, where $a = 3$ and $u = 2x$. Let $2x = 3 \sin\theta$; then $4x^2 = 9 \sin^2\theta$, and

$$9 - 4x^2 = 9 - 9 \sin^2\theta = 9(1 - \sin^2\theta) = 9 \cos^2\theta$$

If $2x = 3 \sin\theta$, $x = \frac{3}{2}\sin\theta$, and $dx = \frac{3}{2}\cos\theta\, d\theta$. Therefore,

$$\int \frac{dx}{9 - 4x^2} = \int \frac{(3/2)\cos\theta\, d\theta}{9\cos^2\theta}$$

$$= \frac{1}{6} \int \frac{d\theta}{\cos\theta} = \frac{1}{6} \int \sec\theta\, d\theta$$

$$= \frac{1}{6} \ln \left| \sec\theta + \tan\theta \right| + C$$

Figure 6.7 illustrates the assumption $2x = 3 \sin\theta$, which implies $\sin\theta = 2x/3$. Thus, $\sec\theta = 3/\sqrt{9 - 4x^2}$, and $\tan\theta = 2x/\sqrt{9 - 4x^2}$.

$$\int \frac{dx}{9 - 4x^2} = \frac{1}{6} \ln \left| \frac{3 + 2x}{\sqrt{9 - 4x^2}} \right| + C$$

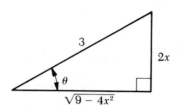

Figure 6.7

Example 4   Find $\int \dfrac{x^2}{\sqrt{1 - x^2}}\, dx$.

Solution   This problem could be solved by using parts.

$$\int \frac{x^2}{\sqrt{1-x^2}}\,dx = \int x \cdot \frac{x}{\sqrt{1-x^2}}\,dx$$
$$= -\frac{1}{2}\int x \cdot \frac{-2x}{\sqrt{1-x^2}}\,dx$$

If we let

$$u = x \quad \text{and} \quad dv = \frac{-2x}{\sqrt{1-x^2}}\,dx$$

then

$$du = dx \quad \text{and} \quad v = 2\sqrt{1-x^2}$$

$$\int \frac{x^2}{\sqrt{1-x^2}}\,dx = -\frac{1}{2}\int x \cdot \frac{-2x}{\sqrt{1-x^2}}\,dx$$
$$= -\frac{1}{2}\left[2x\sqrt{1-x^2} - \int 2\sqrt{1-x^2}\,dx\right] + C$$

The integral $\int \sqrt{1-x^2}\,dx$ now requires a trigonometric substitution. It seems, therefore, that a substitution in the first place would have been a better approach to this problem. Let $x = \sin\theta$; then $x^2 = \sin^2\theta$, and

$$\sqrt{1-x^2} = \sqrt{\cos^2\theta} = \cos\theta$$
$$dx = \cos\theta\,d\theta$$

Therefore,

$$\int \frac{x^2}{\sqrt{1-x^2}}\,dx = \int \frac{\sin^2\theta}{\cos\theta}(\cos\theta)\,d\theta$$
$$= \int \sin^2\theta\,d\theta$$
$$= \frac{1}{2}\int (1 - \cos 2\theta)\,d\theta$$
$$= \frac{1}{2}\left(\theta - \frac{1}{2}\sin 2\theta\right) + C$$

If $x = \sin\theta$, then $\theta = \sin^{-1}x$. Also, $\sin 2\theta = 2\sin\theta\cos\theta$, which the information from Figure 6.8 shows to be $\sin 2\theta = 2x\sqrt{1-x^2}$.

$$\int \frac{x^2}{\sqrt{1-x^2}}\,dx = \frac{1}{2}\left(\theta - \frac{1}{2}\sin 2\theta\right) + C$$

$$= \frac{1}{2} \sin^{-1} x - \left(\frac{1}{4}\right) 2x \sqrt{1 - x^2} + C$$

$$= \frac{1}{2} (\sin^{-1} x - x \sqrt{1 - x^2}) + C$$

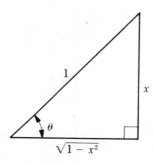

Figure 6.8

Example 5   Evaluate $\int_0^1 \sqrt{3 + x^2} \, dx$.

Solution   Let $x = \sqrt{3} \tan \theta$; then

$$3 + x^2 = 3(1 + \tan^2 \theta) = 3 \sec^2 \theta$$

and

$$\sqrt{3 + x^2} = \sqrt{3} \sec \theta$$
$$dx = \sqrt{3} \sec^2 \theta \, d\theta$$

It is important to note here that the original definite integral had its limits stated in terms of x, that is, from $x = 0$ to $x = 1$. The substitution, $x = \sqrt{3} \tan \theta$, requires either that the limits now be changed to correspond to θ, or that the integral in terms of θ be interpreted as an indefinite integral, then expressed in terms of x and evaluated for the original limits.

Method 1
Change of limits: If $x = 0$, then $x = \sqrt{3} \tan \theta$ implies that

$$0 = \sqrt{3} \tan \theta \qquad \text{and} \qquad \theta = 0$$

If $x = 1$, then $1 = \sqrt{3} \tan \theta$ and

$$\tan \theta = \frac{1}{\sqrt{3}}$$

$$\theta = \frac{\pi}{6}$$

$$\int_0^1 \sqrt{3 + x^2}\, dx = 3 \int_0^{\pi/6} \sec^3 \theta\, d\theta$$

From Example 2, we know that

$$\int \sec^3 \theta\, d\theta = \frac{1}{2}(\sec \theta \tan \theta + \ln|\sec \theta + \tan \theta|)$$

Thus

$$3 \int_0^{\pi/6} \sec^3 \theta\, d\theta = \frac{3}{2}(\sec \theta \tan \theta + \ln|\sec \theta + \tan \theta|) \Big]_0^{\pi/6}$$

$$= \frac{3}{2}\left(\frac{2}{\sqrt{3}} \cdot \frac{1}{\sqrt{3}} + \ln\left|\frac{2}{\sqrt{3}} + \frac{1}{\sqrt{3}}\right|\right) - \frac{3}{2}(0 + \ln 1)$$

$$= \frac{3}{2}\left[\frac{2}{3} + \ln\left(\frac{3}{\sqrt{3}}\right)\right]$$

$$= 1 + \frac{3}{2}\ln(\sqrt{3}) = 1 + \frac{3}{4}\ln 3$$

Method 2
Instead of changing limits, express $3 \int \sec^3 \theta\, d\theta$ in terms of x from Figure 6.9.

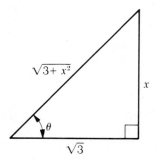

Figure 6.9

$$3 \int \sec^3 \theta\, d\theta = \frac{3}{2}(\sec \theta \tan \theta + \ln|\sec \theta + \tan \theta|)$$

$$= \frac{3}{2}\left(\frac{\sqrt{3 + x^2}}{\sqrt{3}} \cdot \frac{x}{\sqrt{3}} + \ln\left|\frac{\sqrt{3 + x^2}}{\sqrt{3}} + \frac{x}{\sqrt{3}}\right|\right)$$

and evaluate this expression $\Big]_0^1$

$$= \frac{3}{2}\left(\frac{2}{3} + \ln\left|\frac{2}{\sqrt{3}} + \frac{1}{\sqrt{3}}\right|\right) - \frac{3}{2}(0 + \ln 1)$$

$$= 1 + \frac{3}{2}\ln(\sqrt{3}) = 1 + \frac{3}{4}\ln 3$$

It is evident that method 1 is faster, and that a change of limits will often facilitate a given problem. It is sometimes convenient to write $\int_{x=a}^{x=b} u\,du$, which reminds us that a and b are limits for x.

Example 6   Evaluate $\displaystyle\int_2^4 \frac{x\,dx}{2 + x^2}$.

Solution   Let $u = 2 + x^2$, then $du = 2x\,dx$. Changing the limits to evaluate the integral for u, when $x = 2, u = 6$ and when $x = 4, u = 18$. Then

$$\int_2^4 \frac{x\,dx}{2 + x^2} = \frac{1}{2}\int_6^{18} \frac{du}{u} = \frac{1}{2}\ln u\ \Big]_6^{18}$$

$$= \frac{1}{2}(\ln 18 - \ln 6)$$

$$= \frac{1}{2}\ln\frac{18}{6}$$

$$= \frac{1}{2}\ln 3$$

Now that you understand how trigonometric substitution works, and what a tedious process it is to work with, you might take another look at the Table of Integrals, and in particular, at forms 13–21.

6.5 EXERCISES

In Exercises 1–15, integrate and verify your answers whenever practical by using the Table of Integrals.

1. $\displaystyle\int \sqrt{1 - x^2}\,dx$

2. $\displaystyle\int \frac{1}{\sqrt{1 + x^2}}\,dx$

3. $\int \sqrt{4 - x^2} \, dx$

4. $\int \sqrt{4x^2 - 1} \, dx$

5. $\int \dfrac{x + 1}{\sqrt{9 - x^2}} \, dx$

6. $\int \dfrac{\cos \theta}{\sqrt{3 - \sin^2 \theta}} \, d\theta$

7. $\int \dfrac{\sin^2 \theta}{1 + \cos 2\theta} \, d\theta$

8. $\int \dfrac{dx}{x \sqrt{x^2 + 1}}$

9. $\int \dfrac{\sqrt{x^2 + 1}}{x} \, dx$

10. $\int \dfrac{dx}{3x^2 + 2x + 4}$ [*Hint:* Complete the square.]

11. $\int \dfrac{dx}{3 - 2x - x^2}$

12. $\int \dfrac{x^2}{x^2 - 1} \, dx$

13. $\int \dfrac{x^2}{x^2 + 1} \, dx$

14. $\int (x^2 - 1) \, dx$

15. $\int \sqrt{x^2 - 1} \, dx$

In Exercises 16–24, evaluate by using an appropriate change of variables and limits.

16. $\displaystyle\int_1^{3/2} \dfrac{dx}{\sqrt{2x - x^2}}$

17. $\displaystyle\int_0^1 \dfrac{dx}{\sqrt{2 - x^2}}$

18. $\displaystyle\int_{-1}^{\sqrt{3}} \dfrac{x^2 \, dx}{(4 - x^2)^{3/2}}$

19. $\displaystyle\int_0^1 \dfrac{dx}{(x^2 + 1)^2}$

20. $\displaystyle\int_2^4 \dfrac{x \, dx}{x^2 + 3}$

21. $\displaystyle\int_0^2 x^2 \sqrt{1 + x^3} \, dx$

22. $\displaystyle\int_2^4 \dfrac{dx}{\sqrt{x} \, (\sqrt{x} - 1)}$

23. $\displaystyle\int_0^1 \dfrac{dx}{4 - x^2}$

24. $\displaystyle\int_0^2 \dfrac{dx}{x^2 + 4}$

6.6 IMPROPER INTEGRALS

The Fundamental Theorem of Calculus states that if a function f is defined and continuous over the interval $[a, b]$ in its domain, and if F is any antiderivative of f, then

$$\int_a^b f(x) \, dx = F(b) - F(a)$$

What happens if the hypotheses of the Fundamental Theorem are not satisfied? For instance, what about $\int_0^2 dx/x^2$? This integral is not defined at $x = 0$. Or $\int_1^\infty dx/x^2$? The interval $[1, \infty)$ is not closed, since ∞ is not a real number, and x gets larger and larger without bound.

Integrals of the types illustrated above are called *improper integrals,* and they may or may not exist. Let us first examine $\int_1^\infty dx/x^2$. If $\int_1^\infty dx/x^2$ is viewed as the area under the graph of $y = 1/x^2$, which is above the x axis and to the right of $x = 1$, this may be tested by considering the definite integral $\int_1^b dx/x^2$, and then the limit (if it exists) as $b \to \infty$.

$$\int_1^b \frac{dx}{x^2} = -\frac{1}{x}\Big]_1^b = -\frac{1}{b} + 1$$

$$\lim_{b \to \infty} \left(-\frac{1}{b} + 1\right) = 1$$

This limit exists; therefore, $\int_1^\infty dx/x^2 = 1$. The graph of this function (Figure 6.10) shows that the area under the curve over the interval $[1, \infty)$ is actually unbounded, yet the definite integral exists, and thus the area is measurable.

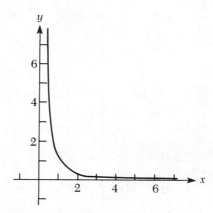

Figure 6.10 Graph of $y = \dfrac{1}{x^2}, \quad x > 0$

The other example, $\int_0^2 dx/x^2$ can be treated in a similar manner, by writing

$$\lim_{b \to 0^+} \int_b^2 \frac{dx}{x^2}$$

Note that $b \to 0^+$, that is, b approaches zero from the right side, and a one-sided limit is desired.

$$\lim_{b \to 0^+} \int_b^2 \frac{dx}{x^2} = \lim_{b \to 0^+} \left[-\frac{1}{x} \right]_b^2 = \lim_{b \to 0^+} \left(-\frac{1}{2} + \frac{1}{b} \right) = \infty$$

Since this limit does not exist, we cannot evaluate $\int_0^2 dx/x^2$.

Listed below are the definitions for improper integrals.

Definition

Let f be a continuous function for all $x \geqslant a$. Then

$$\int_a^\infty f(x) \, dx = \lim_{b \to \infty} \int_a^b f(x) \, dx$$

provided this limit exists.

Definition

Let f be a continuous function for all $x \leqslant a$. Then

$$\int_{-\infty}^a f(x) \, dx = \lim_{b \to -\infty} \int_b^a f(x) \, dx$$

provided this limit exists.

Definition

Let $f(x)$ be a continuous function for all x.

$$\int_{-\infty}^\infty f(x) \, dx = \int_{-\infty}^0 f(x) \, dx + \int_0^\infty f(x) \, dx$$

Example 1   Find the area under the curve $y = 1/(x^2 + 1)$ over the interval $(-\infty, \infty)$, if it exists.

Solution   The graph of the curve is represented in Figure 6.11. By definition,

$$\int_{-\infty}^{\infty} \frac{1}{x^2 + 1} \, dx = \int_{-\infty}^{0} \frac{1}{x^2 + 1} \, dx + \int_{0}^{\infty} \frac{1}{x^2 + 1} \, dx$$

since y is defined for all values of x.

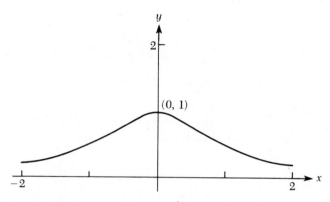

Figure 6.11 Graph of $y = \dfrac{1}{x^2 + 1}$

Clearly, the curve is symmetric about the y axis, and is always above the x axis. We can, therefore, attempt to compute $\lim\limits_{b \to \infty} \int_{0}^{b} dx/(x^2 + 1)$ and if the integral exists, double it.

$$\int_{0}^{b} \frac{dx}{x^2 + 1} = \tan^{-1} x \Big]_{0}^{b} = \tan^{-1} b - \tan^{-1} 0 = \tan^{-1} b$$

$$\lim_{b \to \infty} \tan^{-1} b = \frac{\pi}{2}$$

The area is twice this limit, $2(\pi/2) = \pi$.

Definition

Let f be a function continuous over $[a, b)$.

$$\int_{a}^{b} f(x) \, dx = \lim_{\delta_1 \to 0^+} \int_{a}^{b - \delta_1} f(x) \, dx$$

provided this limit exists.

Definition

Let f be a function continuous over $(a, b]$.

$$\int_a^b f(x)\, dx = \lim_{\delta_2 \to 0^+} \int_{a+\delta_2}^b f(x)\, dx$$

provided this limit exists.

The two definitions stated above lead to the following definition of the integral for a possible discontinuity of the integrand for a value of x between a and b.

Definition

Let f be a function continuous on the closed inverval $[a, b]$ *except* possibly at $x = c$, where $a < c < b$.

$$\int_a^b f(x)\, dx = \lim_{\delta_1 \to 0^+} \int_a^{c-\delta_1} f(x)\, dx + \lim_{\delta_2 \to 0^+} \int_{c+\delta_2}^b f(x)\, dx$$

provided *both* of these limits exist.

Example 2 Evaluate $\displaystyle \int_1^2 \frac{dx}{\sqrt{x-1}}$.

Solution When $x = 1$ the integrand is undefined. Using the definition,

$$\int_1^2 \frac{dx}{\sqrt{x-1}} = \lim_{\delta \to 0^+} \int_{1+\delta}^2 \frac{dx}{\sqrt{x-1}}$$

provided this limit exists

$$\lim_{\delta \to 0^+} \int_{1+\delta}^2 \frac{dx}{\sqrt{x-1}} = \lim_{\delta \to 0^+} 2\sqrt{x-1} \Big]_{1+\delta}^2$$

$$= \lim_{\delta \to 0^+} (2\sqrt{1} - 2\sqrt{\delta}) = 2$$

Example 3   Evaluate $\displaystyle\int_1^2 \frac{dx}{x-1}$.

Solution   On the surface, this example looks very similar to the previous example. When $x = 1$ the integrand is undefined. Therefore, by definition

$$\int_1^2 \frac{dx}{x-1} = \lim_{\delta \to 0^+} \int_{1+\delta}^2 \frac{dx}{x-1}$$

provided this limit exists.

$$\lim_{\delta \to 0^+} \int_{1+\delta}^2 \frac{dx}{x-1} = \lim_{\delta \to 0^+} \ln|x-1| \ \Big]_{1+\delta}^2$$
$$= \lim_{\delta \to 0^+} [\ln 1 - \ln \delta]$$
$$\lim_{\delta \to 0^+} \ln 1 = 0 \quad \text{but} \quad \lim_{\delta \to 0^+} \ln \delta = -\infty$$

Since the second limit does not exist we cannot evaluate the integral, which is therefore meaningless.

Example 4   Find $\displaystyle\int_{-\pi/4}^{\pi/4} \cot \theta \, d\theta$.

Solution   This integrand is undefined for a value of θ in the interval $(-\pi/4, \pi/4)$ rather than at an endpoint. Specifically, $\cot \theta$ is undefined for $\theta = 0$. By definition

$$\int_{-\pi/4}^{\pi/4} \cot \theta \, d\theta = \lim_{\delta_1 \to 0^+} \int_{-\pi/4}^{0-\delta_1} \cot \theta \, d\theta + \lim_{\delta_2 \to 0^+} \int_{0+\delta_2}^{\pi/4} \cot \theta \, d\theta$$

provided both limits exist.

$$\lim_{\delta_1 \to 0^+} \int_{-\pi/4}^{0-\delta_1} \cot \theta \, d\theta = \lim_{\delta_1 \to 0^+} \ln|\sin \theta| \ \Big]_{-\pi/4}^{0-\delta_1}$$
$$= \lim_{\delta_1 \to 0^+} \ln \left| \frac{\sin (0 - \delta_1)}{-\sqrt{2}/2} \right| = -\infty$$

Since this limit does not exist, we need not try to evaluate the second limit, and we cannot evaluate the integral.

6.6 Exercises

In Exercises 1–20, evaluate the integrals, if they exist. Use the Table of Integrals if necessary.

1. $\displaystyle\int_1^\infty \frac{2\,dx}{x^2}$

2. $\displaystyle\int_0^1 \frac{2\,dx}{x^2}$

3. $\displaystyle\int_1^\infty \frac{dx}{x}$

4. $\displaystyle\int_2^\infty \frac{3\,dx}{x^3}$

5. $\displaystyle\int_0^2 \frac{dx}{4-x^2}$

6. $\displaystyle\int_{-2}^0 \frac{dx}{x+2}$

7. $\displaystyle\int_0^\infty e^x\,dx$

8. $\displaystyle\int_0^\infty e^{-x}\,dx$

9. $\displaystyle\int_1^\infty \frac{dx}{x^{3/2}}$

10. $\displaystyle\int_3^\infty \frac{dx}{(1+x)^{3/2}}$

11. $\displaystyle\int_2^9 \frac{dx}{\sqrt{x-2}}$

12. $\displaystyle\int_0^8 \frac{dx}{\sqrt[3]{x-2}}$

13. $\displaystyle\int_{-\infty}^\infty \frac{dx}{4x^2+1}$

14. $\displaystyle\int_{-\infty}^\infty e^{-3x}\,dx$

15. $\displaystyle\int_{-3}^3 \frac{dx}{x^3}$

16. $\displaystyle\int_1^3 \frac{dx}{\sqrt{9-x^2}}$

17. $\displaystyle\int_{-\infty}^0 \frac{x\,dx}{e^{x^2}}$

18. $\displaystyle\int_0^\infty \sqrt{e^x}\,dx$

19. $\displaystyle\int_0^\infty \frac{dx}{\sqrt{e^x}}$

20. $\displaystyle\int_0^1 \frac{\ln x\,dx}{x}$

21. Show that $\displaystyle\int_0^\infty e^{-ax}\,dx = \frac{1}{a}$, $a > 0$.

22. Show that $\displaystyle\int_0^\infty xe^{-ax}\,dx = \frac{1}{a^2}$, $a > 0$ provided $\displaystyle\lim_{b\to\infty} \frac{b}{e^{ab}} = 0$.

23. Consider $\displaystyle\int_0^1 x^n\,dx$. Is this integral improper for any values of n? If so, state these values.

In Exercises 24–30, evaluate the integrals if they exist.

24. $\displaystyle\int_{\pi/4}^{\pi/2} \tan\theta\,d\theta$

25. $\displaystyle\int_0^\pi \frac{d\theta}{1+\cos\theta}$ [*Hint:* Multiply numerator and denominator by $(1-\cos\theta)$.]

26. $\displaystyle\int_{\pi/2}^{\pi} \frac{d\theta}{1 - \sin\theta}$ (See hint in Exercise 25.)

27. $\displaystyle\int_{1}^{2} \frac{dx}{\sqrt{x - 1}}$

28. $\displaystyle\int_{1}^{\infty} \frac{x\,dx}{(1 + x^2)^2}$

29. $\displaystyle\int_{-1}^{2} \frac{dx}{\sqrt[3]{x}}$

30. $\displaystyle\int_{-1}^{1} \frac{dx}{x^2}$

NEW TERMS

Elasticity

Exponential function

Improper integral

Integration by parts

Logarithmic function

Natural logarithm

REVIEW

In Exercises 1–12, find dy/dx.

1. $y = e^{x^2+1}$

2. $y = \dfrac{1}{\sqrt{e^x}}$

3. $y = xe^{x^2}$

4. $y = 3^{2x+5}$

5. $y = 10^{2-x}$

6. $y = \ln(x^2 + x + 1)$

7. $y = x \ln x$

8. $y = \ln\left(\dfrac{x^2 + 1}{3x - 2}\right)$

9. $y = \ln \sqrt[3]{x^4 + 1}$

10. $y = \dfrac{e^x}{\ln e^x}$

11. $y = \log_5 x^2$

12. $y = \log_2(3x + 1)$

In Exercises 13–30, integrate and use the Table of Integrals if necessary.

13. $\int e^{3x}\,dx$

14. $\int xe^{x^2}\,dx$

15. $\int \dfrac{\ln x}{x}\,dx$

16. $\int \dfrac{dx}{x + 4}$

17. $\int \dfrac{x}{x^2 + 4}\,dx$

18. $\int \dfrac{dx}{x^2 + 4}$

19. $\int \ln x\,dx$

20. $\int x3^{x^2}\,dx$

21. $\int e^{2x}\sin x\,dx$

22. $\int \sqrt{4 - x^2}\,dx$

23. $\int \dfrac{dx}{4 - x^2}$

24. $\int \dfrac{\sin \sqrt{x}}{\sqrt{x}} \, dx$

25. $\int x \cos x \, dx$

26. $\int \sin^{-1} x \, dx$

27. $\int x \ln x^2 \, dx$

28. $\int x \ln x \, dx$

29. $\int \dfrac{dx}{x^2 + 4x + 6}$

30. $\int \dfrac{dx}{\sqrt{1 - 4x - x^2}}$

In Exercises 31–40, evaluate by the Fundamental Theorem.

31. $\displaystyle\int_1^5 (x^3 + 2x + 1) \, dx$

32. $\displaystyle\int_0^3 x^2 \sqrt{x^3 + 9} \, dx$

33. $\displaystyle\int_{-2}^2 x^2 e^x \, dx$

34. $\displaystyle\int_1^4 x \ln x \, dx$

35. $\displaystyle\int_1^2 x \, 2^{x^2} \, dx$

36. $\displaystyle\int_0^{\pi/2} \sin^2 \theta \, d\theta$

37. $\displaystyle\int_1^0 \tan^{-1} x \, dx$

38. $\displaystyle\int_0^1 \dfrac{dx}{1 + x^2}$

39. $\displaystyle\int_2^5 \dfrac{dx}{x + 1}$

40. $\displaystyle\int_{-\pi/2}^0 (1 - \sin x) \, dx$

In Exercises 41–45, evaluate each improper integral, if it exists.

41. $\displaystyle\int_0^4 \dfrac{2 \, dx}{\sqrt{x}}$

42. $\displaystyle\int_1^\infty \dfrac{2 \, dx}{\sqrt{x}}$

43. $\displaystyle\int_1^\infty \dfrac{dx}{x^2 + 1}$

44. $\displaystyle\int_1^\infty \dfrac{dx}{x^3}$

45. $\displaystyle\int_1^\infty \dfrac{dx}{x}$

46. For the function defined by $f(x) = e^x \sin x$, find all maximum and/or minimum values over the interval $[-\pi, \pi]$.

47. For what values of x is the function defined in Exercise 46 increasing? Decreasing? Concave upward? Concave downward?

48. Use the information from Exercises 46 and 47 to sketch the graph of $y = e^x \sin x$ over the interval $[-\pi, \pi]$.

49. Evaluate $\displaystyle\int_{-\pi}^\pi e^x \sin x \, dx$. Does this integral correspond to the area under the curve and bounded by the x axis and the line $x = \pi$? Explain.

7

Applications of Integration

The last two chapters have introduced the integral and developed techniques of integration. With these skills at hand, we can now inspect some of the applications of integration. We say *some* because the number is very large, literally hundreds. Many of the applications are in science and engineering. We will hint at these in this chapter in the sections on area and volume. Social science applications are contained in the sections on probability and normal curve areas. Physical scientists will find applications in the discussion of natural growth and decay. The sections on supply and demand functions and on marginal analysis relate to business and economics, as does the section dealing with capital and present value. The last section of this chapter, elementary differential equations, can be applied to all of the above mentioned areas of application and many more.

7.1 AREAS

Finding the areas bounded by plane curves has already been considered in a previous chapter. Specifically, if $y = f(x)$ defines a function such that $f(x) \geq 0$ for all $x \in [a, b]$, then

$$\int_a^b f(x)\, dx$$

equals the area bounded by the graphs of $y = f(x)$, $x = a$, $x = b$, and $y = 0$ (the x axis), provided the integral in question exists.

Example 1   Find the area bounded by $y = \sin x$, $x = 0$, $x = \pi/2$, and $y = 0$.

Solution $\text{Area} = \displaystyle\int_0^{\pi/2} \sin x \, dx = -\cos x \Big]_0^{\pi/2}$

$$= -\cos \frac{\pi}{2} - (-\cos 0)$$

$$= 0 - (-1) = 1$$

The area in this example is illustrated in Figure 7.1. There are any number of variations of this basic result.

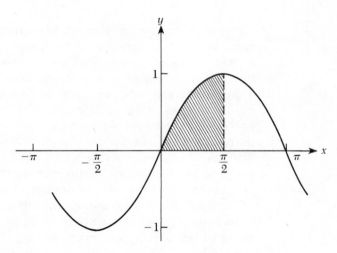

Figure 7.1 $y = \sin x$, $x \in [0, \pi/2]$

Variation

Consider a function f given by $y = f(x)$ where $f(x) \le 0$ for all x such that $x \in [a, b]$. Associated with this function there is a second function given by $y = |f(x)|$, which is always nonnegative over the interval $[a, b]$. This situation is illustrated in Figure 7.2.

It is reasonable to conclude that the two shaded areas of Figure 7.2 are equal. Since $|f(x)|$ is a positive-valued function on the interval, the area of Figure 7.2a and 7.2b is given by

$$\text{Area} = \int_a^b |f(x)| \, dx$$

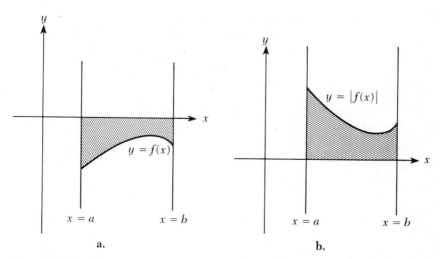

Figure 7.2   **a.** $y = f(x) \leq 0, \quad x \in [a, b]$; **b.** $y = |f(x)| \geq 0, \quad x \in [a, b]$

Example 2   Find the area bounded by $y = -x^2$, $x = 2$, $x = 4$, and the x axis.

Solution   The area is shown in Figure 7.3. This area is given by

$$\int_2^4 \left| -x^2 \right| dx = \int_2^4 x^2 dx = \left. \frac{x^3}{3} \right]_2^4$$
$$= \frac{64}{3} - \frac{8}{3} = \frac{56}{3}$$

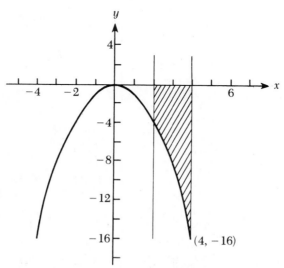

Figure 7.3   The area enclosed by $y = -x^2$, $x = 2$, $x = 4$, and $y = 0$

Variation

Consider a function that is sometimes positive-valued and sometimes negative-valued on an interval $[a, b]$. The graph of such a function is shown in Figure 7.4. Assume that the zeros of this

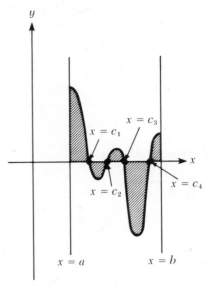

Figure 7.4 A positive/negative-valued function with zeros at $x = c_1, c_2, c_3,$ and c_4

function are known, for example, in the figure: $f(c_1) = f(c_2) = f(c_3) = f(c_4) = 0$. Further suppose that

$$f(x) \geq 0 \quad \text{for} \quad x \in [a, c_1]$$
$$f(x) \leq 0 \quad \text{for} \quad x \in [c_1, c_2]$$
$$f(x) \geq 0 \quad \text{for} \quad x \in [c_2, c_3]$$
$$f(x) \leq 0 \quad \text{for} \quad x \in [c_3, c_4]$$
$$f(x) \geq 0 \quad \text{for} \quad x \in [c_4, b]$$

The total area in this type of problem can be found by breaking down the problem into a number of parts, evaluating the individual areas in turn, and finding the sum of the partial areas. For the above problem, the shaded area of Figure 7.4 is

$$\int_a^{c_1} f(x)\,dx + \int_{c_1}^{c_2} |f(x)|\,dx + \int_{c_2}^{c_3} f(x)\,dx + \int_{c_3}^{c_4} |f(x)|\,dx + \int_{c_4}^{b} f(x)\,dx$$

which becomes

$$\int_a^{c_1} f(x)\,dx - \int_{c_1}^{c_2} f(x)\,dx + \int_{c_2}^{c_3} f(x)\,dx - \int_{c_3}^{c_4} f(x)\,dx + \int_{c_4}^{b} f(x)\,dx$$

Example 3 Find the area bounded by the sine curve and the x-axis between $x = 0$ and $x = 4\pi$.

Solution The area in question is shown in Figure 7.5.

$$\int_0^{\pi} \sin x\,dx - \int_{\pi}^{2\pi} \sin x\,dx + \int_{2\pi}^{3\pi} \sin x\,dx - \int_{3\pi}^{4\pi} \sin x\,dx$$

$$= -\cos x\Big]_0^{\pi} + \cos x\Big]_{\pi}^{2\pi} - \cos x\Big]_{2\pi}^{3\pi} + \cos x\Big]_{3\pi}^{4\pi}$$

$$= -\cos \pi + \cos 0 + \cos 2\pi - \cos \pi - \cos 3\pi$$
$$\qquad\qquad\qquad\qquad + \cos 2\pi + \cos 4\pi - \cos 3\pi$$

$$= -(-1) + 1 + 1 - (-1) - (-1) + 1 + 1 - (-1)$$
$$= 1 + 1 + 1 + 1 + 1 + 1 + 1 + 1 = 8$$

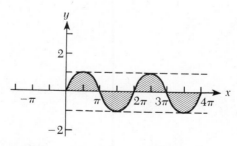

Figure 7.5 $y = \sin x, \quad x \in [0, 4\pi]$

Example 4 Find the area bounded by the x axis, $y = (x - 3)(x - 2)(x + 1)$, $x = 0$, and $x = 4$.

Solution The graph of this function is shown in Figure 7.6. The area is given by

$$\int_0^2 (x^3 - 4x^2 + x + 6)\, dx - \int_2^3 (x^3 - 4x^2 + x + 6)\, dx$$

$$+ \int_3^4 (x^3 - 4x^2 + x + 6)\, dx$$

$$= \frac{x^4}{4} - \frac{4x^3}{3} + \frac{x^2}{2} + 6x \bigg]_0^2 - \left[\frac{x^4}{4} - \frac{4x^3}{3} + \frac{x^2}{2} + 6x \right]_2^3$$

$$+ \left[\frac{x^4}{4} - \frac{4x^3}{3} + \frac{x^2}{2} + 6x \right]_3^4$$

$$= \frac{22}{3} - \left(-\frac{7}{12} \right) + \frac{47}{12} = \frac{88 + 7 + 47}{12} = \frac{142}{12} = \frac{71}{6}$$

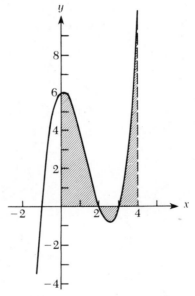

Figure 7.6 $y = (x - 3)(x - 2)(x + 1)$

$$= x^3 - 4x^2 + x + 6$$

Variation

Consider the area bounded by the graphs of $y = f(x)$, $y = g(x)$, $x = a$, and $x = b$ where $g(x) \leqslant f(x)$ for all $x \in [a, b]$. The situation is illustrated in Figure 7.7.

The required area is the difference between the two shaded areas; that is

$$\text{Area} = \int_a^b f(x)\, dx - \int_a^b g(x)\, dx = \int_a^b (f(x) - g(x))\, dx$$

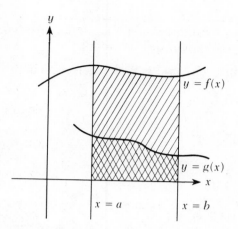

Figure 7.7 $y = f(x), y = g(x)$ with $f(x) \geqslant g(x)$
for all $x \in [a, b]$

Example 5 Find the area bounded by $y = x^2 + 2$, $y = x$, $x = 1$, and $x = 2$ as shown in Figure 7.8.

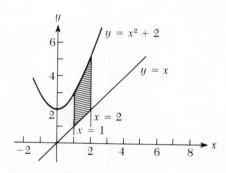

Figure 7.8 $y = x, y = x^2 + 2, x = 1$, and $x = 2$

Solution The area is given by

$$\int_1^2 (x^2 + 2)\, dx - \int_1^2 x\, dx = \int_1^2 (x^2 + 2 - x)\, dx = \frac{x^3}{3} + 2x - \frac{x^2}{2} \Big]_1^2$$

$$= \frac{8}{3} + 4 - 2 - \left(\frac{1}{3} + 2 - \frac{1}{2}\right)$$

$$= \frac{8}{3} + 2 - \frac{1}{3} - 2 + \frac{1}{2}$$

$$= \frac{7}{3} + \frac{1}{2} = \frac{17}{6}$$

Example 6   Find the area bounded by $y = \cos x + 1$, $y = \frac{3}{2}$, $x = 0$, and $x = \pi$ as shown in Figure 7.9.

Solution   The two curves intersect where

$$\cos x + 1 = \frac{3}{2}$$

or

$$\cos x = \frac{1}{2}$$

that is, where

$$x = \frac{\pi}{3} \quad \text{and} \quad y = \frac{3}{2}$$

The area can be found by breaking the problem into two parts. For

$$x \in \left[0, \frac{\pi}{3}\right] \qquad \cos x + 1 \geqslant \frac{3}{2}$$

and for

$$x \in \left[\frac{\pi}{3}, \pi\right] \qquad \frac{3}{2} \geqslant \cos x + 1$$

Thus the area is

$$\int_0^{\pi/3} \left(\cos x + 1 - \frac{3}{2}\right) dx + \int_{\pi/3}^{\pi} \left(\frac{3}{2} - [\cos x + 1]\right) dx$$

$$= \int_0^{\pi/3} \left(\cos x - \frac{1}{2}\right) dx + \int_{\pi/3}^{\pi} \left(\frac{1}{2} - \cos x\right) dx$$

$$= \sin x - \frac{x}{2}\Big]_0^{\pi/3} + \left(\frac{x}{2} - \sin x\right)\Big]_{\pi/3}^{\pi}$$

$$= \sin \frac{\pi}{3} - \frac{\pi}{6} + \left[\frac{\pi}{2} - \left(\frac{\pi}{6} - \sin \frac{\pi}{3}\right)\right]$$

$$= \frac{\sqrt{3}}{2} - \frac{\pi}{6} + \left(\frac{\pi}{2} - \frac{\pi}{6} + \frac{\sqrt{3}}{2}\right) = \sqrt{3} + \frac{\pi}{6}$$

In this example one of the two functions is a constant function.

Variation
Sometimes a problem can be set up and solved with respect to an inverse function more easily than with respect to the given function. This is best illustrated with an example.

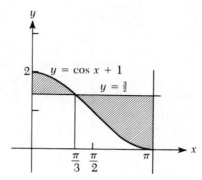

Figure 7.9 $y = \cos x + 1$, $y = \frac{3}{2}$,
$x = 0$, and $x = \pi$

Example 7   Find the area bounded by $y = \ln x$, $x = 0$, $y = 0$, and $y = \ln 4$.

Solution   The area is the shaded portion of Figure 7.10. The area is given by

$$A_1 + A_2 = \ln 4 + \int_1^4 (\ln 4 - \ln x)\, dx$$

This integral is complicated. The problem can be simplified by viewing x as a function of y in the equation $y = \ln x$.

From the definition of $y = \ln x$, it follows that $x = e^y$. Now, suppose we partition the y *axis* from $y = 0$ to $y = \ln 4$. Then the shaded portion of Figure 7.10 has area

$$\int_0^{\ln 4} e^y\, dy = e^y \Big]_0^{\ln 4}$$
$$= e^{\ln 4} - e^0$$
$$= 4 - 1 = 3$$

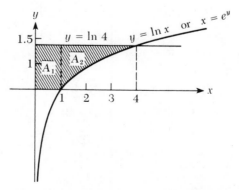

Figure 7.10 $y = \ln x$, $x = 0$, $y = 0$, and $y = \ln 4$

The variations and examples considered above make it clear that the key to solution is determining the exact integrand and limits of integration. Therefore, the sketch of the problem is invaluable. Once a sketch has been made it is usually a simple matter to find the required integral or integrals. It should be observed that in many problems the determination of the nature of the graph of the functions involved may require the applications of the graphing methods using calculus, developed earlier in this text.

7.1 EXERCISES

In Exercises 1–23, find the finite area bounded by the graphs of the given functions. Make a sketch.

1. $y = x^3, \quad x = -3, \quad x = -2, \quad y = 0$
2. $y = \sqrt{x} - 4, \quad x = 1, \quad x = 4, \quad y = 0$
3. $y = x^2 + 2x - 3, \quad x = -2, \quad x = 0, \quad y = 0$
4. $y = x^2 + 2x - 3, \quad x = -4, \quad x = 2, \quad y = 0$
5. $y = x^3 + x, \quad x = 1, \quad y = 0$
6. $y = x^3 + x, \quad x = -1, \quad y = 0$
7. $y = x^3 + x, \quad x = 0, \quad y = 2$
8. $y = (x - 3)(x - 2)(x + 1), \quad x = 0, \quad x = 4, \quad y = 0$
9. $y = \sqrt{x}, \quad y = x$
10. $y = 0, \quad y = x, \quad y = -x + 2, \quad y = x - 4$
11. $y = x^2 + 6, \quad y = x^2 + 2, \quad x = 0, \quad x = 1$
12. $y = x^2 + 2, \quad y = 4 - x^2$
13. $y = x^2 - 6x + 9, \quad y = 4$
14. $y = x^2 + 10x + 25, \quad y = -x^2 - 10x - 25, \quad x = -5, \quad x = 0$
15. $y = \sin x, \quad x = -\pi/2, \quad x = \pi/2, \quad y = 0$
16. $y = (\sin x) - 1, \quad y = 0, \quad x = \pi, \quad x = 4\pi$
17. $y = -e^{-x}, \quad x = \ln 2, \quad x = \ln 5, \quad y = 0$
18. $y = \ln x, \quad y = \ln 3, \quad x = 10, \quad x = 100$
19. $y = \ln x^2, \quad y = \ln 4, \quad x = e$
20. $y = \tan x, \quad x = 2\pi/3, \quad x = \pi, \quad y = 0$
21. $y = \tan^2 x, \quad y = 0, \quad x = \pi/3, \quad x = 0$
22. $y = 0, \quad y = (x - 3)(x - 1)(x - 5)$
23. $y = 0, \quad y = (3x - 2)(x + 2)(x - 2)$

In Exercises 24–27, set up—but do not evaluate—integrals to find the area bounded by the indicated curves.

24. $x = 0$, $y = 0$, $x = 3$, $y = e^{-x^2}$
25. $y = 0$, $y = \sqrt{(x - 2)(x - 3)}$
26. $x^2 + y^2 = 4$, $y \geq 0$
27. $x^{2/3} + y^{2/3} = 8^{2/3}$, $x = 0$, $y \geq 0$

7.2 PROBABILITY

Many applications of integration are made by giving a suitable interpretation to the area under a curve. A very useful application involves viewing such areas as the probability of an outcome of an experiment.

> **Definition**
>
> The *probability* of a specific outcome of an experiment is the fraction of the time that outcome can be expected to take place if the experiment is repeated a very large number of times.

Example 1

Consider the experiment of tossing a coin. The probability of a head is $\frac{1}{2}$ (assuming an honest coin), since half of the time one would expect to get a head.

Example 2

In the experiment of rolling a pair of dice, $\frac{1}{6}$ of the time 7 will come up; thus the probability of a 7 is $\frac{1}{6}$.

An analysis of a given experiment can be achieved by specifying a complete set of *mutually exclusive* outcomes for the experiment together with the probability of each outcome. To require the outcomes to be mutually exclusive is to insist that the occurrence of one outcome precludes the occurrence of any other outcome of the set. The set of outcomes must be extensive enough so that any possible outcome of the experiment falls into one of the outcomes in the set.

Example 3	Consider an experiment with a pair of dice. This experiment can be analyzed with the following table showing a set of mutually exclusive outcomes together with the corresponding distribution of probabilities.

Outcome: total spots	Probability	Outcome: total spots	Probability
2	$\frac{1}{36}$	8	$\frac{5}{36}$
3	$\frac{2}{36} = \frac{1}{18}$	9	$\frac{4}{36} = \frac{1}{9}$
4	$\frac{3}{36} = \frac{1}{12}$	10	$\frac{3}{36} = \frac{1}{12}$
5	$\frac{4}{36} = \frac{1}{9}$	11	$\frac{2}{36} = \frac{1}{18}$
6	$\frac{5}{36}$	12	$\frac{1}{36}$
7	$\frac{6}{36} = \frac{1}{6}$		

These probabilities were calculated by thinking of a pair of dice, one red and one green. Then there are 36 possible outcomes as listed below:

Red	Green	Red	Green	Red	Green
1	1	3	1	5	1
1	2	3	2	5	2
1	3	3	3	5	3
1	4	3	4	5	4
1	5	3	5	5	5
1	6	3	6	5	6
2	1	4	1	6	1
2	2	4	2	6	2
2	3	4	3	6	3
2	4	4	4	6	4
2	5	4	5	6	5
2	6	4	6	6	6

The question is how many of these outcomes result in a total of 2 spots, how many in 3, and so forth. For 2 spots there is only 1 outcome. Thus the probability of a 2 is $\frac{1}{36}$. There are 2 ways to get 3 spots, therefore the probability of 3 spots is $\frac{2}{36}$, and so on.

Notice that the sum of the probabilities is 1. Figure 7.11 illustrates a probability distribution for this experiment. Each possible outcome is indicated by its total number of spots represented on a number line. A series of equal-width columns is then drawn. The height of each column is adjusted so that the area in each column represents the same fraction of the total area of all the columns as the probability of the outcome.

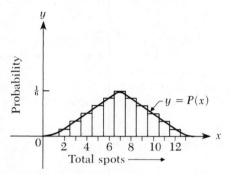

Figure 7.11   Dice probability diagram

This approach is fine in cases like the above where the number of possible outcomes of the experiment is small. However, when the number of outcomes being considered becomes large, it is usually simpler to replace discrete columns with a continuous function that yields corresponding areas when integrated between the bounding values of a given column. For example if $y = P(x)$ is the smooth curve shown in Figure 7.11,

$$\int_{1.5}^{2.5} P(x)\,dx = \frac{1}{36}$$

so that the area corresponding to the outcome "2" will be correct. Similarly,

$$\int_{2.5}^{3.5} P(x)\,dx = \frac{2}{36}$$
$$\int_{3.5}^{4.5} P(x)\,dx = \frac{3}{36}$$

and so forth.

The *probability density function* for the experiment is P. In other words, a probability density function is a continuous function that can be used to replace the discrete descriptions of probability. Then the probabilities can be calculated by integration.

While in many real-life cases the actual nature of such probability density functions is complex, the basic properties of such a function are relatively simple. First, we will be talking about experiments whose outcomes correspond to some set of real numbers. That is, the set of outcomes for the experiment will be represented by a number x, where $x \in [a, b]$. Second, $P(x) \geq 0$ for all $x \in [a, b]$. Third, $\int_a^b P(x)\,dx = 1$; that is, the total area bounded by the probability density function and the x axis must be 1, because the probability that a given result will be one of the possible answers is, of course, 1. In practice almost any nonnegative-valued function f whose integral is finite on a specific interval might represent a probability density function. The factor

$$c = \frac{1}{\int_a^b f(x)\,dx}$$

can be used to adjust the area involved to 1 square unit. That is, if

$$c = \frac{1}{\int_a^b f(x)\,dx}$$

and

$$P(x) = cf(x)$$

then P could be a probability density function, because

$$\int_a^b P(x)\,dx = \int_a^b cf(x)\,dx = c \int_a^b f(x)\,dx$$
$$= \frac{1}{\int_a^b f(x)\,dx} \cdot \int_a^b f(x)\,dx = 1$$

Example 4   Consider the function f such that $f(x) = (\sin x)/2$ on $[0, \pi]$. Within this interval $\sin x \geq 0$, and

$$\int_0^\pi \frac{\sin x}{2}\,dx = \frac{1}{2}\,[-\cos x]_0^\pi$$
$$= \frac{1}{2}\,[-\cos \pi - (-\cos 0)]$$
$$= \frac{1}{2}\,(+1 + 1) = 1$$

Thus, $f(x) = (\sin x)/2$ meets the basic requirements of a probability density function. The area is shown in Figure 7.12.

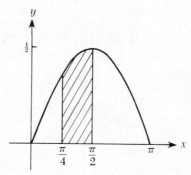

Figure 7.12 $y = \dfrac{\sin x}{2}, \quad x \in \left[\dfrac{\pi}{4}, \dfrac{\pi}{2}\right]$

Example 5   Assume that the outcomes of a certain experiment correspond to real-number values between 0 and π and that the probability density function P for the outcomes is known to be $P(x) = \frac{1}{2}\sin x$. What is the probability that when the experiment is performed an outcome between $\pi/4$ and $\pi/2$ will occur?

Solution   The area is the shaded portion of Figure 7.12. The probability will correspond to the area bounded by the x axis and the probability density function and the lines $x = \pi/4$ and $x = \pi/2$. Specifically, the probability of an outcome between $\pi/4$ and $\pi/2$ is

$$
\begin{aligned}
\int_{\pi/4}^{\pi/2} \frac{1}{2}\sin x \, dx &= \frac{1}{2}(-\cos x)\Big]_{\pi/4}^{\pi/2} \\
&= \frac{1}{2}\left[-\cos\frac{\pi}{2} - \left(-\cos\frac{\pi}{4}\right)\right] \\
&= \frac{1}{2}\left[\cos\frac{\pi}{4}\right] \\
&= \frac{1}{2}\cdot\frac{\sqrt{2}}{2} = \frac{\sqrt{2}}{4} \approx \frac{1.414}{4} = 0.3535
\end{aligned}
$$

Thus approximately 35.35 percent of the time one would expect an experimental outcome between $\pi/4$ and $\pi/2$ for this experiment.

It is often the case that the experimental values in question will correspond to improper integrals.

Example 6

A behavioral psychologist is studying the length of time x it takes a group of individuals to learn a task. Through trial and error it is discovered that the probability density function, P, for any group is given by

$$P(x) = \frac{1}{(x + 1)^2} \quad \text{for} \quad x \in [0, \infty)$$

where x is measured in minutes. What is the probability that any group selected at random will require 3 minutes or more to learn the task?

Solution

The probability of an experimental outcome greater than or equal to 3 will be given by

$$\int_3^\infty \frac{1}{(x + 1)^2} \, dx = \lim_{b \to \infty} \int_3^b \frac{1}{(x + 1)^2} \, dx$$

$$= \lim_{b \to \infty} \left[\frac{-1}{(x + 1)} \right]_3^b$$

$$= \lim_{b \to \infty} \left[\frac{-1}{b + 1} + \frac{1}{3 + 1} \right] = \frac{1}{4}$$

This area is shown in Figure 7.13. Thus for this experiment $\frac{1}{4}$ of the time an outcome greater than or equal to 3 can be expected or $\frac{1}{4}$ of the groups would require 3 minutes or more to learn the task.

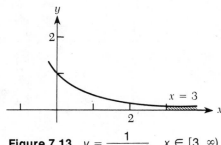

Figure 7.13 $y = \dfrac{1}{(x + 1)^2}, \quad x \in [3, \infty)$

Sometimes the interval of outcomes can extend from $-\infty$ to $+\infty$.

Example 7

Consider an experiment whose outcomes range from $-\infty$ to $+\infty$ and whose probability density function is given by

$$P(x) = \frac{1}{\pi} \frac{1}{x^2 + 1}$$

What is the probability that an outcome corresponding to values between $x = 1$ and $x = \sqrt{3}$ will occur?

Solution   Checking that this function meets the requirements of a probability density function,

$$\int_{-\infty}^{\infty} \frac{1}{\pi} \frac{1}{x^2 + 1} \, dx = \frac{1}{\pi} \lim_{b \to \infty} \int_{-b}^{b} \frac{1}{x^2 + 1} \, dx$$

$$= \frac{1}{\pi} \lim_{b \to \infty} \tan^{-1} x \Big]_{-b}^{b}$$

$$= \frac{1}{\pi} \left\{ \lim_{b \to \infty} [\tan^{-1} b - \tan^{-1}(-b)] \right\}$$

$$= \frac{1}{\pi} \left[\frac{\pi}{2} - \left(-\frac{\pi}{2} \right) \right] = 1$$

Then the probability of an outcome between 1 and $\sqrt{3}$ is given by

$$\frac{1}{\pi} \int_{1}^{\sqrt{3}} \frac{1}{x^2 + 1} \, dx$$

and is shown in Figure 7.14 (page 328).

$$\frac{1}{\pi} \int_{1}^{\sqrt{3}} \frac{1}{x^2 + 1} \, dx = \frac{1}{\pi} \tan^{-1} x \Big]_{1}^{\sqrt{3}}$$

$$= \frac{1}{\pi} [\tan^{-1} \sqrt{3} - \tan^{-1} 1]$$

$$= \frac{1}{\pi} \left[\frac{\pi}{3} - \frac{\pi}{4} \right] = \frac{1}{12}$$

Thus, $\frac{1}{12}$ of the time the described experiment will have an outcome between 1 and $\sqrt{3}$.

The practical applications of probability often invoke calculations similar to those considered here. Usually the problem involves the comparison of an observed experimental outcome with the theoretical probability of its occurrence. For example, it might be known that 95 percent of all patients with a certain medical problem die. Based on this, one can construct a mathematical probability model of what might be expected in a group of 20 patients. Then such a group is given a new treatment. Suppose that all but 2 recover. By assuming that the treatment

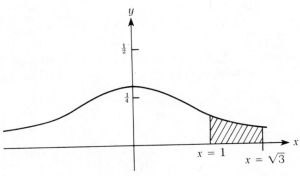

Figure 7.14 $y = \dfrac{1}{\pi} \cdot \dfrac{1}{x^2 + 1}$, $x = 1$, and $x = \sqrt{3}$

was not effective and then using the mathematical model one can calculate the probability that only 2 of the 20 would die. Suppose that this probability comes out 0.00002 percent. That is, 99.99998 percent of the time *more* than 2 patients would be expected to die. It then would appear reasonable to conclude that the treatment was effective.

Naturally we have oversimplified our example to illustrate one possible direction of application of the theory of probability. However, a modern course in probability theory and its application makes use of the calculus tools we have considered here.

7.2 EXERCISES

1. A life insurance table states that the probability of a 70-year-old man living to be 80 is $\frac{4}{9}$. Interpret this according to the definition of probability.

2. A gambler observes that on the average she wins a certain game 60 times out of every 110 times she plays. What is her probability of winning? What is her probability of losing?

3. A single card is drawn from a deck of playing cards. Two possible outcomes are that the card is a red card, or it is an ace. Are these outcomes mutually exclusive? If not, why not?

4. Give an example of three mutually exclusive outcomes of an experiment involving tossing five coins and counting the number of heads.

In Exercises 5–8, assume that an experiment has outcomes corresponding to real-number values between 0 and π with a
probability density function given by $P(x) = \frac{1}{2}\sin x$. Determine
the probability of outcomes of this experiment in the domain
indicated. Sketch the curve and shade the appropriate area.

5. An outcome $x \in \left[\dfrac{\pi}{4}, \dfrac{3\pi}{4}\right]$

6. An outcome $x \in \left[0, \dfrac{\pi}{2}\right]$

7. An outcome x such that $-\dfrac{\pi}{6} \leqslant x - \dfrac{\pi}{2} \leqslant \dfrac{\pi}{6}$

8. An outcome x such that $x - \dfrac{\pi}{2} \geqslant \dfrac{\pi}{3}$ or $x - \dfrac{\pi}{2} \leqslant -\dfrac{\pi}{3}$

In Exercises 9–12, assume that the probability density function
for the time x in minutes it takes a group of individuals to learn
a task is given by

$$P(x) = \frac{1}{(x+1)^2} \qquad x \in [0, \infty)$$

In each case, determine the probability of an outcome in the
domain indicated. Sketch the curve and shade the appropriate
area.

9. An outcome $x \in [1, 6]$ **10.** An outcome $x \in [4, \infty)$

11. An outcome $x \in [0, 1]$ **12.** An outcome $x \in [1, 5]$

In Exercises 13–16, assume that an experiment has outcomes
corresponding to the real numbers with a probability density
function given by

$$P(x) = \frac{1}{\pi}\frac{1}{x^2 + 1}$$

In each case, determine the probability of an outcome in the
domain indicated. Make use of the trigonometric function tables
in the Appendix. Sketch the curve and shade the appropriate
area.

13. An outcome such that $x \in \left[0, \dfrac{1}{\sqrt{3}}\right]$

14. An outcome such that $x \in [-1, 1]$

15. An outcome such that $x \in [-3, 1]$

16. An outcome such that $|x| \leq 2$

17. For the probability density function $P(x) = 1/(x + 1)^2$, $x \geq 0$, determine a value for a such that there is 80 percent probability that $x \in [0, a]$.

18. For the probability density function $P(x) = (1/\pi)1/(x^2 + 1)$, $x \in (-\infty, \infty)$, determine a value for a such that there is a 95 percent chance that $|x| \leq a$.

19. For the probability density function $P(x) = \frac{1}{2} \sin x$, $x \in [0, \pi]$, determine a value for a such that there is a 50 percent probability that

$$-a \leq x - \frac{\pi}{2} \leq a.$$

In Exercises 20 and 21, suppose that the probability that an American male presently 40 years of age will die at age x years is given by the probability density function P such that

$$P(x) = \frac{1}{36,000}(-x^2 + 140x - 4000) \qquad x \in [40, 100]$$

20. What is the probability that an American male, now 40 years old, will die on or before his 60th birthday?

21. Repeat Exercise 20 for the 80th birthday.

In Exercises 22 and 23, assume that a zoologist has established an I.Q. test for monkeys based upon their ability to learn certain tricks. The I.Q.'s are distributed according to the probability density function P such that

$$P(x) = \frac{0.62x}{x^2 + 1} \qquad x \in [10, 50]$$

22. What portion of the monkey population has an I.Q. greater than 40?

23. What is the probability that a monkey selected at random will have an I.Q. less than 20?

Many functions are nonnegative-valued over a specific interval and have integrals that are finite, but not equal to 1. That is, for $x \in [a, b]$, $f(x) \geq 0$ and $\int_a^b f(x)\, dx = A$, a positive finite value.

To find a related function P that does have a unit integral it is only necessary to multiply the original function by the factor

$$\frac{1}{A} = \frac{1}{\displaystyle\int_a^b f(x)\,dx}$$

then $P(x) = \dfrac{1}{A} f(x)$ will be such that

$$\int_a^b P(x)\,dx = \int_a^b \frac{1}{A} f(x)\,dx = \frac{1}{A} \int_a^b f(x)\,dx = \frac{1}{A}(A) = 1$$

In Exercises 24–32, find the related $P(x)$ expression; then find the probability of an outcome within the second range of values indicated.

24. $f(x) = \cos x$ for $x \in \left[-\dfrac{\pi}{2}, \dfrac{\pi}{2}\right]$, an outcome such that $x \in \left[0, \dfrac{\pi}{4}\right]$

25. $f(x) = x^2$ for $x \in [0, 2]$, an outcome such that $x \in [1, 2]$

26. $f(x) = x(x^2 + 1)^2$ for $x \in [1, 2]$, an outcome such that $-\frac{1}{4} \le x - \frac{1}{2} \le \frac{1}{4}$

27. $f(x) = \dfrac{x^2}{(x^3 + 1)^2}$ for $x \in [0, \infty)$, an outcome such that $x \ge \sqrt[3]{3}$

28. $f(x) = xe^{-x^2}$ for $x \in [0, \infty)$, an outcome such that $x \le 1$

29. $f(x) = \dfrac{1}{x^{3/2}}$ for $x \in [1, \infty)$, an outcome such that $x \le 9$

30. $f(x) = \dfrac{1}{(2x)^2 + 1}$ for $x \in (-\infty, \infty)$, an outcome such that $x \in [-\frac{1}{2}, \frac{1}{2}]$

31. $f(x) = \dfrac{1}{x^2 + 2x + 2}$ for $x \in (-\infty, \infty)$, an outcome such that $|x + 1| \le 1$

32. $f(x) = \dfrac{1}{e^x + e^{-x}}$ for $x \in (-\infty, \infty)$, an outcome such that $x \in [\ln 2, \ln 10]$

$$\left[Hint: \quad \frac{1}{e^x + e^{-x}} = \frac{e^x}{e^x(e^x + e^{-x})} = \frac{e^x}{e^{2x} + 1} \right]$$

In Exercises 33–36, for each of the functions and intervals given, determine a value for c such that x = c divides the area involved into two equal size areas.

33. $f(x) = \cos x$ on the interval $\left[0, \dfrac{\pi}{2}\right]$

34. $f(x) = x^2$ on the interval $[0, 4]$

35. $f(x) = \dfrac{1}{\pi}\dfrac{1}{x^2 + 1}$ on the interval $[0, \infty)$

36. $f(x) = \dfrac{1}{e^x + e^{-x}}$ on the interval $(-\infty, \infty)$

In Exercises 37 and 38, determine values for a, b, and c such that x = a, x = b, and x = c divide the area under the curve into four areas of equal size. The numbers a, b, and c are called the quartiles of the distribution.

37. $y = \frac{1}{2}\sin x$ on $[0, \pi]$

38. $y = \dfrac{1}{\pi}\dfrac{1}{x^2 + 1}$ on $(-\infty, \infty)$

7.3 FAMILIES OF CURVES, PARAMETERS, MOMENTS

A large group of real-life probability problems have probability density functions of the same type. That is, they belong to the same family of curves. A *family of curves* is a set of curves having a common characteristic.

Example 1

The curves whose equations can be written in the form $y = Ax + B$ are members of the family of straight lines. By assigning specific values to A and B, a specific member of the family can be found.

Example 2

The curves whose equations are of the form $y = A(x - B)^2$, where $A > 0$, are all members of a family of parabolas which open upward and have vertices lying on the x axis.

The constants in the examples are called *parameters*. By giving specific values to parameters like A and B in these examples, an individual member of a family of curves is specified. There is usually some physical or intuitive reason to guess which family of curves describes a specific problem. To make use of a density function in solving a problem, one must first determine the actual values of the parameters needed to identify the one individual member of the family of curves which best describes the distribution of outcomes. The three most common parameters used in identifying individual members of a family of probability distributions are the *first moment of the distribution about zero,* the *second moment of the distribution about the mean,* and the square root of this latter quantity.

Definition

Let P be a probability density function defined over an interval $[a, b]$. The *first moment of the distribution,* μ, is defined as

$$\mu = \int_a^b xP(x)\, dx$$

provided this integral exists. μ is also called *the average value of the distribution, the expectation of the distribution, the mean of the distribution,* or *the mean of the density function.*

At first glance, μ does not appear to be the usual kind of average: a sum divided by the number of addends. But an example will show that it is just that. Suppose that $y = P(x)$ defines a probability density function for an intelligence test. The test scores range from 40 to 195. That is, the domain of x is from $a = 40$ to $b = 195$. Further, suppose that the standardizing population, the number of people who have taken the test, is 7000 and that test scores are whole numbers only; therefore, scores such as $105\frac{1}{2}$ cannot happen. A diagram such as Figure 7.11 could be made for $y = P(x)$. The width of any rectangle is one unit, and its height is found by substituting the value indicated at the base of the rectangle into $P(x)$. By definition, the area of the ith rectangle is the probability of x_i occurring. In other words, the area of the ith rectangle is the fraction of time that x_i will occur. But the area of the ith rectangle is the product of its width and height: $1 \cdot P(x_i) = P(x_i)$. Therefore, in this example, $y_i = P(x_i)$ gives the probability of score x_i occurring. For instance, $P(120)$ would give the probability of an intelligence score of 120. If $P(120) = 0.1$, then one-tenth of the people taking the test would have a score of 120. Since 7000 have taken the test, 0.1 of 7000, or 700, would score 120. If $K = 7000$, the total number taking the test, then $f(x) = KP(x)$ defines a function that gives the actual number of people receiving each test score. If we apply this information for a score of 120, $f(120) = 7000\,P(120) = 7000(0.1) = 700$. That is, $f(120) = 700$, the number of times the score 120 happened. The expression $f(x_i) \cdot x_i$ is the number of x_i scores times the actual score, x_i. Referring again to the 120 score, $f(x_i) \cdot x_i = (700)(120) = 84,000$, which is the total number of points accumulated by the 700 people who scored 120. Thus, $f(x_i) \cdot x_i$ is the total number of points accumulated by all the people who had an intelligence score of x_i. Therefore, the following expresses the total number of points scored by all 7000 people taking the test.

$$\sum_{i=1}^{n} x_i \cdot f(x_i)$$

The average test score can now be found by dividing the above expression by the number who took the test, 7000. Another way of writing 7000 is

$$\sum_{i=1}^{n} f(x_i)$$

since $f(x_i)$ is the number receiving each score, and the sum of those receiving each individual score must equal the number who took the test. The average test score is the total number of test points divided by the number who took the test:

$$\mu = \frac{\sum\limits_{i=1}^{n} x_i f(x_i)}{\sum\limits_{i=1}^{n} f(x_i)}$$

Multiplying numerator and denominator by $\Delta x/K$ gives

$$\mu = \frac{\sum\limits_{i=1}^{n} x_i \cdot \dfrac{f(x_i)}{K} \Delta x}{\sum\limits_{i=1}^{n} \dfrac{f(x_i)}{K} \Delta x}$$

By definition,

$$f(x_i) = KP(x_i)$$

Solving for $P(x_i)$ gives

$$P(x_i) = \frac{f(x_i)}{K}$$

Substituting this in the last expression for the average gives

$$\mu = \frac{\sum\limits_{i=1}^{n} x_i \cdot P(x_i)\, \Delta x}{\sum\limits_{i=1}^{n} P(x_i)\, \Delta x}$$

The limit of the average as n becomes infinite is

$$\mu = \lim_{n \to \infty} \frac{\sum\limits_{i=1}^{n} x_i \cdot P(x_i)\, \Delta x}{\sum\limits_{i=1}^{n} P(x_i)\, \Delta x} = \frac{\int_a^b xP(x)\, dx}{\int_a^b P(x)\, dx}$$

But $\int_a^b P(x)\, dx = 1$. So the average x score is

$$\frac{\int_a^b xP(x)\, dx}{1} = \int_a^b xP(x)\, dx = \mu$$

Example 3   Let $P(x) = \frac{1}{2}\sin x$ on the interval $[0, \pi]$ (Figure 7.15). Then

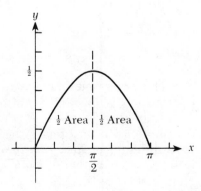

Figure 7.15 $y = \frac{1}{2}\sin x, \quad x \in [0, \pi]$

$$\mu = \int_0^\pi \frac{x \sin x}{2} \, dx$$

$$= \frac{1}{2} \int_0^\pi x \sin x \, dx$$

Using the Table of Integrals,

$$\frac{1}{2} \int_0^\pi x \sin x \, dx = \frac{1}{2} [\sin x - x \cos x]_0^\pi$$

$$= \frac{\sin \pi - \pi(\cos \pi) - (\sin 0 - 0 \cos 0)}{2}$$

$$= \frac{-\pi(-1)}{2} = \frac{\pi}{2}$$

This result agrees with the usual intuitive concept of an average for the symmetric area bounded by $y = \frac{1}{2} \sin x$.

Example 4   Find the first moment of the probability density function

$$P(x) = \frac{1}{\pi} \frac{1}{x^2 + 1} \qquad \text{on } x \in (-\infty, \infty)$$

Solution $\mu = \dfrac{1}{\pi} \displaystyle\int_{-\infty}^{\infty} \dfrac{x}{x^2 + 1} \, dx$

$$= \lim_{b \to \infty} \frac{1}{\pi} \int_{-b}^{b} \frac{x}{x^2 + 1} \, dx$$

$$= \lim_{b \to \infty} \frac{1}{\pi} \left[\frac{1}{2} \ln (x^2 + 1) \right]_{-b}^{b}$$

$$= \lim_{b \to \infty} \frac{1}{2\pi} [\ln (b^2 + 1) - \ln (b^2 + 1)]$$

$$= \lim_{b \to \infty} \frac{1}{2\pi} (0) = 0$$

The first moment, μ, can be thought of as describing where the distribution is located, while σ^2 measures the variability, or the spread of the distribution. Figure 7.16a shows several different distributions, each exactly the same shape, with the same σ^2 but with different means. Figure 7.16b shows several different distributions, each with the same mean but with different variances. Notice that, as the value of σ^2 decreases, the distribution gets tighter.

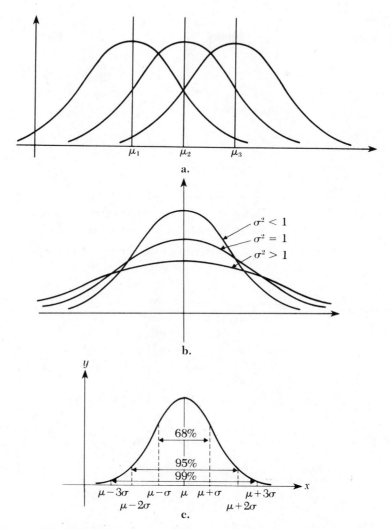

Figure 7.16 **a.** Normal curves with $\sigma^2 = 1$ and various values for μ;
b. normal curves with $\mu = 0$ and various values for σ^2;
c. division into segments of the area under a normal curve

As in the case of μ, the mean of the distribution, the definition of σ^2, the variance of the distribution, differs in formula from the usual expression found in statistics; however, it can be shown in similar manner that the formulas are equivalent,

$$\sigma^2 = \lim_{n \to \infty} \sum_{i=1}^{n} (x_i - \mu)^2 P(x_i)\, \Delta x = \int_a^b (x - \mu)^2 P(x)\, dx$$

Definition

Let P be a probability density function defined on an interval $x \in [a, b]$. Then the *second moment of P around the mean,* σ^2, is defined by

$$\sigma^2 = \int_a^b (x - \mu)^2 P(x) \, dx$$

where μ is the mean of the density function, provided such an integral exists. The symbol σ^2 is read "sigma squared" and is also called *the variance of the distribution*; $\sigma = \sqrt{\sigma^2}$ is called the *standard deviation*, or *root mean squared deviation* of the distribution.

The standard deviation of a distribution is used to divide the area under a *normal* curve (the topic of the next section) into the following segments: 68 percent of the frequencies in the distribution are located within one standard deviation ($\pm\sigma$) to the right and left of μ, 95 percent of the frequencies occur within two standard deviations ($\pm 2\sigma$) from μ, and 99 percent of the frequencies occur within three standard deviations ($\pm 3\sigma$) from μ. This is illustrated in Figure 7.16c.

Example 5   Let $P(x) = \frac{1}{2} \sin x$ on $[0, \pi]$, with $\mu = \pi/2$. Find σ.

Solution $\sigma^2 = \int_0^\pi \left(x - \frac{\pi}{2}\right)^2 \frac{\sin x}{2} \, dx$

Let

$$u = x - \frac{\pi}{2}$$

Thus, $du = dx$. If $x = 0$, then

$$u = -\frac{\pi}{2}$$

and if $x = \pi$, then

$$u = \frac{\pi}{2}$$

Thus,

$$\sigma^2 = \int_{-\pi/2}^{\pi/2} u^2 \; \frac{\sin\left(u + \frac{\pi}{2}\right)}{2} \, du$$

However,

$$\sin\left(u + \frac{\pi}{2}\right) = \cos u$$

thus,

$$\sigma^2 = \frac{1}{2} \int_{-\pi/2}^{\pi/2} u^2 \cos u \; du$$

$$= \frac{1}{2} \left[2u \cos u + (u^2 - 2) \sin u \right]_{-\pi/2}^{\pi/2}$$

$$= \frac{1}{2} \left\{ \left[(2) \frac{\pi}{2} \cos \frac{\pi}{2} + \left(\left(\frac{\pi}{2}\right)^2 - 2 \right) \sin \frac{\pi}{2} \right] \right.$$

$$\left. - \left[2 \left(-\frac{\pi}{2} \right) \cos \left(-\frac{\pi}{2} \right) + \left(\left(-\frac{\pi}{2} \right)^2 - 2 \right) \sin \left(-\frac{\pi}{2} \right) \right] \right\}$$

$$= \left[\left(\frac{\pi}{2} \right)^2 - 2 \right] = \frac{\pi^2}{4} - 2 = \frac{\pi^2 - 8}{4}$$

$$\approx \frac{9.8696 - 8}{4} = \frac{1.8696}{4} = 0.4674$$

Therefore,

$$\sigma = \sqrt{0.4674} \approx 0.6837$$

As one can see from the examples, the calculations of μ, σ^2, and σ are often very complicated. In most cases, these parameters are estimated from observed data rather than calculated directly. However, the concepts of first and second moments have analogies in physical applications that do require direct theoretical calculation.

Example 6

An individual's ability to absorb glucose is determined by administering glucose orally, and then computing the amount absorbed into the blood in a fixed length of time. The probability that an individual will absorb x milliliters of glucose is given by the probability density function P such that

$$P(x) = \frac{3}{8} x^2 \qquad x \in [0, 2]$$

Find the mean and variance for P.

Solution

$$\mu = \int_0^2 \frac{3}{8} x^2 \cdot x \, dx$$

$$= \frac{3}{8} \frac{x^4}{4} \Big]_0^2 = \left(\frac{3}{8}\right)\left(\frac{16}{4}\right) = \frac{3}{2}$$

$$\sigma^2 = \int_0^2 \frac{3}{8} x^2 \left(x - \frac{3}{2}\right)^2 dx$$

$$= \frac{3}{8} \int_0^2 x^2 \left(x^2 - 3x + \frac{9}{4}\right) dx$$

$$= \frac{3}{8} \int_0^2 \left(x^4 - 3x^3 + \frac{9x^2}{4}\right) dx$$

$$= \frac{3}{8} \left[\frac{x^5}{5} - \frac{3x^4}{4} + \frac{3x^3}{4}\right]_0^2 = \frac{3}{20}$$

7.3 EXERCISES

In Exercises 1–10, find μ in each case for the given probability density function.

1. $P(x) = \dfrac{3x^2}{56}$ on the interval $[2, 4]$

2. $P(x) = \dfrac{3(8x - 2x^2)}{64}$ on the interval $[0, 4]$

3. $P(x) = \dfrac{\cos x}{2}$ on the interval $\left[-\dfrac{\pi}{2}, \dfrac{\pi}{2}\right]$

4. $P(x) = \dfrac{3(x^3 - 4x^2 + x + 6)}{22}$ on the interval $[0, 2]$

5. $P(x) = \dfrac{6}{17}(x^2 - x + 2)$ on the interval $[1, 2]$

6. $P(x) = \dfrac{6x^2 - 3x^3}{4}$ on the interval $[-1, 1]$

7. $P(x) = ce^{-x/2}$ on the interval $[-2, 3]$, where $c = \dfrac{-e^{3/2}}{2(1 - e^{5/2})}$

8. $P(x) = \sqrt{2} \sin x$ on the interval $\left[\dfrac{\pi}{2}, \dfrac{3\pi}{4}\right]$

9. $P(x) = \dfrac{4}{\pi(x^2 + 4)}$ on the interval $[-2, 2]$

10. $P(x) = \ln x$ on the interval $[1, e]$

In Exercises 11–15, find μ *and* σ^2 *for the indicated probability density functions.*

11. $P(x) = \dfrac{3}{2} \sqrt{x}$, for [0, 1]   **12.** $P(x) = \dfrac{5x^4}{2}$, for [−1, 1]

13. $P(x) = \dfrac{6}{27} (3x - x^2)$, for [0, 3]

14. $P(x) = \dfrac{3}{2} (x + 1)^2$, for [−2, 0]

15. $P(x) = \dfrac{1}{18} (x - 5)^2$, for [2, 8]

In Exercises 16–20, calculate μ *in each case.*

16. $P(x) = \dfrac{4x^{1/3}}{3}$, for [1, 3]   **17.** $P(x) = |x|$, for [−1, 1]

18. $P(x) = \dfrac{1}{6}$, for [0, 6]   **19.** $P(x) = \dfrac{e^{-x^2}}{\sqrt{2\pi}}$, for [1, ∞)

20. $P(x) = \dfrac{e^x}{9}$, for [1, ln 10]

7.4 NORMAL CURVE AREAS, AREAS WITH TABLES

The one probability density function that has the widest application is the *normal, Gaussian,* or *bell* distribution. The formula for this density is

$$P(x) = \frac{1}{\sigma \sqrt{2\pi}} e^{-\frac{1}{2}\left(\frac{x - \mu}{\sigma}\right)^2}$$

over the range of values $x \in (-\infty, \infty)$. Figure 7.17 shows the graph of one member of this family of densities for the specific values $\mu = 0$ and $\sigma = 1$, where μ and σ^2 are the first and second moments as considered in Section 7.3 and are the identifying parameters for this family. When an experiment has outcomes known to follow this distribution, the outcomes are said to be normally distributed. The density is symmetric about the value $x = \mu$.

Note that the caption for Figure 7.17 says

$$y = \frac{1}{\sqrt{2\pi}} e^{-\frac{1}{2}z^2} \qquad \text{where } z = \frac{x - \mu}{\sigma}, \mu = 0, \sigma = 1$$

The integral

$$\frac{1}{\sqrt{2\pi}} \int_0^{z_1} e^{-\frac{1}{2}z^2}\, dz$$

cannot be integrated by any of the previously introduced methods of integration and is in fact not an elementary integral. It is, however, extremely important. The values of this integral for various values of z_1 are found in a table called a *normal curve area table* or a *normal curve of error table*. The latter name comes from the historical fact that some of the first quantities found to be normally distributed were the errors in repeated measurements of the physical characteristics of an object. Table I, found in the Appendix of this text, is one of these tables.

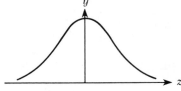

Figure 7.17 $y = \dfrac{1}{\sqrt{2\pi}} e^{-\frac{1}{2}z^2}$ where $z = \dfrac{x - \mu}{\sigma}$,
$\mu = 0$, and $\sigma = 1$

Example 1   Use Table I to evaluate $\dfrac{1}{\sqrt{2\pi}} \displaystyle\int_0^{1.82} e^{-\frac{1}{2}z^2}\, dz$

Solution   In the column headed z, locate 1.8; then move horizontally to the column headed 0.02 to find the number 0.4656. Thus

$$\frac{1}{\sqrt{2\pi}} \int_0^{1.82} e^{-\frac{1}{2}z^2}\, dz = 0.4656$$

Example 2   Use Table I to evaluate $\dfrac{1}{\sqrt{2\pi}} \displaystyle\int_{0.5}^{2.6} e^{-\frac{1}{2}z^2}\, dz$.

Solution   The lower limit of this integral is not zero. However,

$$
\begin{aligned}
\int_{z_1}^{z_2} f(z)\, dz &= \int_{z_1}^{0} f(z)\, dz + \int_0^{z_2} f(z)\, dz \\
&= -\int_0^{z_1} f(z)\, dz + \int_0^{z_2} f(z)\, dz \\
&= \int_0^{z_2} f(z)\, dz - \int_0^{z_1} f(z)\, dz
\end{aligned}
$$

Therefore, we evaluate $\dfrac{1}{\sqrt{2\pi}} \displaystyle\int_{0.5}^{2.6} e^{-\frac{1}{2}z^2} \, dz$ as

$$\frac{1}{\sqrt{2\pi}} \int_{0}^{2.6} e^{-\frac{1}{2}z^2} \, dz - \frac{1}{\sqrt{2\pi}} \int_{0}^{0.5} e^{-\frac{1}{2}z^2} \, dz = 0.4953 - 0.1915 = 0.3038$$

Recalling that we let $z = (x - \mu)/\sigma$, $\mu = 0$, and $\sigma = 1$, we can now show that for any known values of μ and σ, the probability of an experimental outcome between two values x_1 and x_2 can be found by using the normal curve area table, because

$$\frac{1}{\sigma\sqrt{2\pi}} \int_{x_1}^{x_2} e^{-\frac{1}{2}\left(\frac{x-\mu}{\sigma}\right)^2} \, dx = \frac{1}{\sqrt{2\pi}} \int_{z_1}^{z_2} e^{-\frac{1}{2}z^2} \, dz$$

for some z_2 and z_1. Let

$$z = \frac{x - \mu}{\sigma}$$

Then

$$dz = \frac{dx}{\sigma} \quad \text{or} \quad \sigma \, dz = dx$$

When $x = x_1$,

$$z = \frac{x_1 - \mu}{\sigma} = z_1$$

and, when $x = x_2$,

$$z = \frac{x_2 - \mu}{\sigma} = z_2$$

Then

$$\frac{1}{\sigma\sqrt{2\pi}} \int_{x_1}^{x_2} e^{-\frac{1}{2}\left(\frac{x-\mu}{\sigma}\right)^2} \, dx = \frac{1}{\sigma\sqrt{2\pi}} \int_{z_1}^{z_2} e^{-\frac{1}{2}z^2} \sigma \, dz$$

$$= \frac{1}{\sqrt{2\pi}} \int_{z_1}^{z_2} e^{-\frac{1}{2}z^2} \, dz$$

Notice that in the transformed integral the parameters μ and σ no longer appear in the integrand but have been incorporated into the new limits of integration.

Example 3   The results of an experiment are known to be normally distributed with $\mu = 3$ and $\sigma = 2$. What is the probability of an experimental outcome between 3 and 5?

Solution   Since the experimental results are normally distributed,

$$P(x) = \frac{1}{\sigma\sqrt{2\pi}} e^{-\frac{1}{2}\left(\frac{x-\mu}{\sigma}\right)^2}$$

$$= \frac{1}{2\sqrt{2\pi}} e^{-\frac{1}{2}\left(\frac{x-3}{2}\right)^2}$$

The probability of an outcome between 3 and 5 will be given by

$$p = \int_3^5 P(x)\, dx$$

$$= \frac{1}{2\sqrt{2\pi}} \int_3^5 e^{-\frac{1}{2}\left(\frac{x-3}{2}\right)^2}\, dx$$

Letting $z = (x - 3)/2$,

$$p = \frac{1}{\sqrt{2\pi}} \int_0^1 e^{-\frac{1}{2}z^2}\, dz$$

where when $x = 3$, $z = (3 - 3)/2 = 0$; and when $x = 5$, $z = (5 - 3)/2 = 1$. According to Table I,

$$\frac{1}{\sqrt{2\pi}} \int_0^1 e^{-\frac{1}{2}z^2}\, dz = 0.3413$$

The curve and area in question are shown in Figure 7.18. Thus 34.13 percent of the time, a result between 3 and 5 can be expected.

Note that we could have predicted this result in another way. We were told that the distribution was normal, with $\mu = 3$, $\sigma = 2$. From our previous discussion of σ, we know that about 68 percent of the frequencies occur within plus or minus one σ of μ, or approximately 34 percent of the frequencies can be expected between μ and $\mu + \sigma$; thus approximately 34 percent of the time a result between 3 and $3 + 2 = 5$ can be expected.

This example was set up so that the lower limit of the transformed integral was 0. If this is not the case, no difficulty is encountered but a few more calculations arise. Table I yields the shaded area corresponding to a specific z_1 value, as illustrated in Figure 7.18. Notice that the graph of

$$y = \frac{1}{\sqrt{2\pi}} e^{-\frac{1}{2}z^2}$$

is symmetrical about $z = 0$. This means that for any positive value z_1,

$$\frac{1}{\sqrt{2\pi}} \int_{-z_1}^{0} e^{-\frac{1}{2}z^2} dz = \frac{1}{\sqrt{2\pi}} \int_{0}^{z_1} e^{-\frac{1}{2}z^2} dz$$

Using this, together with the usual properties of integrals and a sketch of the specific z values involved, the table can be made to yield almost any desired integral of the type being considered.

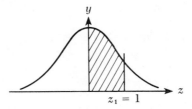

Figure 7.18   Normal curve area, $z_1 = 1$

Example 4   Suppose that the results of an experiment are normally distributed with $\mu = 3$ and $\sigma = 2$. What is the probability of a result between 1 and 4?

Solution   Using

$$z = \frac{x - \mu}{\sigma} = \frac{x - 3}{2}$$

we get

$$p = \int_{1}^{4} P(x)\, dx = \frac{1}{\sqrt{2\pi}} \int_{-1}^{1/2} e^{-\frac{1}{2}z^2} dz$$

When $x = 1$,

$$z = \frac{1 - 3}{2} = \frac{-2}{2} = -1$$

and when $x = 4$,

$$z = \frac{4 - 3}{2} = \frac{1}{2}$$

The area corresponding to p is shown in Figure 7.19. However,

$$\frac{1}{\sqrt{2\pi}} \int_{-1}^{1/2} e^{-\frac{1}{2}z^2} dz = \frac{1}{\sqrt{2\pi}} \int_{-1}^{0} e^{-\frac{1}{2}z^2} dz + \frac{1}{\sqrt{2\pi}} \int_{0}^{1/2} e^{-\frac{1}{2}z^2} dz$$

$$\frac{1}{\sqrt{2\pi}} \int_{-1}^{0} e^{-\frac{1}{2}z^2} dz = \frac{1}{\sqrt{2\pi}} \int_{0}^{+1} e^{-\frac{1}{2}z^2} dz = 0.3413$$

$$\frac{1}{\sqrt{2\pi}} \int_{0}^{1/2} e^{-\frac{1}{2}z^2} dz = 0.1915$$

These values are from Table I. Thus,

$$p = 0.3413 + 0.1915 = 0.5328$$

indicating a 53.28 percent chance of an outcome between 1 and 4.

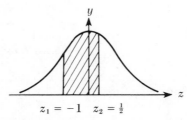

Figure 7.19   Normal curve area,
$z_1 = -1$ and $z_2 = \frac{1}{2}$

Example 5

Human I.Q.'s (Intelligence Quotients), derived from certain psychological tests, have a mean of 100 and a standard deviation of 10. I.Q. scores are normally distributed. What is the probability that a person selected at random will have an I.Q. between 105 and 120?

Solution $p = \displaystyle\int_{105}^{120} P(x)\, dx$ where $P(x) = \dfrac{1}{10\sqrt{2\pi}} e^{-\frac{1}{2}\left(\frac{x-100}{10}\right)^2}$

Letting $z = (x - 100)/10$,

$$p = \frac{1}{\sqrt{2\pi}} \int_{1/2}^{2} e^{-\frac{1}{2}z^2}\, dz$$

This area is shown in Figure 7.20. To use the table to evaluate the integral we need only observe that

$$\frac{1}{\sqrt{2\pi}} \int_{0}^{2} e^{-\frac{1}{2}z^2}\, dz = \frac{1}{\sqrt{2\pi}} \int_{0}^{1/2} e^{-\frac{1}{2}z^2}\, dz + \frac{1}{\sqrt{2\pi}} \int_{1/2}^{2} e^{-\frac{1}{2}z^2}\, dz$$

Thus,

$$\frac{1}{\sqrt{2\pi}} \int_{1/2}^{2} e^{-\frac{1}{2}z^2}\, dz = \frac{1}{\sqrt{2\pi}} \int_{0}^{2} e^{-\frac{1}{2}z^2}\, dz - \frac{1}{\sqrt{2\pi}} \int_{0}^{1/2} e^{-\frac{1}{2}z^2}\, dz$$
$$= 0.4772 - 0.1915 = 0.2857$$

Thus, there is a 0.2857 probability of a result between 105 and 120. That is, about 29 percent of the population has an I.Q. between 105 and 120.

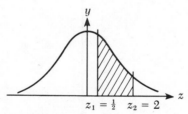

Figure 7.20 Normal curve area,
$z_1 = \frac{1}{2}$ and $z_2 = 2$

Example 6

An experimental result is known to be normally distributed with $\mu = 3$ and $\sigma = 2$. Find a value for a such that 95 percent of the time results between $\mu - a$ and $\mu + a$ can be expected.

Solution We know that

$$\frac{1}{2\sqrt{2\pi}} \int_{3-a}^{3+a} e^{-\frac{1}{2}\left(\frac{x-3}{2}\right)^2}\, dx = 0.9500$$

Using the usual transformation $z = (x - 3)/2$, this becomes

$$\frac{1}{\sqrt{2\pi}} \int_{-a/2}^{a/2} e^{-\frac{1}{2}z^2}\, dz = 0.9500$$

because $x = 3 + a,$ $\qquad z = \dfrac{3 + a - 3}{2} = \dfrac{a}{2}$

and $x = 3 - a,$ $\qquad z = \dfrac{3 - a - 3}{2} = -\dfrac{a}{2}$

The area corresponding to this integral is shown in Figure 7.21. Since the function is symmetrical,

$$\frac{1}{\sqrt{2\pi}} \int_0^{a/2} e^{-\frac{1}{2}z^2}\, dz = \frac{1}{2}(0.9500) = 0.475$$

According to Table I, a z_1 value such that

$$\frac{1}{\sqrt{2\pi}} \int_0^{z_1} e^{-\frac{1}{2}z^2}\, dz = 0.475$$

is

$$z_1 = 1.96$$

Thus,

$$\frac{a}{2} = 1.96$$
$$a = 3.92$$

Thus 95 percent of the time, an experimental result between $3 - 3.92$ and $3 + 3.92$, or -0.92 and 6.92, can be expected. We note that approximately 95 percent of all outcomes occur between $\mu - 2\sigma$ and $\mu + 2\sigma$.

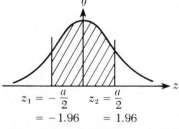

Figure 7.21 Normal curve area enclosing
95 percent of the total area

7.4 EXERCISES

In Exercises 1–10, use the normal curve area table to evaluate:

1. $\dfrac{1}{\sigma\sqrt{2\pi}} \displaystyle\int_3^7 e^{-\frac{1}{2}\left(\frac{x-\mu}{\sigma}\right)^2} dx,$

if $\mu = 3, \sigma = 2$

2. $\dfrac{1}{2\sqrt{2\pi}} \displaystyle\int_{-2}^3 e^{-\frac{1}{2}\left(\frac{x-3}{2}\right)^2} dx$

3. $\dfrac{1}{\sqrt{2\pi}} \displaystyle\int_{-3}^{-1} e^{-\frac{1}{2}\left(\frac{x+1}{1}\right)^2} dx$

4. $\dfrac{1}{\sqrt{2\pi}} \displaystyle\int_{-1}^{3.2} e^{\frac{-(x+1)^2}{2}} dx$

5. $\dfrac{1}{2\sqrt{2\pi}} \displaystyle\int_1^4 e^{-\frac{1}{2}\left(1-\frac{x}{2}\right)^2} dx$

6. $\dfrac{1}{4\sqrt{2\pi}} \displaystyle\int_2^{5.1} e^{-\frac{x^2}{16}\left(\frac{1}{2}\right)} dx$

7. $\dfrac{1}{\sigma\sqrt{2\pi}} \displaystyle\int_{-1.1}^{2.2} e^{-\frac{x^2}{2\sigma^2}} dx,$

if $\sigma = 1.1$

8. $\dfrac{1}{(2.1)\sqrt{2\pi}} \displaystyle\int_{-2}^{2.1} e^{-\frac{x^2}{2(2.1)^2}} dx$

9. $\dfrac{1}{\sqrt{2\pi}} \displaystyle\int_1^\infty e^{-\frac{1}{2}\left(\frac{x+1}{1}\right)^2} dx$

10. $\dfrac{1}{\sqrt{2\pi}} \displaystyle\int_2^\infty e^{\frac{-(x-1)^2}{2}} dx$

In Exercises 11 and 12, if I.Q.'s are normally distributed with $\mu = 100$ and $\sigma = 10$, find:

11. The portion of the population with I.Q. 80 to 120

12. The portion of the population with I.Q. 180 or over

In Exercises 13 and 14, assume that the life of a top-quality light bulb is normally distributed with an average life of 1000 hours and a standard deviation of 50 hours.

13. What is the probability that a bulb will last for 2000 hours or more?

14. If a bulb is selected at random, what is the probability that it will have a life of 900 hours or less?

In Exercises 15 and 16, suppose that the height of California women is normally distributed with $\mu = 65$ inches and $\sigma = 4$ inches.

15. What portion of California women are 6 feet tall or shorter?

16. If a California woman is selected at random, what is the probability that she is from 5 to 6 feet tall?

In Exercises 17–21, take into account that normal curve probabilities can be used to approximately evaluate probabilities that are not normally distributed. For example, if n honest coins are tossed, the resulting distribution of heads is approximately normally distributed with $\mu = n(\tfrac{1}{2}) = n/2$ and $\sigma = \sqrt{n(\tfrac{1}{4})} = \sqrt{n}/2$. Assume that in an experiment 400 coins are tossed. Then the probability of getting between 190 and 210 heads is approximately

$$\frac{1}{\sigma\sqrt{2\pi}} \int_{189.5}^{210.5} e^{-\frac{1}{2}\left(\frac{x-\mu}{\sigma}\right)^2} dx$$

but $n/2 = 200$ and $\sqrt{n}/2 = 20/2 = 10$. Therefore,

$$p = \frac{1}{10\sqrt{2\pi}} \int_{189.5}^{210.5} e^{-\frac{1}{2}\left(\frac{x-200}{10}\right)^2} dx$$

which can be evaluated using a normal curve numerical table.

17. Evaluate this integral. Why were the limits 189.5 and 210.5 used?
18. Find the probability of more than 210 heads.
19. Find the probability of fewer than 205 heads.
20. Find the probability of between 205 and 210 heads.
21. Find the probability of not fewer than 180 heads and not more than 201 heads.

In Exercises 22–27, use the normal curve values to determine, if possible, a value of z_1 or x_1 so that each of the integrals has the value indicated.

22. a. $\dfrac{1}{\sqrt{2\pi}} \displaystyle\int_{0}^{z_1} e^{-\frac{1}{2}z^2} dz = 0.37$

  b. $\dfrac{1}{\sqrt{2\pi}} \displaystyle\int_{z_1}^{0} e^{-\frac{1}{2}z^2} dz = 0.46$

23. a. $\dfrac{1}{\sqrt{2\pi}} \displaystyle\int_{0}^{z_1} e^{-\frac{1}{2}z^2} dz = 0.425$

  b. $\dfrac{1}{\sqrt{2\pi}} \displaystyle\int_{z_1}^{0} e^{-\frac{1}{2}z^2} dz = 0.321$

24. a. $\dfrac{1}{\sqrt{2\pi}} \displaystyle\int_{-\infty}^{z_1} e^{-\frac{1}{2}z^2} dz = 0.75$

  b. $\dfrac{1}{\sqrt{2\pi}} \displaystyle\int_{z_1}^{\infty} e^{-\frac{1}{2}z^2} dz = 0.60$

25. a. $\dfrac{1}{\sqrt{2\pi}} \displaystyle\int_{-z_1}^{z_1} e^{-\frac{1}{2}z^2} dz = 0.60$

b. $\dfrac{1}{\sqrt{2\pi}} \displaystyle\int_{-z_1}^{z_1} e^{-\frac{1}{2}z^2}\, dz = 0.50$

26. **a.** $\dfrac{1}{3\sqrt{2\pi}} \displaystyle\int_{2}^{x_1} e^{-\frac{1}{2}\left(\frac{x-2}{3}\right)^2}\, dx = 0.40$

b. $\dfrac{1}{3\sqrt{2\pi}} \displaystyle\int_{-1}^{x_1} e^{-\frac{1}{2}\left(\frac{x+1}{3}\right)^2}\, dx = 0.25$

27. **a.** $\dfrac{1}{\sqrt{2\pi}} \displaystyle\int_{0}^{z_1} e^{-\frac{1}{2}z^2}\, dz = 0.61$

b. $\dfrac{1}{\sqrt{2\pi}} \displaystyle\int_{-z_1}^{\infty} e^{-\frac{1}{2}z^2}\, dz = 1.2$

7.5 APPLICATIONS TO CAPITAL AND PRESENT VALUE

Many models of economic problems can be constructed using the concepts of calculus. For example, suppose that one wishes to determine the total amount of revenue or profit from a certain source over a period of time. Further, suppose that a function, f, gives the rate of revenue as a function of time, t, in years.

That is to say, $f(t)$ is an income of dollars per year, or any amount of money per unit of time. Then the total amount of revenue or income received over a given period of time is the rate, $f(t)$, multiplied by the period of time of earning. For the purpose of analysis let us assume that f is continuous.

Thus, if the period of time involved is t' years, the total revenue would be approximated by

$$R \approx \sum_{i=1}^{n} f(t_i)\, \Delta t$$

where $0 \le t_1 \le t_2 \le t_3 \le \cdots \le t_n = t'$, $\Delta t = t'/n$, and the revenue on each subinterval is approximated by $f(t_i)\, \Delta t$. If we take a limiting case, as n tends to ∞ and Δt tends to zero for each subinterval,

$$R = \lim_{n \to \infty} \sum_{i=1}^{n} f(t_i)\, \Delta t$$

The total revenue, R, fits the limit definition of an integral. Thus,

$$R = \int_{0}^{t'} f(t)\, dt$$

Example 1	Suppose that the revenue rate in dollars from a certain source is known to be given by

$$f(t) = (t + 1)^{-3/2}(3000) \text{ dollars per year}$$

where t is given in years. What is the total revenue expected from this source over the next 18 months?

Solution The revenue is given by

$$
\begin{aligned}
\int_0^{3/2} 3000(t + 1)^{-3/2} \, dt &= 3000(-2)(t + 1)^{-1/2} \Big]_0^{3/2} \\
&= -6000 \left[\left(\frac{3}{2} + 1 \right)^{-1/2} - (1)^{-1/2} \right] \\
&= -6000 \left[\left(\frac{5}{2} \right)^{-1/2} - 1 \right] \\
&= -6000 \left[\left(\frac{2}{5} \right)^{+1/2} - 1 \right] \\
&= 6000 \left[1 - \sqrt{\frac{2}{5}} \right] \text{ dollars} \approx \$2205.27
\end{aligned}
$$

This type of problem leads in turn to a problem of a more complex nature. Suppose a business firm knows that it will need a certain sum of money at some later time. It then might ask how much would have to be invested today at a fixed interest rate in order to have this required amount at the later date. We need not consider the actual amount, but rather ask how much the firm would have to invest now to have $1 at the later time. Suppose it invested $1 at a rate r per year. We need only consider the $1 figure, as any other dollar amount can be found by multiplying the $1 figure by the actual amount. For example, if $1 yields $1.05, then $7300 yields $7300(1.05) = \$7665$.

At the end of one year the business would have $(1 + r)$ dollars. In two years it would have

$$(1 + r) + (1 + r)r = (1 + r)(1 + r) = (1 + r)^2 \text{ dollars}$$

and so forth. In general, there would be $(1 + r)^t$ dollars if the interest were compounded yearly.

If the amount were compounded n times per year, the $1 would become

$$\left(1 + \frac{r}{n} \right)^{tn} \text{ dollars} = \left(1 + \frac{r}{n} \right)^{(n/r)(tr)} = \left[\left(1 + \frac{r}{n} \right)^{n/r} \right]^{tr}$$

Now consider what happens as n tends to infinity, that is, the number of times interest is compounded becomes large.

$$\lim_{n\to\infty} \left(1 + \frac{r}{n}\right)^{n/r} = \lim_{x\to 0} (1 + x)^{1/x} = e$$

where $r/n = x$. This limit is sometimes used as a definition of e. Thus,

$$\lim_{n\to\infty} \left(\left[1 + \frac{r}{n}\right]^{n/r}\right)^{tr} = e^{tr}$$

One dollar today becomes e^{tr} dollars in t years at an interest rate r when interest is compounded continuously. However, the original problem was what amount a becomes \$1 in t years. Using the result for \$1,

$$ae^{tr} = 1$$
$$a = e^{-tr}$$

In other words, \$1 t years in the future has a *present value* of e^{-tr} dollars.

Example 2
$

How much would one have to invest at 6 percent interest in order to have \$5000 in 30 months?

Solution $t = \dfrac{5}{2}$ years $r = 0.06$

Thus,

$$5000e^{-\frac{5}{2}(0.06)} = 5000e^{-0.15} = 5000(0.86071) = \$4303.55$$

The ideas of present value and future revenue can be combined. That is, we can formulate an expression for the total present value or worth of future revenue. Consider the actual amount of a future revenue. Assuming that the rate of revenue is given by a function f, the actual amount of revenue over a short period of time Δt would be $f(t)\,\Delta t$. The present value of this is $e^{-rt}f(t)\,\Delta t$ if the interest rate is r.

The total present value of this revenue over an interval of time from 0 to t' would be given by

$$\int_0^{t'} e^{-rt} f(t)\, dt$$

This integral is called the *capital value* of the future income.

Example 3

$

Consider a constant income of $100 per month over a period covering the next 5 years. What is the capital value of this income if the interest rate is 5 percent?

Solution   The capital value is given by

$$\int_0^{t'} e^{-rt} f(t) \, dt$$

where $t' = 5$, $f(t) = \$100$ per month or $1200 per year, and $r = 0.05$. Therefore, we have

$$
\begin{aligned}
\int_0^5 e^{-0.05t} 1200 \, dt &= 1200 \int_0^5 e^{-0.05t} \, dt \\
&= 1200 \left(\frac{e^{-0.05t}}{-0.05} \right) \Big]_0^5 \\
&= \frac{-1200}{0.05} [e^{-0.25} - e^0] \\
&= 24{,}000[1 - e^{-0.25}] = 24{,}000[1 - 0.7788] \\
&= 24{,}000[0.2212] = \$5308.80
\end{aligned}
$$

This means that a total of $5(1200) = \$6000$ will have been invested; that amount—plus the accrued interest at the end of 5 years—is equivalent to a lump sum of $5308.80 invested today at the same rate of interest.

7.5 EXERCISES

$ In Exercises 1–6, for the given revenue function and the indicated number of years, find the expected total revenue.

1. $f(t) = (t + 1)^{-2/3}(2000)$,   for 3 years

2. $f(t) = (t^2 + 1)^{\frac{1}{2}}t(1000)$,   for 4 years

3. $f(t) = (t + 2)^2(500)$,   for 2 years

4. $f(t) = \dfrac{1}{t + 1}(250)$,   for 2.5 years

5. $f(t) = \dfrac{1}{t^2 + 5t + 6}(500)$,   for 10 years

6. $f(t) = |\sin[(t + 2)\pi](1000)|$,   for 3 years

$ In Exercises 7–10, assume interest is compounded continuously.

7. How much would one have to invest at 6 percent to have $4000 in 24 months?

8. How much would one have to have invested at 3 percent 4 years ago to have $9500 today?

9. How much would one have to have invested at 7 percent $3\frac{1}{2}$ years ago to have $102,000 today?

10. How much would one have to invest at 12.5 percent to have $1.5 million in 3 years?

⑨ *In Exercises 11–20, a revenue function is given. Find the capital value of the indicated income at the given time in the future with the given interest rate.*

11. $f(t) = \$150/$month for 3 years, the interest rate is 5 percent

12. $f(t) = \$200/$month for 4 years, the interest rate is 4 percent

13. $f(t) = \$250/$month for 5 years, the interest rate is 8 percent

14. $f(t) = \$500/$year for 10 years, the interest rate is 7 percent

15. $f(t) = \$400t/$year for 8 years, the interest rate is 4.5 percent

16. $f(t) = e^{rt}(t^2 + 2)1000$ dollars/year, the interest rate r is 2 percent, for 4 years

17. $f(t) = e^{rt}(t^3)500$ dollars/year, the interest rate r is 5 percent, for 4 years

18. $f(t) = e^{rt}(\ln(t + 1))1500$ dollars/year, for 5 years if the interest rate r is 3.5 percent

19. $f(t) = e^{rt}(\sin^2 \pi t)1200$ dollars/year, for 6 years if the interest rate r is 6 percent

20. $f(t) = e^{rt}\dfrac{1}{t + 1}2000$ dollars/year, for 10 years if the interest rate r is 6 percent

7.6 MARGINAL ANALYSIS

In one form, integration may be regarded as the recovery of a function when the derivative of the function is known. In many applications of calculus to business and economics, this is the type of problem encountered. When considering the behavior of an economic function through its derivative, one is performing *marginal analysis*. For example, let $p = f(x)$ describe the price at which each unit of a commodity can be sold if x represents the number of units sold. p is called a demand function. The total revenue R produced by such sales would be $R = px = xf(x)$.

Definition

If R is the total revenue produced by the sale of x number of units, then the revenue from the sale of the ith unit sold is said to be the *marginal revenue,* and is defined as dR/dx, the first derivative of R with respect to x.

Example 1
$

If the revenue from the sale of some automotive parts is computed according to

$$R = \frac{1}{10} x^2 - 8x$$

then the total revenue from the sale of 100 units is

$$R = \frac{1}{10} \cdot 100^2 - 8 \cdot 100 = \$200$$

and the total revenue from the sale of 200 units is

$$R = \frac{1}{10} \cdot 200^2 - 8 \cdot 200 = \$2400$$

The marginal revenue from the sale of the 100th unit is computed by evaluating dR/dx for $x = 100$:

$$\frac{dR}{dx} = \frac{1}{5}x - 8 = \$12$$

whereas the marginal revenue of the 200th unit is $\frac{1}{5}(200) - 8 = \$32$.

Example 2
$

Let the marginal revenue relative to the sales of a certain item be given by

$$\frac{dR}{dx} = 4x^2 - 3x + 2$$

What is the total revenue produced by the sale of 5 of these items?

Solution   Since $dR/dx = 4x^2 - 3x + 2$,

$$dR = (4x^2 - 3x + 2)\,dx$$

and

$$R = \int (4x^2 - 3x + 2)\, dx = \frac{4x^3}{3} - \frac{3x^2}{2} + 2x + C$$

However, clearly $R = 0$ if $x = 0$; thus

$$0 = \frac{4}{3}(0)^3 - \frac{3}{2}(0)^2 + 2(0) + C$$
$$C = 0$$

Then,

$$R = \frac{4x^3}{3} - \frac{3x^2}{2} + 2x$$

If $x = 5$,

$$R = \frac{4}{3}(125) - \frac{3}{2}(25) + 2(5) = \frac{500}{3} - \frac{75}{2} + 10$$
$$= \frac{1000 - 225 + 60}{6} = \frac{835}{6}$$

Definition

Let $y = f(x)$ describe the total cost of producing x items of some commodity. Then dy/dx, or y', is the *marginal cost of producing* an item.

Example 3
$

If the marginal cost of producing a certain item is

$$y' = 3 + x + \frac{1}{4x + 1}$$

what is the cost of producing one item if there is a fixed cost of $4?

Solution

$$y = \int y'\, dx = \int \left(3 + x + \frac{1}{4x + 1}\right) dx$$
$$= 3x + \frac{x^2}{2} + \frac{1}{4}\ln|4x + 1| + C$$

Since the fixed cost is $4, $y = 4$ when $x = 0$; hence

$$y = 4 = 3(0) + \frac{0^2}{2} + \frac{1}{4} \ln|4(0) + 1| + C$$

Thus

$$C = 4$$

and

$$y = 3x + \frac{x^2}{2} + \frac{1}{4} \ln|4x + 1| + 4$$

When $x = 1$,

$$y = 3 + \frac{1}{2} + \frac{1}{4} \ln 5 + 4 = \frac{15}{2} + \frac{1}{4} \ln 5 \approx 7.90$$

So the cost of producing one item is approximately $7.90.

Definition

In a certain mathematical model of the economy, the total consumption, c, is described as a function of the total national income x. That is, $c = f(x)$. The *marginal propensity to consume* is dc/dx. If s stands for the total national amount in savings, then $s = g(x)$, that is, the total amount in savings is a function of total income. The *marginal propensity to save* is ds/dx.

If we assume that $x = c + s$, that is, that total income equals the income consumed plus the income saved, then

$$1 = \frac{dc}{dx} + \frac{ds}{dx}$$

or

$$\frac{dc}{dx} = 1 - \frac{ds}{dx}$$

The total physical output of a number of workers or machines is a function of the number of workers or machines.

Definition

If P is a measure of the output of x workers or machines, then
$$P = f(x)$$
The *marginal physical productivity* is dP/dx.

Example 4
§

Suppose that the marginal physical productivity for lumberjacks is given by

$$P' = 100 - 0.08x$$

where P is in thousands of board feet per day. How many lumberjacks would be required to cut 1,000,000 board feet of lumber per day?

Solution   Since $dP/dx = 100 - 0.08x$, then $dP = (100 - 0.08x)\, dx$.

$$P = \int (100 - 0.08x)\, dx = 100x - 0.04x^2 + C$$

However, if $x = 0$, $P = 0$ (no workers, no production). Thus,

$$P = 100x - 0.04x^2$$

The problem is to determine x when $P = 1000$.

$$1000 = 100x - 0.04x^2$$
$$4x^2 - 10,000x + 100,000 = 0$$

$$x = \frac{10,000 \pm \sqrt{100,000,000 - 4(4)(100,000)}}{8}$$

$$= \frac{10,000 \pm \sqrt{98,400,000}}{8} = \frac{10,000 \pm 9919.7}{8}$$

$$= \frac{80.3}{8} \quad \text{or} \quad \frac{19,919.7}{8}$$

$$= 10.04 \quad \text{or} \quad 2489.96$$

Hence 11 lumberjacks would be required to assure 1,000,000 board feet of production; the 2489.96 solution is valid, but, you must admit, not practical.

In addition to the above examples, economists also consider such concepts as *marginal demand* for commodities, the *marginal rate of substitution*, the *marginal efficiency of investment*, and so forth. The basic approach in each case follows the pattern considered here. The marginal rate is the derivative in each case, and as such, has interpretations as indicated in earlier chapters. Perhaps the most important one relates to a single item. This means, for instance, that if $c = f(x)$ represents the total cost of producing and selling x items, then $c' = df/dx$, the marginal cost, when evaluated for $x = 15$ represents the cost of producing and selling the 15th item.

7.6 EXERCISES

1. How does the analysis of marginal revenue relate to the total revenue found by integration in Section 7.5?

2. The total cost of producing dog collars is given by $C(x) = x^2 - 3x$. Find the marginal cost of producing 50 units.

In Exercises 3–5, suppose the price of data-wacks is estimated to be $p(x)$, where $p(x) = 1000 - 4x$ and x is the number of data-wacks sold.

3. Find the revenue function.

4. Find the marginal revenue function.

5. What is the marginal revenue from the sale of the 100th data-wack?

In Exercises 6–9, for the given marginal revenue, find the total revenue produced by the sales of the indicated number of items.

6. $R'(x) = 3x^2 - 6x + 8$; $R = 0$ when $x = 0$;   10 items

7. $R'(x) = x^3 + 2x^2 - x + 52$; $R = 0$ when $x = 0$;   15 items

8. $R'(x) = x\sqrt{x^2 + 2}$; $R = 0$ when $x = 0$;   4 items

9. $R'(x) = x^2 + 5x - 12$; $R = 0$ when $x = 0$;   10 items

10. If the marginal propensity to consume as a function of income is $dc/dx = 4e^x + 2x$, find the analytic expression for consumption and savings as a function of income.

In Exercises 11–14, the marginal cost of producing an item is given, together with the fixed cost. Find the cost of producing the indicated number of items.

11. $y' = x^2 + \dfrac{4x}{2x^2 + 5}$,   with a fixed cost of 50;   20 items

12. $y' = 4x^3 - 2x^2 + 2$,   with a fixed cost of 100;   3 items

13. $y' = x^{1/3} - \dfrac{1}{2}x^{1/2}$,   with a fixed cost of 30;   10 items

14. $y' = x\sqrt{4 + 2x^2}$,   with a fixed cost of 300;   100 items

In Exercises 15 and 16, if $C(x)$ defines the cost of producing x items, and $R(x)$ the revenue from the sale of the same x items, $R(x) - C(x)$ is an expression estimating the profit. Under the indicated conditions find the profit generated by the given number of items.

15. $C'(x) = x^2 + 2$, $R'(x) = 2x + 4$ with $R(0) = 0$ and $C(0) = 4$. Find the profit in 2 items.

16. $R'(x) = x^3$ with $R(0) = 0$, $C'(x) = 4x$ with $C(0) = 0$. Find the profit in 10 items.

17. The marginal physical productivity for x lumberjacks is given by $P' = 23 - 0.06x$, where P is in thousands of board feet per day. How many lumberjacks would be required to cut 2,000,000 board feet per day?

18. The marginal physical productivity for x used car salespersons is given by $P' = 2 - 0.1x$, where P is in cars per day. A company wishes to sell 15 cars per day. How many salespersons should be used?

19. The marginal efficiency of investment is the derivative of the yield of investment with respect to the amount of investment. If the investment in farm equipment in thousands of dollars is given by x, and the yield in tons of sugar beets is y, with $y = f(x)$, then if $y' = 4x^2/100$ find the yield for an investment of $75,000, if there is a yield of 2 tons with no actual cash investment.

7.7 SUPPLY AND DEMAND FUNCTIONS

One straightforward application of integration is related to the economic concepts of supply and demand and the concepts of a consumers' or producers' surplus. In a certain economic model, the price, y, at which a commodity sells is functionally related to the quantity, x, which can be sold at a given price. A sample of the type of relationship being considered is illustrated in the graph of Figure 7.22. Notice that as smaller and smaller amounts of the items are demanded a higher and higher price can be expected, while as more and more of the item can be sold one would expect a lower price. The function involved is called the *demand function*.

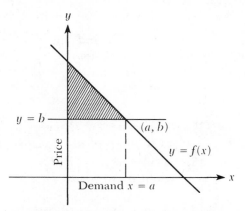

Figure 7.22 Demand function and consumers' surplus (shaded)

In this model let us suppose that in actuality an amount $x = a$ is being sold at a price $b = f(a)$. Then on paper any consumer who would have been willing to pay a higher price for the item "saves" funds or has a surplus. The total surplus involved, considering the whole range of possible values involved, is called the *consumers' surplus*, and corresponds to the shaded area of Figure 7.22. Here integration can be used to evaluate the surplus. The total area beneath the graph of $y = f(x)$, and above $y = 0$, between $x = 0$ and $x = a$ is given by

$$\int_0^a f(x)\, dx$$

The shaded area, which corresponds to the consumers' surplus, is this area minus the rectangular area bounded by $x = 0$, $x = a$, $y = 0$, and $y = b$. Hence the consumers' surplus C is given by

$$C = \int_0^a f(x)\, dx - ab$$

Example 1
[$]

If the demand function is $y = 4 - x^2$ and there is 1 unit actually sold, find the consumers' surplus C.

Solution   The problem is illustrated in Figure 7.23; C is given by

$$C = \int_0^1 (4 - x^2)\, dx - 1(3)$$

$$= 4x - \frac{x^3}{3}\Big]_0^1 - 3 = 4 - \frac{1}{3} - 3 = \frac{2}{3}$$

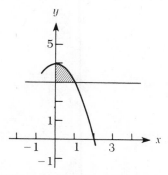

Figure 7.23   Demand function $y = 4 - x^2$

Example 2   If the demand function is given by $y = \sqrt{25 - x^2}$, find the consumers' surplus C if the actual selling price is $y = 3$.

Solution   If $y = 3$, then the corresponding number of items sold is given by

$$3 = \sqrt{25 - x^2}$$
$$9 = 25 - x^2$$
$$x^2 = 16$$
$$x = 4$$

The problem is illustrated in Figure 7.24.

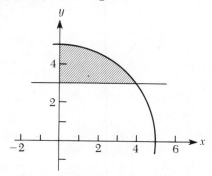

Figure 7.24   Demand function $y = \sqrt{25 - x^2}$,
selling price $y = 3$,
consumers' surplus (shaded)

$$C = \int_0^4 \sqrt{25 - x^2}\, dx - 4(3)$$

Using the Table of Integrals,

$$\int_0^4 \sqrt{25 - x^2}\, dx = \frac{x}{2} \sqrt{25 - x^2} + \frac{25}{2} \sin^{-1}\left(\frac{x}{5}\right)\Big]_0^4$$

$$= 2\sqrt{25 - 16} + \frac{25}{2} \sin^{-1}\frac{4}{5}$$

$$= 6 + 12.5 \sin^{-1} 0.8$$

$$= 6 + 12.5(0.925) = 6 + 11.6 = 17.6$$

Therefore,

$$C = 17.6 - 12 = 5.6$$

If we consider the producer of goods rather than the consumer we can develop the idea of a producers' surplus. Again, our two variables are price and quantity. This time, however, the quantity involved is the number of items that would be produced if they could be sold at a certain price. Here y represents price and x the number of items produced. The problem is illustrated in Figure 7.25.

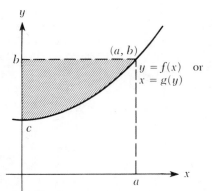

Figure 7.25   Supply function $y = f(x)$, producers' surplus (shaded)

Notice that there is a minimum price which would be required before any items are produced, and as the price paid for items increases, so does the supply. A function used in this manner is called a *supply function*. Suppose that the actual price at which an item is being sold is b, with $b = f(a)$, a the amount being sold. Then any manufacturer or supplier who had expected to supply the item at a lower price has, on paper, made a gain. Figure 7.25 demonstrates this gain by the shaded area below $y = b$ but above $y = f(x)$. The total of this gain is called the *producers' surplus*, and can be evaluated analytically by integration. The producers' surplus, P, is given by

$$P = ab - \int_0^a f(x)\, dx$$

Example 3

Suppose that in Figure 7.25 $y = f(x) = 4 + x^2$ with $a = 5$ and hence $b = 29$. Find the producers' surplus.

Solution

The producers' surplus P is given by

$$P = (29)(5) - \int_0^5 (4 + x^2)\, dx$$

$$= 145 - \left[4x + \frac{x^3}{3} \right]_0^5$$

$$= 145 - \left(20 + \frac{125}{3} \right)$$

$$= 125 - \frac{125}{3} = \frac{250}{3}$$

The functional relation being considered, the producers' surplus, can be evaluated by using integration with respect to y rather than x. If $y = f(x)$ can be solved for x as a function of y as in $x = g(y)$, then the producers' surplus is given by

$$P = \int_c^b g(y)\, dy \qquad \text{where } 0 = g(c)$$

Example 4

Using $f(x) = 4 + x^2$, with $x = 5$ and $y = 29$ as the actual number of items involved and the same price as in Example 3, evaluate the producers' surplus by finding $x = g(y)$.

Solution

We have $y = f(x) = 4 + x^2$. Hence,

$$y - 4 = x^2$$
$$x = \sqrt{y - 4} \qquad \text{with } y \geq 4$$

When $x = 0$, $y = 4$, hence the producers' surplus P is given by

$$P = \int_{4}^{29} \sqrt{y-4}\,dy = \left. \frac{2(y-4)^{3/2}}{3} \right]_{4}^{29}$$

$$= \frac{2(29-4)^{3/2}}{3} = \frac{2(25)^{3/2}}{3} = \frac{2(5)^3}{3}$$

$$= \frac{2(125)}{3} = \frac{250}{3}$$

Under the assumption that the marketplace acts under pure competition, the actual price an item would reach would be determined when supply equals demand. If supply and demand were characterized by supply and demand functions, this would occur when these two functions yielded the same price.

Example 5
$

Assume that in a market under pure competition the supply function for a certain item is given by $g(x) = 3x + 3$, while the demand function for the same item is given by $f(x) = 13 - x^2$. Determine the price at which the item would be sold and the consumers' and producers' surplus.

Solution

The graphs of these supply and demand functions are shown in Figure 7.26. The point of intersection (a, b) can be found by setting $f(x)$ equal to $g(x)$.

$$3x + 3 = 13 - x^2$$
$$x^2 + 3x - 10 = 0$$
$$x = -5 \quad \text{or} \quad x = 2$$

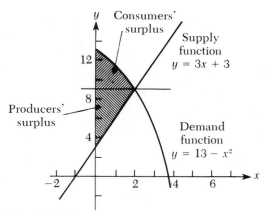

Figure 7.26

Realistically, $x = 2$ is the number of items sold. The corresponding y value of price is $3(2) + 3 = 9$ units. Hence, the point (a, b) is $(2, 9)$. The surpluses can then be found by integration. The producers' surplus P is given by

$$P = (2)(9) - \int_0^2 (3x + 3)\, dx$$
$$= 18 - \left[\frac{3x^2}{2} + 3x\right]_0^2$$
$$= 18 - [6 + 6] = 18 - 12 = 6$$

The consumers' surplus C is given by

$$C = \int_0^2 (13 - x^2)\, dx - (2)(9)$$
$$= 13x - \frac{x^3}{3}\Big]_0^2 - 18 = 26 - \frac{8}{3} - 18$$
$$= 8 - \frac{8}{3} = \frac{16}{3}$$

7.7 EXERCISES

1. Two integrals were given for finding the producers' surplus. One involved the supply function $y = f(x)$, and the other a function $x = g(y)$. Only one integral for the consumers' surplus was given in the first example. Use the demand function in the first example of this section and find an integral for the consumers' surplus involving x as a function of y.

In Exercises 2–7, for the given demand functions find the consumers' surplus for the number of items or price given. Make a sketch in each case.

2. $f(x) = 9 - x^2$, number of items $= 2$
3. $f(x) = 64 - x^3$, number of items $= 3$
4. $f(x) = \sqrt{16 - x^2}$, number of items $= 2$
5. $f(x) = \sqrt{25 - x^2}$, number of items $= 4$
6. $f(x) = 10 - \frac{1}{2}x$, price $= 5$
7. $f(x) = 30 - x^3$, price $= 3$

In Exercises 8–13, for each of the given supply functions find the producers' surplus if the actual number of items is as given. Make a sketch in each case.

8. $f(x) = 1 + x^3$, number of items = 4
9. $f(x) = 2 + \ln(x + 1)$, number of items = 4
10. $f(x) = 3x^2 + 2$, number of items = 10
11. $f(x) = xe^x + 1$, number of items = 3
12. $f(x) = x^2 - 6x + 9$, number of items = 6
13. $f(x) = 4x + x^2$, number of items = 30

In Exercises 14–18, in each case f gives the demand function and g gives the supply function. Assume that the market behaves under pure competition and find the producers' and consumers' surplus and the price at which the items would be sold.

14. $f(x) = 14 - x^2$
    $g(x) = 4x + 2$

15. $f(x) = 10 - x^2$
    $g(x) = 8x + 1$

16. $f(x) = 27 - x^{2/3}$
    $g(x) = 2x + 7$

17. $f(x) = 4 - \sqrt{x}$
    $g(x) = \sqrt{x} + 2$

18. $f(x) = 8e^{-x}$
    $g(x) = 2e^x$

7.8 NATURAL GROWTH AND DECAY

Many real-world problems of population density or mathematically similar ideas lead to problems involving integration.

Example 1

By definition, radioactive substances are those elements that naturally break down into other elements, releasing energy as they do. The rate at which such a substance decays is proportional to the mass of the material present. If A is the amount present then $dA/dt = -kA$, where k is positive and constant. Hence,

$$\frac{dA}{A} = -k\, dt$$

$$\int \frac{dA}{A} = \int -k\, dt \qquad (1)$$

This type of equation is known as a *differential equation*. A further discussion of this topic is to be found in Section 7.10. Integrating both sides of (1),

$$\ln A = -kt + C$$
$$A = e^{-kt+C} = e^C e^{-kt}$$

Letting $e^C = A_0$,

$$A = A_0 e^{-kt}$$

Notice that when $t = 0$, $A = A_0$; hence A_0 is the amount of substance present when $t = 0$.

Definition

The *half-life*, t_h, of a radioactive substance is the time necessary for a given amount of the substance to decay to one-half of the original amount.

Example 2

Since $A = A_0 e^{-kt}$, what is the relationship between the half-life t_h of a substance and the constant k?

Solution   By definition when $t = t_h$, $A = \frac{1}{2}A_0$. Thus,

$$\frac{1}{2}A_0 = A_0 e^{-kt_h}$$

$$\frac{1}{2} = e^{-kt_h}$$

$$\ln\left(\frac{1}{2}\right) = -kt_h$$

$$-\ln 2 = -kt_h$$

$$kt_h = \ln 2$$

$$t_h = \frac{\ln 2}{k}$$

$$k = \frac{\ln 2}{t_h}$$

Example 3

The half-life of radium is 1590 years. Find an equation describing the amount of radium present as a function of time. Then determine how long it would take a given sample of radium to decay to $\frac{1}{6}$ of its present amount.

Solution   The half-life, $t_h = 1590$; hence

$$k = \frac{\ln 2}{1590}$$

Thus,

$$A = A_0 e^{\frac{-\ln 2}{1590} t}$$

Finding the time necessary for A to equal $A_0/6$,

$$\frac{A_0}{6} = A_0 e^{\frac{-\ln 2}{1590} t}$$

$$\frac{1}{6} = e^{\frac{-\ln 2}{1590} t}$$

$$\ln \frac{1}{6} = \frac{-\ln 2}{1590} t$$

$$-\ln 6 = \frac{-\ln 2}{1590} t$$

$$t = \frac{1590(\ln 6)}{\ln 2} = \frac{1590(1.792)}{0.693} = 4112 \text{ years}$$

For the social or biological scientist one of the most useful applications of the equations describing radioactive decay is the dating of materials. While many radioactive substances are used in this process, the most common is ^{14}C, carbon-14, a radioactive isotope of carbon. Carbon-14 is produced through the action of cosmic rays striking the nitrogen in the earth's upper atmosphere. For many centuries, thousands in fact, the amount of ^{14}C manufactured has exactly balanced the amount lost through radioactive decay, so that the amount present in the atmosphere remains constant. Living things—plants and animals—constantly replace the carbon in their bodies, so that the amount of ^{14}C present in a living organism matches that normally found in the atmosphere. When a living organism dies this replacement process stops, and no new ^{14}C is introduced into the cell structure of the organism. By measuring the amount of ^{14}C remaining in an organic specimen, the date of its death can be determined.

Example 4   Since the half-life of ^{14}C is 5570 years, how old is an organic object
if it has $\frac{1}{10}$ of the normal amount of ^{14}C present?

Solution   Since ^{14}C is a radioactive substance, it obeys the radioactive decay principle described by the equation

$$A = A_0 e^{-kt} \qquad \text{where } k = \frac{\ln 2}{t_h} = \frac{\ln 2}{5570}$$

The problem is to find t such that

$$A = \frac{1}{10} A_0$$

Therefore,

$$\frac{A_0}{10} = A_0 e^{\frac{-\ln 2}{5570} t}$$

$$\frac{1}{10} = e^{\frac{-\ln 2}{5570} t}$$

$$\ln \frac{1}{10} = \frac{-\ln 2}{5570} t$$

$$-\ln 10 = \frac{-\ln 2}{5570} t$$

$$t = \frac{5570 \ln 10}{\ln 2} = \frac{5570(2.303)}{0.693} = 18{,}510 \text{ years}$$

The mathematical model for unlimited population growth is essentially the same as the one for radioactive decay. The reason is that the rate of population growth is proportional to the size of the population just as the rate of radioactive decay is proportional to the mass of the material present. The difference is that radioactive substances decrease in size while populations generally increase in size.

Example 5   If the population of the earth was 3.9 billion in 1975, and is increasing at the rate of 2 percent per year, when will the population reach 50 billion?

Solution   If the population is P and the rate of change of the population, dP/dt, is proportional to P, then

$$\frac{dP}{dt} = kP = 0.02P$$

$$\frac{dP}{P} = 0.02 \, dt$$

$$\int \frac{dP}{P} = \int 0.02 \, dt$$
$$\ln P = 0.02t + C$$
$$P = e^{0.02t+C} = e^C e^{0.02t}$$
$$P = P_0 e^{0.02t} \qquad (P_0 \text{ is the population when } t = 0)$$

Since we are measuring time from 1975, $P_0 = 3.9$ (in billions). Hence,

$$P = 3.9 e^{0.02t}$$

Setting $P = 50$ billion,

$$50 = 3.9 e^{0.02t}$$
$$e^{0.02t} = \frac{50}{3.9}$$
$$0.02t = \ln \frac{50}{3.9}$$
$$t = \frac{\ln(50/3.9)}{0.02} = \frac{\ln 50 - \ln 3.9}{0.02}$$
$$= \frac{3.91 - 1.36}{0.02} \approx 128 \text{ years}$$

Hence the population will reach 50 billion in 1975 + 128 years, or 2103 A.D.

More complex models of growth and decay assume that the rate of growth also depends on a limiting factor. For example, in a closed growth system the rate of growth slows down as the total population reaches some limiting factor.

Example 6

Assume that the earth cannot support a population greater than 20 billion persons, and that the rate of population growth is proportional to how close the world population is to this limiting value. What is the mathematical expression describing the world population as a function of time?

Solution   If P is the world population, then according to the described model

$$\frac{dP}{dt} = k(20 - P)$$

where k is positive to assure that dP/dt is positive. Then

$$\frac{dP}{20 - P} = k\, dt$$

$$\int \frac{dP}{20 - P} = \int k\, dt$$

$$-\ln(20 - P) = kt + C \qquad (P < 20)$$

$$20 - P = e^{-(kt+C)} = e^{-kt}e^{-C}$$

Let $e^{-C} = B$. Then

$$20 - P = Be^{-kt}$$

where B and k are positive. Thus

$$P = 20 - Be^{-kt}$$

What does B represent? Let us assume that $P = 3.9$ billion in 1975. If we let $t = 0$ in 1975,

$$3.9 = 20 - Be^{-k \cdot 0}$$
$$3.9 = 20 - B$$
$$B = 16.1$$

and B is the difference between the world population at $t = 0$ and the 20-billion limit.

The equations developed in this section have many interpretations other than those given. For example, the same equations describe radioactive decay and unlimited population growth. The equation developed above for limited population growth also describes investment amounts when a company wishes to invest toward a certain total investment, and the rate of investment is proportional to how close the total investment is to the desired limit. Several other similar mathematical models are considered in the exercises.

7.8 EXERCISES

1. A 20-gram sample of radium with a half-life of 1590 years is refined and stored for future use. How much radium will remain at the end of 15.9 years? How much will remain at the

end of 15,900 years? How much will remain at the end of 159,000 years?

2. An organic sample is found to have $\frac{1}{5}$ of the normal amount of ^{14}C. How old is it? (The half-life of ^{14}C is 5570 years.)

3. An organic sample is found to have 0.01% of the normal amount of ^{14}C. How old is it?

4. Carbon-14 dating is said to be accurate only over a period of about 70,000 years. What factors contribute to this limitation?

5. A radioactive isotope of potassium is also used for dating. This isotope of potassium has a half-life of approximately 1,300,000,000 years. If a sample containing this isotope is found to have $\frac{2}{3}$ of the original amount of radioactive potassium, how old is it?

6. Referring to the potassium isotope of Exercise 5, a sample is found to have $\frac{1}{5}$ of the original amount of radioactive potassium. How old is the sample?

In Exercises 7 and 8, assume that the population of the earth was 3.9 billion in 1975 and population growth is unlimited.

7. If the earth's population is growing at the rate of 2 percent per year, when will a population of 60 billion be reached?

8. If the earth's population has been growing at a rate of 2 percent per year, estimate the world population in 1900, 1800, and 1500.

9. If India's population was 600 million in 1970 and was growing at a 5 percent annual rate, and China's population was 750 million and growing at an annual rate of 4 percent, when will the population of the two be equal? Why is this model of population unrealistic?

In Exercises 10–12, assume, as in Example 6, that the world cannot support a population of more than 20 billion. If in the analysis $k = 0.02$ and in 1975 the world's population was 3.9 billion, then:

10. When will the world population reach 10 billion?

11. When will the world population reach 19 billion?

12. Show that after suitable values of P_0 and k have been found $\lim_{t \to \infty} P = 20$.

13. Suppose that the rate of bacteria growth in a culture is proportional to the bacteria present. If in the first hour of growth the culture grows to $1\frac{1}{2}$ times its original amount, how long will it take to double the original amount?

14. If the culture in Exercise 13 grows at a rate such that it doubles the original amount in 2 days, how long would it take to triple the original amount?

15. Assume a bacteria culture growing in such a way that it doubles the original amount in t days. Discuss the parallels between this number t and the half-life of a radioactive substance.

16. Some scientists discover a new radioactive substance. They discover that their 1 gram sample is decaying at a rate of 0.004 gram per day. What is the half-life of this substance?

17. Repeat Exercise 16 assuming that the substance is decaying at the rate of 0.05 gram per day.

18. Ten grams of a substance is stored. If after 1 year the substance has decayed to 9 grams, could it be radium?

In Exercises 19–21, assume that the world population follows a pattern of inhibited growth where the population P is such that

$$\frac{dP}{dt} = kP(20 - P) \qquad \text{where P is in billions}$$

19. Show that

$$P = \frac{20P_0}{P_0 + (20 - P_0)e^{-20kt}}$$

satisfies this equation if P_0 is the population when $t = 0$.

20. Using the result of Exercise 19 and $k = 0.02$, determine when the world population will reach 15 billion if the population in 1975 is 3.9 billion.

21. According to the results of Exercise 19, $k = 0.02$ and $P_0 = 3.9$ billion in 1975; when will the world's population reach 20 billion?

7.9 VOLUMES

Integration can be used to find volumes as well as areas. Consider a solid as shown in Figure 7.27. Assume that along an axis through the solid the cross-sectional area perpendicular to the axis at each point on the axis is known. The volume can be estimated by partitioning the portion of the axis within the solid. Let the

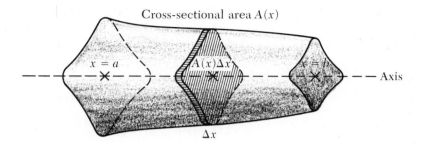

Figure 7.27   Volume by slicing

points along the axis be identified by values of x, and let $A(x)$ give the cross-sectional area perpendicular to the axis at each point on the axis from $x = a$ through $x = b$. Let ΔV represent the volume between $x = x_i$ and $x = x_i + \Delta x$. Then ΔV is approximately $A(x_i)\Delta x$. V, the total volume, is then estimated by

$$V = \sum_{i=1}^{n} A(x_i)\,\Delta x$$

Letting the divisions become finer and finer, that is, finding the limit of the indicated sum as the values of Δx tend toward zero, gives

$$V = \lim_{n \to \infty} \sum_{i=1}^{n} A(x_i)\,\Delta x = \int_a^b A(x)\,dx$$

Example 1   Find the volume of the right circular cone shown in Figure 7.28.

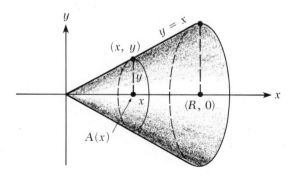

Figure 7.28   Volume by known cross-sectional area

Solution First we find an expression for the cross-sectional area as a function of x. Note that a cross-section of the cone is a circle of radius y. Then its area is πy^2. However, the circle touches the line $y = x$ so $\pi y^2 = \pi x^2$ and the cross-sectional area is $A(x) = \pi x^2$. The volume is

$$V = \int_0^R \pi x^2 \, dx$$

$$= \frac{\pi}{3} [x^3]_0^R$$

$$= \frac{1}{3} \pi R^3$$

A large family of volume problems of this type can be solved by using the concept of a solid of revolution. Consider the area bounded by $y = f(x)$, $x = a$, $x = b$, and $y = 0$ where $f(x) \geqslant 0$ for all $x \in [a, b]$. Such an area is shown in Figure 7.29. If this

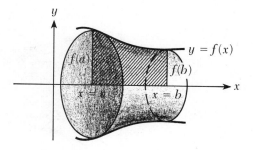

Figure 7.29   A solid of revolution

area is revolved about the x axis, the resulting solid is called a *solid of revolution*. Due to the manner in which the solid has been generated, cross-sectional areas taken perpendicular to the x axis are clearly circles of radius $f(x)$. Hence the cross-sectional area as a function of x is given by

$$A(x) = \pi [f(x)]^2$$

The volume is

$$V = \int_a^b A(x) \, dx = \int_a^b \pi [f(x)]^2 \, dx$$

Example 2   Find the volume generated when the graph of $y = \sin x$ between $x = 0$ and $x = \pi/2$ is revolved around the x axis.

Solution
$$V = \int_0^{\pi/2} \pi[\sin x]^2 \, dx$$
$$= \pi \int_0^{\pi/2} \frac{1 - \cos 2x}{2} \, dx$$
$$= \pi \left[\frac{x}{2} - \frac{\sin 2x}{4} \right]_0^{\pi/2}$$
$$= \pi \left[\frac{\pi}{4} - \frac{\sin \pi}{4} - \left(0 - \frac{\sin 0}{4} \right) \right] = \frac{\pi^2}{4} \text{ cubic units}$$

Sometimes the volume integral involves an improper integral. This in turn can lead to paradoxes involving infinity.

Example 3   The area bounded by $y = 0$, $y = 1/x^{2/3}$, and $x = 1$ is revolved about the x axis, as in Figure 7.30. Find the volume thus generated.

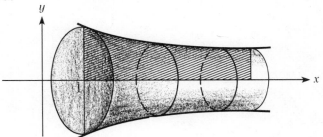

Figure 7.30   Graph of $y = x^{-2/3}$ revolved around the x axis

Solution   From above,

$$V = \int_1^{\infty} \pi \left(\frac{1}{x^{2/3}} \right)^2 dx$$
$$= \pi \lim_{b \to \infty} \int_1^b \frac{1}{(x^{4/3})} \, dx$$
$$= \pi \lim_{b \to \infty} -3x^{-1/3} \big]_1^b$$
$$= \pi \lim_{b \to \infty} \left(\frac{-3}{b^{1/3}} + 3 \right) = 3\pi$$

The volume is 3π cubic units.

The paradox arises when we consider the size of the area that was revolved. The area in question is shown in Figure 7.30. Integrating to find the area one gets

$$A = \int_1^\infty \frac{1}{x^{2/3}} \, dx$$

$$= \lim_{b \to \infty} \int_1^b \frac{1}{x^{2/3}} \, dx$$

$$= \lim_{b \to \infty} \, [3x^{1/3}]_1^b = \lim_{b \to \infty} \, [3b^{1/3} - 3]$$

$$= \infty \qquad \text{undefined}$$

In short, the volume is finite, but the cross-sectional area is infinite!

Example 4 The region bounded by $y = \sqrt{x}$ and $y = x^2$ is revolved around the x axis. Find the volume generated (see Figure 7.31).

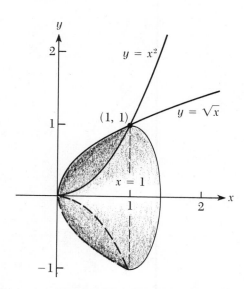

Figure 7.31 Volume formed by revolving the shaded region bounded by $y = \sqrt{x}$ and $y = x^2$ around the x axis

Solution The volume can be found by taking the difference in volumes generated when $y = \sqrt{x}$ and $y = x^2$ are revolved separately.

$$V = \int_0^1 \pi(\sqrt{x})^2 \, dx - \int_0^1 \pi(x^2)^2 \, dx$$
$$= \pi \int_0^1 (x - x^4) \, dx$$
$$= \pi \left[\frac{x^2}{2} - \frac{x^5}{5} \right]_0^1$$
$$= \pi \left[\frac{1}{2} - \frac{1}{5} \right] = \frac{3\pi}{10} \text{ cubic units}$$

7.9 EXERCISES

In Exercises 1–10, $A(x)$ gives the cross-sectional area of a solid perpendicular to the x axis over the indicated interval. Find the volume.

1. $x = 0$ to $x = 1$, $A(x) = 3x^2$

2. $x = 0$ to $x = 1$, $A(x) = \dfrac{x^3}{(x^4 + 1)^3}$

3. $x = 2$ to $x = 5$, $A(x) = \dfrac{x - 1}{x + 1}$

4. $x = 1$ to $x = 3$, $A(x) = \left(2x - \dfrac{1}{2x} \right)^2$

5. $x = 1$ to ∞, $A(x) = \dfrac{1}{x\sqrt{x}}$

6. $x = 1$ to $x = e^3$, $A(x) = \dfrac{\ln x}{x}$

7. $x = \dfrac{\pi}{2}$ to $x = \dfrac{2\pi}{3}$, $A(x) = \dfrac{\sin x}{1 - \cos x}$

8. $x = 0$ to $x = \sqrt{\dfrac{\pi}{2}}$, $A(x) = x \sin 3x^2$

9. $x = \ln \dfrac{\pi}{4}$ to $x = 0$, $A(x) = e^x \sin e^x$

10. $x = 2\pi$ to $x = 4\pi$, $A(x) = |\sin x|$

In Exercises 11–21, the indicated area is to be revolved around the x axis. Find the volume generated. Sketch the area involved.

11. $y = \sqrt{x}$, $x = 0$, $x = 4$, $y = 0$

12. $y = \dfrac{1}{x + 1}$, $x = 0$, $x = 8$, $y = 0$

13. $x = 3$, $y = 0$, $y = x + 2$, $x = 0$

14. $x = y^2$, $x = 0$, $y = 2$ **15.** $x = -y^2$, $x = 0$, $y = 3$

16. $y^2 = x^3$, $x = 4$, $y = 0$

17. $y = 0$, $y = \cos x$ between $x = -\dfrac{\pi}{2}$ and $x = \dfrac{\pi}{2}$

18. $y = 0$, $x = 0$, $y = \dfrac{1}{(x + 1)^{1/3}}$ [*Hint*: Graph extends to ∞.]

19. $y = 0$, $y = e^{-x}$, $x = 0$, $x = \ln 2$

20. $y = 0$, $y = \dfrac{1}{x}$, $x = 1$ [*Hint*: Graph extends to ∞.]

21. $y = 0$, $y = e^{-x}$, $x = 0$

22. Find the area of the region described in Exercise 18.

23. Find the area of the region described in Exercise 20.

24. Find the area of the region described in Exercise 21.

25. If, instead of y being a function of x, x was given as a function of y, find an integral which would be used to find the volume generated when the region bounded by $x = f(y)$, $x = 0$, $y = c$, and $y = d$ is revolved around the y axis.

In Exercises 26–30, note that the method we have used can be generalized to find the volume generated when regions are revolved around lines other than the x axis. In these problems find the volume generated when the area is revolved about the specified line.

26. The region bounded by $y = 0$, $x = 4$, and $y = \sqrt{x}$ is revolved around the line $y = -1$.

27. The region bounded by $y = 0$, $x = 4$, and $y = \sqrt{x}$ is revolved around the line $y = 2$.

28. The region bounded by $y = 1$, $x = 1$, $x = 4$, and $y = \sqrt{x}$ is revolved around the line $y = 1$.

29. The region bounded by $y = 0$, $y = \sqrt{x}$, and $x = 4$ is revolved around the line $x = 4$.

30. The region bounded by $y = 0$, $y = \sqrt{x}$, $x = 1$, and $x = 4$ is revolved around the line $x = 0$.

7.10 ELEMENTARY DIFFERENTIAL EQUATIONS

Many of the applications of integration considered in this chapter have been specific cases of a more general concept known as a *differential equation.* A differential equation is an equation involving an unknown function and its derivatives. The objective is to recover the missing function. We have done this in many of the examples of previous sections.

Example 1

In Section 7.6 we considered the following problem: Let the marginal revenue relative to sales of a certain object be given by

$$\frac{dR}{dx} = 4x^2 - 3x + 2$$

What is the total revenue produced by the sales of 5 items?

Solution

This is a very simple type of differential equation. The unknown function is R. It lends itself to direct solution by the method known as *separation of variables*. In this type of problem, using differential notation, the expressions involved can be *separated* in such a way that integration can be used to recover the missing function. Specifically,

$$dR = (4x^2 - 3x + 2)\, dx$$

Both sides of this equation can be integrated. We then have

$$\int dR = \int (4x^2 - 3x + 2)\, dx$$

Integrating:

$$R = \frac{4x^3}{3} - \frac{3x^2}{2} + 2x + C$$

The problem is completed by finding an appropriate value for C. In this case we reasoned that revenue was zero when $x = 0$. Therefore

$$0 = \frac{4}{3}(0)^3 - \frac{3}{2}(0)^2 + 2(0) + C$$
$$C = 0$$

and

$$R = \frac{4}{3}x^3 - \frac{3}{2}x^2 + 2x$$

When $x = 5$,

$$R = \frac{835}{6} \approx 139.17$$

Many simple differential equations can be solved by the method of separation of variables.

Example 2   Find y if $dy/dx = 3x^2y$.

   Solution   Separate the variables. Specifically, if

$$\frac{dy}{dx} = 3x^2y$$

then

$$\frac{dy}{y} = 3x^2\,dx$$

Then

$$\int \frac{dy}{y} = \int 3x^2\,dx$$
$$\ln y = x^3 + C$$

We can solve for y by stating both sides of the equation as a power of e. That is,

$$e^{\ln y} = e^{x^3 + C}$$

However, $e^{\ln y} = y$ and $e^{x^3 + C} = e^{x^3}e^C$. Letting $e^C = C_1$ (since C is a constant, e^C is a constant), we have

$$y = C_1 e^{x^3}$$

The specific value of C_1 will depend on the conditions particular to the problem. Solutions in which constants like C_1 remain are considered *general solutions*. When the constant is given a value to meet specific conditions, we have a *particular solution*.

Example 3 If $x = \frac{1}{2}$ when $t = 0$ and

$$\frac{dx}{dt} = 1 + 2x$$

find x when $t = 1$.

Solution To solve the problem we must find a particular solution to the differential equation.

$$\frac{dx}{dt} = 1 + 2x$$

Then

$$\frac{dx}{1 + 2x} = dt$$

$$\int \frac{dx}{1 + 2x} = \int dt$$

$$\frac{1}{2} \ln|1 + 2x| = t + C$$

$$\ln|1 + 2x| = 2t + C_1 \qquad (C_1 = 2C)$$

$$e^{\ln|1+2x|} = e^{2t+C_1}$$

$$1 + 2x = e^{2t}e^{C_1}$$

$$= C_2 e^{2t} \qquad (C_2 = e^{C_1})$$

$$2x = C_2 e^{2t} - 1$$

$$x = C_3 e^{2t} - \frac{1}{2} \qquad \left(C_3 = \frac{C_2}{2}\right)$$

When $t = 0$, $x = \frac{1}{2}$; thus

$$\frac{1}{2} = C_3 e^0 - \frac{1}{2}$$

$$1 = C_3$$

Therefore,

$$x = e^{2t} - \frac{1}{2}$$

When $t = 1$,

$$x = e^2 - \frac{1}{2} \approx 6.89$$

Problems can also involve trigonometric functions.

Example 4   Find how y and x are related if $\sin x \sin y \, dx = -\cos x \cos y \, dy$.

Solution   If $\sin x \sin y \, dx = -\cos x \cos y \, dy$, then

$$\frac{\sin x}{\cos x} \, dx = \frac{-\cos y}{\sin y} \, dy$$

$$\int \frac{\sin x}{\cos x} \, dx = - \int \frac{\cos y}{\sin y} \, dy$$

$$-\ln |\cos x| + C = -\ln |\sin y|$$

Then

$$e^{\ln|\cos x| + C} = e^{\ln|\sin y|}$$

$$e^C \cos x = \sin y$$

$$\sin y = C_1 \cos x \qquad (C_1 = e^C)$$

This answer is in *implicit* form.

Example 5   Sociologists use differential equations to describe how information is spread through a population by means of mass media. Suppose that a population is made up of P individuals and a certain piece of information is being disseminated. Let $y = f(t)$ be the number of people who have heard the information after t units of time. Then $P - y$ are the people who have not heard the information. The rate that people are learning the information is proportional to the number who have not yet heard it. That is, $y' = k(P - y)$. Solve the differential equation for y, the number of people that have heard the information after t units of time.

Solution

$$y' = k(P - y)$$

$$\frac{dy}{dt} = k(P - y) \qquad \text{separate the variables}$$

$$\frac{dy}{P - y} = k \, dt \qquad \text{integrate both sides}$$

$$\int \frac{dy}{P - y} = k \int dt$$

$$-\ln (P - y) = kt + C$$

$$\ln (P - y) = -kt - C \tag{1}$$

However, when $t = 0$, $y = 0$ and substitution gives:

$$\ln P = -C$$

Therefore equation 1 becomes

$$\ln (P - y) = -kt + \ln P$$

From properties of logarithms,

$$e^{-kt+\ln P} = P - y$$
$$e^{-kt}e^{\ln P} = P - y$$

But $e^{\ln P} = P$, so

$$Pe^{-kt} = P - y \qquad \text{solving for } y$$
$$y = P - Pe^{-kt}$$
$$f(t) = y = P(1 - e^{-kt})$$

and $f(t)$ is the number of individuals who have heard a piece of information after t units of time.

Naturally not all differential equations are separable. In fact, there is an extensive body of theory and practice for solving differential equations. It will be worthwhile to consider another method. An equation of the type

$$\frac{dy}{dx} + P(x)y = Q(x) \tag{2}$$

where $P(x)$ and $Q(x)$ are functions of x, is called a *linear differential equation of the first degree*. Such equations can be solved by using an *integrating factor*, that is, a special factor which, when multiplied by each expression in the equation, converts the equation to a solvable form.

Example 6   Consider the equation

$$\frac{dy}{dx} - \frac{2}{x} y = 2x^3 \tag{3}$$

Solution   This is in the form of Equation 2 with $P(x) = -2/x$ and $Q(x) = 2x^3$. If we multiply Equation 3 by x^{-2}, we have

$$x^{-2} \frac{dy}{dx} - 2x^{-3}y = 2x \tag{4}$$

On examining the left side of Equation 4, we find that this is a *product* derivative. That is,

$$\frac{d}{dx}(x^{-2}y) = x^{-2}\frac{dy}{dx} - 2x^{-3}y$$

So Equation 4 becomes

$$d(x^{-2}y) = 2x \, dx$$

Integrating both sides,

$$\int d(x^{-2}y) = \int 2x \, dx$$
$$x^{-2}y = x^2 + C$$
$$y = x^4 + Cx^2 \qquad \text{a solution}$$

That is nice! But where did the x^{-2} come from? Consider Equation 2 in the abstract:

$$\frac{dy}{dx} + P(x)y = Q(x)$$

We want to select $R(x)$ such that when we examine

$$R(x)\frac{dy}{dx} + R(x)P(x)y = R(x)Q(x)$$

the left side is a product derivative. That is,

$$R(x)\frac{dy}{dx} + R(x)P(x)y = \frac{d}{dx}[R(x)y] \tag{5}$$

What can we choose for $R(x)$? Consider

$$\frac{d}{dx}[R(x)y] = R(x)\frac{dy}{dx} + \frac{d}{dx}[R(x)]y$$

Then equating this with the left side of Equation 5,

$$R(x) \frac{dy}{dx} + \frac{d}{dx}[R(x)]y = R(x) \frac{dy}{dx} + R(x)P(x)y$$

To reach our objective, we have to choose $R(x)$ so that

$$\frac{d}{dx}[R(x)] = R(x)P(x)$$

Then

$$\frac{d[R(x)]}{R(x)} = P(x) dx$$

Integrating,

$$\ln[R(x)] = \int P(x) dx + C$$
$$e^{\ln[R(x)]} = e^{\int P(x) dx + C}$$
$$R(x) = e^{\int P(x) dx} \cdot e^C$$
$$= C_1 e^{\int P(x) dx}$$

Therefore, if we choose $R(x) = C_1 e^{\int P(x) dx}$, where C_1 is an arbitrary constant, we can use this as an integrating factor.

Example 7   Show that x^{-2} is the factor determined by the above analysis for the differential equation

$$\frac{dy}{dx} - \frac{2}{x} y = 2x^3$$

Solution   We have $P(x) = -2/x$; therefore

$$\int P(x) dx = \int \frac{-2}{x} dx = -2 \ln x = \ln x^{-2}$$
$$e^{\int P(x) dx} = e^{\ln x^{-2}} = x^{-2}$$

The factor is

$$C_1 e^{\int P(x) dx} = C_1 x^{-2}$$

Letting $C_1 = 1$ (C_1 is arbitrary) we have x^{-2} as our factor.

Example 8   Solve $dy/dx = e^{-2x} - y$.

Solution   Putting the problem in the form of Equation 2 above, we have

$$\frac{dy}{dx} + y = e^{-2x}$$

Therefore, $P(x) = 1, Q(x) = e^{-2x}$.

$$\int P(x)\,dx = \int 1\,dx = x$$
$$C_1 e^{\int P(x)\,dx} = C_1 e^x$$

Letting $C_1 = 1$,

$$C_1 e^x = e^x$$

We can then solve the equation by multiplying by e^x.

$$\frac{dy}{dx} e^x + ye^x = e^{-x}$$

The left side is the derivative of $e^x y$. Therefore

$$e^x y = \int e^{-x}\,dx$$
$$= -e^{-x} + C$$
$$y = \frac{-e^{-x} + C}{e^x}$$
$$= Ce^{-x} - e^{-2x}$$

Example 9   Solve

$$\frac{dy}{dx} + (\cot x)y = \cos x$$

Solution   This is in the required form of Equation 2, with $P(x) = \cot x$, $Q(x) = \cos x$.

$$\int P(x)\,dx = \int \cot x\,dx = \int \frac{\cos x}{\sin x}\,dx$$
$$= \ln(\sin x)$$
$$e^{\int P(x)\,dx} = e^{\ln(\sin x)} = \sin x$$

Let $C_1 = 1$. The original equation becomes

$$\sin x\, \frac{dy}{dx} + (\cos x)y = \sin x \cos x \qquad \text{with } R(x) = \sin x$$
$$\frac{d}{dx}[(\sin x)\cdot y] = \sin x \cos x$$

or

$$(\sin x) \cdot y = \int \sin x \cos x \, dx$$

$$(\sin x)y = \frac{\sin^2 x}{2} + C$$

$$y = \frac{\sin x}{2} + \frac{C}{\sin x}$$

$$= \frac{1}{2} \sin x + C \csc x$$

Example 10

💲

The money spent by a corporation for advertising must have an effect upon the corporation's net profits. That is, if p is profit and x dollars are spent on advertising then $p = f(x)$ for some function f. Under certain conditions profits are related to advertising expenditure by

$$\frac{dp}{dx} + p = 20 - x$$

Find p as a function of x.

Solution

The differential equation is in the form of Equation 2 with $P(x) = 1$ and $Q(x) = 20 - x$. Then

$$\int P(x) \, dx = \int 1 \, dx = x$$

and

$$e^{\int P(x)\,dx} = e^x$$

Multiplying the differential equation by e^x gives

$$e^x \frac{dp}{dx} + e^x p = 20e^x - xe^x$$

Integrating both sides,

$$e^x p = 20e^x - \int xe^x \, dx + C$$

The integral of xe^x can be found using the Table of Integrals.

$$e^x p = 20e^x - (x - 1)e^x + C$$

$$p = 20 - (x - 1) + \frac{C}{e^x}$$

$$p = 21 - x + Ce^{-x}$$

7.10 EXERCISES

In Exercises 1–12, solve the differential equations using separation of variables. Solutions can be left in implicit form, that is, without solving for y.

1. $\dfrac{dy}{dx} = (x + 3)^2$

2. $\dfrac{dy}{dx} = \sin x$

3. $\dfrac{dy}{dx} = x^2 - 2x + 4$

4. $x^2\,dx = y\,dy$

5. $(y^2 + 1)\,dx + 2xy\,dy = 0$

6. $\dfrac{dy}{dx} = e^{x-y}$

7. $xe^y\,dx + e^x\,dy = 0$

8. $x(1+y^2)dx + y(1+x^2)dy = 0$

9. $\sqrt{xy}\,\dfrac{dy}{dx} = 1$

10. $y \sin x\,dx = \cos x\,dy$

11. $y\,dx - \sqrt{1 - x^2}\,dy = 0$

12. $\dfrac{dy}{dx} - 2x = \dfrac{2x}{y}$

In Exercises 13–16, find the indicated particular solution for y as a function of x.

13. $y\,dy = 2x\,dx,\quad y = 0$ when $x = 1$

14. $x^2\,dy = (1 - x)y\,dx,\quad y = 1$ when $x = \frac{1}{2}$

15. $\dfrac{dy}{dx} = -\cos 2x,\quad y = 1$ when $x = \dfrac{\pi}{2}$

16. $\dfrac{dy}{dx} = e^x,\quad y = 0$ when $x = 0$

In Exercises 17–22, solve the linear differential equations.

17. $\dfrac{dy}{dx} + 2y = e^{-x}$

18. $\dfrac{dy}{dx} + \dfrac{2}{x}y = x + 2$

19. $\dfrac{dy}{dx} + (\tan x)y = \cos^2 x$

20. $x\,dy + y\,dx = \sin x\,dx$

21. $y\,dx + x\,dy = x\,dx$

22. $\dfrac{dy}{dx} - 2xy = x$

In Exercises 23 and 24, find particular solutions.

23. $\dfrac{dy}{dx} - 2y = e^{3x},\quad y = 2$ when $x = 0$

24. $\dfrac{dy}{dx} + \dfrac{3y}{x} = 4,\quad y = 5$ when $x = 1$

25. It was shown in Example 5 that in a population of P individuals, information dispensed by mass media will reach y people in t units of time where

$$y = P(1 - e^{-kt})$$

Find k in the above equation if the population consists of 2000 persons and 100 people knew of the information after 1 hour.

26. If net profit, P, is related to dollars, x, spent on advertising, by $P = 21 - x - Ce^{-x}$ and a profit of $50 will result when no advertising is done then what is the value of C in the above equation?

27. Newton's Law of Cooling states that an object placed in air will cool at a rate that is proportional to the difference between the temperature of the object and that of the air around it. An object of temperature 40°C is placed in air that is maintained at 10°C. After 20 minutes the temperature of the object has dropped to 30°C. What is the temperature of the object after 40 minutes in the air?

28. An infectious disease is introduced into a population of P individuals. After t units of time y individuals have the disease. An epidemiologist determines that the rate at which more individuals are catching the disease is proportional to the number who are not infected. Find an expression for y as a function of t.

NEW TERMS

Capital value	Present value
Demand function	Probability
First moment	Probability density function
Half-life	Second moment
Margin	Standard deviation
Mutually exclusive	Supply function
Normal curve	z-score

REVIEW

In Exercises 1–8, find the area bounded by the given graphs. Sketch the curves.

1. $y = x^2 + 4x + 4$, $x = 0, x = 3, y = 0$
2. $y = (\sqrt{x} - 4)^2$, $x = 1, x = 0, y = 0$
3. $y = -xe^{-x^2}$, $x = 0, x = \sqrt{\ln 3}, y = 0$
4. $y = \dfrac{1}{x} \ln x$, $x = 1, x = 2, y = 0$
5. $y = x + \sin x$, $y = x, x = 0, x = \dfrac{\pi}{6}$
6. $y = x^2 + 3$, $y = 2, x = 0, x = 1$
7. $y = \sin x$, $y = -x - 1, x = \dfrac{\pi}{4}, x = \dfrac{\pi}{2}$
8. $y = \tan^{-1} x$, $y = -e^x, x = 0, x = 1$
9. State the definition of a probability.
10. An alternate definition of the probability of an outcome of an experiment that can turn out in n ways, all of which are considered equally likely, is the ratio of the number of these ways favorable to the outcome over n. Compare this definition with the fraction of the time definition used in the text.
11. Why is the area under a probability density curve chosen to be equal to 1?

In Exercises 12 and 13, consider an experiment that has outcomes which fall between −1 and 1 with a probability density function given by

$$P(x) = \frac{\pi}{4} \cos\left(\frac{\pi x}{2}\right)$$

12. Sketch the curve described by $P(x)$ and find the probability of an experimental outcome between 0 and $\frac{1}{2}$.
13. Find the probability of an experimental outcome between $-\frac{1}{3}$ and $+\frac{1}{6}$.

In Exercises 14 and 15, consider an experiment that has outcomes corresponding to the real numbers, with a probability density function given by

$$P(x) = \frac{1}{\pi} \frac{1}{x^2 + 1}$$

14. What is the probability of an outcome corresponding to a real number greater than 1?

15. What is the probability of an outcome corresponding to a real number between $-\frac{1}{2}$ and $\frac{1}{3}$?

In Exercises 16 and 17, assume that the probability that x accidents will occur next year on a stretch of interstate freeway is given by the probability density function

$$P(x) = 0.05e^{-0.05x}$$

16. What is the probability that there will be 20 or fewer accidents next year?

17. What is the probability that there will be 100 or more accidents next year?

18. Consider the function $f(x) = 2x^2/(x^3 + 1)^2$ on the interval $x \in [0, \infty)$. Find a related $P(x)$, probability density function.

In Exercises 19 and 20, for the given probability density functions find μ and σ^2.

19. $P(x) = \frac{4}{3}x^{1/3}$ on the interval $x \in [0, 1]$.

20. $P(x) = \frac{3}{16}(x + 1)^2$ on the interval $x \in [-3, 1]$.

In Exercises 21–24, find the indicated normal curve areas using the normal curve area table.

21. $\dfrac{1}{3\sqrt{2\pi}} \displaystyle\int_{-4}^{4} e^{-\frac{1}{2}\left(\frac{x-1}{3}\right)^2} dx$

22. $\dfrac{1}{5\sqrt{2\pi}} \displaystyle\int_{0}^{3} e^{-\frac{1}{2}\left(\frac{x}{5}\right)^2} dx$

23. $\dfrac{1}{2\sqrt{2\pi}} \displaystyle\int_{0}^{3} e^{-\frac{1}{2}\left(\frac{x-1.5}{2}\right)^2} dx$

24. $\dfrac{1}{3\sqrt{2\pi}} \displaystyle\int_{3}^{\infty} e^{-\frac{1}{2}\left(\frac{x-3}{3}\right)^2} dx$

25. In an experiment involving tossing 10,000 coins, what is the normal curve approximation of the probability of tossing more than 5025 heads?

In Exercises 26 and 27, determine a value for z_1 so that the given integral has the value indicated.

26. $\dfrac{1}{\sqrt{2\pi}} \displaystyle\int_{0}^{z_1} e^{-\frac{1}{2}z^2} dz = 0.41$

27. $\dfrac{1}{\sqrt{2\pi}} \displaystyle\int_{-z_1}^{z_1} e^{-\frac{1}{2}z^2} dz = 0.28$

In Exercises 28–31, for the given revenue function and the indicated number of years, find the expected total revenue.

28. $f(t) = \dfrac{1}{(t + 2)^2}$ (500), for 3 years

29. $f(t) = \dfrac{1}{t+1} (1000),$ for 4 years

30. $f(t) = \dfrac{1}{t^2+1} (1500),$ for 6 years

31. $f(t) = \left[\dfrac{\sin (\pi t)}{\pi} + t \right] 100,$ for 10 years

32. How much would one have to invest at 5.5 percent to have $120,000 five years in the future?

33. How much would have had to be invested at 2.5 percent four years ago to have $15,000 today?

In Exercises 34–37, a revenue function is given. Find the capital value at the given interest rate of the given income at the time in the future indicated.

34. $f(t) = \$225$ per month, the interest rate is 6 percent; 4 years in the future.

35. $f(t) = \$400$ per month, the interest rate is 5 percent; 3 years in the future.

36. $f(t) = \$100t$ per month, the interest rate is 3 percent; 10 years in the future.

37. $f(t) = e^{rt} \dfrac{1}{t+1} (1000)$ dollars per year for 5 years with an interest rate $r = 2$ percent.

In Exercises 38–41, f gives a demand function and g gives a supply function for a certain item. Assume that the market for this item functions under pure competition and find the producers' and consumers' surplus and the price of the item.

38. $f(x) = 7 - x^2$
    $g(x) = x + 1$

39. $f(x) = 40 - x^2$
    $g(x) = 2x + 5$

40. $f(x) = \cos x + 2$
    $g(x) = \dfrac{2x}{\pi} + 1$

41. $f(x) = 100 - x^2$
    $g(x) = (x - 2)^2$

42. If the marginal revenue $R'(x) = x^2 + 2x$ for a certain item, with $R = 0$ when $x = 0$, find the total revenue produced from the sale of four items.

43. If the marginal revenue $R'(x) = (e^{-x}/4) + 8x$, with $R = 0$ when $x = 0$, find the total revenue from the sale of six items.

44. If the marginal cost of a certain item is $c'(x) = x + (e^{-x}/2)$, with a fixed cost of 3, find the cost of producing four items.

45. If the marginal cost of a certain item is $c'(x) = x^{1/4} - (x^{1/5}/2)$, with a fixed cost of 3, what is the cost of producing ten items?

46. The marginal physical productivity of pretzel makers is given by $P' = (102 - 0.4x)$, where P is in thousands of pretzels per day. How many workers would be required to produce 1,000,000 pretzels per day?

47. A 10-gram sample of radium with a half-life of 1590 years is refined and stored. At some later time the sample is opened and found to contain 8 grams of radium. How long has the sample been stored?

48. The half-life of uranium, ^{235}U, is 7.1×10^8 years. If 30 pounds of pure ^{235}U are stored, how long would it require for this to decay to 22.8 pounds of ^{235}U?

49. Assume that a population is not limited and is growing at a rate of k percent per year. If the population is 205 million in 1970, what value of k will lead to a population of 250 million by 2000?

In Exercises 50–53, a cross-sectional area function is given for a certain volume. Find the volume over the indicated interval.

50. $A(x) = \dfrac{1}{x} - \dfrac{1}{x^2}$, $x = 1$ to $x = 4$

51. $A(x) = \ln x$, $x = e^2$ to $x = e^3$

52. $A(x) = x \sin (2x^2)$, $x = 0$ to $x = \sqrt{\dfrac{\pi}{2}}$

53. $A(x) = \tan x$, $x = 0$ to $x = 1$

In Exercises 54–57, the area bounded by the given graphs is revolved about the indicated line. Find the volume generated.

54. $x = 1$, $x = 4$, $y = 0$, $y = \sqrt{x} - \dfrac{1}{\sqrt{x}}$;   about the x axis

55. $x = 1$, $x = 4$, $y = 0$, $y = \sqrt{x} - \dfrac{1}{\sqrt{x}}$;
about the line $y = -2$

56. $x = 1$, $x = 2$, $y = e^x$, $y = 0$;   about the x axis

57. $x = 1$, $x = 2$, $y = e^x$, $y = 0$;   about the y axis

58. A drinking glass is designed by taking the area bounded by $y = 2$, $x = 0$, $y = 0$, $x = 5$, $y = \sqrt{x - 1}$ (all in inches) and revolving it around the x axis. If glass weighs 0.93 ounce per cubic inch and costs $0.27 per pound, what is the cost of the material in such a glass?

59. A glass bud vase is made by revolving the area bounded by $y = 0$, $y = (x/5) + 1.5$, $y = \sqrt[3]{x}$, $x = -0.2$, and $x = 5$ around the x axis. How much would the vase weigh if the glass used weighs 0.93 ounce per cubic inch?

In Exercises 60–63, solve the differential equations.

60. $\dfrac{dy}{dx} = \dfrac{x}{y^2}$

61. $y\,\dfrac{dy}{dx} = x^2 e^{y^2}$

62. $\dfrac{dy}{dx} + 2xy = 2x$

63. $x\,\dfrac{dy}{dx} = y + x^3$

8

Functions of Several Variables

Chapter 7 completed our discussion of calculus of functions of a single variable. Such functions are important, but they are not capable of modeling many real-life situations. Some phenomena are functions of more than one variable. For example, the outcome of a partisan election in a certain precinct may depend on the party preference of the voters of the precinct, on the popularity of the candidate, and on the weather on election day. In business, the productivity of an individual worker is affected by rate of work, the availability of the most effective tools, and the availability of capital and land. The purpose of this chapter is to extend the functional concept to such cases and to extend the ideas of a derivative and integral.

The first two sections introduce functions of two independent variables, their graphs, and limits. Elementary linear programming is introduced in Section 8.3, using graphing techniques. The next three sections apply the derivative to functions of two or more variables. Derivatives in this case are called *partial derivatives* and are used in maximum–minimum problems and in a chain rule for functions of several variables; they are also used to develop a differential, called the *total differential*. Sections 8.7 and 8.8 apply integration to functions of two variables. This technique is called *multiple integration*. The last section of this chapter generalizes partial differentiation and multiple integration to functions of many variables.

8.1 FUNCTIONS OF SEVERAL VARIABLES; LIMITS

The extension of the function concept to functions of several variables is straightforward enough. The three basic ingredients of a function—the domain, the range, and the function rule—remain, but the domain must be modified to allow two or more independent variables to be involved.

The actual details of such an extension can be carried out in steps: first, a specific extension to functions of two independent variables, and then a generalized extension to functions of n independent variables.

Definition

A *function of two variables* is a rule of correspondence which associates each member from a set of ordered pairs called the *domain* with one and only one number from a set called the *range*.

Notice that this definition parallels the original definition of a function of a single variable. We will use the notation $z = f(x, y)$ to indicate that the image under the functional rule of the pair (x, y) is z.

Example 1   Let $f(x, y) = x^2 + y^2$. Calculate values of $f(x, y)$ if the domain is $\{(x, y) \mid x^2 + y^2 \leq 1\}$.

Solution   Specifically,

$$f\left(\frac{1}{2}, \frac{1}{4}\right) = \left(\frac{1}{2}\right)^2 + \left(\frac{1}{4}\right)^2 = \frac{1}{4} + \frac{1}{16} = \frac{5}{16}$$

$$f\left(0, -\frac{1}{3}\right) = 0^2 + \left(-\frac{1}{3}\right)^2 = \frac{1}{9}$$

$$f(0, 0) = 0$$

$f(3, -2)$ is meaningless, since $(3, -2)$ is not in the domain of the function.

Example 2   Does $D = \{(x, y) \mid x \in [0, 1], \ y \in [0, 1]\}$, $f(x, y) = z$ such that $z^2 + x^2 + y^2 = 1$ define a function?

Solution D and f fail to define a function. Consider $f(0, 0)$; then $z^2 + 0^2 + 0^2 = 1$. Thus, z could be ± 1, and the images under f are not unique; hence no function has been defined.

As with a function of a single variable, the definition of a function is applied by convention to a compact notational form. The notation $z = f(x, y)$ will be used to indicate that the variable z is a real-valued function of two independent variables x and y. The domain of the function is assumed to be the set of all ordered pairs (x, y) for which the expression is defined and real. The range of the function is the set of image values.

Example 3 $z = f(x, y) = \dfrac{\sqrt{4 - x^2}}{y + 2}$; find the domain of f and the values of z for a few values of x and y.

Solution   The domain of the function is

$$\{(x, y)\,|\,x \in [-2, 2] \quad \text{and} \quad y \neq -2\}$$

The restriction on x is needed to ensure that $\sqrt{4 - x^2}$ is a real number, and that on y is necessary to exclude division by zero. The functional notation applies in the usual way.

$$f(1, 3) = \frac{\sqrt{4 - 1}}{3 + 2} = \frac{\sqrt{3}}{5}$$

$$f(0, 8) = \frac{\sqrt{4 - 0}}{8 + 2} = \frac{\sqrt{4}}{10} = \frac{2}{10} = \frac{1}{5}$$

$$f(2, 7) = \frac{\sqrt{4 - 4}}{7 + 2} = 0$$

$$f(2, 12) = \frac{\sqrt{4 - 4}}{12 + 2} = 0$$

$$f(w, r) = \frac{\sqrt{4 - w^2}}{r + 2}$$

$$f(x^2, y^3) = \frac{\sqrt{4 - (x^2)^2}}{y^3 + 2}$$

Example 4 $z = f(x, y) = \sin\left(\dfrac{1}{x + y}\right)$; find the values of z for particular values of x and y.

Solution   The domain of this function is $\{(x, y)\,|\,x \neq -y\}$. The range of this function is the range of the sine function: $z \in [-1, 1]$

$$f(0, 1) = \sin\left(\frac{1}{0 + 1}\right) = \sin(1) \approx 0.84$$

$$f\left(\frac{2}{\pi}, 0\right) = \sin\left(\frac{1}{\frac{2}{\pi} + 0}\right) = \sin\left(\frac{\pi}{2}\right) = 1$$

$$f(w, y) = \sin\left(\frac{1}{w + y}\right)$$

$$f(w^2 + 1, k) = \sin\left(\frac{1}{(w^2 + 1) + k}\right)$$

Since the domains of the functions being considered are sets of ordered pairs, it is natural to visualize these domains by finding their graphs in an xy plane.

Example 5   Graph the domain of the function f such that

$$D = \{(x, y)\,|\,x^2 + y^2 \leq 1\}$$
$$f(x, y) = x^2 + y^2$$

Solution   The graph of D is shown in Figure 8.1.

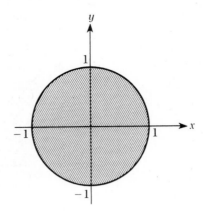

Figure 8.1 $D = \{(x, y)\,|\,x^2 + y^2 \leq 1\}$

Example 6   Graph the domain of f such that $z = f(x, y) = \dfrac{\sqrt{4 - x^2}}{y + 2}$.

Solution   It is shown in Figure 8.2.

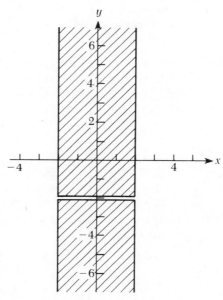

Figure 8.2 Domain of f: $z = f(x, y) = \dfrac{\sqrt{4 - x^2}}{y + 2}$

Example 7 Graph the domain of $z = f(x, y) = \sin\left(\dfrac{1}{x + y}\right)$.

Solution It is shown in Figure 8.3.

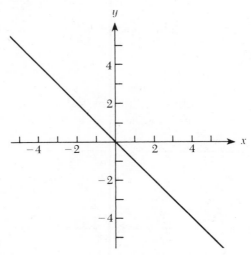

Figure 8.3 Domain of f: $z = f(x, y) = \sin\left(\dfrac{1}{x + y}\right)$; all points in the plane except those on the line $x = -y$

Example 8   Determine the domain and construct its graph for the function defined by $z = f(x, y) = \ln x$.

Solution   Here, z is a function of two variables, yet the defining expression for z omits reference to y. In this case, the domain would be $\{(x, y) \mid 0 < x\}$, which does not restrict the value of y. Thus,

$$
\begin{aligned}
f(1, 7) &= \ln 1 = 0 \\
f(1, 3) &= \ln 1 = 0 \\
f(e, e) &= \ln e = 1 \\
f(e, 7) &= \ln e = 1 \\
f(e, y) &= \ln e = 1
\end{aligned}
$$

Still, f can be thought of as a function of y in the same way the constant function

$$f(x) = 3$$

is viewed as a function of x with

$$
\begin{aligned}
f(2) &= 3 \\
f(0) &= 3 \qquad \text{and so forth}
\end{aligned}
$$

The graph of the domain of this function is the right half of an xy plane, and is shown in Figure 8.4.

The concept of a limit extends to a function of two variables.

$$\lim_{(x,\, y) \to (x_0,\, y_0)} f(x, y) = L$$

indicates that the value of $f(x, y)$ approaches (or is at) L as the point corresponding to (x, y) is taken nearer and nearer to (x_0, y_0).

Example 9 $\displaystyle \lim_{(x,\, y) \to (0,\, 0)} \frac{\sqrt{4 - x^2}}{y + 2} = 1$

Example 10 $\displaystyle \lim_{(x,\, y) \to (4/\pi,\, 0)} \sin\left(\frac{1}{x + y}\right) = \frac{1}{\sqrt{2}}$ $\left(\text{that is, } \sin\dfrac{\pi}{4}\right)$

Example 11 $\displaystyle \lim_{(x,\, y) \to (0,\, 0)} \sin\left(\frac{1}{x + y}\right)$ is not defined

---

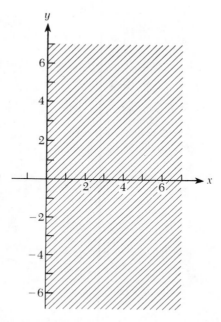

Figure 8.4 Domain of f such that $f(x, y) = \ln x$

The solutions to the examples above were found in an informal manner. In order to provide a formal definition for such limits, it is necessary to assign a specific meaning to the terms "approaches" and "nearer" as used above. The quantity $\left| f(x, y) - L \right|$ would provide a measure of how close $f(x, y)$ is to L for a specific pair (x, y) but the question of how near (x, y) is to (x_0, y_0) and how this is to be measured is a more complex one. The distance formula would provide one measure of this. That is, $\sqrt{(x - x_0)^2 + (y - y_0)^2}$ could be used. One could then say that

$$\lim_{(x, y) \to (x_0, y_0)} f(x, y) = L$$

provided that we can guarantee that $\left| f(x, y) - L \right|$ is as small as we want whenever $\sqrt{(x - x_0)^2 + (y - y_0)^2}$ is small enough.

However, while $\sqrt{(x - x_0)^2 + (y - y_0)^2}$ is a natural measure of how close (x, y) is to (x_0, y_0), it is mathematically easier to measure the closeness of (x, y) to (x_0, y_0) with two quantities, $\left| x - x_0 \right|$ and $\left| y - y_0 \right|$. The advantage of these measures lies in the fact that they can be considered separately. That is, the behavior of the function can be examined when x is fixed and $\left| y - y_0 \right|$ becomes small, or y is fixed, and $\left| x - x_0 \right|$ becomes small. Geometrically

$\sqrt{(x - x_0)^2 + (y - y_0)^2}$ involves a circular area about the point (x_0, y_0) while $|x - x_0|$, $|y - y_0|$ relates to a rectangular region about (x_0, y_0).

Figure 8.5 illustrates the geometry. The definition might take on two forms.

1. *Circular:* $\lim\limits_{(x, y) \to (x_0, y_0)} f(x, y) = L$

if and only if given a positive number ε there exists a δ such that $|f(x, y) - L| < \varepsilon$ whenever

$$0 < \sqrt{(x - x_0)^2 + (y - y_0)^2} < \delta$$

2. *Rectangular:* $\lim\limits_{(x, y) \to (x_0, y_0)} f(x, y) = L$

if and only if given a positive number ε there exists a δ such that $|f(x, y) - L| < \varepsilon$ whenever

$$0 < |x - x_0| < \delta \quad \text{and} \quad 0 < |y - y_0| < \delta$$

Because of the complex nature of the formal application of limits in these cases, we will restrict our study to an informal application such as the examples above.

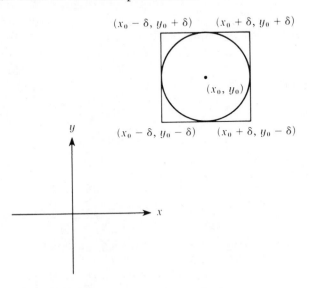

Figure 8.5 $\lim\limits_{(x, y) \to (x_0, y_0)} f(x, y) = L$

8.1 EXERCISES

1. Give three examples of real-life quantities whose value depends on two or more independent quantities.

In Exercises 2–10, for each D and f given find, if possible, the functional images of $(0, 0)$, $(1, 1)$, $(2, 2)$, $(4, 1)$, $(2x, 3)$, (u, v), and $(x^2 + y^2, x)$. Graph D in each case.

2. $D = \{(x, y) \mid x^2 + y^2 \leqslant 25\}$
   $$f(x, y) = \frac{x^2 + y^2}{25}$$

3. $D = \{(x, y) \mid |x - 2| \leqslant 2, \quad |y - 3| \leqslant 3\}$
   $$f(x, y) = \frac{\sqrt{x^2 + 4}}{\sqrt{y + 1}}$$

4. $D = \{(x, y) \mid |x - 3| < 3, \quad |y - 3| < 3\}$
   $$f(x, y) = \sqrt{18 - (x - 3)^2 - (y - 3)^2}$$

5. $D = \{(x, y) \mid xy = 1\}$
   $$f(x, y) = \frac{1}{x^2 y^2}$$

6. $D = \{(x, y) \mid xy = 0\}$
   $$f(x, y) = \ln(xy + 1)$$

7. $D = \{(x, y) \mid x, y \text{ real numbers}\}$
   $$f(x, y) = x^3 + 3x^2 y + 3xy^2 + y^3$$

8. $D = \{(x, y) \mid x, y \text{ real numbers}\}$
   $$f(x, y) = (x + y)^3$$

9. $D = \{(x, y) \mid x = 2, \quad y^2 \leqslant 1\}$
   $$f(x, y) = \cos\left(\pi x - \frac{\pi}{2} y\right)$$

10. $D = \left\{(x, y) \mid |x| \leqslant \frac{\pi}{2}, \quad |y| \leqslant \frac{\pi}{2}\right\}$
    $$f(x, y) = (\sin x)(\cos y)$$

In Exercises 11–24, for each functional expression determine the domain of the function and graph that domain.

11. $f(x, y) = \dfrac{1}{(x + 2)(y - 3)}$

12. $f(x, y) = \dfrac{y^2 + 2}{x^2 - 5x + 6}$

13. $f(x, y) = (x + y)^3$

14. $f(x, y) = \sqrt{25 - x^2 - y^2}$

15. $f(x, y) = \sqrt[3]{-x^2 - y^2}$

16. $f(x, y) = (x + 2)(x + 3)(y + 2)(y + 3)$

17. $f(x, y) = \sqrt{xy}$ **18.** $f(x, y) = e^{\ln x} \cdot e^{\ln y}$

19. $f(x, y) = \dfrac{\sqrt{e^x + e^y}}{\sqrt{1 - e}}$ **20.** $f(x, y) = \cos\left(\dfrac{x + y}{2}\right)$

21. $f(x, y) = \dfrac{(x^2 + 4)(y^2 - 9)}{\sin(xy)}$

22. $f(x, y)$ such that $[f(x, y)] + x^2 + y^2 = 9$ (greatest integer)

23. $f(x, y)$ such that $f(x, y) + x - 2y^2 = 4$

24. $f(x, y)$ such that $f(x, y) = 0$

In Exercises 25–38, evaluate the limits.

25. $\lim\limits_{(x,\, y)\to(1,\, 2)} (x + y)^3$

26. $\lim\limits_{(x,\, y)\to(0,\, 1)} (x + 2)(x + 3)(y + 2)(y + 3)$

27. $\lim\limits_{(x,\, y)\to(-1,\, 1)} \dfrac{1}{(x + 2)(y + 1)}$ **28.** $\lim\limits_{(x,\, y)\to(0,\, 0)} \dfrac{x^2 + y^2}{2(x + y + 1)}$

29. $\lim\limits_{(x,\, y)\to(-1,\, 4)} \left(\dfrac{|x - 3|}{|y + 2|}\right)^2$ **30.** $\lim\limits_{(x,\, y)\to(4,\, 3)} \dfrac{\sqrt{x^2 + y^2}}{2}$

31. $\lim\limits_{(x,\, y)\to(4,\, 4)} \dfrac{y^2 + 2}{x^2 - 5x + 8}$ **32.** $\lim\limits_{(x,\, y)\to(-1,\, -1)} e^{x^2 + y^2}$

33. $\lim\limits_{(x,\, y)\to(1,\, -1)} e^{\ln x} e^{\ln(-y)}$ **34.** $\lim\limits_{(x,\, y)\to(e,\, e)} \dfrac{\ln x}{\ln y + 1}(x^2 + y^2)$

35. $\lim\limits_{(x,\, y)\to(0,\, 0)} \cos\left(\dfrac{x + y}{2}\right)$ **36.** $\lim\limits_{(x,\, y)\to(0,\, 5)} \dfrac{\sin x}{x} y^2$

37. $\lim\limits_{(x,\, y)\to(\pi/2,\, \pi/4)} \dfrac{\sin x \sin y}{\sqrt{2}}$ **38.** $\lim\limits_{(x,\, y)\to(0,\, 0)} y \sin\left(\dfrac{1}{x}\right)$

39. $\lim\limits_{(x,\, y)\to(0,\, 0)} \sin x \cot x$

40. Provide an informal definition for $w = f(x, y, z)$ and

$$\lim\limits_{(x,\, y,\, z)\to(x_0,\, y_0,\, z_0)} f(x, y, z) = L$$

41. Find

$$\lim\limits_{(x,\, y,\, z)\to(1,\, 0,\, 1)} \left\{\dfrac{x^2 + y^2 + z^2}{\sqrt{3}}\right\}$$

42. Find $f\left(7, 4, -\dfrac{2}{\pi}\right)$ if

$$f(x, y, z) = \frac{x^2 - 2xy + \sin z}{\sqrt{x^2 + y^2 + (\pi z)^2}}$$

In Exercises 43–48, consider

$$\lim_{(x, y)\to(1, 1)} \frac{x^2(x^2 + 1)}{y^2} = 2$$

Now consider taking this limit along some "path" or curve passing through (1, 1), for example, $x = y$ or $y = \sin(\pi x/2)$.

$$\lim_{\substack{(x, y)\to(1, 1) \\ \text{along the path} \\ x = y}} \frac{x^2(x^2 + 1)}{y^2} = \lim_{x\to 1} \frac{x^2(x^2 + 1)}{x^2} = 2$$

$$\lim_{\substack{(x, y)\to(1, 1) \\ \text{along the path} \\ y = \sin(\pi x/2)}} \frac{x^2(x^2 + 1)}{y^2} = \lim_{x\to 1} \frac{x^2(x^2 + 1)}{\sin^2(\pi x/2)} = 2$$

However, $\displaystyle\lim_{(x, y)\to(0, 0)} \dfrac{y(x^2 + 1)}{x}$ *is undefined. Consider, then*

$$\lim_{\substack{(x, y)\to(0, 0) \\ \text{along the path} \\ y = x^2}} \frac{y(x^2 + 1)}{x} = \lim_{x\to 0} \frac{x^2(x^2 + 1)}{x} = \lim_{x\to 0} x(x^2 + 1) = 0$$

whereas

$$\lim_{\substack{(x, y)\to(0, 0) \\ \text{along the path} \\ y = \sin x}} \frac{y(x^2 + 1)}{x} = \lim_{x\to 0} \frac{\sin x(x^2 + 1)}{x} = 1$$

Thus, one gets different values along different paths. In each case below find the limit along the given path.

43. $\displaystyle\lim_{\substack{(x, y)\to(0, 0) \\ \text{along the path} \\ x = y}} \frac{y(x^2 + 1)}{x}$

44. $\displaystyle\lim_{\substack{(x, y)\to(2, 2) \\ \text{along the path} \\ x = y}} \frac{x - 2}{y^2 - 4}$

45. $\lim\limits_{\substack{(x,\,y)\to(0,\,0) \\ \text{along the path} \\ y\,=\,\sin x}} \dfrac{x^2(x+2)}{y^2}$
    46. $\lim\limits_{\substack{(x,\,y)\to(2,\,2) \\ \text{along the path} \\ x\,=\,y}} \dfrac{x^2-4}{2y-4}$

47. $\lim\limits_{\substack{(x,\,y)\to(2,\,2) \\ \text{along the path} \\ y\,=\,x^2/2}} \dfrac{x^2-4}{2y-4}$
    48. $\lim\limits_{\substack{(x,\,y)\to(2,\,2) \\ \text{along the path} \\ y\,=\,2/x^2}} \dfrac{x^2-4}{2y-4}$

8.2 THREE-DIMENSIONAL COORDINATE GEOMETRY

If two-dimensional coordinate geometry provides a useful tool for visualizing the domains of functions of two independent variables, it is reasonable to assume that a three-dimensional coordinate geometry could provide a visualization of such functions as a whole. While there are many schemes which could be used to construct such a three-dimensional system, the most straightforward one makes use of a standard rectangular Cartesian coordinate system in a plane, with the third coordinate measuring perpendicular distances to the plane. Such a system is illustrated in Figure 8.6. By constructing the third or z axis perpendicular to the xy plane

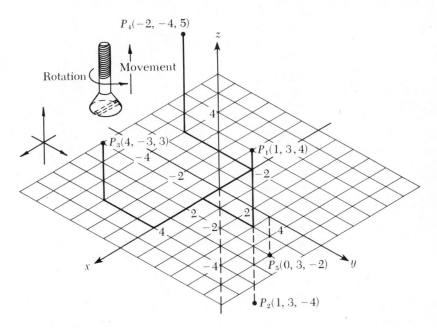

Figure 8.6 A three-dimensional coordinate system

at the origin of the xy system, each point in space can be assigned an identifying set of coordinates. An ordered triple then can serve as the coordinates of the point.

The coordinate system shown in Figure 8.6 is considered *right-handed* because of the positive direction assigned to the z axis and the y axis. Consider a standard right-handed screw (a normal wood screw for example) placed along the z axis, as illustrated in Figure 8.6. Assume the slot of the screw is parallel to the x axis. Now visualize the direction this screw would advance if the slot were rotated 90° to be parallel to the y axis. The direction the screw moves determines the positive direction to be assigned to the z axis. Points identified by $P;(x, y, z)$ are located by using the x and y values to identify the location of the point relative to the xy plane and then using the value of z to tell how far above or below the xy plane the point lies.

Example 1

Consider the points $P_1;(1, 3, 4)$, $P_2;(1, 3, -4)$, $P_3;(4, -3, 3)$, $P_4;(-2, -4, 5)$, and $P_5;(0, 3, -2)$. Each of these points has been plotted in Figure 8.6.

Defining the three coordinate axes, in turn, defines three coordinate planes. These are the original xy plane, or the $z = 0$ plane, the xz plane where $y = 0$, and the yz plane where $x = 0$. The coordinates of a point (x, y, z) can be thought of as the perpendicular distances to the three coordinate planes. The three coordinate planes divide space into eight octants. The octant where all three coordinates of a point are positive is known as the first octant. There is no set pattern used in identifying the other seven octants.

The formula used to calculate the distance between two points in a plane when their coordinates are known has its counterpart for two points in space. If $P_1;(x_1, y_1, z_1)$ and $P_2;(x_2, y_2, z_2)$ are two points, whose coordinates are given relative to a three-dimensional coordinate system, and P_1P_2 denotes the distance between P_1 and P_2, then:

$$P_1P_2 = \sqrt{(x_2 - x_1)^2 + (y_2 - y_1)^2 + (z_2 - z_1)^2}$$

Example 2

Find the distance between $P_1;(2, 3, 1)$ and $P_2;(4, 1, 5)$.

Solution

$P_1;(2, 3, 1)$ and $P_2;(4, 1, 5)$ are shown in Figure 8.7.

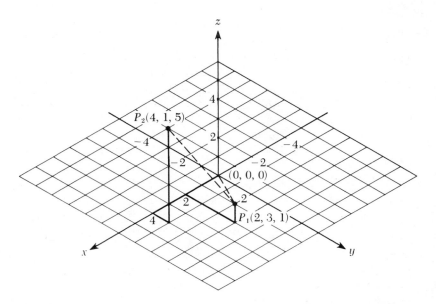

Figure 8.7 The distance formula for P_1;(2, 3, 1) and P_2;(4, 1, 5)

$$P_1P_2 = \sqrt{(4-2)^2 + (1-3)^2 + (5-1)^2}$$
$$= \sqrt{2^2 + (-2)^2 + 4^2} = \sqrt{4 + 4 + 16}$$
$$= \sqrt{24} = 2\sqrt{6}$$

Example 3   Find the distance from P_1;(2, 3, 1) to O;(0, 0, 0).

Solution   The distance is shown in Figure 8.7.

$$P_1O = \sqrt{(2-0)^2 + (3-0)^2 + (1-0)^2}$$
$$= \sqrt{4 + 9 + 1} = \sqrt{14}$$

The definition of the graph of a function of two variables defined by $z = f(x, y)$ follows the same form as the definition of the graph of a function of a single variable in a two-dimensional coordinate system.

Definition

The *graph* of a function defined by $z = f(x, y)$ is the set of points P;(x, y, z) whose coordinates satisfy the defining functional equation.

Example 4

Consider $z = f(x, y) = \sqrt{9 - x^2 - y^2}$. The domain of this function is the set of all values of x and y such that $x^2 + y^2 \leq 9$. This condition is necessary to assure that the value of z will be real. In the xy plane of a three-dimensional coordinate system this domain corresponds to a circular region bounded by the circle $x^2 + y^2 = 9$; a circle of a radius 3 with its center at the origin. This region is shown in Figure 8.8. Now examine the defining equation

$$z = \sqrt{9 - x^2 - y^2}$$

This is equivalent to

$$z^2 = 9 - x^2 - y^2$$

with the further condition that z must be nonnegative. This in turn is equivalent to

$$x^2 + y^2 + z^2 = 9$$
$$\sqrt{x^2 + y^2 + z^2} = 3$$

A comparison of this result with the distance formula indicates that a point $P;(x, y, z)$, where x, y, and z satisfy the original equation, must lie three units from the origin and above the xy plane. Thus, $z = \sqrt{9 - x^2 - y^2}$ represents a hemisphere of radius 3 with center at the origin.

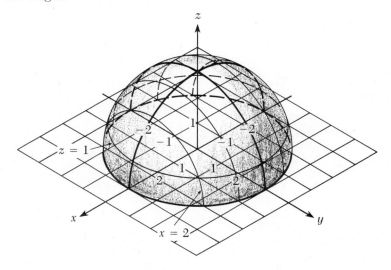

Figure 8.8 Graph of $z = \sqrt{9 - x^2 - y^2}$

In the example we have a function that generated a graph with a recognizable geometric form. In order to analyze more complex, less prearranged problems, more graphing tools are needed. One such tool is the trace of a graph. This concept is the three-dimensional extension of the idea of the points of intersection of two graphs in two dimensions.

Definition

The graph of an equation in three dimensions is called a *surface*.

Definition

The curve formed when two surfaces intersect is called the *trace* of one surface on the other.

Consider all of the points satisfying an equation like $z = 3$. All of these points would lie in a plane parallel to the xy plane and 3 units above it. In general $z = c$, a constant, represents the equation of a plane parallel to the xy plane and $|c|$ units above or below it. Similarly $x = c$ or $y = c$ represent planes parallel to the yz or xz planes. In order to visualize the nature of the graph of an equation like $z = f(x, y)$, it is useful to examine the traces the graph makes with the planes $x = $ a constant, $y = $ a constant, or $z = $ a constant.

Example 5   Draw the traces of the surface corresponding to $z = \sqrt{9 - x^2 - y^2}$ and $z = c$, a constant plane.

Solution   If $z = 1$, then

$$1 = \sqrt{9 - x^2 - y^2}$$
$$1 = 9 - x^2 - y^2$$
$$x^2 + y^2 = 8$$

Thus the trace of the surface defined by $z = \sqrt{9 - x^2 - y^2}$ in the plane $z = 1$ is a circle with radius $\sqrt{8}$ and center on the z axis. The circle is shown in Figure 8.8. Similarly the trace of the surface with the plane $z = 2$ is given by $2 = \sqrt{9 - x^2 - y^2}$, which reduces to $x^2 + y^2 = 5$, a circle with radius $\sqrt{5}$ also centered on the z axis. In fact, all of the traces which exist with planes whose equations are of the form $z = c$, a constant, are circles centered along the z axis. Now consider traces of the surface with planes parallel to the yz plane, that is, those planes whose equations are of the type $x = c$, a constant. With the plane $x = 2$ the trace is given by

$$z = \sqrt{9 - 4 - y^2} = \sqrt{5 - y^2}$$

This is a semicircle of radius $\sqrt{5}$, and it is shown in Figure 8.8. All of the traces with $x = c$, a constant plane, will be semicircles. In a like manner, all of the traces with y equal to a constant plane will be semicricles. The overall conclusion from examination of the traces is that the surface is a hemispherical dome of radius 3 centered at the origin.

Example 6

Using traces, sketch the graph in the first octant of the surface defined by $z = 4 - x - y$.

Solution

Consider the traces this surface makes with each of the coordinate planes in turn. With $z = 0$, the xy plane, the trace is given by $0 = 4 - x - y$, which is equivalent to $y = 4 - x$, a straight line. With $y = 0$, the xz plane, the trace is given by $z = 4 - x$. With $x = 0$, the yz plane, the trace is described by $z = 4 - y$.

These traces are shown in Figure 8.9. The traces with the planes $z = 1$, $z = 2$, $z = 3$, $x = 1$, $x = 2$, $x = 3$, $y = 1$, $y = 2$, and $y = 3$ are also shown in Figure 8.9. The surface $z = 4 - x - y$ is a plane.

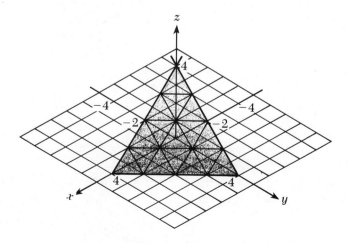

Figure 8.9 The first octant graph of $z = 4 - x - y$

Example 7   Using traces, sketch the graph of the surface $z = x^2 + y^2$.

Solution   In this case the trace of the surface with the xy plane is a point; $0 = x^2 + y^2$, whose only solution is $x = 0$ and $y = 0$. Thus the point $(0, 0, 0)$ is on the surface. The surface does not have a trace with any $z = c$ planes where c is negative; $x^2 + y^2$ is never negative. With the plane $z = 1$ the trace is given by $1 = x^2 + y^2$. This is a circle with radius 1 centered on the z axis. The planes $z = 2, z = 3$, etc., each yield traces that are circles centered on the z axis. In the plane $x = 0$, the trace is the parabola $z = y^2$. The trace in the $x = 1$ plane is the raised parabola $z = y^2 + 1$. These trace curves are shown in Figure 8.10. The graph appears to be bowl shaped. This surface is called a paraboloid because of the parabolic traces it produces.

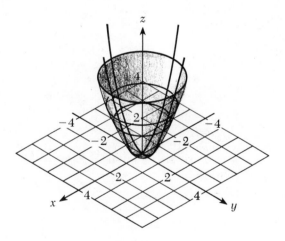

Figure 8.10   Graph of $z = x^2 + y^2$

Example 8   Graph the surface $y = x^2$.

Solution   The equation $y = x^2$ is independent of z since it contains no z term. When $z = 0$, in the xy plane, $y = x^2$ graphs as a well-known parabola. But in the $z = 1$ plane $y = x^2$ also has a parabolic graph. In fact, $y = x^2$ has the same graph for any value of z. The surface is graphed in Figure 8.11.

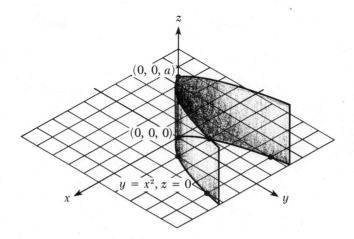

Figure 8.11   The surface $y = x^2$ (a parabolic cylinder)

8.2   EXERCISES

In Exercises 1–5, plot each of the pairs of points indicated in a three-dimensional coordinate system and calculate the distance between them.

1. $A;(0, 1, 3)$, $B;(6, -1, 2)$ **2.** $A;(1, -1, 4)$, $B;(3, 2, 7)$

3. $A;(-1, -1, 4)$, $B(1, 1, 4)$ **4.** $A;(3, 2, 7)$, $B;(-3, 1, -6)$

5. $A;(-4, -5, -4)$, $B;(-4, -5, -2)$

In Exercises 6–10, each equation describes a plane in three dimensions. Graph the traces each plane forms with the coordinate planes.

6. $z = x + y + 2$ **7.** $z = 4 + 3x - 2y$

8. $z = -x - y + 3$ **9.** $12z = -2x - 3y + 24$

10. $x + y = 3$

In Exercises 11–19, using traces, sketch each of the surfaces.

11. $z = \sqrt{25 - x^2 - y^2}$ **12.** $z = 25 - \dfrac{x^2}{9} - \dfrac{y^2}{16}$

13. $x^2 + y^2 + z^2 = 16$ **14.** $z = x^2 - y$

15. $z = y^2 - x + 2$ **16.** $z = y^2$

17. $z = \sin y$ **18.** $z = \sqrt{9 - (x - 3)^2 - (y - 2)^2}$

19. $z = e^{x+y}$

20. Consider the paraboloid described by $4z = x^2 + y^2$. Show, using the distance formula, that any point on this surface is the same distance from the point $(0, 0, 1)$ as it is from the plane $z = -1$.

21. Find all of the points on the z axis that are exactly 5 units from the point $(1, 2\sqrt{2}, 0)$.

In Exercises 22–24, assuming that the description of a left-handed coordinate system follows in form that of a right-handed coordinate system, plot each of the points indicated in a left-handed system.

22. Plot $A;(1, 2, -3)$, $B;(0, 1, 3)$, $C;(-3, 1, 5)$.

23. Plot $A;(-1, -1, 4)$, $B;(-4, -5, 3)$, $C;(-2, -6, 8)$.

24. Sketch $z = \sqrt{9 - x^2 - y^2}$.

8.3 LINEAR PROGRAMMING

If $z = f(x, y) = ax + by + c$ for a, b, and c constants, then $f(x, y)$ will graph in three dimensions as a plane. Such a plane in general has no maximum or minimum value for z. However, in a number of practical problems the allowable values of x and y are restricted to a small region in the xy plane, and the plane

$$z = f(x, y) = ax + by + c$$

will have a maximum and minimum value over this region. Finding such maximum and minimum values is called linear programming. Since its discovery in the early 1940s linear programming has grown steadily in its application to economics. We will present the basic theory for the three-dimensional case. However, linear programming can be extended to n-dimensional space and can include a number of sophisticated techniques for finding maxima or minima.

Figure 8.12 shows a shaded region in the xy plane and the portion of the plane $z = f(x, y) = ax + by + c$ above the region. The maximum and minimum values of z on the plane and above the region will occur when the coordinates of one of the vertices of the region are substituted into $ax + by + c$. In this case the maximum

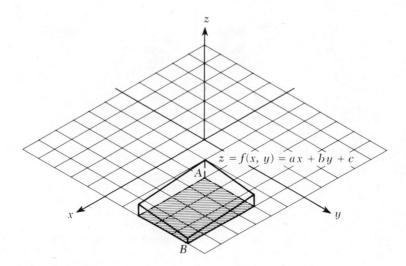

Figure 8.12 Maximum or minimum values of $z = f(x, y) = ax + by + c$ over a region in the xy plane

value of z occurs above the point A and the minimum occurs above point B.

When the function to be maximized or minimized is linear (graphs as a plane), the maximum or minimum value will always occur at one of the vertices of the region in the xy plane. Therefore it is not necessary to graph the plane. Simply graph the region in the xy plane and test its vertices to discover maximum or minimum values above the region.

Example 1   Find the maximum value of z and the minimum value of z for $z = 4x + 3y$ given the following conditions on x and y:

$$x \geqslant 0 \qquad y \geqslant 0$$
$$x + y \leqslant 6 \qquad 2x + y \leqslant 8$$

Solution   The intersection of the constraints on x and y is the shaded region in Figure 8.13. Now we test the four vertices of the region, $(0, 0)$, $(4, 0)$, $(2, 4)$, and $(0, 6)$, in $z = 4x + 3y$ to find the maximum and minimum values of z.

$$z = f(0, 0) = 4(0) + 3(0) = 0$$
$$z = f(4, 0) = 4(4) + 3(0) = 16$$
$$z = f(2, 4) = 4(2) + 3(4) = 20$$
$$z = f(0, 6) = 4(0) + 3(6) = 18$$

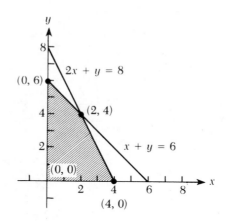

Figure 8.13   The region bounded by
$x \geqslant 0, y \geqslant 0, x + y \leqslant 6$,
and $2x + y \leqslant 8$

Therefore the maximum value of z is 20 and occurs when $x = 2$ and $y = 4$, and the minimum value of z is 0 and occurs when $x = y = 0$.

Example 2
💲

A company produces two types of food supplements, called types A and B. Type A yields a profit of \$10.00 per ounce and type B yields a profit of \$20.00 per ounce. The supplements are manufactured in two different departments. Department I requires 2 hours to manufacture one ounce of A and 3 hours to manufacture one ounce of B. Department II requires 1 hour to manufacture a single ounce of A and 4 hours to manufacture an ounce of B. How many ounces of A and B should be produced to give maximum profit if Department I has 20 hours of time available for production and Department II has 15 hours available?

Solution

Let x be the number of ounces of supplement A produced and y be the number of ounces of B produced. The profit, P, is

$$P = 10x + 20y$$

We will graph the constraints on x and y, and then test the coordinates of the vertices thus discovered in the profit function to find which gives maximum profit. It is clear that x and y must be nonnegative, that is,

$$x \geqslant 0 \qquad \text{and} \qquad y \geqslant 0$$

The information on production in Departments I and II is given in the following table.

	Hours available	Hours required to produce 1 ounce	
		A	B
Dept. I	20	2	3
Dept. II	15	1	4

Using the data from Department I,

$$2x + 3y \leqslant 20$$

The information on Department II implies

$$x + 4y \leqslant 15$$

Thus, the constraints on x and y are

$$x \geqslant 0$$
$$y \geqslant 0$$
$$2x + 3y \leqslant 20$$
$$x + 4y \leqslant 15$$

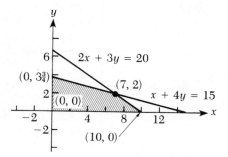

Figure 8.14 The region bounded by $x \geqslant 0$, $y \geqslant 0$, $2x + 3y \leqslant 20$, and $x + 4y \leqslant 15$

These are graphed in Figure 8.14. The coordinates of the vertices are substituted into the profit function $P = 10x + 20y$ to give:

For $(0, 0)$ $P = 0$
For $(10, 0)$ $P = 10(10) + 20(0) = \$100$
For $(7, 2)$ $P = 10(7) + 20(2) = \$110$
For $(0, 3\frac{3}{4})$ $P = 10(0) + 20(3\frac{3}{4}) = \$75$

The maximum profit of \$110 will occur when 7 ounces of supplement A and 2 ounces of supplement B are produced.

The linear programming presented here involves a function of x and y that is linear and that we wish to maximize or minimize based upon certain restrictions on x and y. The next example will, at first glance, appear to be more complicated than this. It will appear as if the income function is a function of three variables. However, the example is such that one of the variables can be represented as a function of the other two.

Example 3

$

During the spring season a resort operator rents bicycles for \$12 per day, skidoos for \$40 per day, and motorcycles for \$25 per day. He has storage space for a combined total of 50 bicycles, skidoos, and motorcycles. In order to satisfy certain regular customers he must have at least 10 bicycles, but he will never need more than 20. He is certain to rent 5 skidoos every day while the greatest demand for skidoos was 35 in one day. How many bicycles, skidoos, and motorcycles should he have for maximum income?

Solution

Let m be the number of motorcycles, b the number of bicycles, and s the number of skidoos. We will use the fact that $b + m + s = 50$ to solve for m in terms of b and s. In this way, the income can be represented as a function of just two variables. The income, I, is

$$I = 12b + 40s + 25m$$

Substituting $m = 50 - b - s$,

$$\begin{aligned} I &= 12b + 40s + 25(50 - b - s) \\ &= 12b + 40s + 1250 - 25b - 25s \\ &= 1250 - 13b + 15s \end{aligned}$$

We choose to write m in terms of b and s in the income function, because most of the information in the problem concerned skidoos and bicycles. The constraints on b and s are:

$$b \in [10, 20] \qquad s \in [5, 35] \qquad b + s \leqslant 50$$

These constraints are graphed in Figure 8.15.

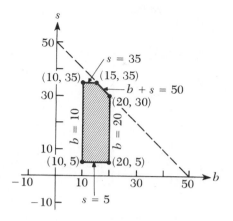

Figure 8.15 The region bounded by $10 \leqslant b \leqslant 20$, $5 \leqslant s \leqslant 35$, and $b + s \leqslant 50$

Testing the coordinates of the vertices in the income function $I = 1250 - 13b + 15s$ yields:

For (10, 5) $I = 1250 - 13(10) + 15(5)$
             $= 1250 - 130 + 75 = \$1195$
For (20, 5) $I = 1250 - 13(20) + 15(5)$
             $= 1250 - 260 + 75 = \$1065$
For (20, 30) $I = 1250 - 13(20) + 15(30)$
             $= 1250 - 260 + 450 = \$1440$
For (15, 35) $I = 1250 - 13(15) + 15(35)$
             $= 1250 - 195 + 525 = \$1580$
For (10, 35) $I = 1250 - 13(10) + 15(35)$
             $= 1250 - 130 + 525 = \$1645$

The maximum income is $1645 and occurs when there are 10 bicycles, 35 skidoos, and 5 motorcycles.

8.3 EXERCISES

1. Find the maximum value of z for $z = 2x + 5y$ given:
   $x \geqslant 2$, $y \geqslant 1$, $x + y \leqslant 8$
2. Find the maximum value of z for $z = 2x - 5y$ if:
   $x \geqslant 2$, $y \geqslant 2$, $x + y \leqslant 10$
3. Find the minimum value of P for $P = 2x + 4y$ if:
   $x \geqslant 0$, $y \geqslant 0$, $x + y \geqslant 8$, $2x + 3y \geqslant 19$

4. Find the minimum value of P for $P = 2x + 6y - 8$ if:
$0 \leqslant x \leqslant 10$, $y \geqslant 0$, $2x + y \geqslant 16$, $x + 5y \geqslant 26$

5. Find the maximum and minimum values of z for $z = 6x + 3y$
if: $x + y \leqslant 6$, $x - y \leqslant 2$, $2y - 3x \leqslant 12$

6. Find the minimum value of z for $z = 4x - y$ if:
$x - y \geqslant -1$, $3x + 2y \geqslant 17$, $x + 4y \geqslant 9$

7. Find the maximum and minimum values of I for $I = 3x + 10y$
if: $x + 2y \leqslant 6$, $x - y \leqslant 6$, $x \geqslant -6$, $y \leqslant 1$

8. Find the maximum and minimum values of I for $I = x + y$
if: $x + 3y \geqslant 9$, $2x - y \geqslant 4$, $y \geqslant 0$, $x \leqslant 15$

9. Find the maximum value of P for $P = 6x - 6y$ if:
$3x + y \leqslant 15$, $x + 3y \leqslant 15$, $x + y \leqslant 7$, $x \geqslant 1$, $y \geqslant 1$

10. Find the maximum value of P for $P = 16x + 4y$ if:
$6x + 5y \leqslant 30$, $x + 2y \leqslant 10$, $2x + y \leqslant 8$, $x \geqslant 0$, $y \geqslant 0$

💲 11. A company produces vitamin A and vitamin B. The profit on
vitamin A is $150 a pound and vitamin B yields a profit of $200
per pound. Vitamin A takes 5 hours per pound to manufacture
and 3 hours per pound to package. Vitamin B requires 6 hours
per pound to manufacture and 2 hours per pound to package.
A total of 60 hours of manufacturing time is available, while
only 30 hours of packaging time is available. How many
pounds of vitamins A and B should be manufactured to maximize profits?

💲 12. A special diet requires a daily intake of at least 10 ounces of
protein and 14 ounces of carbohydrate. The protein and carbohydrate are to come from two grain mixtures. Mixture I contains 3 ounces of protein and 6 ounces of carbohydrate per
pound and costs $2.00 per pound. Mixture II contains 1 ounce
of protein and 5 ounces of carbohydrate per pound and costs
$1.50 per pound. How much of mixtures I and II should be
purchased to satisfy the dietary requirements at minimum
cost?

💲 13. Two types of hammers, claw and sledge, go through three
distinct manufacturing phases: stamping, assembly, and
finishing. The time, in hours, required for each phase is given
below:

	Stamping	Assembly	Finishing
Claw	$\frac{1}{4}$	$\frac{1}{4}$	1
Sledge	$\frac{1}{4}$	$\frac{1}{2}$	$\frac{1}{2}$

The times available are 20 hours for stamping, 25 hours for assembly, and 50 hours for finishing. What is the maximum profit that can be made under these conditions if a claw hammer makes $1.00 in profit and a sledge hammer $1.50?

14. Two types of television antennas, UHF and VHF, go through two manufacturing phases, fabricating and assembly. Fabrication takes 3 hours for each UHF antenna and 6 hours for each VHF antenna. Assembly times are 6 hours for a UHF and 3 hours for a VHF antenna. The fabricating shop and the assembly plant each have 2100 hours of time available. How many antennas of each type should be manufactured to maximize profits if the net profit is $40 on each UHF antenna and $50 on each VHF antenna?

15. A pharmacist has 100 units of protein, 160 units of bulk additive, and 400 units of vitamins in his store. He makes two health food mixtures with these ingredients. Mixture A requires 1 unit of protein, 2 units of bulk, and 2 units of vitamin and sells for $100 per bag. Mixture B consists of 1 unit of protein, 1 unit of bulk and 5 units of vitamin and sells for $110 per bag. How many bags of each type of mixture should the pharmacist make for maximum revenue?

16. Two partners are setting up a business to sell television sets, stereos, and portable dishwashers. They want a combined total of 100 of all three items. To satisfy certain customers they must have at least 10 stereos but they cannot sell more than 30 in the immediate future (which is all they are worried about). They want fewer than 60 television sets but the number of these must equal or exceed the number of stereos. Their investment averages $250 for a television set, $150 for a stereo, and $100 for a dishwasher. How many of each item should they order to minimize their investment?

17. Suppose that you wish to give a party. You have $250 to spend on liquid refreshments, which, due to the idiosyncrasies of your guests, must be beer and Scotch whiskey. You need at least 750 drinks. Beer costs $0.60 a quart, is 4 percent alcohol, and a quart contains 4 drinks. Scotch costs $7.00 a quart, is 50 percent alcohol, and each quart contains 20 drinks. How many quarts of beer and Scotch should be purchased so that the party will have maximum alcoholic content?

8.4 PARTIAL DERIVATIVES

There are two possible approaches to defining a derivative of a function of two variables. One is to make an exact analogy to the derivative of a function of a single variable.

Consider $z = f(x, y)$ with a specific point, (x_0, y_0), in the domain of the function. Then one can define the derivative of z at (x_0, y_0) as

$$\lim_{(\Delta x, \Delta y) \to (0, 0)} \left[\frac{f(x_0 + \Delta x, y_0 + \Delta y) - f(x_0, y_0)}{\sqrt{\Delta x^2 + \Delta y^2}} \right]$$

where $\sqrt{\Delta x^2 + \Delta y^2}$ measures the change in the independent variables from (x_0, y_0) to $(x_0 + \Delta x, y_0 + \Delta y)$. At first glance this approach seems reasonable enough.

In reality it is quite complex and not nearly as useful as another approach known as partial derivatives.

Definition

Let $z = f(x, y)$ define a function of x and y. Then

$$\frac{\partial z}{\partial x} = f_x(x, y) = \lim_{\Delta x \to 0} \frac{f(x + \Delta x, y) - f(x, y)}{\Delta x}$$

$$\frac{\partial z}{\partial y} = f_y(x, y) = \lim_{\Delta y \to 0} \frac{f(x, y + \Delta y) - f(x, y)}{\Delta y}$$

provided such a limit exists at the point (x, y) in the domain of the function. $\partial z / \partial x$ is called the *partial derivative of z with respect to x* and $\partial z / \partial y$ is called the *partial derivative of z with respect to y*.

This definition of the partial derivatives in effect creates a derivative of a function of two variables by treating one of the variables as a constant. To partially differentiate with respect to x, one need only treat y as a constant and apply the usual differentiation rules.

Example 1   If $z = x^2y$, find $\partial z/\partial x$.

    Solution

$$\frac{\partial z}{\partial x} = \lim_{\Delta x \to 0} \frac{(x + \Delta x)^2 y - x^2 y}{\Delta x}$$

$$= \lim_{\Delta x \to 0} y \left[\frac{(x + \Delta x)^2 - x^2}{\Delta x} \right]$$

$$= y \lim_{\Delta x \to 0} \left[\frac{(x + \Delta x)^2 - x^2}{\Delta x} \right]$$

However,

$$\lim_{\Delta x \to 0} \left[\frac{(x + \Delta x)^2 - x^2}{\Delta x} \right] = \frac{d}{dx}(x^2) \quad \text{or} \quad 2x$$

by definition. Thus,

$$\frac{\partial z}{\partial x} = y2x = 2xy$$

Example 2   If $z = \dfrac{y + 2}{x}$, find $\dfrac{\partial z}{\partial x}$

    Solution

$$\frac{\partial z}{\partial x} = \lim_{\Delta x \to 0} \frac{\dfrac{y + 2}{x + \Delta x} - \dfrac{y + 2}{x}}{\Delta x}$$

$$= \lim_{\Delta x \to 0} \frac{(y + 2)x - (y + 2)(x + \Delta x)}{x(x + \Delta x)\Delta x}$$

$$= \lim_{\Delta x \to 0} (y + 2) \frac{(x - x - \Delta x)}{x(x + \Delta x)\Delta x}$$

$$= \lim_{\Delta x \to 0} \frac{(y + 2)(-\Delta x)}{x(x + \Delta x)(\Delta x)}$$

$$= \lim_{\Delta x \to 0} \frac{(y + 2)(-1)}{x(x + \Delta x)} = (y + 2)\frac{(-1)}{x^2}$$

Notice if we treat y as a constant, then

$$\frac{d}{dx}\left(\frac{y + 2}{x} \right) = (y + 2)\frac{d}{dx}\left(\frac{1}{x} \right)$$

$$= (y + 2)\left(\frac{-1}{x^2} \right)$$

$$= -\frac{(y + 2)}{x^2}$$

as above.

Example 3   If $z = \sin(x + y)$, find $\partial z/\partial x$.

Solution $\dfrac{\partial z}{\partial x} = \lim\limits_{\Delta x \to 0} \dfrac{\sin[(x + \Delta x) + y] - \sin(x + y)}{\Delta x}$

$= \lim\limits_{\Delta x \to 0} \left\{ \dfrac{\sin(x + \Delta x)\cos y + \cos(x + \Delta x)\sin y - \sin x \cos y - \cos x \sin y}{\Delta x} \right.$

$= \lim\limits_{\Delta x \to 0} \left\{ \left[\dfrac{\sin(x + \Delta x) - \sin x}{\Delta x} \right] \cos y + \left[\dfrac{\cos(x + \Delta x) - \cos x}{\Delta x} \right] \sin y \right\}$

However,

$$\lim_{\Delta x \to 0} \left[\frac{\sin(x + \Delta x) - \sin x}{\Delta x} \right] = \frac{d}{dx}(\sin x) = \cos x$$

and

$$\lim_{\Delta x \to 0} \left[\frac{\cos(x + \Delta x) - \cos x}{\Delta x} \right] = \frac{d}{dx}(\cos x) = -\sin x$$

Thus,

$$\frac{\partial z}{\partial x} = \cos x \cos y - \sin x \sin y = \cos(x + y)$$

Without supplying a formal proof, it is reasonable to conclude that the mechanics of partial differentiation follow the usual differentiation rules, treating other independent variables as constants.

Example 4 $\dfrac{\partial}{\partial x}(x^2 + y^2) = 2x$

Example 5 $\dfrac{\partial}{\partial x}(xy^3) = y^3$

Example 6 $\dfrac{\partial}{\partial x}(e^{x^2 y^3}) = e^{x^2 y^3} \cdot 2xy^3$

Example 7 $\dfrac{\partial}{\partial x} \sqrt{x^2 + y^2} = \dfrac{1}{2 \sqrt{x^2 + y^2}} (2x) = \dfrac{x}{\sqrt{x^2 + y^2}}$

Example 8 $\dfrac{\partial}{\partial x} (x^3 + 3x^2y + 3xy^2 + y^3) = 3x^2 + 6xy + 3y^2$

Example 9 $\dfrac{\partial}{\partial x} [\sin (x^2 + y^2)] = \cos (x^2 + y^2)(2x)$

Example 10 $\dfrac{\partial}{\partial x} (x^3 + 3xy + y^3) = 3x^2 + 3y$

The obvious parallel exists for partial differentiation with respect to y or any other variable.

Example 11 $\dfrac{\partial}{\partial y} (x^2 + y^2) = 2y$

Example 12 $\dfrac{\partial}{\partial y} (x^3 + 3x^2y + 3xy^2 + y^3) = 3x^2 + 6xy + 3y^2$

Example 13 $\dfrac{\partial}{\partial y} (x^2y) = x^2$

Example 14 $\dfrac{\partial}{\partial y} (xy^3) = 3xy^2$

Example 15 $\dfrac{\partial}{\partial y} \sqrt{x^2 - y^2} = \dfrac{1}{2 \sqrt{x^2 - y^2}} (-2y) = \dfrac{-y}{\sqrt{x^2 - y^2}}$

Example 16 $\dfrac{\partial}{\partial y} (x^3 + 3xy + y^4) = 3x + 4y^3$

Example 17 $\dfrac{\partial}{\partial y} \sin (x^2 + y^2) = \cos (x^2 + y^2)(2y)$

An alternative notation is

$$\frac{\partial}{\partial y} f(x, y) = f_y(x, y)$$

Just as the regular differentiation process can be repeated any number of times, so can the process of partial differentiation.

Example 18   Let $z = x^5 y^6 = f(x, y)$. Find the third partial derivative of z with respect to x and y.

Solution $\dfrac{\partial z}{\partial x} = f_x(x, y) = 5x^4 y^6$

$\dfrac{\partial^2 z}{\partial x^2} = f_{xx}(x, y) = \dfrac{\partial}{\partial x}\left(\dfrac{\partial z}{\partial x}\right) = \dfrac{\partial}{\partial x}(5x^4 y^6) = 20x^3 y^6$

$\dfrac{\partial^3 z}{\partial x^3} = f_{xxx}(x, y) = 60x^2 y^6$

$\dfrac{\partial z}{\partial y} = f_y(x, y) = 6x^5 y^5$

$\dfrac{\partial^2 z}{\partial y^2} = f_{yy}(x, y) = \dfrac{\partial}{\partial y}\left(\dfrac{\partial z}{\partial y}\right) = \dfrac{\partial}{\partial y}(6x^5 y^5) = 30x^5 y^4$

$\dfrac{\partial^3 z}{\partial y^3} = f_{yyy}(x, y) = 120x^5 y^3$

With partial differentiation, a new type of higher-order derivative can be introduced. Since any partial derivative is usually still a function of both of the original variables, the second or higher derivative can be found with respect to a variable other than the first one used.

Example 19   If $z = f(x, y) = x^5 y^6$, find $f_{yyx}(x, y)$.

Solution $$\frac{\partial z}{\partial x} = 5x^4 y^6$$

$$\frac{\partial}{\partial y}\left(\frac{\partial z}{\partial x}\right) = \frac{\partial}{\partial y}(5x^4 y^6) = 30x^4 y^5$$

$$\frac{\partial}{\partial y}\left(\frac{\partial}{\partial y}\left(\frac{\partial z}{\partial x}\right)\right) = 150x^4 y^4 = f_{yyx}(x, y)$$

These types of partial derivatives are called *mixed partial derivatives*. If $z = f(x, y)$,

$$\frac{\partial^2 z}{\partial y \, \partial x} = \frac{\partial}{\partial y} \left(\frac{\partial z}{\partial x} \right) = f_{yx}(x, y)$$

$$\frac{\partial^2 z}{\partial x \, \partial y} = \frac{\partial}{\partial x} \left(\frac{\partial z}{\partial y} \right) = f_{xy}(x, y)$$

Thus, there are two mixed second-order partial derivatives. The concept extends to six mixed third-order partial derivatives.

$$\frac{\partial^3 z}{\partial x \, \partial y^2} = \frac{\partial}{\partial x} \left(\frac{\partial^2 z}{\partial y^2} \right) = f_{xyy}(x, y)$$

$$\frac{\partial^3 z}{\partial y \, \partial x \, \partial y} = \frac{\partial}{\partial y} \left(\frac{\partial^2 z}{\partial x \, \partial y} \right) = f_{yxy}(x, y)$$

$$\frac{\partial^3 z}{\partial y^2 \, \partial x} = \frac{\partial^2}{\partial y^2} \left(\frac{\partial z}{\partial x} \right) = f_{yyx}(x, y)$$

$$\frac{\partial^3 z}{\partial y \, \partial x^2} = \frac{\partial}{\partial y} \left(\frac{\partial^2 z}{\partial x^2} \right) = f_{yxx}(x, y)$$

$$\frac{\partial^3 z}{\partial x \, \partial y \, \partial x} = \frac{\partial}{\partial x} \left(\frac{\partial^2 z}{\partial y \, \partial x} \right) = f_{xyx}(x, y)$$

$$\frac{\partial^3 z}{\partial x^2 \, \partial y} = \frac{\partial^2}{\partial x^2} \left(\frac{\partial z}{\partial y} \right) = f_{xxy}(x, y)$$

Example 20 If $z = x^7 y^{7/2}$, find the third-order mixed partial derivatives of z.

Solution

$$\frac{\partial z}{\partial x} = 7x^6 y^{7/2}$$

$$\frac{\partial z}{\partial y} = \frac{7}{2} x^7 y^{5/2}$$

$$\frac{\partial^2 z}{\partial x^2} = 42x^5 y^{7/2}$$

$$\frac{\partial^2 z}{\partial y^2} = \frac{35}{4} x^7 y^{3/2}$$

$$\frac{\partial^2 z}{\partial y \, \partial x} = \frac{\partial}{\partial y} \left(\frac{\partial z}{\partial x} \right) = \frac{\partial}{\partial y} (7x^6 y^{7/2}) = \frac{49}{2} x^6 y^{5/2}$$

$$\frac{\partial^2 z}{\partial x \, \partial y} = \frac{\partial}{\partial x} \left(\frac{\partial z}{\partial y} \right) = \frac{\partial}{\partial x} \left(\frac{7}{2} x^7 y^{5/2} \right) = \frac{49}{2} x^6 y^{5/2}$$

$$\frac{\partial^3 z}{\partial y \, \partial x^2} = \frac{\partial}{\partial y} \left(\frac{\partial^2 z}{\partial x^2} \right) = \frac{\partial}{\partial y} (42x^5 y^{7/2}) = \frac{294}{2} x^5 y^{5/2}$$

$$\frac{\partial^3 z}{\partial x \, \partial y \, \partial x} = \frac{\partial}{\partial x} \left(\frac{\partial^2 z}{\partial y \, \partial x} \right) = \frac{\partial}{\partial x} \left(\frac{49}{2} x^6 y^{5/2} \right) = \frac{294}{2} x^5 y^{5/2}$$

$$\frac{\partial^3 z}{\partial x^2\, \partial y} = \frac{\partial^2}{\partial x^2}\left(\frac{\partial z}{\partial y}\right) = \frac{\partial}{\partial x}\left(\frac{\partial}{\partial x}\left(\frac{7}{2}\,x^7 y^{5/2}\right)\right)$$

$$= \frac{\partial}{\partial x}\left(\frac{49}{2}\,x^6 y^{5/2}\right) = \frac{294}{2}\,x^5 y^{5/2}$$

$$\frac{\partial^3 z}{\partial x\, \partial y^2} = \frac{\partial}{\partial x}\left(\frac{\partial^2 z}{\partial y}\right) = \frac{\partial}{\partial x}\left(\frac{35}{4}\,x^7 y^{3/2}\right) = \frac{245}{4}\,x^6 y^{3/2}$$

$$\frac{\partial^3 z}{\partial y\, \partial x\, \partial y} = \frac{\partial}{\partial y}\left(\frac{\partial^2 z}{\partial x\, \partial y}\right) = \frac{\partial}{\partial y}\left(\frac{49}{2}\,x^6 y^{5/2}\right) = \frac{245}{4}\,x^6 y^{3/2}$$

$$\frac{\partial^3 z}{\partial y^2\, \partial x} = \frac{\partial}{\partial y}\left(\frac{\partial^2 z}{\partial y\, \partial x}\right) = \frac{\partial}{\partial y}\left(\frac{49}{2}\,x^6 y^{5/2}\right) = \frac{245}{4}\,x^6 y^{3/2}$$

From this example, it appears that

$$\frac{\partial^2 z}{\partial x\, \partial y} = \frac{\partial^2 z}{\partial y\, \partial x}$$

$$\frac{\partial^3 z}{\partial x^2\, \partial y} = \frac{\partial^3 z}{\partial x\, \partial y\, \partial x} = \frac{\partial^3 z}{\partial y\, \partial x^2}$$

$$\frac{\partial^3 z}{\partial y^2\, \partial x} = \frac{\partial^3 z}{\partial y\, \partial x\, \partial y} = \frac{\partial^3 z}{\partial x\, \partial y^2}$$

This result is, in fact, always true provided the partial derivatives are continuous. Proofs of this can be found in more advanced courses. In general, the order of differentiation does not affect the result in finding a mixed partial derivative. We might also mention that different authors may take f_{xy} to mean $\dfrac{\partial}{\partial x}\left(\dfrac{\partial f}{\partial y}\right)$ while others see it as $\dfrac{\partial}{\partial y}\left(\dfrac{\partial f}{\partial x}\right)$.

Example 21   Find $\dfrac{\partial^3}{\partial x^2\, \partial y}\,(x^2 y^3 - \sin \sqrt{y^{3/2} - \ln y})$.

Solution   Rather than following the indicated order of differentiation, it is easier to proceed as follows:

$$\frac{\partial}{\partial x}\,(x^2 y^3 - \sin \sqrt{y^{3/2} - \ln y}) = 2xy^3$$

$$\frac{\partial^2}{\partial x^2}\,(x^2 y^3 - \sin \sqrt{y^{3/2} - \ln y}) = 2y^3$$

$$\frac{\partial^3}{\partial x^2\, \partial y}\,(x^2 y^3 - \sin \sqrt{y^{3/2} - \ln y}) = 6y^2$$

Example 22   If $z = \ln (x^2 + y)$, find $\dfrac{\partial^2 z}{\partial x \, \partial y}$

Solution $\dfrac{\partial z}{\partial x} = \dfrac{1}{x^2 + y} \, (2x)$

$\dfrac{\partial^2 z}{\partial y \, \partial x} = \dfrac{-2x}{(x^2 + y)^2} = \dfrac{\partial^2 z}{\partial x \, \partial y}$

An alternate solution is:

$$\dfrac{\partial z}{\partial y} = \dfrac{1}{x^2 + y}$$

$$\dfrac{\partial^2 z}{\partial x \, \partial y} = \dfrac{\partial^2 z}{\partial y \, \partial x} = \dfrac{-1}{(x^2 + y)^2} \, (2x) = \dfrac{-2x}{(x^2 + y)^2}$$

8.4 EXERCISES

In Exercises 1–16, find $\partial z/\partial x$ and $\partial z/\partial y$ in each case.

1. $z = x^2 + 2xy + y^2$

2. $z = (x + 2)(y + 3)$

3. $z = \dfrac{x + y}{2}$

4. $z = x^2 + 3xy^3 - 4y$

5. $z = x^{1/5} + y^{1/5} - \sqrt{xy}$

6. $z = \dfrac{1}{x^2 y^2}$

7. $z = \dfrac{y^2 + 2}{x^2 - 5x + 8}$

8. $z = \dfrac{\sqrt{x^2 - y^2}}{2xy}$

9. $z = 25 - (x - y)^4 + (y - 1)^4$

10. $z = 2xy - 5y^2 - 2x^2 + 4x + 4y - 4$

11. $z = 2x + xy + y^2 + 3y + 5$ **12.** $z = \ln (x^2 y)$

13. $z = e^{x^2 + 2xy}$

14. $z = \cos \left(\dfrac{x + y}{2} \right)$

15. $z = \sin (x^2 + 2 \cos y)$

16. $z = e^{\ln x} e^{\ln y}$

In Exercises 17–21, find the four second partial derivatives.

17. $z = \dfrac{1}{x^2 y^2}$

18. $z = 25 - (x - y)^4 + (y - 1)^4$

19. $z = \dfrac{y^2 + 2}{x^2 - 5x + 8}$

20. $z = 2x + 2y + y^2 + 3x^2 + 5$

21. $z = x^2 + xy + y^2 + 3x - 3y + 4$

In Exercises 22–27, find the indicated third partial derivatives.

22. $f(x, y) = x^9y^{-4/3}$,   find f_{xxx}, f_{xyy}

23. $f(x, y) = \ln(x^2y^2)$,   find f_{yyy}, f_{xyx}

24. $f(x, y) = \sin(x^2 + y^2)$,   find f_{yyx}, f_{xxy}

25. $f(x, y) = e^{xy}$,   find $f_{xxy}, f_{xyx}, f_{yxx}$

26. $f(x, y) = 2x^3 + 2y^3$,   find $f_{xyy}, f_{yxy}, f_{yyx}$

27. By first differentiating with respect to y, find

$$\frac{\partial^3}{\partial x^2\,\partial y}\,[x^2y^2 - \sin(\sqrt{y^{3/2} - \ln y})]$$

In Exercises 28 and 29, note that for a given function z of two variables, the Laplacian operator, ∇^2 (read "del" squared), is defined as

$$\nabla^2 z = \frac{\partial^2 z}{\partial x^2} + \frac{\partial^2 z}{\partial y^2}$$

A function $z = f(x, y)$ is called harmonic if it satisfies Laplace's equation $\nabla^2 z = 0$. Show that the functions indicated are harmonic.

28. $z = x^2 - y^2$ **29.** $z = \ln\sqrt{x^2 + y^2}$

In Exercises 30 and 31, find all values (x, y) that satisfy the simultaneous conditions

$$\frac{\partial z}{\partial x} = 0 \quad and \quad \frac{\partial z}{\partial y} = 0$$

30. $z = 2x + 2y + y^2 + 3x^2 + 5$

31. $z = 2xy - 5y^2 - 2x^2 + 4x + 4y - 4$

In Exercises 32–36, $\partial z/\partial x$ and $\partial z/\partial y$ are given. Use antidifferentiation with respect to x holding y constant and antidifferentiation with respect to y holding x constant and find z. [Hint: $\frac{\partial}{\partial x} f(y) = 0$ and $\frac{\partial}{\partial y} f(x) = 0$; thus, $\frac{\partial}{\partial x}[f(x, y) + g(y)] = \frac{\partial}{\partial x} f(x, y)$ and $\frac{\partial}{\partial y}[f(x, y) + h(x)] = \frac{\partial}{\partial y} f(x, y).$]

32. $\dfrac{\partial z}{\partial x} = 12x^2y^2$

   $\dfrac{\partial z}{\partial y} = 8x^3y$

33. $\dfrac{\partial z}{\partial x} = y\cos(xy) - 5x^4$

   $\dfrac{\partial z}{\partial y} = x\cos(xy) + 3y^2$

34. $\dfrac{\partial z}{\partial x} = \dfrac{1}{x}$

$\dfrac{\partial z}{\partial y} = \dfrac{1}{y}$

35. $\dfrac{\partial z}{\partial x} = y + \dfrac{y}{x}$

$\dfrac{\partial z}{\partial y} = x + \ln xy + 1$

36. $\dfrac{\partial z}{\partial x} = \dfrac{1}{2\sqrt{x - y^2}}$

$\dfrac{\partial z}{\partial y} = \dfrac{-y}{\sqrt{x - y^2}}$

8.5 EXTREMA FOR FUNCTIONS OF TWO VARIABLES —MAXIMUM AND MINIMUM

One of the most useful applications of differentiation was finding the extrema (maximum and minimum values) of a function. Naturally, one wonders about a similar application involving partial differentiation and a function of two variables. The trace of a graph in three dimensions can assist in visualizing the role of the partial derivative in such a problem.

Consider the trace of a surface $z = f(x, y)$ in the plane $y = c$, where c is a constant (see Figure 8.16). The trace is a plane curve in the $y = c$ plane. In this plane, z is a function of x only. Now $\partial z/\partial x$ is found by treating y as a constant, which is true in the $y = c$ plane, and $\partial z/\partial x$ acts like the ordinary derivative of z with

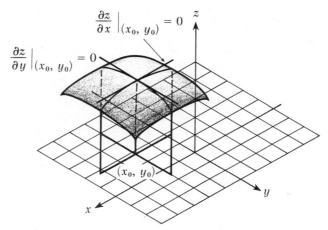

Figure 8.16 Graph of $z = f(x, y)$, with $f(x_0, y_0)$ an extremum (maximum)

respect to x in the $y = c$ plane. It is reasonable to conclude that at any value of x, $\partial z/\partial x$ yields the slope of the line in the $y = c$ plane tangent to the trace curve in the $y = c$ plane. Now suppose (x_0, y_0) corresponds to an extremum of $z = f(x, y)$, and

$$\frac{\partial z}{\partial x}\bigg|_{(x_0, y_0)}$$

[read, "$\partial z/\partial x$ evaluated at (x_0, y_0)"] and

$$\frac{\partial z}{\partial y}\bigg|_{(x_0, y_0)}$$

[read, "$\partial z/\partial y$ evaluated at (x_0, y_0)"] both exist. Since $f(x_0, y_0)$ is an extremum of z, it must also be an extremum for the trace $z = f(x, y)$ in the $y = y_0$ plane. Thus $\dfrac{\partial z}{\partial x}\bigg|_{(x_0, y_0)}$ must equal zero. If the roles of x and y are changed in the preceding discussion, one can conclude that if $f(x_0, y_0)$ is an extremum,

$$\frac{\partial z}{\partial x}\bigg|_{(x_0, y_0)} = 0$$

and

$$\frac{\partial z}{\partial y}\bigg|_{(x_0, y_0)} = 0$$

These are necessary, but not sufficient, conditions for $f(x_0, y_0)$ to be an extremum of z, for, as we know in the ordinary regular derivative case, the fact the derivative of a function is zero at a specific value of the independent variable does not assure a maximum or minimum function value. This happens when the curve only flattens and does not change direction, as $f(x) = x^3$ does at $x = 0$. A later example will demonstrate a second way in which a zero for both first partials can fail to be an extremum of the function.

Example 1   Find the extrema of

$$z = (x - 1)^2 - (y - 2)^2$$

Solution $\dfrac{\partial z}{\partial x} = 2(x - 1)$ $\dfrac{\partial z}{\partial y} = -2(y - 2)$

Clearly the only possibility for an extremum occurs at $(1, 2)$. Checking second partials,

$$\left.\frac{\partial^2 z}{\partial x^2}\right|_{(1,\, 2)} = 2 \qquad \left.\frac{\partial^2 z}{\partial y^2}\right|_{(1,\, 2)} = -2$$

Our conclusion is that the trace of z in the $y = 2$ plane has a minimum at $(1, 2)$ while the trace of the surface in the $x = 1$ plane has a maximum at $(1, 2)$. Thus, $(1, 2)$ cannot correspond to either a maximum or minimum value of z. This point is called a saddle point, and is illustrated in Figure 8.17.

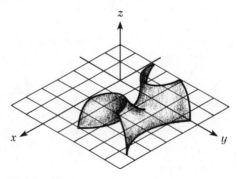

Figure 8.17 Graph of $z = (x - 1)^2 - (y - 2)^2$
(a hyperbolic paraboloid)

Example 2   Find the extrema of the function

$$z = 2xy - 5y^2 - 2x^2 + 4x + 4y - 4$$

Solution $\dfrac{\partial z}{\partial x} = 2y - 4x + 4 \qquad \dfrac{\partial z}{\partial y} = 2x - 10y + 4$

Setting these equal to zero one can find the only possibilities for extrema.

$$\begin{aligned} 2x - 10y + 4 &= 0 \\ -4x + 2y + 4 &= 0 \end{aligned}$$

Thus,

$$\begin{aligned} 2x - 10y + 4 &= 0 \\ -20x + 10y + 20 &= 0 \\ \hline -18x + 24 &= 0 \\ x &= \frac{4}{3} \end{aligned}$$

And,

$$
\begin{aligned}
4x - 20y + 8 &= 0 \\
-4x + 2y + 4 &= 0 \\
\hline
-18y + 12 &= 0 \\
y &= \frac{2}{3}
\end{aligned}
$$

Conclusion: If an extremum for z exists, it must occur at $(\frac{4}{3}, \frac{2}{3})$.

Does $(\frac{4}{3}, \frac{2}{3})$ in the example correspond to a maximum or minimum value of z? Perhaps the second partial derivatives can test the result.

$$
\frac{\partial^2 z}{\partial x^2} = -4 \qquad \frac{\partial^2 z}{\partial y^2} = -10 \qquad \frac{\partial^2 z}{\partial x\, \partial y} = 2
$$

The fact that $\partial^2 z/\partial x^2 = -4$ would indicate that the point $(\frac{4}{3}, \frac{2}{3}, z)$ corresponds to a maximum of the trace curve in the $y = \frac{2}{3}$ plane, and because $\partial^2 z/\partial y^2 = -10$, one can conclude that $(\frac{4}{3}, \frac{2}{3}, z)$ is also a maximum of the trace of the surface in the $x = \frac{4}{3}$ plane. However, $\partial^2 z/(\partial x\, \partial y) = 2$, a positive value. What effect does this fact have on the nature of z at $(\frac{4}{3}, \frac{2}{3})$? For the moment, let us assume that we do have a maximum. The value of the maximum is

$$
\begin{aligned}
f\left(\frac{4}{3}, \frac{2}{3}\right) &= 2\left(\frac{4}{3}\right)\left(\frac{2}{3}\right) - 5\left(\frac{2}{3}\right)^2 - 2\left(\frac{4}{3}\right)^2 + 4\left(\frac{4}{3}\right) + 4\left(\frac{2}{3}\right) - 4 \\
&= \frac{16}{9} - \frac{20}{9} - \frac{32}{9} + \frac{16}{3} + \frac{8}{3} - 4 \\
&= \frac{16 - 20 - 32 + 48 + 24 - 36}{9} \\
&= \frac{88 - 88}{9} = 0
\end{aligned}
$$

Is 0 actually the largest value of z? To make the application of extrema of functions of two variables useful, a simple test is required which will indicate whether a pair of values (x_0, y_0) which produce simultaneous zeros of the first partial derivatives are maxima or minima of $f(x, y)$. Advanced calculus supplies just such a test, which we will use without proof.

Definition

For a given function defined by $z = f(x, y)$,

$$\Delta = \frac{\partial^2 z}{\partial x^2} \cdot \frac{\partial^2 z}{\partial y^2} - \left(\frac{\partial^2 z}{\partial x \, \partial y}\right)^2$$

and $\Delta|_{(x_0, \, y_0)}$ denotes Δ evaluated at $x = x_0$, $y = y_0$.

Theorem 8.1   (The Δ test) Let $z = f(x, y)$ define a function of two variables, and $(x_0, \, y_0)$ such that

$$\frac{\partial z}{\partial x}\bigg|_{(x_0, \, y_0)} = 0 \qquad \frac{\partial z}{\partial y}\bigg|_{(x_0, \, y_0)} = 0$$

Then if $\Delta|_{(x_0, \, y_0)} > 0$ and $\partial^2 z/\partial x^2 < 0$ or $\partial^2 z/\partial y^2 < 0$, $f(x_0, \, y_0)$ is a maximum value of z. If $\Delta|_{(x_0, \, y_0)} > 0$ and $\partial^2 z/\partial x^2 > 0$ or $\partial^2 z/\partial y^2 > 0$, then $f(x_0, \, y_0)$ is a minimum value of z. If $\Delta|_{(x_0, \, y_0)} < 0$, then $f(x_0, \, y_0)$ is neither a maximum nor minimum value for z. If $\Delta|_{(x_0, \, y_0)} = 0$, the Δ test fails.

Notice that if $\partial^2 z/\partial x^2$ and $\partial^2 z/\partial y^2$ differ in sign, then $\Delta < 0$ and we have neither a maximum nor a minimum. When $\Delta = 0$ the test fails, meaning we cannot tell with this test if the point is a maximum or minimum. Some other method must be used in these cases.

Example 3   If $z = 2xy - 5y^2 - 2x^2 + 4x + 4y - 4$, then $(\tfrac{4}{3}, \tfrac{2}{3})$ corresponds to the only possible extremum for z. Apply the Δ test.

Solution $\dfrac{\partial^2 z}{\partial x^2} = -4 \qquad \dfrac{\partial^2 z}{\partial y^2} = -10 \qquad \dfrac{\partial^2 z}{\partial x \, \partial y} = 2$

Thus,

$$\Delta\bigg|_{\left(\frac{4}{3}, \frac{2}{3}\right)} = (-4)(-10) - (2)^2 = 36$$

Since $\partial^2 z/\partial x^2 < 0$, this must correspond to a relative maximum. Thus, $f(\tfrac{4}{3}, \tfrac{2}{3}) = 0$ is the maximum value of z.

Example 4   If $z = (x - 1)^2 - (y - 2)^2$ as in a previous example, $(1, 2)$ is the only possibility for an extremum. What does the Δ test tell us about this function?

Solution $\dfrac{\partial^2 z}{\partial x^2} = 2$ $\dfrac{\partial^2 z}{\partial y^2} = -2$ $\dfrac{\partial^2 z}{\partial x\,\partial y} = 0$

Thus,

$$\Delta\big|_{(1,\,2)} = (2)(-2) - (0)^2 = -4$$

This indicates that z has no maxima or minima.

Example 5   Find the extrema of $f(x,\,y) = z = 7x^2 + y^2 - 5xy - 3x + 6y - 2$.

Solution $\dfrac{\partial z}{\partial x} = 14x - 5y - 3$ $\dfrac{\partial z}{\partial y} = 2y - 5x + 6$

Setting these equal to zero and solving,

$$14x - 5y - 3 = 0$$
$$-5x + 2y + 6 = 0$$

or

$$28x - 10y - 6 = 0$$
$$\underline{-25x + 10y + 30 = 0}$$
$$3x + 24 = 0$$
$$x = -8$$

Thus,

$$-5(-8) + 2y + 6 = 0$$
$$40 + 2y + 6 = 0$$
$$2y = -46$$
$$y = -23$$

Also,

$$\dfrac{\partial^2 z}{\partial x^2} = 14 \qquad \dfrac{\partial^2 z}{\partial y^2} = 2 \qquad \dfrac{\partial^2 z}{\partial x\,\partial y} = -5$$

Therefore,

$$\Delta\big|_{(-8,\,-23)} = (14)(2) - (-5)^2 = 3$$

Since $\partial^2 z/\partial x^2 = 14$, a positive value, we conclude that

$$f(-8, -23) = 7(-8)^2 + (-23)^2 - 5(-8)(-23) - 3(-8) + 6(-23) - 2$$
$$= 448 + 529 - 920 + 24 - 138 - 2$$
$$= 1001 - 1060 = -59$$

is a minimum value for z.

Example 6 Find any extrema of

$$z = f(x, y) = y \ln x - x$$

Solution

$$\frac{\partial z}{\partial x} = \frac{y}{x} - 1 \qquad\qquad \frac{\partial^2 z}{\partial x^2} = \frac{-y}{x^2}$$

$$\frac{\partial z}{\partial y} = \ln x \qquad\qquad \frac{\partial^2 z}{\partial y^2} = 0$$

$$\frac{\partial^2 z}{\partial x \, \partial y} = \frac{1}{x}$$

Setting $\partial z/\partial x$ and $\partial z/\partial y$ equal to zero,

$$\frac{y}{x} - 1 = 0$$

$$\frac{y - x}{x} = 0$$

$$x = y \qquad\qquad \ln x = 0$$
$$x = 1$$

Thus the only possibility for an extremum is at $(1, 1)$.

$$\Delta\big|_{(1, 1)} = \left(\frac{-1}{1^2}\right)(0) - \left(\frac{1}{1}\right)^2 = -1$$

There are no maxima or minima.

Example 7 Find the maximum and minimum values, if any, for

$$z = f(x, y) = 2x^4 - x^2 + 3y^2$$

Solution

$$\frac{\partial z}{\partial x} = 8x^3 - 2x \qquad\qquad \frac{\partial^2 z}{\partial x^2} = 24x^2 - 2$$

$$\frac{\partial z}{\partial y} = 6y \qquad\qquad \frac{\partial^2 z}{\partial y^2} = 6$$

$$\frac{\partial^2 z}{\partial x \, \partial y} = 0$$

Setting $\partial z/\partial x$ and $\partial z/\partial y$ equal to zero,

$$
\begin{array}{cc}
8x^3 - 2x = 0 & 6y = 0 \\
2x(4x^2 - 1) = 0 & y = 0 \\
x = 0 \quad \text{or} \quad x = \pm \dfrac{1}{2} &
\end{array}
$$

Thus the criteria yield three possible points to test: $(0, 0)$, $(\frac{1}{2}, 0)$, and $(-\frac{1}{2}, 0)$.

$$
\Delta = (24x^2 - 2)(6) - 0
$$

$$
\Delta\big|_{(0, 0)} = -12 \quad \text{no maximum or minimum}
$$

$$
\Delta\big|_{(\frac{1}{2}, 0)} = \left[24\left(\frac{1}{4} \right) - 2 \right](6) = 24
$$

$$
\frac{\partial^2 z}{\partial x^2}\bigg|_{(\frac{1}{2}, 0)} = (24x^2 - 2) = 6 - 2 = 4
$$

Thus z is minimum at $(\frac{1}{2}, 0)$.

$$
\Delta\big|_{(-\frac{1}{2}, 0)} = 24 \qquad \text{and} \qquad \frac{\partial^2 z}{\partial x^2}\bigg|_{(-\frac{1}{2}, 0)} = 4
$$

indicating $(-\frac{1}{2}, 0)$ also yields a relative minimum.

$$
f\left(\frac{1}{2}, 0 \right) = 2\left(\frac{1}{2} \right)^4 - \left(\frac{1}{2} \right)^2 + 3(0)^2 = \frac{2}{16} - \frac{1}{4} = -\frac{1}{8}
$$

$$
f\left(-\frac{1}{2}, 0 \right) = 2\left(-\frac{1}{2} \right)^4 - \left(-\frac{1}{2} \right)^2 + 3(0)^2 = -\frac{1}{8}
$$

Example 8
$

Suppose that a long range economic model yields the conclusion that the return from a certain type of investment will be determined by two factors, one measured by x and a second by y, with R, the return, given by

$$
R = ke^{-(x-2)^2-(y-3)^2}
$$

where k is a positive constant. What values of x and y will produce a maximum return?

Solution $R = ke^{-(x-2)^2} \cdot e^{-(y-3)^2}$

Therefore,

$$
\frac{\partial R}{\partial x} = 2k[-(x - 2)]e^{-(x-2)^2} \cdot e^{-(y-3)^2}
$$

$$\frac{\partial R}{\partial y} = -2k(y-3)e^{-(x-2)^2} \cdot e^{-(y-3)^2}$$

$$\frac{\partial^2 R}{\partial x^2} = -2k[(x-2)2[-(x-2)]e^{-(x-2)^2} + e^{-(x-2)^2}]e^{-(y-3)^2}$$

$$\frac{\partial^2 R}{\partial y^2} = -2k[(y-3)2[-(y-3)]e^{-(y-3)^2} + e^{-(y-3)^2}]e^{-(x-2)^2}$$

$$\frac{\partial^2 R}{\partial x \, \partial y} = 4k(x-2)(y-3)e^{-(x-2)^2}e^{-(y-3)^2}$$

When $\partial R/\partial x$ and $\partial R/\partial y$ are set equal to zero,

$$-2k(x-2)e^{-(x-2)^2}e^{-(y-3)^2} = 0 \qquad (x = 2)$$
$$-2k(y-3)e^{-(x-2)^2}e^{-(y-3)^2} = 0 \qquad (y = 3)$$

Thus the only possible maximum or minimum is $(2, 3)$.

$$\begin{aligned}
\Delta\big|_{(2,\,3)} &= \{-2k[(x-2)2(-(x-2))e^{-(x-2)^2} + e^{-(x-2)^2}]e^{-(y-3)^2}\} \\
&\quad \times \{-2k[(y-3)2(-(y-3))e^{-(y-3)^2} + e^{-(y-3)^2}]e^{-(x-2)^2}\} \\
&\quad - [4k(x-2)(y-3)e^{-(x-2)^2}e^{-(y-3)^2}]^2 \\
&= 4k^2
\end{aligned}$$

Since

$$\Delta\big|_{(2,\,3)} > 0 \qquad \text{and} \qquad \frac{\partial^2 R}{\partial x^2}\bigg|_{(2,\,3)} = -2k < 0$$

$(2, 3)$ corresponds to a maximum value for R.

$$R\big|_{(2,\,3)} = ke^{-(2-2)^2-(3-3)^2} = k$$

8.5 EXERCISES

In Exercises 1–16, apply the Δ test to determine the extrema, if possible.

1. $z = f(x, y) = x^3 - 12xy + 8y^3$
2. $z = f(x, y) = x^2 - y^2 - 2xy - 4x$
3. $z = f(x, y) = x^2 + xy + y^2 - 3x + 2$
4. $z = f(x, y) = x^2 - xy + y^2 - x - y$
5. $z = f(x, y) = x^3 - 12x + y^2$
6. $z = f(x, y) = x^3 - 6x - 6xy + 6y + 3y^2$
7. $z = f(x, y) = x^3 + y^3 - 9xy$

8. $z = f(x, y) = 4 - x^{2/3} - y^{2/3}$

9. $z = f(x, y) = \dfrac{1}{x} + xy - \dfrac{8}{y}$

10. $z = f(x, y) = x^4 + y^4 - 4xy$

11. $z = f(x, y) = x^2 + 9 + 6x \cos y$

12. $z = f(x, y) = 25 + (x - y)^4 + (x - 1)^4$

13. $z = f(x, y) = xy(4 - x - y)$

14. $z = f(x, y) = 4 \ln x + e^y - x - y$

15. $z = f(x, y) = \dfrac{xy}{2} - \ln (x^2 + y^2)^{1/2}$

16. $z = f(x, y) = \sin x + \sin y$

17. In a certain economic model, the cost, z, of a given item is related to two factors, measured in x and y, by the equation $z = x^2 + y^2 - \ln xy$, where $x > 0$ and $y > 0$. Determine the values of x and y which will correspond to a minimum value of z.

18. The profit P from the sales of a certain commodity is related to cost, production factors, advertising costs, and the like, by the equation

$$P = -4a^2 - 2b^2 - ab + 18b + 20a$$

where a and b are variables describing actual costs in thousands of dollars. Determine how much should be invested in each factor to maximize profit.

19. A sociological analysis indicates that a parameter measuring the crime rate, R, depends on the amount spent on welfare, w, as measured in hundreds of millions of dollars, and on the amount spent on prisons, p, as measured in hundreds of millions of dollars.

$$R = w^3 + 2p^3 - 6w + 6p - 6wp$$

How much should be spent on each type of program to yield a minimum crime rate parameter?

20. The profit P as a function of the prices, x and y, of two competing items is given by

$$P = 65 - 2(x - 7)^2 - 4(y - 7)^2 - 4x - 2y + 14$$

Determine the values of x and y which maximize profits.

21. In an election campaign for United States Senator, a political scientist discovers that there is a relation between the number of extra votes received (over the normal) and the amount spent on television by both parties in the Los Angeles and San Francisco areas. How much money should be spent in each place to maximize votes if x is the amount spent in Los Angeles (in hundred thousands of dollars) and y is the amount spent in San Francisco (in hundred thousands of dollars), and the number of extra votes V is given by

$$V = -x^3 + 2xy - y^2$$

where V is in hundred thousands.

22. An open rectangular box is to be constructed to hold 1 cubic meter of volume. How should it be constructed to minimize the amount of construction material?

23. A maker of candy uses television and national magazines for advertising. The manufacturer knows that profit P is related to the amount T spent on television ads and the amount M spent on magazine ads by the equation

$$P = TM(12 - 4T - 3M)$$

where P, T, and M are in hundred thousands of dollars. If the equation is valid, how much should be spent on each format to maximize profit?

24. A transportation specialist discovers that a "resistance" parameter describes commuter resistance to moving from auto transport to rapid transit. The key variables are average daily cost d and average daily parking cost P. These are related to the resistance parameter R by the equation

$$R = P^3 - 12Pd + 8d^2$$

What values for P and d would produce a minimum resistance parameter?

In Exercises 25–28, note that the method known as least squares is used to find the line $y = mx + b$ that best fits a set of experimental data, $(x_1, y_1), (x_2, y_2), \ldots, (x_n, y_n)$. If the line fits the data perfectly, then for any pair, (x_i, y_i), $y_i = mx_i + b$. In general, however, y_i will not be exactly $mx_i + b$. Thus, y_i will deviate from $mx_i + b$ by $y_i - (mx_i + b)$. We wish to choose m and b so that all *the deviations will be minimum. Some of the deviations will be positive and others will be negative. If we take the squares of the deviations, they will all be positive or zero. Thus a negative deviation will count as much as a positive one. Since the experimental data are known, if we add the squares of all the deviations, we will have a function of two variables, m and b, which we wish to minimize. That is, the line y = mx + b that best fits is the one for which*

$$f(m, b) = \sum_{i=1}^{n} [y_i - (mx_i + b)]^2$$

is minimum. Find such a line for the data given below.

25. $(0, 0), (4, 3), (2, 2)$ **26.** $(1, 2), (0, 1), (6, -2)$

27. $(2, 0), (0, 1), (-3, 2)$ **28.** $(0, 5), (2, 1), (-1, 3)$

8.6 THE CHAIN RULE AND TOTAL DIFFERENTIALS

Two very useful concepts related to the differentiation of a function of a single variable were the chain rule and the differential. Both of these ideas have parallels related to the functions of two variables. Suppose z defines a function of two variables, x and y, while x and y, in turn, are functions of two more variables, u and v. That is,

$$z = f(x, y) \qquad x = h(u, v) \qquad y = g(u, v)$$

Then the *chain rule for partial differentiation* states

$$\frac{\partial z}{\partial u} = \frac{\partial z}{\partial x} \cdot \frac{\partial x}{\partial u} + \frac{\partial z}{\partial y} \cdot \frac{\partial y}{\partial u}$$

$$\frac{\partial z}{\partial v} = \frac{\partial z}{\partial x} \cdot \frac{\partial x}{\partial v} + \frac{\partial z}{\partial y} \cdot \frac{\partial y}{\partial v}$$

Example 1   If $z = 3x^2 + 2y$, $x = u^2 v$, and $y = u^2 - v^3$, find $\partial z/\partial u$ and $\partial z/\partial v$.

Solution $\dfrac{\partial z}{\partial x} = 6x$ $\dfrac{\partial z}{\partial y} = 2$

$\dfrac{\partial x}{\partial u} = 2uv$ $\dfrac{\partial x}{\partial v} = u^2$

$\dfrac{\partial y}{\partial u} = 2u$ $\dfrac{\partial y}{\partial v} = -3v^2$

Thus,

$$\frac{\partial z}{\partial u} = (6x)(2uv) + 2(2u) = 12(u^2v)uv + 4u = 12u^3v^2 + 4u$$

$$\frac{\partial z}{\partial v} = (6x)(u^2) + 2(-3v^2) = 6(u^2v)u^2 - 6v^2 = 6u^4v - 6v^2$$

In the above example it is possible to substitute for x and y and directly compute the partials for u and v. Such substitution is not always desirable but in this case we use it to show that the chain rule and ordinary partial differentiation give the same result. Substituting $x = u^2v$ and $y = u^2 - v^3$ in $z = 3x^2 + 2y$ gives

$$z = 3(u^2v)^2 + 2(u^2 - v^3)$$
$$= 3u^4v^2 + 2u^2 - 2v^3$$

Then the partials are

$$\frac{\partial z}{\partial u} = 12u^3v^2 + 4u \qquad \frac{\partial z}{\partial v} = 6u^4v - 6v^2$$

Note these results agree with those of Example 1.

Example 2   If $z = 3x^2y$, $x = 4u^2/v$, and $y = \ln(u/v)$, find $\partial z/\partial u$ and $\partial z/\partial v$.

Solution

$$\frac{\partial z}{\partial x} = 6xy \qquad\qquad \frac{\partial z}{\partial y} = 3x^2$$

$$\frac{\partial x}{\partial u} = \frac{8u}{v} \qquad\qquad \frac{\partial x}{\partial v} = \frac{-4u^2}{v^2}$$

$$\frac{\partial y}{\partial u} = \frac{1}{u} \qquad\qquad \frac{\partial y}{\partial v} = -\frac{1}{v}$$

Therefore,

$$\frac{\partial z}{\partial u} = \frac{\partial z}{\partial x}\frac{\partial x}{\partial u} + \frac{\partial z}{\partial y}\frac{\partial y}{\partial u}$$

$$= 6xy\left(\frac{8u}{v}\right) + 3x^2\left(\frac{1}{u}\right)$$

$$= 6\left(\frac{4u^2}{v}\right)\ln\left(\frac{u}{v}\right)\left(\frac{8u}{v}\right) + 3\left(\frac{4u^2}{v}\right)^2\frac{1}{u}$$

$$= 192\frac{u^3}{v^2}\ln\left(\frac{u}{v}\right) + 48\frac{u^3}{v^2}$$

$$\frac{\partial z}{\partial v} = \frac{\partial z}{\partial x}\frac{\partial x}{\partial v} + \frac{\partial z}{\partial y}\frac{\partial y}{\partial v}$$

$$= 6xy\left(\frac{-4u^2}{v^2}\right) + (3x^2)\left(-\frac{1}{v}\right)$$

$$= 6\left(\frac{4u^2}{v}\right)\ln\left(\frac{u}{v}\right)\left(\frac{-4u^2}{v^2}\right) + 3\left(\frac{4u^2}{v}\right)^2\left(-\frac{1}{v}\right)$$

$$= -96\frac{u^4}{v^3}\ln\left(\frac{u}{v}\right) - 48\frac{u^4}{v^3} = -48\left[2\ln\left(\frac{u}{v}\right) + 1\right]\frac{u^4}{v^3}$$

Example 3   If $z = x^2 \cos 2y$ and $x = \sqrt{u^2 + v^2}$ and $y = \tan^{-1}(u/v)$, find $\partial z/\partial u$ and $\partial z/\partial v$.

Solution
$$\frac{\partial z}{\partial x} = 2x \cos 2y \qquad\qquad \frac{\partial z}{\partial y} = -2x^2 \sin 2y$$

$$\frac{\partial x}{\partial u} = \frac{u}{\sqrt{u^2 + v^2}} = \frac{u}{x} \qquad\qquad \frac{\partial x}{\partial v} = \frac{v}{\sqrt{u^2 + v^2}} = \frac{v}{x}$$

$$\frac{\partial y}{\partial u} = \frac{1}{1 + \left(\frac{u}{v}\right)^2} \cdot \frac{1}{v} = \frac{\frac{1}{v}}{\frac{v^2 + u^2}{v^2}} = \frac{v}{u^2 + v^2} = \frac{v}{x^2}$$

$$\frac{\partial y}{\partial v} = \frac{1}{1 + \left(\frac{u}{v}\right)^2}\left(\frac{-u}{v^2}\right) = \frac{-u}{v^2 + u^2} = \frac{-u}{x^2}$$

Thus,

$$\frac{\partial z}{\partial u} = \frac{\partial z}{\partial x}\frac{\partial x}{\partial u} + \frac{\partial z}{\partial y}\frac{\partial y}{\partial u}$$

$$= (2x \cos 2y)\left(\frac{u}{x}\right) + (-2x^2 \sin 2y)\left(\frac{v}{x^2}\right)$$

$$= 2u \cos 2y - 2v \sin 2y$$

$$\frac{\partial z}{\partial v} = \frac{\partial z}{\partial x}\frac{\partial x}{\partial v} + \frac{\partial z}{\partial y}\frac{\partial y}{\partial v}$$

$$= (2x \cos 2y)\left(\frac{v}{x}\right) + (-2x^2 \sin 2y)\left(\frac{-u}{x^2}\right)$$

$$= 2v \cos 2y + 2u \sin 2y$$

Now consider the case when x and y, instead of being functions of two variables, are functions of a single variable t. That is,

$$z = f(x, y) \qquad x = h(t) \qquad y = g(t)$$

Then $\partial x/\partial t$ and dx/dt are equal, as are $\partial y/\partial t$ and dy/dt, and z may be considered a function of t alone. The chain rule becomes

$$\frac{\partial z}{\partial t} = \frac{dz}{dt} = \frac{\partial z}{\partial x} \cdot \frac{dx}{dt} + \frac{\partial z}{\partial y} \cdot \frac{dy}{dt}$$

Example 4   If $z = x^2 + y^2$, $x = t^4$, and $y = t^3$, find dz/dt.

Solution $\dfrac{\partial z}{\partial x} = 2x \qquad \dfrac{\partial z}{\partial y} = 2y$

$\dfrac{dx}{dt} = 4t^3 \qquad \dfrac{dy}{dt} = 3t^2$

$\begin{aligned} \frac{dz}{dt} &= \frac{\partial z}{\partial x} \cdot \frac{dx}{dt} + \frac{\partial z}{\partial y} \cdot \frac{dy}{dt} \\ &= 2x \cdot 4t^3 + 2y \cdot 3t^2 \\ &= 8xt^3 + 6yt^2 \end{aligned}$

Substituting for x and y,

$\begin{aligned} &= 8xt^3 + 6yt^2 \\ &= 8t^4 t^3 + 6t^3 t^2 \\ &= 8t^7 + 6t^5 \end{aligned}$

If we had initially replaced x and y in the above example, then

$$z = (t^4)^2 + (t^3)^2 = t^8 + t^6$$

And the derivative of z with respect to t can be directly computed.

$$\frac{dz}{dt} = 8t^7 + 6t^5$$

Note that this is the same result obtained in the example. Of course it is not always convenient to replace x and y as was done above.

The *total differential* can then be defined as follows.

Definition

If $z = f(x, y)$, and dx and dy are differential changes in x and y, then

$$dz = f_x\, dx + f_y\, dy = \frac{\partial z}{\partial x}\, dx + \frac{\partial z}{\partial y}\, dy$$

Example 5   If $z = 4x^3y^2$, find dz.

   Solution $dz = (12x^2y^2)\, dx + (8x^3y)\, dy$

In the case of the function of a single variable defined by $y = f(x)$, dy is interpreted as the change in y along the tangent line to $y = f(x)$, if x is changed by the amount dx. The actual change in y along $y = f(x)$ is given by $\Delta y = f(x + dx) - f(x)$, and if dx is small, Δy is approximated by dy. In a more complete development of three-dimensional analytic geometry than has been presented here, it would be necessary to supply an appropriate definition of the plane tangent to a surface $z = f(x, y)$ at a pair of values (x, y). If x and y are changed by an amount dx and dy, then dz corresponds to the change in z along that tangent plane. The actual change in z along the surface is given by $\Delta z = f(x + dx, y + dy) - f(x, y)$. As in the function of a single variable, Δz is approximated by dz if dx and dy are small.

Example 6   If $z = f(x, y) = 2 \ln x - y$, $dx = 0.3$, $dy = 0.4$, $x = 1$, and $y = 2$, calculate Δz and dz.

   Solution $\Delta z = f(x + dx, y + dy) - f(x, y)$
$= f(1.3, 2.4) - f(1, 2)$
$= 2 \ln 1.3 - 2.4 - (\ln 1 - 2)$
$= 2 \ln 1.3 - 0.4$
$= 0.525 - 0.4 = 0.125$

using a calculator with natural log function, or

$$0.524 - 0.4 = 0.124$$

using the log tables in the Appendix.

$$dz = f_x\, dx + f_y\, dy$$
$$= \frac{2}{x}\, dx - dy$$
$$= 2(0.3) - 0.4 = 0.6 - 0.4 = 0.2$$

Note that dz and Δz differ by less than 0.1.

Example 7   Suppose $z = f(x, y) = \sqrt{1 + xy}$. Estimate Δz when $x = 4$, $y = 2$, $dx = 0.1$, and $dy = 0.3$.

Solution

$$dz = \frac{y}{2\sqrt{1+xy}}dx + \frac{x}{2\sqrt{1+xy}}dy$$

$$= \frac{2}{2\sqrt{1+(2)4}}(0.1) + \frac{4}{2\sqrt{1+(2)4}}(0.3)$$

$$= \frac{0.1}{3} + \frac{0.6}{3} = \frac{0.7}{3} = \frac{7}{30} \approx 0.23$$

8.6 EXERCISES

In Exercises 1–10, calculate the indicated derivatives.

1. $z = x^2 + y^2$, $\quad x = u \ln v$, $\quad y = v \ln u$; $\quad$ calculate $\dfrac{\partial z}{\partial u}, \dfrac{\partial z}{\partial v}$

2. $z = x^2 + xy + y^2$, $\quad x = u + v$, $\quad y = u - v$; $\quad$ calculate $\dfrac{\partial z}{\partial u}, \dfrac{\partial z}{\partial v}$

3. $z = e^{x+y} - e^{-x-y}$, $\quad x = \ln u$, $\quad y = \ln v$; $\quad$ calculate $\dfrac{\partial z}{\partial u}, \dfrac{\partial z}{\partial v}$

4. $z = \tan^{-1}x + \sqrt{1-y^2}$, $\quad x = u^2$, $\quad y = v - 1$; $\quad$ calculate $\dfrac{\partial z}{\partial u}, \dfrac{\partial z}{\partial v}$

5. $z = e^{x/y}$, $\quad x = u + 2v$, $\quad y = 2u - v$; $\quad$ calculate $\dfrac{\partial z}{\partial u}, \dfrac{\partial z}{\partial v}$

6. $z = \cos(xy)$, $\quad x = u^2v$, $\quad y = e^{uv}$; $\quad$ calculate $\dfrac{\partial z}{\partial u}, \dfrac{\partial z}{\partial v}$

7. $z = \dfrac{x}{y} + y$, $\quad x = t^2 - 1$, $\quad y = t - 1$; $\quad$ calculate $\dfrac{dz}{dt}$

8. $z = \tan^{-1}\dfrac{r}{\theta}$, $\quad r = x + \sin y$, $\quad \theta = x - \sin y$; $\quad$ calculate $\dfrac{\partial z}{\partial x}, \dfrac{\partial z}{\partial y}$

9. $w = u^2 + ve^u$, $\quad u = x \sin x$, $\quad v = x \cos x$; $\quad$ find $\dfrac{dw}{dx}$

10. $w = x^2y + y^2x$, $\quad x = \ln\left(\dfrac{u}{v}\right)$, $\quad y = \ln(uv)$; $\quad$ calculate $\dfrac{\partial w}{\partial u}, \dfrac{\partial w}{\partial v}$

In Exercises 11–15, calculate dz.

11. $z = (x + 2y)^2$ $\qquad\qquad$ **12.** $z = \ln(x - \ln y)$

13. $z = x^2 + 2xy + 2y^2 + x + y$ $\qquad$ **14.** $z = e^{-x-y}$

15. $z = \sin^2(xy) + \cos^2(xy)$

16. If $z = 3x^2 + 2y$, $x = u^2v$, and $y = u^2 - v^3$, verify that

$$\frac{\partial z}{\partial u} = 12u^3v^2 + 4u \qquad \frac{\partial z}{\partial v} = 6u^4v - 6v^2$$

by replacing x and y with their expressions in u and v and calculating $\partial z/\partial u$ and $\partial z/\partial v$ directly.

17. If $z = 3x^2y$, $x = 4u^2/v$, and $y = \ln(u/v)$, verify by substitution and direct calculation that

$$\frac{\partial z}{\partial u} = 192\,\frac{u^3}{v^2}\ln\left(\frac{u}{v}\right) + 48\,\frac{u^3}{v^2} \qquad \frac{\partial z}{\partial v} = -48\left[2\ln\left(\frac{u}{v}\right) + 1\right]\frac{u^4}{v^3}$$

18. If $z = x^2\cos 2y$, $x = \sqrt{u^2 + v^2}$, and $y = \tan^{-1}(u/v)$ as in Example 3, verify by direct substitution and calculation that $\partial z/\partial u = 2u\cos 2y - 2v\sin 2y$ and $\partial z/\partial v = 2v\cos 2y + 2u\sin 2y$.

19. For $z = \sqrt{1 + xy}$ with $x = 4$, $y = 2$, $dx = 0.1$, $dy = 0.3$ as in Example 7, calculate Δz, the actual value of the change in z.

20. If two sides of a triangle are measured and found to be 3 ± 0.01 inches and 9 ± 0.03 inches, and the angle between the sides is known to be exactly $60°$, estimate the area of the triangle and set limits on the accuracy of the estimate. [*Hint:* Find the area of the triangle, A, as a function of the lengths of the sides and calculate dA.]

21. Two sides of a parallelogram are measured and found to be 8 ± 0.01 centimeters, and 7 ± 0.02 centimeters and one interior angle of the parallelogram is known to be exactly $\pi/6$ radian. Estimate the area of this parallelogram and set limits on the accuracy of the estimate.

22. A tin can is known to be 25 ± 0.02 centimeters in height with a circular top 2.5 ± 0.1 centimeters in radius. Estimate the volume of this can and place limits on this estimate.

\$ 23. For the can in Exercise 22 the material in the top and bottom costs \$0.006 per square centimeter, while the wall material costs \$0.004 per square centimeter. Estimate the cost of materials in the can and place limits on that estimate.

In Exercises 24–26, if $y = f(x)$ and $w = g(x)$, the following rules for differentials hold:

$$dy = f'(x)\,dx \qquad dw = g'(x)\,dx$$
$$d(y + w) = dy + dw \qquad d(y \cdot w) = y\,dw + w\,dy$$
$$d\left(\frac{y}{w}\right) = \frac{w\,dy - y\,dw}{w^2}$$

Show that if $z = f(x, y)$ and $w = g(x, y)$, then:

24. $d(z + w) = dz + dw$ **25.** $d(z \cdot w) = z\, dw + w\, dz$

26. $d\left(\dfrac{z}{w}\right) = \dfrac{w\, dz - z\, dw}{w^2}$

8.7 REPEATED INTEGRATION

The process of integration can also be extended to functions of more than one variable, and, just as there were two approaches to the extension of differentiation, there are two approaches to the extension of integration. One approach involves the extension of the basic concept of an integral by translating the interval of integration into a region of integration. This approach will be considered in the next section. The second approach involves a repeated application of the usual integration process, treating one of the variables as a constant.

Example 1   Let $z = x^2 y^3$, and evaluate $\int_0^4 z\, dx$, where y is treated as a constant.

Solution
$$\int_0^4 z\, dx = \int_0^4 x^2 y^3\, dx = \frac{x^3 y^3}{3}\Big]_0^4$$
$$= \frac{4^3 y^3}{3} - \frac{0^3 y^3}{3} = \frac{64}{3} y^3$$

In this example, the final result is still a function of y. Therefore, it can be integrated with respect to y.

Example 2   Evaluate $\int_1^3 \int_0^4 z\, dx\, dy$, where $z = x^2 y^3$, and $\int_0^4 z\, dx$ is evaluated treating y as a constant.

Solution
$$\int_1^3 \int_0^4 z\, dx\, dy = \int_1^3 \left(\int_0^4 x^2 y^3\, dx \right) dy$$
$$= \int_1^3 \frac{64}{3} y^3\, dy$$
$$= \frac{64}{3} \left[\frac{y^4}{4} \right]_1^3 = \frac{64}{3} \left(\frac{1}{4}\right) [3^4 - 1^4]$$
$$= \frac{16}{3}(81 - 1) = \frac{16}{3}(80) = \frac{1280}{3}$$

In general, if $z = f(x, y)$, then

$$\int_a^b \int_c^d z \, dx \, dy = \int_a^b \left(\int_c^d z \, dx \right) dy$$

where the inner integral is evaluated by treating y as a constant.

Example 3   Evaluate $\int_0^5 \int_1^e \dfrac{y}{x} \, dx \, dy = I$.

Solution $\int_1^e \dfrac{y}{x} \, dx = y[\ln x]_1^e = y[\ln e - \ln 1] = y[1 - 0] = y$

Thus,

$$I = \int_0^5 y \, dy = \frac{y^2}{2} \Big]_0^5 = \frac{25}{2}$$

In the preceding examples and the ones that follow, we will assume z is sufficiently well-behaved that no improper integration is involved.

Since the inner integration treats y as a constant, the limits of integration could contain y, and thus depend on y. The mechanics are straightforward.

Example 4   Evaluate $\int_4^5 \int_0^{y^2} xy \, dx \, dy = I$.

Solution $\int_0^{y^2} xy \, dx = \dfrac{x^2 y}{2} \Big]_0^{y^2} = \dfrac{(y^2)^2 y}{2} - \dfrac{0^2 y}{2} = \dfrac{y^4 y}{2} = \dfrac{y^5}{2}$

Thus,

$$I = \int_4^5 \frac{y^5}{2} \, dy = \frac{y^6}{12} \Big]_4^5$$

$$= \frac{5^6}{12} - \frac{4^6}{12} = \frac{1}{12}[15{,}625 - 4096] = \frac{11{,}529}{12} = \frac{3843}{4}$$

Example 5   Evaluate $I = \int_0^2 \int_{2y-4}^0 \dfrac{2x - 1}{y + 1} \, dx \, dy$.

Solution
$$\int_{2y-4}^{0} \frac{2x-1}{y+1} \, dx = \frac{1}{y+1} \left[x^2 - x \right]_{2y-4}^{0}$$

$$= \frac{1}{y+1} \left[0 - \{(2y-4)^2 - (2y-4)\} \right]$$

$$= \frac{-1}{y+1} \left[(2y-4)^2 - (2y-4) \right]$$

$$= \frac{-1}{y+1} \left[4y^2 - 16y + 16 - 2y + 4 \right]$$

$$= \frac{-1}{y+1} \left[4y^2 - 18y + 20 \right]$$

$$= \frac{-2}{y+1} \left[2y^2 - 9y + 10 \right]$$

$$= -2 \left[2y - 11 + \frac{21}{y+1} \right]$$

Therefore,

$$I = -2 \int_{0}^{2} \left(2y - 11 + \frac{21}{y+1} \right) dy$$
$$= -2 \left[y^2 - 11y + 21 \ln (y+1) \right]_{0}^{2}$$
$$= -2 \left[4 - 22 + 21 \ln 3 - (0 - 0 + 21 \ln 1) \right]$$
$$= -2 \left[-18 + 21 \ln 3 \right] = 6 \left[6 - 7 \ln 3 \right]$$

There is no reason why the inner integration must be with respect to x. That is,

$$\int_{a}^{b} \int_{c}^{d} f(x, y) \, dy \, dx$$

is evaluated with respect to y by treating x as a constant in the innermost integration. In this case, c and d can be functions of x.

Example 6 Evaluate $I = \int_{0}^{2} \int_{0}^{\sqrt{x+1}} \sqrt{x+1} \, y^4 \, dy \, dx$.

Solution
$$\int_{0}^{\sqrt{x+1}} \sqrt{x+1} \, y^4 \, dy = \sqrt{x+1} \, \frac{y^5}{5} \Big]_{0}^{\sqrt{x+1}}$$

$$= \frac{\sqrt{x+1} \, (\sqrt{x+1})^5}{5}$$

$$= \frac{(\sqrt{x+1})^6}{5} = \frac{(x+1)^3}{5}$$

Therefore,

$$I = \int_0^2 \frac{(x+1)^3}{5} \, dx = \frac{(x+1)^4}{20} \Big]_0^2$$

$$= \frac{3^4}{20} - \frac{1^4}{20} = \frac{81-1}{20} = \frac{80}{20} = 4$$

8.7 EXERCISES

In Exercises 1–20, evaluate each repeated integral.

1. $\displaystyle\int_{-1}^2 \int_0^4 (2y - x + 4) \, dx \, dy$ **2.** $\displaystyle\int_0^4 \int_{-1}^2 (2y - x + 4) \, dy \, dx$

3. $\displaystyle\int_1^4 \int_{-2}^2 xy \, dy \, dx$ **4.** $\displaystyle\int_0^{16} \int_0^y \sqrt{16 - x} \, dx \, dy$

5. $\displaystyle\int_0^{10} \int_{x-5}^x xy \, dy \, dx$ **6.** $\displaystyle\int_0^3 \int_0^{x^2-1} (x - 1) \, dy \, dx$

7. $\displaystyle\int_0^3 \int_{2y}^{y^2} (y - 2x) \, dx \, dy$ **8.** $\displaystyle\int_1^3 \int_0^{\ln x} e^y \sqrt{(x-1)^2 + 4} \, dy \, dx$

9. $\displaystyle\int_0^{\ln 2} \int_0^{\ln 2} e^{x+y} \, dx \, dy$ **10.** $\displaystyle\int_{-5}^5 \int_0^{\sqrt{25-x^2}} dy \, dx$

11. $\displaystyle\int_{-1}^1 \int_0^{1-y^2} (x + y) \, dx \, dy$ **12.** $\displaystyle\int_0^1 \int_0^{1-y^2} (x + y) \, dx \, dy$

13. $\displaystyle\int_{-3}^3 \int_y^{2y} \sqrt{xy - y^2} \, dx \, dy$ **14.** $\displaystyle\int_0^2 \int_{\sqrt{x}}^{(x^2+4)/2} xy \, dy \, dx$

15. $\displaystyle\int_{-1}^1 \int_{-\sqrt{1-y^2}}^{+\sqrt{1-y^2}} (x + y) \, dx \, dy$ **16.** $\displaystyle\int_0^2 \int_{x/2}^{\sqrt{x+2}-1} dy \, dx$

17. $\displaystyle\int_0^2 \int_0^{\sqrt{4-y^2}} xy \sqrt{4 - x^2} \, dx \, dy$ **18.** $\displaystyle\int_0^\pi \int_{\sin x}^1 y \, dy \, dx$

19. $\displaystyle\int_{\pi/2}^\pi \int_0^{1-\sin y} x \, dx \, dy$ **20.** $\displaystyle\int_{\pi/2}^\pi \int_0^{4\cos x} 2y \, dy \, dx$

In Exercises 21 and 22, consider

$$I_1 = 2 \int_0^1 \int_0^x xy^2 \, dy \, dx \qquad I_2 = \int_{-1}^1 \int_0^x xy^2 \, dy \, dx$$

21. Does $\displaystyle I_2 = \int_{-1}^1 \int_0^x xy^2 \, dy \, dx = \frac{I_1}{2}$?

22. Does $I_2 = I_1$?

23. In general, if $a < b < c$ and $z = f(x, y)$, is the following equation true?

$$\int_a^c \int_{f_1(x)}^{f_2(x)} z\ dy\ dx = \int_a^b \int_{f_1(x)}^{f_2(x)} z\ dy\ dx + \int_b^c \int_{f_1(x)}^{f_2(x)} z\ dy\ dx$$

24. In general, if b is exactly halfway between a and c, does

$$\int_a^c \int_{f_1(x)}^{f_2(x)} z\ dy\ dx = 2 \int_b^c \int_{f_1(x)}^{f_2(x)} z\ dy\ dx?$$

25. If $\int_0^2 \int_0^{x_2} e^{y/x}\ dy\ dx = e^2 - 1$, what is $\int_0^{x_2} \int_0^2 e^{y/x}\ dx\ dy$?

8.8 MULTIPLE INTEGRATION

Recall that the integral of a function of a single variable was defined as a Riemann sum and illustrated using area. The integral of a function of two variables can be analogously defined and illustrated using volume.

In Figure 8.18, we have a region A in the xy plane, partitioned

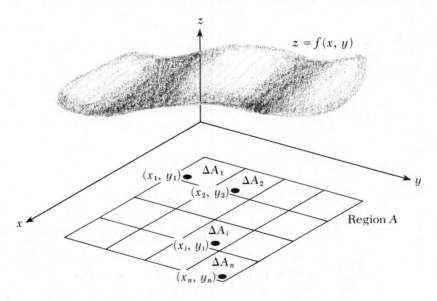

Figure 8.18 Region A

into n subregions, of area ΔA_1, ΔA_2, ΔA_3, . . . , ΔA_n. The ith subregion has area ΔA_i. Let (x_i, y_i) be any point in the ith subregion and let $z = f(x, y)$ be defined for all points in A. Then the integral of $f(x, y)$ over A is defined.

$$\int_A f(x, y) \, dA = \lim_{\substack{n \to \infty \\ |\Delta A_i| \to 0}} \sum_{i=1}^{n} f(x_i, y_i) \, \Delta A_i$$

provided such a limit exists, where $|\Delta A_i|$ denotes the largest area of any of the subregions of the partition being considered.

If $f(x, y)$ is positive for all points of region A, then the integral can be interpreted as the volume of the space above region A and below the surface $z = f(x, y)$. In Figure 8.19, we choose (x_i, y_i) at a

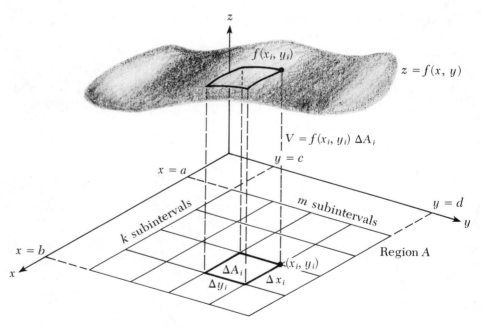

Figure 8.19 Volume above ΔA_i

corner of ΔA_i. The rectangular solid with base ΔA_i and height $f(x_i, y_i)$ is an estimate of the volume above ΔA_i and below $f(x, y)$. The limit of the sum of all such volumes as their number becomes infinite is the volume above A and below $f(x, y)$. That is,

$$V = \lim_{\substack{|\Delta A_i| \to 0 \\ n \to \infty}} \sum_{i=1}^{n} f(x_i, y_i) \, \Delta A_i = \int_A f(x, y) \, dA$$

The key to evaluating the above expression and, therefore, computing the volumes below surfaces is repeated integration. Consider a region A, in Figure 8.19, bounded by the lines $x = a$, $x = b$, $y = c$, and $y = d$; a, b, c, and d constants; $a < b$ and $c < d$. Suppose that the interval along the x axis is divided into k subintervals, while that along the y axis is divided into m subintervals.

Let Δx_i be the length of the ith subinterval along the x axis, and Δy_j be the length of the jth subinterval along the y axis. The area A has then been divided into $k \times m$ subregions. The areas of these subregions are $\Delta x_1 \Delta y_1$, $\Delta x_1 \Delta y_2$, $\Delta x_1 \Delta y_3$, . . . , $\Delta x_1 \Delta y_m$, $\Delta x_2 \Delta y_1$, . . . , $\Delta x_2 \Delta y_m$, . . . , $\Delta x_k \Delta y_1$, . . . , $\Delta x_k \Delta y_m$. Let (x_i, y_j) be a point in the ijth subregion. Following the definition of the multiple integral, form a Riemann sum.

Sum
$$= [f(x_1, y_1) \, \Delta x_1 \, \Delta y_1 + f(x_1, y_2) \, \Delta x_1 \, \Delta y_2 + \cdots + f(x_1, y_m) \, \Delta x_1 \, \Delta y_m]$$
$$+ [f(x_2, y_1) \, \Delta x_2 \, \Delta y_1 + f(x_2, y_2) \, \Delta x_2 \, \Delta y_2 + \cdots + f(x_2, y_m) \, \Delta x_2 \, \Delta y_m]$$
$$\vdots$$
$$+ [f(x_k, y_1) \, \Delta x_k \, \Delta y_1 + f(x_k, y_2) \, \Delta x_k \, \Delta y_2 + \cdots + f(x_k, y_m) \, \Delta x_k \, \Delta y_m]$$

Using summation notation,

$$\text{Sum} = \sum_{j=1}^{m} f(x_1, y_j) \, \Delta x_1 \, \Delta y_j + \sum_{j=1}^{m} f(x_2, y_j) \, \Delta x_2 \, \Delta y_j + \cdots$$
$$+ \sum_{j=1}^{m} f(x_k, y_j) \, \Delta x_k \, \Delta y_j$$

Using a double summation notation,

$$\text{Sum} = \sum_{i=1}^{k} \left[\sum_{j=1}^{m} f(x_i, y_j) \, \Delta y_j \, \Delta x_i \right] = \sum_{i=1}^{k} \left[\sum_{j=1}^{m} f(x_i, y_j) \, \Delta y_j \right] \Delta x_i$$

If $\int_A f(x, y) \, dA$ exists, it must be given by the limit of the sum as k and m tend to infinity. Specifically,

$$\int_A f(x, y) \, dA = \lim_{\substack{k \to \infty \\ |\Delta x| \to 0}} \left[\lim_{\substack{m \to \infty \\ |\Delta y| \to 0}} \left\{ \sum_{i=1}^{k} \left[\sum_{j=1}^{m} f(x_i, y_j) \, \Delta y_j \right] \Delta x_i \right\} \right]$$

However, the limit as m tends to infinity does not affect the sum over k; thus

$$\int_A f(x, y) \, dA = \lim_{\substack{k \to \infty \\ |\Delta x| \to 0}} \sum_{i=1}^{k} \left[\lim_{\substack{m \to \infty \\ |\Delta y| \to 0}} \sum_{j=1}^{m} f(x_i, y_j) \, \Delta y_j \right] \Delta x_i$$

If one compares the inner limit/sum combination with the definition of an ordinary integral, realizing that x_i is fixed as the sum is taken over j, and m tends to infinity, then

$$\lim_{\substack{m \to \infty \\ |\Delta y| \to 0}} \sum_{j=1}^{m} f(x_i, y_j) \, \Delta y_j = \int_c^d f(x_i, y) \, dy$$

That is, this limit is the integral of $f(x, y)$ with respect to y, holding x at a constant value of x_i. Then

$$\int_A f(x, y) \, dA = \lim_{\substack{k \to \infty \\ |\Delta x| \to 0}} \sum_{i=1}^{k} \left[\int_c^d f(x_i, y) \, dy \right] \Delta x_i$$

However, $\int_c^d f(x_i, y) \, dy$ is a function of x_i, or x in the ordinary sense; therefore

$$\int_A f(x, y) \, dA = \int_a^b \left[\int_c^d f(x, y) \, dy \right] dx$$

This is a repeated integral; specifically, it is a repeated integral whose limits of integration correspond to the boundaries of the region of the multiple integral.

Example 1   Find the volume below the surface $z = 3x^2 + 3y^2$ and above the region bounded by $x = 0$, $x = 1$, $y = 0$, and $y = 1$.

Solution   The surface and region are graphed in Figure 8.20. The required volume V is such that

$$V = \int_A f(x, y)\, dA = \int_0^1 \int_0^1 (3x^2 + 3y^2)\, dy\, dx$$
$$= \int_0^1 \left[3x^2 y + y^3\right]_0^1 dx$$
$$= \int_0^1 (3x^2 + 1)\, dx$$
$$= x^3 + x\Big]_0^1 = 2 \text{ cubic units}$$

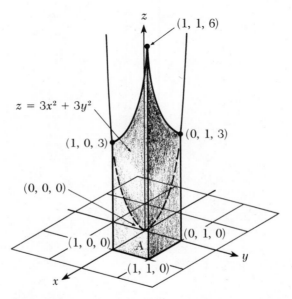

Figure 8.20 Volume above A and below $z = 3x^2 + 3y^2$

The above discussion does not constitute a proof, only a reasonable argument for the result. It also applies to a more general result involving regions with boundaries other than constants. Consider a function $f(x, y)$ and a region A bounded by $y = g_2(x)$, $y = g_1(x)$, $x = a$, $x = b$ with $g_2(x) > g_1(x)$ for all $x \in [a, b]$. Such a region is shown in Figure 8.21. Then

$$\int_A f(x, y)\, dA = \int_a^b \int_{g_1(x)}^{g_2(x)} f(x, y)\, dy\, dx$$

This result is also a very reasonable one. The inner integration involves holding x constant. A typical constant value for x is shown as a dashed line in Figure 8.21. Within the region, and along the dashed line, y varies from $g_1(x)$ to $g_2(x)$. To complete integration over the region, one then needs to integrate the inner result with respect to x from $x = a$ to $x = b$.

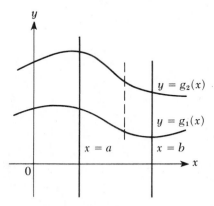

Figure 8.21   The region bounded by $y = g_1(x)$,
$y = g_2(x)$, $x = a$, and $x = b$

Example 2

Consider the region A bounded by $y = x$ above and $y = x^2$ below, and by $x = \frac{1}{2}$ on the left and $x = \frac{3}{4}$ on the right. This region is shown in Figure 8.22. If $f(x, y) = xy^2$, evaluate $\int_A f(x, y)\, dA$.

Solution   From Figure 8.22,

$$\int_A f(x, y)\, dA = \int_{1/2}^{3/4} \int_{x^2}^{x} xy^2\, dy\, dx$$

$$= \int_{1/2}^{3/4} \frac{xy^3}{3} \Big]_{x^2}^{x} dx$$

$$= \int_{1/2}^{3/4} \left(\frac{x^4}{3} - \frac{x^7}{3} \right) dx$$

$$= \frac{1}{3} \left[\frac{x^5}{5} - \frac{x^8}{8} \right]_{1/2}^{3/4}$$

$$= \frac{1}{3} \left[\frac{1}{5} \left(\frac{3}{4} \right)^5 - \frac{1}{8} \left(\frac{3}{4} \right)^8 \right] - \frac{1}{3} \left[\frac{1}{5} \left(\frac{1}{2} \right)^5 - \frac{1}{8} \left(\frac{1}{2} \right)^8 \right]$$

$$= \frac{1}{15} \cdot \frac{3^5}{2^{10}} - \frac{1}{(3)2^3} \cdot \frac{3^8}{2^{16}} - \frac{1}{3(5)} \cdot \frac{1}{2^5} + \frac{1}{(3)2^3} \cdot \frac{1}{2^8}$$

$$= \frac{3^5 \cdot 2^9 - 3^8(5) - 2^{14} + (5)2^8}{3(5)(2)^{19}} = \frac{76,507}{7,864,320}$$

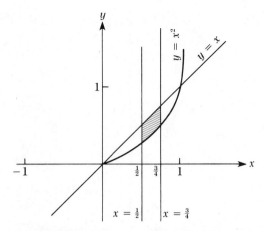

Figure 8.22 The region bounded by $y = x^2$, $y = x$, $x = \frac{1}{2}$, and $x = \frac{3}{4}$

Repeated integration can be used to find volumes over regions that are not four-sided.

Example 3 Find the volume in the first octant bounded by the coordinate planes and plane $z = f(x, y) = 4 - x - y$.

Solution See Figure 8.23. The volume is given by

$$V = \int_0^4 \int_0^{4-x} f(x, y) \, dy \, dx$$

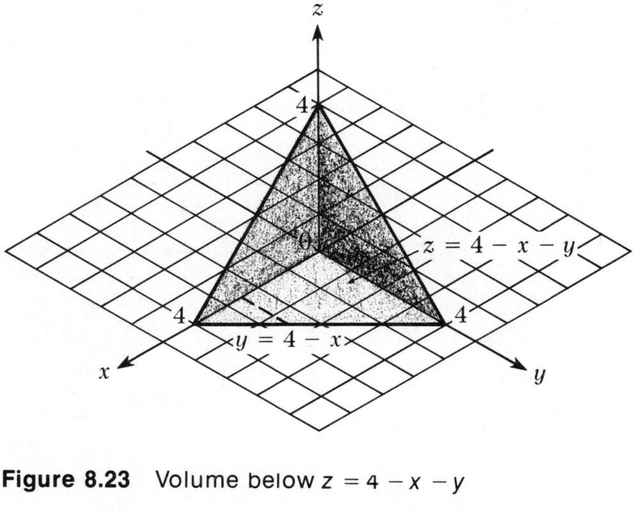

Figure 8.23 Volume below $z = 4 - x - y$

The inner integration is done holding x constant. A typical x value is shown with a dashed line in the figure. Along this line, y varies from $y = 0$ to $y = 4 - x$. In the outer integral, x varies from $x = 0$ to $x = 4$.

$$
\begin{aligned}
V &= \int_0^4 \int_0^{4-x} (4 - x - y)\, dy\, dx \\
&= \int_0^4 \left[4y - xy - \frac{y^2}{2} \right]_0^{4-x} dx \\
&= \int_0^4 \left[4(4 - x) - x(4 - x) - \frac{1}{2}(4 - x)^2 \right] dx \\
&= \int_0^4 \left[16 - 4x - 4x + x^2 - \frac{1}{2}(16 - 8x + x^2) \right] dx \\
&= \int_0^4 \left[\frac{x^2}{2} - 4x + 8 \right] dx \\
&= \frac{x^3}{6} - 2x^2 + 8x \Big]_0^4 \\
&= \frac{64}{6} - 32 + 32 = \frac{32}{3} \text{ cubic units}
\end{aligned}
$$

The solid of Figure 8.23 is a pyramid. The volume of a pyramid is $V = 1/3\, Bh$ where B is the area of its base and h its height. From Figure 8.23, the base of the pyramid is a triangle of area 8 square units and its height is 4 units. Therefore the volume is

$$
V = \frac{1}{3}\, Bh = \frac{1}{3} \cdot 8 \cdot 4 = \frac{32}{3} \text{ cubic units}
$$

The result here agrees with that found by integration.

Example 4 If $f(x, y) = 1$, then $\int_A 1\, dA$ will yield the area of A. Find the area of the region bounded by $y = x^2$ and $y = 9 - x^2$.

Solution The area in question is shown in Figure 8.24. The points of intersection are given by

$$
\begin{aligned}
x^2 &= 9 - x^2 \\
2x^2 &= 9
\end{aligned}
$$

$$
x = \pm \frac{3}{\sqrt{2}} \qquad y = \frac{9}{2}
$$

The area is given by

$$\int_A dA = \int_{-3/\sqrt{2}}^{3/\sqrt{2}} \int_{x^2}^{9-x^2} dy \, dx$$

Since $\int dy = y$,

$$\int_A dA = \int_{-3/\sqrt{2}}^{3/\sqrt{2}} y \bigg]_{x^2}^{9-x^2} dx$$

$$= \int_{-3/\sqrt{2}}^{3/\sqrt{2}} (9 - 2x^2) \, dx$$

Since $\int (9 - 2x^2) \, dx = 9x - \dfrac{2x^3}{3}$,

$$\int_A dA = 9x - \frac{2x^3}{3} \bigg]_{-3/\sqrt{2}}^{3/\sqrt{2}}$$

$$= \left[(9)\frac{3}{\sqrt{2}} - \frac{2(3/\sqrt{2})^3}{3} \right] - \left[-(9)\frac{3}{\sqrt{2}} + \frac{2(3/\sqrt{2})^3}{3} \right]$$

$$= 2 \left(\frac{27}{\sqrt{2}} - \frac{9}{\sqrt{2}} \right) = \frac{2}{\sqrt{2}} (18) = \frac{36}{\sqrt{2}} = 18\sqrt{2}$$

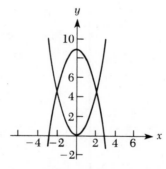

Figure 8.24 The area bounded by
$y = x^2$ and $y = 9 - x^2$

When the region is bounded by curves where x is given as a function of y, only minor changes must be made in the method.

Example 5   Find the area bounded by $y = 0$, $x = e^y$, and $x = 2$.

Solution   This region is shown in Figure 8.25. This time a dashed line has been used to indicate a constant y value. Notice that for a constant

y, x varies from e^y to 2.

$$\int_A dA = \int_0^{\ln 2} \int_{e^y}^{2} dx \, dy$$

$$= \int_0^{\ln 2} x \Big]_{e^y}^{2} dy = \int_0^{\ln 2} (2 - e^y) \, dy$$

$$= 2y - e^y \Big]_0^{\ln 2}$$

$$= 2 \ln 2 - e^{\ln 2} - (0 - e^0)$$

$$= 2 \ln 2 - 2 + 1 = 2 \ln 2 - 1$$

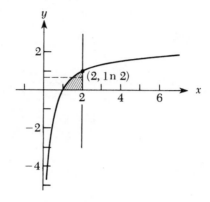

Figure 8.25   The region bounded by $y = 0$,
$y = \ln x$, and $x = 2$

8.8   EXERCISES

In Exercises 1–26, for each defined function and region given, find $\int_A f(x, y) \, dA.$

1. $f(x, y) = 3x^2y + y$, A bounded by $x = 0$, $x = 1$, $y = 1$, $y = 3$

2. $f(x, y) = x^2y$, A bounded by $y = 0$, $y = 3$, $x = 0$, $x = 4$

3. $f(x, y) = (x + y)^2$, A bounded by $x = -1$, $x = 1$, $y = -1$, $y = 1$

4. $f(x, y) = (x + 1)^2(y + 2)^2$, A bounded by $x = 1$, $x = 2$, $y = 2$, $y = 4$

5. $f(x, y) = \dfrac{x}{y}$, A bounded by $x = 0$, $x = 1$, $y = 1$, $y = e$

6. $f(x, y) = xy$, A bounded by $y = 0, x = y^{1/2}, x = 2 - y^{1/2}$

7. $f(x, y) = e^x e^y$, A bounded by $y = 0, y = 4, x = 0, x = 4$

8. $f(x, y) = xy(x + y)$, A bounded by $x = 1, x = 2, y = \pi,$
$y = 2\pi$

9. $f(x, y) = x + y$, A bounded by $y = 0, y = \dfrac{x}{2}, x = 0, x = 6$

10. $f(x, y) = \dfrac{y^2}{x}$, A bounded by $y = 1, y = x, x = 1, x = e$

11. $f(x, y) = 2y$, A bounded by $y = 0, y = \cos x, x = 0, x = \dfrac{\pi}{6}$

12. $f(x, y) = x$, A bounded by $y = 2x, y = x^2$

13. $f(x, y) = e^x \cdot e^y$, A bounded by $x = 0, x = \ln y, y = \ln 4$

14. $f(x, y) = 1$, A bounded by $y = 1, y = e^x, x = 0, x = 2$

15. $f(x, y) = 1$, A bounded by $x = \ln y, x = 2, y = 1$

16. $f(x, y) = x \sin y$, A bounded by $y = 0, y = x, x = 0, x = \dfrac{\pi}{2}$

17. $f(x, y) = \dfrac{x^2}{1 + y^2}$, A bounded by $y = -1, y = 1, x = 0, x = 2$

18. $f(x, y) = 2x^2y - x + 2$, A bounded by $y = \dfrac{x}{3}, \quad y = 4 - x,$
$x = -1, x = 0$

19. $f(x, y) = 2xy - y^2 + 1$, A bounded by $x = 2y + 2,$
$x = 3 - 3y, y = 0$

20. $f(x, y) = 2y$, A bounded by $y = +9 - x^2, y = 0$

21. $f(x, y) = 2xy$, A bounded by $y = x^2 + 4, y = 3x + 2,$
$x = -2, x = 1$

22. $f(x, y) = 1$, A bounded by $y = x^2, y = x + 2$

23. $f(x, y) = 1$, A bounded by $y = \sqrt{25 - x^2}, y = 0$

24. $f(x, y) = 1$, A bounded by $x = -y, y = x - x^2$

25. $f(x, y) = 1$, A bounded by $y = \sqrt{x + 4}, y = \dfrac{3}{5}x, y = 0$

26. $f(x, y) = 2x$, A bounded by $x = 0, x = 2 \sin y \cos y, y = 0,$
$y = \dfrac{\pi}{3}$

Multiple integrals are required to change the order of repeated integration.

$$\int_a^b \int_{f(x)}^{g(x)} f(x, y) \, dy \, dx = \int_A f(x, y) \, dA$$

where A depends on the expressions for $f(x)$ and $g(x)$. Once A has been found, a second repeated integral can be found, with reversed order of integration but the same value, by finding new limits of integration for $\int_A f(x, y)\, dA$. For example, consider $\int_0^{\ln 2} \int_{e^y}^{2} dx\, dy$ as shown in Figure 8.25; A is bounded by $x = e^y$, $y = 0$, and $x = 2$. Therefore we could attempt to set up integration with respect to y first, holding x constant. If we are considering a value within the region with x constant, it is clear that y varies from 0 to its value along the curve. Here $x = e^y$; however we want y as a function of x. That means that we will have to view the curve as $y = \ln\ x$. Then:

$$\int_0^{\ln 2} \int_{e^y}^{2} dx\, dy = \int_A dA = \int_1^2 \int_0^{\ln x} dy\, dx$$

where A is the region bounded by $y = 0$, $x = 2$, $y = \ln\ x$, or $x = e^y$. In Exercises 27–31, find a second repeated integral with reversed order of integration but the same value.

27. $\int_0^1 \int_{y^3}^{y^2} dx\, dy$

28. $\int_0^3 \int_0^{\sqrt{9-x^2}} dy\, dx$

29. $\int_{-1}^1 \int_{-1}^{y} \sqrt{1 + y^2}\, dx\, dy$

30. $\int_0^{\sqrt{2}} \int_{x^2-1}^{1} x\ \ln\ x\, dy\, dx$

31. $\int_{-1}^4 \int_0^{2y} e^{-x^2}\, dx\, dy$

In Exercises 32–39, if $z = f(x, y) \geqslant 0$ for all (x, y) in a region A, then $\int_A z\, dA$ yields the volume bounded by the region A in the xy plane and the surface $z = f(x, y)$ with walls vertical lines on the boundaries of A. In each case, find the volume indicated using double integration.

32. The volume bounded by $z = 0$, $z = 4$, $x = 1$, $x = 3$, $y = 2$, and $y = 5$

33. The volume in the first octant bounded by the plane $z = -x - y + 3$

34. The volume bounded by $z = 25 - x^2 - y^2$ and $z = 0$

35. The volume bounded by $z = 0$, $z = \dfrac{1}{\sqrt{2\pi}} e^{-x^2/2}$, $y = 0$, and $y = 1$

36. The volume bounded by $z = 0$, $z = x + y$, $y = 0$, $x = 0$, $x = 1$, and $y = 1$

37. The volume bounded by $z = e^{x+y}$, $z = 0$, $x = 0$, $x = \ln\ 2$, $y = 0$, and $y = \ln\ 3$

38. The volume bounded by $z = x^2 + y^2 + 2$, $z = 0$, $x = -1$, $x = 1$, $y = -1$, and $y = 1$

39. The volume bounded by $z = xy$, $y \geq 0$, $x \geq 0$, and $x^2 + y^2 = 1$

8.9 FUNCTIONS OF MORE THAN TWO VARIABLES

The concept of a function extends to functions of more than two variables. In a realistic sense it is probable that the more variables we can handle, the more useful the approach, as most real phenomena are controlled by a large number of factors. Functions such as

$$w = f(x, y, z) = xy^3z$$

$$\frac{\partial w}{\partial x} \qquad \frac{\partial w}{\partial y} \qquad \frac{\partial w}{\partial z} \qquad \frac{\partial^4 w}{\partial x\, \partial y^2\, \partial z}$$

$$\int_1^3 \int_2^5 \int_7^1 xy^3z \, dz \, dx \, dy$$

can be evaluated following the patterns we have established.

Example 1 If $w = f(x, y, z)$ defines a function of three variables x, y, and z, and

$$f(x, y, z) = \frac{x^2 - xy}{z}$$

find $f(1, 2, 1), f(-1, 4, 3)$, and $f(2, 2, 2)$.

Solution
$$f(1, 2, 1) = \frac{1^2 - 1(2)}{1} = \frac{1 - 2}{1} = -1$$

$$f(-1, 4, 3) = \frac{(-1)^2 - (-1)(4)}{3} = \frac{1 + 4}{3} = \frac{5}{3}$$

$$f(2, 2, 2) = \frac{(2)^2 - 2(2)}{2} = \frac{4 - 4}{2} = \frac{0}{2} = 0$$

Example 2 If $w = f(x, y, z) = e^x \ln y + z$, find $f(0, e, 1)$, $f(\ln 3, 1, 2)$, and $f(2, 2, \ln 2)$.

Solution
$$f(0, e, 1) = e^0 \ln e + 1 = 1 \cdot 1 + 1 = 2$$
$$f(\ln 3, 1, 2) = e^{\ln 3} \cdot \ln 1 + 2 = 3 \cdot 0 + 2 = 0 + 2 = 2$$
$$f(2, 2, \ln 2) = e^2 \ln 2 + \ln 2$$

Example 3   If $w = f(x, y, z)$ defines a function of three variables x, y, and z, and

$$f(x, y, z) = \frac{x \sin (\pi y)}{z^3}$$

find $f(1, \frac{1}{2}, 8)$, $f(0, \frac{1}{4}, -1)$, and $f(-1, -\frac{1}{2}, 2)$.

Solution $f(1, \frac{1}{2}, 8) = \dfrac{1 \sin (\pi/2)}{8^3} = \dfrac{1(1)}{512} = \dfrac{1}{512}$

$f(0, \frac{1}{4}, -1) = \dfrac{(0) \sin (\pi/4)}{(-1)^3} = 0$

$f(-1, -\frac{1}{2}, 2) = \dfrac{(-1) \sin (-\pi/2)}{2^3} = \dfrac{-1(-1)}{8} = \dfrac{1}{8}$

Example 4   If $w = \dfrac{x^2 - xy}{z}$, find $\partial w/\partial x$, $\partial w/\partial y$, $\partial w/\partial z$.

Solution   As before, each partial derivative is found by treating all but one variable as constants.

$$\frac{\partial w}{\partial x} = \frac{2x - y}{z} \qquad \frac{\partial w}{\partial y} = \frac{-x}{z} \qquad \frac{\partial w}{\partial z} = \frac{-(x^2 - xy)}{z^2}$$

Example 5   If $w = e^x \ln y + z$, find $\partial w/\partial x$, $\partial w/\partial y$, and $\partial w/\partial z$.

Solution $\dfrac{\partial w}{\partial x} = e^x \ln y \qquad \dfrac{\partial w}{\partial y} = \dfrac{e^x}{y} \qquad \dfrac{\partial w}{\partial z} = 1$

Example 6   If $w = f(x, y, z) = \dfrac{x \sin (\pi y)}{z^3}$, find $\partial w/\partial x$, $\partial w/\partial y$, and $\partial w/\partial z$.

Solution $\dfrac{\partial w}{\partial x} = \dfrac{\sin (\pi y)}{z^3} \qquad \dfrac{\partial w}{\partial y} = \dfrac{\pi x \cos (\pi y)}{z^3} \qquad \dfrac{\partial w}{\partial z} = \dfrac{-3x \sin (\pi y)}{z^4}$

Example 7   Evaluate the three-fold repeated integral

$$J = \int_0^1 \int_{-2}^{3x} \int_0^{xy^2} z^2 \, dz \, dy \, dx$$

Solution $\displaystyle\int_0^{xy^2} z^2\,dz = \frac{z^3}{3}\Big]_0^{xy^2} = \frac{(xy^2)^3}{3} = \frac{x^3y^6}{3}.$

$$J = \int_0^1 \int_{-2}^{3x} \frac{x^3y^6}{3}\,dy\,dx$$

However,

$$\int_{-2}^{3x} \frac{x^3y^6}{3}\,dy = \frac{x^3y^7}{21}\Big]_{-2}^{3x}$$

$$= \frac{x^3}{21}[(3x)^7 - (-2)^7]$$

$$= \frac{x^3}{21}[2187x^7 + 128]$$

Thus,

$$J = \int_0^1 \left[\frac{2187}{21}x^{10} + \frac{128}{21}x^3\right]dx$$

$$= \frac{2187}{(21)11}x^{11} + \frac{128}{(21)4}x^4\Big]_0^1$$

$$= \frac{2187}{231} + \frac{32}{21}\left(\frac{11}{11}\right)$$

$$= \frac{2187 + 352}{231} = \frac{2539}{231}$$

We have seen how the chain rule extends to the more complex case of partial derivatives. That is, if $w = f(x, y, z)$, $x = h(u, v)$, $y = g(u, v)$, and $z = j(u, v)$, then

$$\frac{\partial w}{\partial u} = \frac{\partial w}{\partial x}\cdot\frac{\partial x}{\partial u} + \frac{\partial w}{\partial y}\cdot\frac{\partial y}{\partial u} + \frac{\partial w}{\partial z}\cdot\frac{\partial z}{\partial u}$$

Example 8   If $w = x^2 + y^2 + z^2$, $x = uv^3$, $y = vu^4$, and $z = u^5v^5$, find $\partial w/\partial v$.

Solution $\displaystyle\frac{\partial w}{\partial v} = \frac{\partial w}{\partial x}\cdot\frac{\partial x}{\partial v} + \frac{\partial w}{\partial y}\cdot\frac{\partial y}{\partial v} + \frac{\partial w}{\partial z}\cdot\frac{\partial z}{\partial v}$

$= 2x(3uv^2) + 2y(u^4) + 2z(5u^5v^4)$

It is often useful to find a general approach to various related concepts. For example, consider n independent variables denoted by $x_1, x_2, x_3, \ldots, x_n$. Then $z = f(x_1, x_2, \ldots, x_n)$ indicates z is a function of these n variables, and one can consider partial derivatives of z with respect to each of these variables.

$$\frac{\partial z}{\partial x_1} \quad \text{or} \quad \frac{\partial z}{\partial x_7} \quad \text{or} \quad \frac{\partial^2 z}{\partial x_3^2} \quad \text{or} \quad \frac{\partial^2 z}{\partial x_1 \, \partial x_7} \quad \text{or} \quad \frac{\partial z}{\partial x_i}$$

or it may be useful to consider an n-fold repeated integral. It may be denoted by

$$\int_{a_1}^{b_1} \int_{a_2}^{b_2} \int_{a_3}^{b_3} \cdots \int_{a_n}^{b_n} f(x_1, x_2, \ldots, x_n) \, dx_n \, dx_{n-1} \cdots dx_2 \, dx_1$$

where a_n and b_n may be functions of $x_1 \ldots x_{n-1}$, and a_{n-1} and b_{n-1} may be functions of $x_1 \ldots x_{n-2}$, and so forth. Clearly, the outermost integration must have constant limits, whereas a_2 and b_2, at their worst, are functions only of x_1.

The summation notation provides a compact form of the chain rule. Suppose $z = f(x_1, x_2, \ldots, x_n)$ while $x_i = g(y_1, \ldots, y_k)$, that is, each x_i is a function of k variables y_1 to y_k. Then

$$\frac{\partial z}{\partial y_i} = \sum_{j=1}^{n} \frac{\partial z}{\partial x_j} \frac{\partial x_j}{\partial y_i} \qquad (i = 1, 2, \ldots, k)$$

As obscure as these ideas may seem, they do have many useful applications. For example, one encounters n-fold repeated integrals, and an n-variable chain rule in advanced mathematical statistics. Specifically, the mathematical expression for the statistical expectation with respect to a six-tuple random valued phenomenon takes the form

$$\int_{-\infty}^{\infty} \int_{-\infty}^{\infty} \int_{-\infty}^{\infty} \int_{-\infty}^{\infty} \int_{-\infty}^{\infty} \int_{-\infty}^{\infty} g(x_1, x_2, x_3, x_4, x_5, x_6)$$
$$\cdot f(x_1, x_2, x_3, x_4, x_5, x_6) \, dx_1 \, dx_2 \, dx_3 \, dx_4 \, dx_5 \, dx_6$$

where $f(x_1, x_2, x_3, x_4, x_5, x_6)$ is a probability density function, and $g(x_1, x_2, x_3, x_4, x_5, x_6)$ is any continuous function.

8.9   EXERCISES

In Exercises 1–14, for the given defined functions find the indicated partial derivatives.

1. $f(x, y, z) = x^2 y^3 z^4$; $\quad \dfrac{\partial f}{\partial x}, \dfrac{\partial^2 f}{\partial x^2}, \dfrac{\partial f}{\partial y}, \dfrac{\partial^2 f}{\partial x \, \partial y}$

2. $f(x, y, z) = x^2 + y^2 - z^3$; $\dfrac{\partial f}{\partial x}$, $\dfrac{\partial f}{\partial y}$, $\dfrac{\partial^2 f}{\partial x^2}$, $\dfrac{\partial^2 f}{\partial x \, \partial z}$

3. $f(x, y, z) = \dfrac{xy^2}{z}$; $\dfrac{\partial f}{\partial x}$, $\dfrac{\partial f}{\partial y}$, $\dfrac{\partial f}{\partial z}$, $\dfrac{\partial^2 f}{\partial z^2}$

4. $f(x, y, z) = \dfrac{z^2}{xy}$; $\dfrac{\partial f}{\partial z}$, $\dfrac{\partial^2 f}{\partial z^2}$, $\dfrac{\partial^2 f}{\partial x \, \partial y}$, $\dfrac{\partial^3 f}{\partial x \, \partial y \, \partial z}$

5. $f(x, y, z) = x^2 \ln y - y \ln z + zy$; $\dfrac{\partial f}{\partial x}$, $\dfrac{\partial^2 f}{\partial x^2}$, $\dfrac{\partial f}{\partial z}$, $\dfrac{\partial^2 f}{\partial x \, \partial y}$

6. $f(x, y, z) = \dfrac{\sqrt{x^2 + y^2 + z^2}}{x}$; $\dfrac{\partial f}{\partial x}$, $\dfrac{\partial f}{\partial y}$, $\dfrac{\partial f}{\partial z}$, $\dfrac{\partial^2 f}{\partial z^2}$

7. $f(x, y, z) = e^{\frac{x+y+z}{2}}$; $\dfrac{\partial f}{\partial x}$, $\dfrac{\partial f}{\partial y}$, $\dfrac{\partial f}{\partial z}$, $\dfrac{\partial^3 f}{\partial x \, \partial y \, \partial z}$

8. $f(x_1, x_2, x_3, x_4, x_5) = \sin (x_1 + x_2 - x_3 - x_4 + x_5)$;

$\dfrac{\partial^2 f}{\partial x_1 \, \partial x_2}$, $\dfrac{\partial^2 f}{\partial x_4^2}$, $\dfrac{\partial f}{\partial x_5}$

9. $f(u, v, w) = \dfrac{u^2 - v^2}{w}$; $\dfrac{\partial f}{\partial u}$, $\dfrac{\partial f}{\partial v}$, $\dfrac{\partial f}{\partial w}$, $\dfrac{\partial^3 f}{\partial w^3}$

10. $f(x, y, z) = \sin (x^2 - y + z)$; $\dfrac{\partial f}{\partial x}$, $\dfrac{\partial f}{\partial y}$, $\dfrac{\partial f}{\partial z}$, $\dfrac{\partial^2 f}{\partial x \, \partial y}$

11. $f(x, y, z, w) = \ln (xyzw)$; $\dfrac{\partial f}{\partial x}$, $\dfrac{\partial f}{\partial y}$, $\dfrac{\partial f}{\partial z}$, $\dfrac{\partial f}{\partial w}$, $\dfrac{\partial^2 f}{\partial x \, \partial y}$

12. $f(x, y, z, w) = \tan^{-1} x - \sin y + \sin^{-1} z + w^3$;

$\dfrac{\partial^2 f}{\partial x \, \partial y}$, $\dfrac{\partial^2 f}{\partial x \, \partial z}$, $\dfrac{\partial^2 f}{\partial x \, \partial w}$, $\dfrac{\partial^2 f}{\partial y \, \partial z}$

13. $f(x_1, x_2, x_3, x_4, x_5) = \displaystyle\sum_{i=1}^{5} x_i^2 + \sin (x_1 x_2)$; $\dfrac{\partial f}{\partial x_2}$, $\dfrac{\partial^2 f}{\partial x_3^2}$, $\dfrac{\partial^2 f}{\partial x_1 \, \partial x_3}$

14. $f(r, \theta, \psi) = r \tan^{-1} \left(\dfrac{\theta}{\psi} \right)$; $\dfrac{\partial f}{\partial r}$, $\dfrac{\partial f}{\partial \theta}$, $\dfrac{\partial f}{\partial \psi}$, $\dfrac{\partial^2 f}{\partial r^2}$

In Exercises 15–23, find the indicated repeated integrals.

15. $\displaystyle\int_0^1 \int_1^2 \int_2^3 2 \, xyz \, dy \, dz \, dx$

16. $\displaystyle\int_0^1 \int_1^2 \int_2^3 2 \, xyz \, dz \, dy \, dx$

17. $\displaystyle\int_0^2 \int_0^{2w} \int_0^{2z} \int_0^{2y} 2x \, dx \, dy \, dz \, dw$

18. $\displaystyle\int_0^{2\pi} \int_0^5 \int_{-\sqrt{25-r^2}}^{\sqrt{25-r^2}} r \, dz \, dr \, d\theta$

19. $\int_{-\pi/2}^{\pi/2} \int_{0}^{2\pi} \int_{0}^{5} r^2 \sin \psi \, dr \, d\theta \, d\psi$

20. $\int_{1}^{2} \int_{0}^{y^2} \int_{1}^{e^2} \frac{1}{x} \, dx \, dz \, dy$

21. $\int_{0}^{2\pi} \int_{0}^{2\pi} \int_{0}^{2\pi} \sin x \sin y \cos z \, dz \, dy \, dx$

22. $\int_{1}^{2} \int_{\sqrt{\pi/4}}^{\sqrt{\pi/2}} \int_{0}^{x} \sin x^2 \, dy \, dx \, dz$

23. $\underbrace{\int_{0}^{3} \int_{0}^{3} \int_{0}^{3} \cdots \int_{0}^{3}}_{n \text{ times}} dx_1 \, dx_2 \cdots dx_n$

In Exercises 24 and 25, if $w = x^2yz^3$, $x = u^2v^2 - u$, $y = \dfrac{u}{v}$, *and* $z = \dfrac{u^2 + v^2}{u}$, *find:*

24. $\dfrac{\partial w}{\partial u}$ **25.** $\dfrac{\partial w}{\partial v}$

In Exercises 26 and 27, if $w = \sin x^2 \sin y \cos z$, $x = u^2v - 1$, $y = \dfrac{u}{v} \pi$, *and* $z = \sqrt{u^2 - v^2}$, *find:*

26. $\dfrac{\partial w}{\partial u}$ **27.** $\dfrac{\partial w}{\partial v}$

In Exercises 28–30, if $z = e^{\sum_{i=1}^{5} x_i}$, $x_1 = y_1$, $x_2 = y_1 + y_2^2$, $x_3 = \sum_{i=1}^{3} y_b^i$, $x_4 = \sum_{i=1}^{4} y_b^i$ *and* $x_5 = \sum_{i=1}^{5} y_b^i$ *find:*

28. $\dfrac{\partial z}{\partial y_1}, \dfrac{\partial z}{\partial y_2}$ **29.** $\dfrac{\partial z}{\partial y_3}, \dfrac{\partial z}{\partial y_5}$ **30.** $\dfrac{\partial^2 z}{\partial y_1^2}$

NEW TERMS

Hyperbolic paraboloid   Partial derivative
Multiple integration   Surface
Paraboloid   Trace

REVIEW

1. Give three examples of everyday quantities whose value depends on the value of two independent quantities.

2. What conventions normally apply to the domain of a function of two variables defined by an equation when the domain of the function is not specified?

In Exercises 3–7, for each defining expression, determine the domain of the function and sketch that domain in an xy plane.

3. $f(x, y) = \sqrt{25 - x^2 - y^2}$ 4. $f(x, y) = \ln xy$

5. $f(x, y) = \dfrac{e^x}{e^y}(x^2 + y^2)$ 6. $f(x, y) = \sin\left(\dfrac{1}{x + y}\right) \cdot e^{x+y}$

7. $f(x, y) = e^{\frac{1}{x + y}}$

In Exercises 8–10, find each of the functional images indicated.

8. $f(x, y) = \dfrac{e^x}{e^y}(x^2 + y^2)$;   find $f(1, 1)$, $f(3, \ln 2)$, $f(0, 1)$, $f(-1, -2), f(x, x)$

9. $f(x, y) = \sin\left(\dfrac{1}{x + y}\right) \cdot e^{x+y}$; find $f\left(0, \dfrac{2}{\pi}\right)$, $f\left(\dfrac{4}{\pi}, \dfrac{2}{\pi}\right)$, $f(1, -1)$

10. $f(x, y) = \dfrac{|x| \, |y - x|}{|x| - 2}$;   find $f(1, 1), f(2, 1), f(0, 0), f(x, x)$, $f(w, w)$

In Exercises 11–19, using traces, sketch each of the following surfaces in three dimensions.

11. $z = x^2 + y^2$ 12. $z = \sin x + 1 - y$

13. $z = 36 - (x - 2)^2 - (y - 2)^2$ 14. $z = e^{(x^2+y^2)/2}$

15. $z = x^2 - y^2$ 16. $z = (x - y)x^2$

17. $z = x^2 + 4x + 4 + y^2 + 6y + 9$ 18. $z = x + 2y - 3$

19. $z = -x - y - 4$

In Exercises 20–24, note that an alternate to the three-dimensional rectangular coordinates in a plane is cylindrical coordinates. These are defined by using polar coordinates in the xy plane rather than the usual coordinates, and retaining z as a measure of the distance above or below the xy plane. Then

$$x = r \cos \theta \qquad y = r \sin \theta \qquad z = z$$

where (x, y, z) *are the rectangular coordinates of a point, and* (r, θ, z) *are the cylindrical coordinates. See Figure 8.26.*

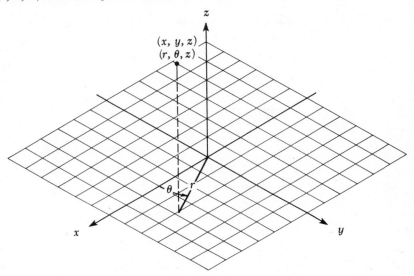

Figure 8.26 Cylindrical coordinates

20. Plot the following points in cylindrical coordinates: $(1, 0, 1)$, $(1, \pi, 1)$, $(3, 2\pi, 2)$, $(-2, \pi, 0)$, $\left(3, \dfrac{\pi}{4}, 1\right)$, $\left(3, -\dfrac{\pi}{4}, 1\right)$, and $\left(3, \dfrac{\pi}{3}, 1\right)$.

21. What is the three-dimensional surface described by the cylindrical coordinate equation $r = 5$?

22. Describe the surface defined by the cylindrical coordinate equation $\theta = \pi/4$.

23. Describe the surface defined by the cylindrical coordinate equation $z = \sqrt{25 - r^2}$.

24. Find the cylindrical coordinate equation for the surface described by the rectangular coordinate equation

$$z = \sqrt{25 - (x - 5)^2 + y^2}$$

In Exercises 25–30, find the indicated partial derivatives.

25. $z = xe^y + ye^x$; find $\dfrac{\partial z}{\partial x}, \dfrac{\partial z}{\partial y}, \dfrac{\partial^2 z}{\partial x \, \partial y}$

26. $z = \dfrac{x^2 - y^2}{x^2 + y^2}$; find $\dfrac{\partial^3 z}{\partial x \, \partial y^2}$

27. $z = x^2 + 3y^2 x$; find $\dfrac{\partial^4 z}{\partial x^4}$

28. $z = \ln\left(\dfrac{x}{y}\right)$, $x = e^u$, and $y = e^{-v}$; find $\dfrac{\partial^2 z}{\partial u^2}$

29. $z = x^2 + y^2$, $x = \sin u \cos v$, and $y = \cos u \cos v$; find $\dfrac{\partial z}{\partial u}$, $\dfrac{\partial z}{\partial v}$

30. $z = \sin\left(\dfrac{x}{y}\right)$, $y = u$, and $x = u \cos v$; find $\dfrac{\partial z}{\partial u}$, $\dfrac{\partial z}{\partial v}$

31. If $x = r \cos \theta$, $y = r \sin \theta$, and $z = f(x, y)$, show that

$$\left(\frac{\partial z}{\partial r}\right)^2 + \frac{1}{r^2} \left(\frac{\partial z}{\partial \theta}\right)^2 = \left(\frac{\partial z}{\partial x}\right)^2 + \left(\frac{\partial z}{\partial y}\right)^2$$

32. Recalling Laplace's equation for a function of two variables, that is,

$$\frac{\partial^2 z}{\partial x^2} + \frac{\partial^2 z}{\partial y^2} = 0$$

show that

$$z = e^x \sin y \qquad \text{and} \qquad z = e^y \cos x$$

both satisfy Laplace's equation.

33. Show that if $z = f_1(x, y)$ and $z = f_2(x, y)$ both satisfy Laplace's equation, $z = f_1(x, y) + f_2(x, y)$ satisfies Laplace's equation.

In Exercises 34 and 35, note that if $F(x, y) = 0$ implicitly defines y as a function of x, it can be shown that

$$\frac{dy}{dx} = \frac{-\partial F / \partial x}{\partial F / \partial y}$$

Find dy/dx in each case.

34. $F(x, y) = x^2 + y^2 - xy = 0$ **35.** $F(x, y) = \sin(x + y) = 0$

In Exercises 36–41, find the maxima and minima of the indicated functions.

36. $f(x, y) = 2xy^2 - x^3 + 3y$

37. $f(x, y) = \cos x - \sin y$, $x \in [0, 2\pi]$, $y \in [0, 2\pi]$

38. $f(x, y) = x^2 y^2 - y$

39. $f(x, y) = 8x^3 - 12x - 12xy + 6y + 3y^2$

40. $f(x, y) = \dfrac{x^4}{16} + 16y^4 - 4xy$ **41.** $f(x, y) = x^2 + 4y^2 - 4x$

In Exercises 42–44, for the given function with $dx = \frac{1}{2}$ and $dy = -\frac{1}{3}$, calculate dz at the indicated point.

42. $z = x^2y^2 - y$ at $(2, 1)$

43. $z = \sin x - \cos y$ at $\left(\dfrac{\pi}{2}, \dfrac{\pi}{3}\right)$

44. $z = 2xy^2 - x^3 + 3y$ at $(2, -1)$

In Exercises 45–50, evaluate each double integral.

45. $\displaystyle\int_0^{\pi/2} \int_1^2 \frac{1}{x}\,dx\,dy$ **46.** $\displaystyle\int_1^3 \int_1^4 (x^2 + y^2)\,dy\,dx$

47. $\displaystyle\int_{-1}^2 \int_0^{x^2} 2y\,dy\,dx$ **48.** $\displaystyle\int_0^2 \int_1^{\ln 2} xe^y\,dy\,dx$

49. $\displaystyle\int_0^1 \int_0^2 xe^{xy}\,dy\,dx$ **50.** $\displaystyle\int_0^{\pi} \int_0^{\pi/2} \sin x \cos y\,dx\,dy$

In Exercises 51–53, find the $\displaystyle\int_A f(x, y)\,dA$ for the $f(x, y)$ and A indicated.

51. $f(x, y) = y$, A bounded by $y = x^2,\ x = y^2$

52. $f(x, y) = 3x^2 - xy$, A bounded by $x = 2y,\ x = 2y^2$

53. $f(x, y) = 1$, A bounded by $y = x^2,\ y = x + 2$

In Exercises 54–56, for the given defined functions, find the indicated partial derivatives.

54. $w = f(x, y, z) = \ln x - \ln y + \ln z + \ln \dfrac{xy}{z}$;   find

$\dfrac{\partial^2 w}{\partial z^2},\ \dfrac{\partial^2 w}{\partial x^2},\ \dfrac{\partial^2 w}{\partial y^2}$

55. $u = f(x, y, z, w),\ x = v^2 - r^{-2},\ y = vr^3, z = \dfrac{v^2}{r^2}, w = \dfrac{1}{v^2}$;   find

$\dfrac{\partial u}{\partial v},\ \dfrac{\partial u}{\partial r}$

56. $f(x) = \displaystyle\sum_{i=1}^{100} \frac{1}{i^2}\, x^{i^2-1}$;   find $\dfrac{df(x)}{dx}$

In Exercises 57–59, evaluate the indicated triple integrals.

57. $\displaystyle\int_0^{2\pi}\int_0^1\int_0^{\sqrt{1-x^2}} xz\,dz\,dx\,dy$ **58.** $\displaystyle\int_0^5\int_0^{\sqrt{25-z^2}}\int_0^{5-z} x\,dy\,dx\,dz$

59. $\displaystyle\int_{-1}^1\int_0^{1-x^2}\int_0^y xy\,dz\,dy\,dx$

60. A farmer wishes to ship tomatoes, lettuce, and onions to the city. Eight hundred boxes can be loaded on a truck, and at least 200 boxes of tomatoes, at least 100 boxes of lettuce, and at least 200 boxes of onions must be shipped. If the profit on tomatoes is $2 per box, on lettuce is $3 per box, and on onions is $1 per box, how should the truck be loaded for maximum profit?

61. A newspaper containing 60 pages is to contain at least 30 pages of news articles and at most 15 pages of advertising. Photographs are limited to one-fourth the number of pages for news articles. If the cost per page is $30 per page of photographs, $50 per page of news articles, and $30 per page of advertisements, what should the composition of the newspaper be for minimum cost?

9
Sequences and Series

The material covered in the first eight chapters is a survey, an introduction to calculus and its applications. One key must be provided to expand these concepts of applied calculus. That key is a fundamental working knowledge of infinite sequences and series. The important result we seek is the representation of a complicated function as an infinite sum of simple terms. This result is achieved in the last three sections of this chapter with the discussions of power series and Taylor's series. The theory used to develop these series is based upon sequences and series and their convergence or divergence, as presented in the first sections of this chapter. Two important tools of mathematics are used extensively in the latter parts of the chapter. These are factorials and L'Hôspital's rules.

9.1 SEQUENCES, CONVERGENCE, AND DIVERGENCE

The common characteristic of the functions considered in this chapter will be their domain of definition. The nature of the functional rule will be secondary. In each case, the domain will be taken from the set of integers.

Definition

A real-valued function whose domain of definition is a subset of integers is called a *sequence*.

Example 1   Let $f(n) = 1/n$, with a domain $n = $ 1, 2, 3, 4, 5, 6, 7, 8, or 9. Find the values of the function over its domain.

Solution $f(1) = \dfrac{1}{1} = 1$ $f(2) = \dfrac{1}{2}$ $f(3) = \dfrac{1}{3}$

$f(4) = \dfrac{1}{4}$ $f(5) = \dfrac{1}{5}$ $f(6) = \dfrac{1}{6}$ and so forth

Example 2   Let $f(n) = n^2$, with a domain of the nonnegative integers. Find the value of the function for $n = $ 3, 1000, and 2.7.

Solution $f(3) = 3^2 = 9$ $f(1000) = 1000^2 = 1,000,000$

$f(2.7)$ is undefined, because 2.7 is not in the domain of f.

Because of the special nature of the domain of a sequence, a special notation is usually adopted for image values. For a given sequence, the image of an integer n will be denoted by a_n or b_n, that is, by means of a subscript rather than the usual $f(n)$ or $g(n)$ notation. The sequence as a whole, that is the totality of the image values, will be indicated by $\{a_n\}$ or $\{b_n\}$. The individual values are called the terms of the sequence.

Example 3   List according to the above notation the two sequences given as examples above.

Solution $a_n = 1/n$, with $n = $ 1, 2, 3, 4, 5, 6, 7, 8, 9; thus,

$$a_1 = \dfrac{1}{1} = 1 \qquad a_2 = \dfrac{1}{2} \qquad a_3 = \dfrac{1}{3} \qquad a_4 = \dfrac{1}{4} \quad \text{and so forth}$$

$b_n = n^2$, n any nonnegative integer:

$$b_0 = 0^2 = 0 \qquad b_1 = 1^2 = 1 \qquad b_2 = 2^2 = 4 \quad \cdots$$
$$b_{1000} = 1000^2 = 1,000,000 \quad \text{and so forth}$$

For our discussion $\{a_n\}$ will be assumed to be an infinite sequence, that is, a sequence with an infinite number of terms. The domain of $\{a_n\}$ will be assumed to be the nonnegative integers unless otherwise indicated.

Definition

Let $\{a_n\}$ be a sequence. If there exists a number L such that $\lim\limits_{n\to\infty} a_n$ exists and is equal to L, then L is the limit of the sequence.

Definition

If $\{a_n\}$ has a limit, then it is said to *converge*; otherwise $\{a_n\}$ is said to *diverge*.

Example 4   Let $\{a_n\}$ be a sequence such that $a_n = 1/n$. Is $\{a_n\}$ convergent?

Solution $\lim\limits_{n\to\infty} a_n = \lim\limits_{n\to\infty} \dfrac{1}{n} = 0$

Thus, $\{a_n\}$ is a convergent sequence.

Example 5   Let $\{a_n\}$ be a sequence such that $a_n = n^2$. Is this sequence convergent?

Solution $\lim\limits_{n\to\infty} a_n = \lim\limits_{n\to\infty} n^2 = \infty$

Therefore, $\{a_n\}$ is a divergent sequence.

Example 6   Let $\{b_n\}$ be such that $b_n = \dfrac{n}{n+1}$. Is this sequence convergent?

Solution $\lim\limits_{n\to\infty} b_n = \lim\limits_{b\to\infty} \dfrac{n}{n+1}$

$\qquad = \lim\limits_{n\to\infty} \dfrac{n/n}{(n/n) + (1/n)}$

$\qquad = \lim\limits_{n\to\infty} \dfrac{1}{1 + (1/n)} = 1$

Therefore, $\{b_n\}$ is convergent.

From the examples, it seems that sequences are defined by describing the functional rule, that is, the nature of the general term of the sequence $\{a_n\}$. This is the usual method used in defining ordinary functions. While in general this is the way sequences will be defined, a second, equally useful way, makes use of their special domain. Instead of giving the form of a general term of the sequence, one gives the specific value of one term, usually the first term of the sequence, and then describes how successive terms are related. That is, one describes how the $(i + 1)$th term is related to the ith term. This type of definition is called a *recursive definition*.

Example 7 Consider the sequence $\{a_n\}$, where $a_1 = 3$ and $a_{i+1} = a_i + 2$. What are the first five terms of this sequence?

Solution $a_1 = 3$
$a_2 = a_1 + 2 = 3 + 2 = 5$
$a_3 = a_2 + 2 = 5 + 2 = 7$
$a_4 = a_3 + 2 = 7 + 2 = 9$
$a_5 = a_4 + 2 = 9 + 2 = 11$

Often, when a sequence is defined recursively it is possible to recover the more usual functional rule-defining equation.

Example 8 Given $\{a_n\}$, $a_1 = 3$, and $a_{i+1} = a_i + 2$, find the functional rule-defining equation and test for convergence.

Solution $a_1 = 3$
$a_2 = a_1 + 2 = 3 + 2$
$a_3 = a_2 + 2 = 3 + 2 + 2$
$a_4 = a_3 + 2 = 3 + 2 + 2 + 2$
$a_5 = a_4 + 2 = 3 + 2 + 2 + 2 + 2$

It follows that

$$a_n = 3 + \underbrace{2 + 2 + \cdots + 2}_{n-1}$$
$$= 3 + 2(n - 1) = 2n + 1$$

Therefore,

$$\lim_{n \to \infty} a_n = \lim_{n \to \infty} (2n + 1) = \infty$$

Thus, $\{a_n\}$ is divergent.

A sequence of the type just considered, where each term is the previous term plus a fixed amount, is called an *arithmetic sequence*. A second type of sequence given a special name is the *geometric sequence*. In a geometric sequence, each term in the sequence is a constant factor times the previous term.

Example 9 Consider $\{a_n\}$ where $a_1 = 1$ and $a_{i+1} = a_i(\frac{1}{2})$. Is this geometric sequence convergent?

Solution $a_1 = 1$

$a_2 = \dfrac{1}{2}$

$a_3 = a_2 \left(\dfrac{1}{2}\right) = \left(\dfrac{1}{2}\right)^2$

$a_4 = a_3 \left(\dfrac{1}{2}\right) = \left(\dfrac{1}{2}\right)^2 \left(\dfrac{1}{2}\right) = \left(\dfrac{1}{2}\right)^3$

$a_5 = a_4 \left(\dfrac{1}{2}\right) = \left(\dfrac{1}{2}\right)^3 \left(\dfrac{1}{2}\right) = \left(\dfrac{1}{2}\right)^4$

It seems

$$a_n = \left(\dfrac{1}{2}\right)^{n-1}$$

$$\lim_{n \to \infty} a_n = \lim_{n \to \infty} \left(\dfrac{1}{2}\right)^{n-1} = \lim_{n \to \infty} \dfrac{1}{2^{n-1}}$$

$$= \lim_{n \to \infty} \dfrac{2}{2^n} = 0$$

This sequence is convergent.

Example 10 If $\{a_n\}$ is such that $a_1 = 1$ and $a_{i+1} = 3a_i$, is $\{a_n\}$ convergent?

Solution $a_1 = 1$
$a_2 = a_1(3) = 3$
$a_3 = a_2(3) = 3^2$
$a_4 = a_3(3) = (3^2)3 = 3^3$

Therefore,

$$a_n = 3^{n-1}$$
$$\lim_{n \to \infty} a_n = \lim_{n \to \infty} 3^{n-1} = \infty$$

and $\{a_n\}$ is divergent.

9.1 EXERCISES

In Exercises 1–10, a sequence $\{a_n\}$ is given. Find values for a_1, a_2, a_3, and a_{20}, and determine if the sequence is convergent or divergent.

1. $a_n = \dfrac{n + 3}{n^2}$

2. $a_n = \dfrac{\cos n\pi}{n^2}$

3. $a_n = \left(1 + \dfrac{1}{n}\right)^{1/n}$

4. $a_n = \dfrac{(-1)^{n-2}}{\ln (n + 1)}$

5. $a_n = \dfrac{1}{\sqrt{n + 1} - \sqrt{n}}$

6. $a_n = \dfrac{\sqrt{n} - \sqrt{n + 2}}{n}$

7. $a_n = \dfrac{3n^3 - 1}{2n^3 + 2}$

8. $a_n = n \sin n\pi$

9. $a_n = \left(\dfrac{1}{2}\right)^n 2^{n-2}$

10. $a_n = \dfrac{e^n}{n}$

In Exercises 11–15, a sequence has been defined by giving a_{i+1} in terms of a_i. Find a_n in each case, and determine if the indicated sequence is convergent.

11. $a_{i+1} = a_i \left(\dfrac{1}{4}\right), \quad a_1 = 1$

12. $a_{i+1} = a_i \dfrac{1}{i + 1}, \quad a_1 = 1$

13. $a_{i+1} = a_i + 2, \quad a_1 = 1$

14. $a_{i+1} = a_i + 1, \quad a_1 = 1$

15. $a_{i+1} = a_i + \dfrac{1}{i + 1}, \quad a_1 = 1$

In Exercises 16–25, the sum of the first n terms of a sequence is given. Determine the form of the general term by finding the difference between $\sum_{i=1}^{n-1} a_i$ and $\sum_{i=1}^{n} a_i$ of the sequence and determine if the sequence is convergent.

16. $\displaystyle\sum_{i=1}^{n} a_i = \dfrac{n}{2n + 1}$

17. $\displaystyle\sum_{i=1}^{n} a_i = \dfrac{n}{3n + 4}$

18. $\displaystyle\sum_{i=1}^{n} a_i = \frac{1}{n+1}$ **19.** $\displaystyle\sum_{i=1}^{n} a_i = \frac{n(n+1)}{2}$

20. $\displaystyle\sum_{i=1}^{n} a_i = \ln(n+1)$ **21.** $\displaystyle\sum_{i=1}^{n} a_i = 2\left[1 - \left(\frac{1}{2}\right)^n\right]$

22. $\displaystyle\sum_{i=1}^{n} a_i = \frac{(1-3^n)}{-2}$ **23.** $\displaystyle\sum_{i=1}^{n} a_i = 3 + 4(n-1)$

24. $\displaystyle\sum_{i=1}^{n} a_i = \frac{n^2}{n+1}$ **25.** $\displaystyle\sum_{i=1}^{n} a_i = \frac{n}{n+1}$

26. Present an argument supporting the hypothesis that all arithmetic sequences $\{a_n\}$ where $a_{i+1} = a_i + c$, $c \neq 0$, must be divergent.

27. Consider a general geometric sequence $\{a_n\}$ with $a_1 = a$, $a_{i+1} = ra_i$, where r is the constant ratio between terms. Determine what conditions must be placed on r to assure that the sequence is convergent.

In Exercises 28–31, consider $\{a_n\}$ and $\{b_n\}$, two convergent sequences. Discuss the convergence or divergence of the sequences formed from $\{a_n\}$ and $\{b_n\}$. Justify your answer in each case.

28. $\{c_n\}$ where $c_n = a_n + b_n$

29. $\{c_n\}$ where $c_n = a_n \cdot b_n$

30. $\{c_n\}$ where $c_n = \dfrac{a_n}{b_n}$

31. $\{c_n\}$ where $c_n = ka_n$, k a constant

In Exercises 32–35, consider $\{a_n\}$ and $\{b_n\}$, two divergent sequences. Discuss the convergence and divergence of the sequence formed from $\{a_n\}$ and $\{b_n\}$. Justify your answer.

32. $\{c_n\}$ where $c_n = a_n - b_n$

33. $\{c_n\}$ where $c_n = a_n \cdot b_n$

34. $\{c_n\}$ where $c_n = \dfrac{a_n}{b_n}$

35. $\{c_n\}$ where $c_n = ka_n$, k a constant $\neq 0$

9.2 SERIES

An idea very closely related to the concept of a sequence is that of a series. A series is nothing more than the sum formed from the terms of a sequence.

Example 1   List a few of the series formed from the sequence $\{a_n\}$ with $a_n = 1/2^n$.

Solution   Some of the series which can be formed from this sequence are:

$$\sum_{i=1}^{3} a_i = a_1 + a_2 + a_3 = \frac{1}{2} + \frac{1}{2^2} + \frac{1}{2^3}$$

$$\sum_{i=1}^{10} a_i = a_1 + a_2 + a_3 + a_4 + a_5 + a_6 + a_7 + a_8 + a_9 + a_{10}$$

$$= \frac{1}{2} + \frac{1}{2^2} + \frac{1}{2^3} + \frac{1}{2^4} + \frac{1}{2^5} + \frac{1}{2^6} + \frac{1}{2^7} + \frac{1}{2^8} + \frac{1}{2^9} + \frac{1}{2^{10}}$$

$$\sum_{k=3}^{7} a_{2k+1} = a_{2(3)+1} + a_{2(4)+1} + a_{2(5)+1} + a_{2(6)+1} + a_{2(7)+1}$$

$$= a_7 + a_9 + a_{11} + a_{13} + a_{15}$$

$$= \frac{1}{2^7} + \frac{1}{2^9} + \frac{1}{2^{11}} + \frac{1}{2^{13}} + \frac{1}{2^{15}}$$

$$\sum_{i=1}^{1000} a_i = a_1 + a_2 + a_3 + \cdots + a_{1000}$$

$$= \frac{1}{2} + \frac{1}{2^2} + \frac{1}{2^3} + \cdots + \frac{1}{2^{1000}}$$

Since we were interested in infinite sequences, it is natural to consider infinite series, that is, a series with an infinite number of terms.

Example 2   Consider the sequence $\{a_n\}$ with $a_n = 1/2^n$. The corresponding infinite series is

$$a_1 + a_2 + a_3 + \cdots + a_i + a_{i+1} + \cdots$$

$$= \frac{1}{2} + \frac{1}{2^2} + \frac{1}{2^3} + \cdots + \frac{1}{2^i} + \frac{1}{2^{i+1}} + \cdots$$

Definition

For a given infinite sequence $\{a_n\}$, the corresponding infinite series will be denoted by

$$\sum_{i=1}^{\infty} a_i = \lim_{n \to \infty} \sum_{i=1}^{n} a_i$$

If this limit exists, the series is said to be *convergent*. Otherwise the series is said to be *divergent*.

Example 3   Consider a geometric sequence $\{a_n\}$ with $a_1 = a > 0$ and $a_{i+1} = ra_i = ar^i$. The corresponding geometric series is given by

$$\sum_{i=1}^{\infty} a_i = \sum_{i=1}^{\infty} ar^{i-1} = a + ar + ar^2 + \cdots$$

Under what conditions is this series convergent?

Solution   To determine if this series is convergent, consider

$$S_n = \sum_{i=1}^{n} a_i = a + ar + ar^2 + \cdots + ar^{n-1}$$

where S_n is called a *partial sum* of the series. A trick can be used to find a compact form for S_n.

Since

$$S_n = a + ar + ar^2 + \cdots + ar^{n-1}$$

and

$$rS_n = ar + ar^2 + ar^3 + \cdots + ar^n$$

then

$$S_n - rS_n = a - ar^n = a(1 - r^n)$$

because all of the inner terms cancel. Therefore,

$$S_n(1 - r) = a(1 - r^n)$$
$$S_n = \frac{a(1 - r^n)}{1 - r}$$

and

$$\sum_{i=1}^{\infty} a_i = \lim_{n \to \infty} \sum_{i=1}^{n} a_i = \lim_{n \to \infty} S_n$$
$$= \lim_{n \to \infty} \frac{a(1 - r^n)}{1 - r}$$

If $|r| < 1$, then $\lim_{n \to \infty} r^n = 0$, and

$$\lim_{n \to \infty} S_n = \lim_{n \to \infty} \frac{a(1 - r^n)}{1 - r} = \frac{a}{1 - r}$$

and the series is convergent.

If $|r| > 1$, then $\lim_{n \to \infty} r^n$ does not exist, and

$$\lim_{n \to \infty} S_n = \lim_{n \to \infty} \frac{a(1 - r^n)}{1 - r}$$

does not exist; therefore $\sum_{i=1}^{\infty} a_i$ is divergent.

If $|r| = 1$, then $r = +1$ or $r = -1$. If $r = 1$,

$$S_n = \sum_{i=1}^{n} a_i = na$$

and

$$\lim_{n \to \infty} S_n = \lim_{n \to \infty} na = \infty$$

If $r = -1$,

$$S_n = \sum_{i=1}^{n} a(-1)^i = 0 \quad \text{or} \quad a$$

depending on whether n is odd or even. In either case, $\lim_{n \to \infty} S_n$ does not exist.

Therefore, we can conclude that a geometric series $\sum_{i=1}^{\infty} a_i$ with $a_i = ar^{i-1}$ is convergent if $|r| < 1$ and divergent if $|r| \geq 1$. Further, if $\sum_{i=1}^{\infty} a_i$ is convergent, it converges to $a/(1 - r)$. That is,

$$\sum_{i=1}^{\infty} a_i = \frac{a}{1 - r}$$

where a is the first term and r the common factor.

Example 4   Is $\sum_{i=1}^{\infty} \frac{1}{2^i}$ convergent?

Solution   Yes, it is a geometric series with $|r| = \left|\frac{1}{2}\right| = \frac{1}{2} < 1$. Furthermore, since the first term is $\frac{1}{2^1} = \frac{1}{2}$,

$$\sum_{i=1}^{\infty} \frac{1}{2^i} = \frac{1/2}{1 - (1/2)} = \frac{1/2}{1/2} = 1$$

A discussion of series may be viewed as the study of two sequences, the original sequence whose terms are the terms of the series, and the sequence of partial sums. That is, $\{a_n\}$ leads to the series $\sum_{i=1}^{\infty} a_i$, which in turn leads to the second sequence $\{S_n\}$

where

$$S_1 = a_1$$
$$S_2 = a_1 + a_2 = S_1 + a_2$$
$$S_3 = a_1 + a_2 + a_3 = S_2 + a_3$$
$$S_4 = a_1 + a_2 + a_3 + a_4 = S_3 + a_4$$

.

.

.

$$S_n = \sum_{i=1}^{n} a_i = S_{n-1} + a_n$$

.

.

.

The definition for convergence can be restated in terms of the convergence of a sequence.

Definition

$\sum_{i=1}^{\infty} a_i$ *converges* if its sequence of partial sums, $\{S_n\}$, where

$S_n = \sum_{i=1}^{n} a_i$, converges; otherwise $\sum_{i=1}^{\infty} a_i$ *diverges.*

The difference in the two definitions is only one of notation, not of concept.

Example 5   Test the series $\sum_{i=1}^{\infty} i$ for convergence.

Solution $S_n = \sum_{i=1}^{n} i = 1 + 2 + 3 + 4 + \cdots + n$

Using the standard method for finding this sum,

$$S_n = \underbrace{1 + 2 + 3 + 4 + \cdots + (n - 1) + n}_{n \text{ terms}}$$

or, writing in reversed order,

$$S_n = n + (n - 1) + (n - 2) + \cdots + 2 + 1$$

Therefore, adding left and right sides,

$$2S_n = 1 + n + [2 + (n - 1)] + [3 + (n - 2)] + \cdots$$
$$+ [(n - 1) + 2] + [n + 1]$$

$$\underbrace{\qquad\qquad\qquad\qquad\qquad\qquad\qquad}_{n \text{ terms}}$$

$$= \underbrace{[1 + n] + [n + 1] + [n + 1] + \cdots + [n + 1] + [n + 1]}_{n \text{ terms}}$$

$$= n(n + 1)$$

Then, dividing both sides by 2,

$$S_n = \frac{n(n + 1)}{2}$$

Then,

$$\lim_{n \to \infty} S_n = \lim_{n \to \infty} \frac{n(n + 1)}{2} = \infty$$

Therefore, $\{S_n\}$ is a divergent sequence, and $\sum_{i=1}^{\infty} i$ is a divergent series.

Although we will consider a number of tests for the convergence and divergence of a series in later sections, a simple test for divergence can be considered here.

Test

Consider $\sum_{i=1}^{\infty} a_i$. If $\lim_{n \to \infty} a_n \neq 0$, the series is divergent. If $\lim_{n \to \infty} a_n = 0$, the series may be either convergent or divergent; in other words, the test fails to give any information.

We wish to show that it is impossible to have a convergent infinite series with $\lim_{n \to \infty} a_n \neq 0$. We assume $\lim_{n \to \infty} a_n \neq 0$ and the series *converges*. Thus

$$\lim_{n \to \infty} S_n = L$$

and

$$\lim_{n \to \infty} S_{n-1} = L$$

also. (It does not matter if we move the index on each partial sum "down" one.) Therefore

$$\lim_{n \to \infty} (S_n - S_{n-1}) = \lim_{n \to \infty} S_n - \lim_{n \to \infty} S_{n-1} = L - L = 0$$

However,

$$S_n - S_{n-1} = a_n$$

Thus,

$$\lim_{n \to \infty} (S_n - S_{n-1}) = \lim_{n \to \infty} a_n = 0$$

This contradicts the assumption that $\lim_{n \to \infty} a_n \neq 0$. Therefore the series diverges.

The conclusion that a necessary condition for convergence is that the nth term of a series tend to zero as n tends to infinity is reasonable, but we will see examples where $\lim_{n \to \infty} a_n = 0$, yet the series is divergent. The series $\sum_{i=1}^{\infty} (1/i) = 1 + \frac{1}{2} + \frac{1}{3} + \frac{1}{4} + \cdots$, called the *harmonic series*, is such an example. It will be shown to be divergent in Section 9.3.

Some examples of the use of the test follow.

Example 6   What is the $\lim_{n \to \infty} a_n$ of the convergent ($|r| < 1$) geometric series below?

$$\sum_{i=1}^{\infty} \frac{1}{2^i} \qquad \left(\text{that is,} \quad a_n = \frac{1}{2^n} \right)$$

Solution $\lim_{n \to \infty} a_n = \lim_{n \to \infty} \frac{1}{2^n} = 0$

Example 7   What is the $\lim_{n \to \infty} a_n$ of the divergent geometric series ($|r| > 1$) below?

$$\sum_{i=1}^{\infty} 2^i \qquad (\text{that is,} \quad a_n = 2^n)$$

Solution $\lim_{n \to \infty} a_n = \lim_{n \to \infty} 2^n = \infty$

Example 8   Are all arithmetic series divergent?

Solution   In any arithmetic series, that is, one formed from an arithmetic sequence $\{a_n\}$ with $a_1 = a$ and $a_{i+1} = a_i + b$, where b is a nonzero constant,

$$a_n = a + b(n - 1)$$

and

$$\lim_{n \to \infty} a_n = \lim_{n \to \infty} \{a + b(n - 1)\} = \pm\infty$$

depending on the sign of b. Thus all arithmetic series are divergent.

Example 9   Is $\displaystyle\sum_{i=1}^{\infty} \frac{i}{i + 1}$ divergent?

Solution
$$a_n = \frac{n}{n + 1}$$

and

$$\lim_{n \to \infty} a_n = \lim_{n \to \infty} \frac{n}{n + 1} = 1$$

Thus,

$$\sum_{i=1}^{\infty} a_i = \sum_{i=1}^{\infty} \frac{i}{i + 1}$$

is divergent.

Example 10   Is the series $\displaystyle\sum_{i=2}^{\infty} \frac{1}{\ln i}$ convergent?

Solution
$$a_n = \frac{1}{\ln n}$$

$$\lim_{n \to \infty} a_n = \lim_{n \to \infty} \frac{1}{\ln n} = 0$$

Thus,

$$\sum_{i=2}^{\infty} \frac{1}{\ln i}$$

may be either convergent or divergent. This test provides no conclusions about this series.

Notice that in all the preceding discussions, all these notations have the same meaning.

$$\lim_{i \to \infty} a_i = \lim_{n \to \infty} a_n = \lim_{k \to \infty} a_k \quad \text{and so forth}$$

$$\sum_{i=1}^{\infty} a_i = \sum_{n=1}^{\infty} a_n = \sum_{l=2}^{\infty} a_{l-1} = \sum_{k=7}^{\infty} a_{k-6} = \sum_{k=1}^{\infty} a_k \quad \text{and so forth}$$

This change of index is a useful device and will be used on occasion. The index of the sum, or the variable in the limit process, is a "dummy" and can be changed.

In the case of some series, we will know S_n, that is, $\sum_{i=1}^{n} a_i$, rather than a specific form for a_n. It is usually a simple matter to recover a_n, since

$$S_n = S_{n-1} + a_n \qquad \text{or} \qquad a_n = S_n - S_{n-1}$$

Example 11   Suppose $S_n = \sum_{i=1}^{n} a_i = \dfrac{n}{n + 1}$; find a_n.

Solution $a_n = S_n - S_{n-1}$

$$S_n = \frac{n}{n + 1} \qquad\qquad S_{n-1} = \frac{n - 1}{n - 1 + 1} = \frac{n - 1}{n}$$

Therefore,

$$\begin{aligned}
a_n &= \frac{n}{n + 1} - \frac{n - 1}{n} \\
&= \frac{n^2 - (n - 1)(n + 1)}{n(n + 1)} \\
&= \frac{n^2 - n^2 + 1}{n(n + 1)} = \frac{1}{n(n + 1)}
\end{aligned}$$

The infinite series is

$$\sum_{i=1}^{\infty} \frac{1}{i(i + 1)}$$

which is convergent to 1, since

$$\lim_{n\to\infty} S_n = \lim_{n\to\infty} \frac{n}{n+1} = 1$$

In summary, here are three important facts from this section:

1. The geometric series

$$\sum_{i=1}^{\infty} ar^{i-1} = a + ar + ar^2 + ar^3 + \cdots = \frac{a}{1-r}$$

converges if $|r| < 1$

2. The harmonic series

$$\sum_{i=1}^{\infty} \frac{1}{i} = 1 + \frac{1}{2} + \frac{1}{3} + \cdots \qquad \text{diverges}$$

3. If $\lim_{n\to\infty} a_n \neq 0$, then $\sum_{i=1}^{\infty} a_i$ diverges.

9.2 EXERCISES

In Exercises 1–6, for each series determine if $\lim_{n\to\infty} a_n = 0$ and draw any possible conclusions about the convergence or divergence of the series.

1. $\displaystyle\sum_{i=1}^{\infty} \frac{i}{i+1}$ 　　**2.** $\displaystyle\sum_{i=1}^{\infty} \frac{1}{\sqrt{i}}$ 　　**3.** $\displaystyle\sum_{i=1}^{\infty} \frac{1}{4^i}$

4. $\displaystyle\sum_{i=0}^{\infty} (-1)^i$ 　　**5.** $\displaystyle\sum_{i=0}^{\infty} \left(\frac{3}{2}\right)^i$ 　　**6.** $\displaystyle\sum_{i=0}^{\infty} i(-1)^i$

In Exercises 7–12, an expression has been given for $\displaystyle\sum_{i=1}^{n} a_i$ Use this expression to determine if $\displaystyle\sum_{i=1}^{\infty} a_i$ is convergent.

7. $\displaystyle\sum_{i=1}^{n} a_i = \frac{2n}{6n+1}$ 　　**8.** $\displaystyle\sum_{i=1}^{n} a_i = \frac{n^3 - 2}{n^3}$ 　　**9.** $\displaystyle\sum_{i=1}^{n} a_i = \frac{n^2}{2n+1}$

10. $\displaystyle\sum_{i=1}^{n} a_i = 2^n$ 　　**11.** $\displaystyle\sum_{i=1}^{n} a_i = -3(1 - 2^n)$ 　　**12.** $\displaystyle\sum_{i=1}^{n} a_i = \ln(n+1)$

In Exercises 13–20, an expression has been given for $S_n = \sum_{i=1}^{n} a_i$. Find the specific form for a_n.

13. $S_n = \dfrac{n^2}{n+1}$ **14.** $S_n = \dfrac{1}{2^n}$ **15.** $S_n = \dfrac{5n}{4n+1}$

16. $S_n = \dfrac{n^2}{n^2+2}$ **17.** $S_n = \dfrac{n}{n+1}$ **18.** $S_n = \ln\left(\dfrac{1}{n+1}\right)$

19. $S_n = 3^n$ **20.** $S_n = \dfrac{n+1}{n-1}$

Any repeating decimal can be written as an infinite series. For example,

$$0.23232323\ldots = 23\left(\frac{1}{100} + \frac{1}{10,000} + \frac{1}{1,000,000} + \cdots\right)$$

$$= 23\sum_{i=1}^{\infty} \frac{1}{10^{2i}} = 23\sum_{i=1}^{\infty}\left(\frac{1}{10^2}\right)^i$$

However, $\displaystyle\sum_{i=1}^{\infty}\left(\frac{1}{10^2}\right)^i$ is a convergent geometric series, which converges to

$$\frac{1/10^2}{1 - (1/10^2)} = \frac{1/10^2}{99/10^2} = \frac{1}{99}$$

Therefore,

$$23\sum_{i=1}^{\infty} \frac{1}{10^{2i}} = \frac{23}{99}$$

or

$$0.2323\ldots = \frac{23}{99}$$

In Exercises 21–26, use this method to convert the following repeating decimals to rational numbers.

21. $0.3333333\ldots$ **22.** $1.41414141\ldots$

23. $3.142857142857142857\ldots$ **24.** $-0.371371371\ldots$

25. $10.371371371\ldots$ **26.** $0.4444244424442\ldots$

27. Let R_n denote the remainder of a series if the series is truncated after n terms. That is,

$$R_n = \sum_{i=n+1}^{\infty} a_i$$

If $\displaystyle\sum_{i=1}^{\infty} a_i$ is convergent, what value must $\displaystyle\lim_{n\to\infty} R_n$ have? Justify your answer.

In Exercises 28 and 29, suppose that a ball is dropped from a height of 100 feet. Each time it hits the ground it bounces two-thirds of the height from which it fell.

28.   Find an infinite series representing the distance the ball travels as it bounces.

29.   Based on your analysis of the infinite series of Exercise 28, how far will the ball travel before it comes to rest?

In Exercises 30–33, consider two convergent infinite series

$$\sum_{i=1}^{\infty} a_i \qquad and \qquad \sum_{i=1}^{\infty} b_i$$

30.   What are $\displaystyle\lim_{n\to\infty} a_n$ and $\displaystyle\lim_{n\to\infty} b_n$?

31.   Is the series $\displaystyle\sum_{i=1}^{\infty} c_i$, where $c_i = a_i + b_i$ for all i, convergent? Justify your answer.

32.   Is the series $\displaystyle\sum_{i=1}^{\infty} c_i$, where $c_i = a_i \cdot b_i$, convergent?

33.   Is the series $\displaystyle\sum_{i=1}^{\infty} \left(\frac{a_i}{b_i}\right)$ convergent? Justify your answer.

In Exercises 34–36, consider two divergent series $\displaystyle\sum_{i=1}^{\infty} a_i$ and $\displaystyle\sum_{i=1}^{\infty} b_i$.

34.   What are $\displaystyle\lim_{n\to\infty} a_n$ and $\displaystyle\lim_{n\to\infty} b_n$?

35.   If $c_i = a_i - b_i$, is $\displaystyle\sum_{i=1}^{\infty} c_i$ convergent or divergent? Justify your answer.

36.   If $c_i = \dfrac{a_i}{b_i}$, is $\displaystyle\sum_{i=1}^{\infty} c_i$ convergent or divergent? Justify your answer.

37.   How does the convergence or divergence of $\displaystyle\sum_{i=1}^{\infty} a_i$ relate to that of $\sum\limits_{i=7}^{\infty} a_i$, $\sum\limits_{i=100,000}^{\infty} a_i$, and $\sum\limits_{i=m}^{\infty} a_i$, where m is any positive integer?

38. A car rolling down a hill moves 2 feet in the first second, 5 feet in the second second, and 8 feet in the third second. At this rate, how far will the car move in 30 seconds?

39. Mr. Jones starts a regular savings plan by depositing $1000 in the Get-Rich Savings and Loan Company. At the beginning of the second year his account is credited with $53 interest, so he deposits only $947. At the beginning of the third year he is credited with $106 and deposits $894. At the beginning of the fourth year he is credited with $159, and deposits $841. If he continues to receive interest and deposit in this manner, how much total interest will he have received by the end of the tenth year?

9.3 TESTS FOR CONVERGENCE AND DIVERGENCE

The definition of a convergent infinite series requires pre-knowledge of the limit of the series. On the other hand, many of the uses of infinite series involve the evaluation of a numerical quantity, whose decimal representation is otherwise unknown, as the limit of a convergent infinite series. To make infinite series useful it will be necessary to consider some tests which can tell if an infinite series is convergent without knowing the limit of the series.

The tests in this section will be structured for series with positive terms only. However, they will also take care of series with all negative terms, for they would then apply to the positive-termed series found by factoring out a common factor of -1. Second, it is clear that the convergence or divergence of an infinite series does not depend on the values of any finite number of finite-valued terms. If one ignores the first term, the first thousand terms, or the first million terms, the series still converges or diverges on the basis of the behavior of the remaining infinite number of terms. Further, we may assume

$$\lim_{n \to \infty} a_n = 0$$

for, from the previous section, we know that without this condition the series must be divergent.

We can construct a graphical representation of an infinite series, which can help us visualize the relationships between the tests and the series. Suppose we construct a graph of the series by constructing a series of rectangular columns. Each column will be 1 unit wide and a_i units high; the left-hand side of the column is at the integer i. Such a graph is illustrated in Figure 9.1. Clearly, if $\sum_{i=1}^{\infty} a_i$ exists, it must equal the total area of the columns.

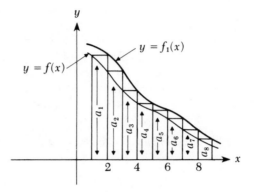

Figure 9.1 The integral test

The fact $\sum_{i=1}^{\infty} a_i$ corresponds to an area suggests that we should be able to use some features of integration, our area-finding tool, to test our series for convergence or divergence. But what do we integrate, and after we are done, what does the result tell us?

Consider an infinite series $\sum_{i=1}^{\infty} a_i$ and a function f such that $f(i) = a_i$ for all integers i. Further suppose that f is continuous for $x \geq 1$. The graph of f in Figure 9.1 compares f to the rectangular columns of $\sum_{i=1}^{\infty} a_i$. The graph of f passes through $(1, a_1), (2, a_2), \ldots,$ (i, a_i), the upper left-hand corner of each column. Now consider a second function $f_1(x) = f(x - 1)$, where f is defined the same as before. In this case, $f_1(2) = f(1) = a_1, f_1(3) = f(2) = a_2$, and so forth. That is, $f_1(i) = a_{i-1}$ for any integer $i \geq 2$. The graph of f_1 is shown in Figure 9.1 and passes through the upper right-hand corners of the columns corresponding to $\sum_{i=2}^{\infty} a_i$. Consider the two improper integrals.

$$\int_2^\infty f(x)\,dx \qquad \text{and} \qquad \int_2^\infty f_1(x)\,dx$$

Both of these exist, or fail to exist together, since f_1 is only f shifted to the right. However, $\sum\limits_{i=2}^\infty a_i$ is trapped between these integrals. That is,

$$\int_2^\infty f(x)\,dx \le \sum_{i=2}^\infty a_i \le \int_2^\infty f_1(x)\,dx$$

Therefore, it is reasonable to conclude that if the integrals converge, the series $\sum\limits_{i=2}^\infty a_i$ must also, and if $\sum\limits_{i=2}^\infty a_i$ diverges, so do the integrals. However, $\sum\limits_{i=1}^\infty a_i$ and $\sum\limits_{i=2}^\infty a_i$ differ only by a_1; therefore they must converge or diverge together. The summary of this result takes the form of the integral test for convergence and divergence.

The Integral Test

Consider an infinite series $\sum\limits_{i=1}^\infty a_i$ and a function f such that for some integer n, $f(i) = a_i$ for all integers $i \ge n$. Then $\int_n^\infty f(x)\,dx$ and $\sum\limits_{i=1}^\infty a_i$ converge or diverge together. (It should be noted that they do not necessarily converge to the same value.)

Stating that the function need only match the terms of the series from some integer n on, and need not match the series for a finite number of terms preceding the nth, is only a restatement, in an alternative form, of the fact that the series' convergence or divergence is not affected by the behavior of a finite number of terms. The preceding discussion does not constitute a formal proof, and it is only presented to make the result seem reasonable.

Example 1   Consider the convergent geometric series $\sum\limits_{i=1}^\infty (1/2^i)$. Apply the integral test to this series.

Solution   We have $a_i = 1/2^i$; therefore the appropriate function is

$$f(x) = \frac{1}{2^x} = 2^{-x}$$

The improper integral is then

$$\int_1^\infty 2^{-x}\,dx = \lim_{b\to\infty} \int_1^b 2^{-x}\,dx$$

$$= \lim_{b\to\infty} \frac{2^{-x}}{-\ln 2}\bigg]_1^b$$

$$= \lim_{b\to\infty} \left[\frac{1}{-\ln 2}\left(\frac{1}{2^b} - \frac{1}{2}\right)\right]$$

$$= \frac{1}{-\ln 2}\left(-\frac{1}{2}\right) = \frac{1}{2\ln 2}$$

The integral exists. Therefore the Integral Test verifies the convergence of the series. The series does not, however, converge to $1/(2\ln 2)$. In fact, it converges to 1.

Example 2   Consider the harmonic series $\sum_{i=1}^\infty (1/i)$. Test this for convergence using the integral test.

Solution   Let $f(x) = 1/x$. The improper integral corresponding to this series is

$$\int_1^\infty \frac{1}{x}\,dx = \lim_{b\to\infty} \int_1^b \frac{1}{x}\,dx$$

$$= \lim_{b\to\infty} [\ln x]_1^b$$

$$= \lim_{b\to\infty} [\ln b - \ln 1]$$

$$= \lim_{b\to\infty} [\ln b] = \infty$$

This harmonic series must diverge.

Example 3   Consider the series $\sum_{i=1}^\infty (1/i^p)$. This series is called a "p series." Determine the restrictions on p required to make the series convergent.

Solution   If $p = 1$, then the series is the harmonic series, which we know to be divergent. If $p \neq 1$, then we may apply the Integral Test, with $f(x) = x^{-p}$.

$$\int_1^\infty f(x)\,dx = \lim_{b\to\infty} \int_1^b x^{-p}\,dx$$

$$= \lim_{b\to\infty} \frac{x^{-p+1}}{-p+1}\bigg]_1^b$$

$$= \lim_{b\to\infty} \left[\frac{1}{1-p}(b^{-p+1} - 1)\right]$$

Whether or not this limit exists depends on the sign of $-p + 1$. If $-p + 1 < 0$, that is, $p > 1$, then

$$\lim_{b \to \infty} (b^{-p+1}) = 0$$

while if $-p + 1 > 0$, that is, $p < 1$,

$$\lim_{b \to \infty} (b^{-p+1}) = \infty$$

Correspondingly, if $p > 1$, then the integral exists and $\sum_{i=1}^{\infty} (1/i^p)$ converges, and if $p \leq 1$, the integral fails to exist and $\sum_{i=1}^{\infty} (1/i^p)$ diverges. In conclusion, $\sum_{i=1}^{\infty} (1/i^p)$ converges if $p > 1$ and diverges if $p \leq 1$.

Although the Integral Test compares the series in question with an integral, it is just as reasonable to construct a test that compares an unknown series with a known series.

Comparison Test I

Consider an infinite series $\sum_{i=1}^{\infty} a_i$ with $a_i \geq 0$ and $a_{i+1} \leq a_i$. Suppose that there exists a second series $\sum_{i=1}^{\infty} b_i$ such that $b_i \geq a_i$ for all integers i. Then if $\sum_{i=1}^{\infty} b_i$ converges, so does $\sum_{i=1}^{\infty} a_i$.

A graph similar to Figure 9.1 can be used to convince us of the validity of the test. Consider Figure 9.2; $\sum_{i=1}^{\infty} b_i$ represents the total area of the b columns, whereas $\sum_{i=1}^{\infty} a_i$ represents that of the a columns, $a_i \leq b_i$. If the total area of the b columns is finite, that is, if $\sum_{i=1}^{\infty} b_i$ converges, then the total area of the a columns must also be finite; that is, $\sum_{i=1}^{\infty} a_i$ converges.

A similar figure would lead to the conclusion that a comparison can be made to prove divergence. The construction of the figure is left as an exercise.

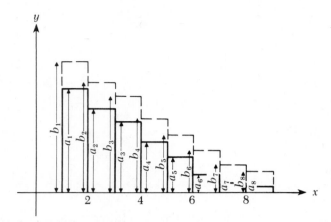

Figure 9.2   Comparison Test I

Comparison Test II

Consider an infinite series $\sum\limits_{i=1}^{\infty} a_i$. Suppose that there exists another infinite series $\sum\limits_{i=1}^{\infty} b_i$ with $b_i \le a_i$ for all integers i. Then if $\sum\limits_{i=1}^{\infty} b_i$ diverges, so does $\sum\limits_{i=1}^{\infty} a_i$.

The key to the use of the comparison tests is that one must decide which way a series goes before deciding which test to use. If the series tested seems to be convergent, then one tries to find a known convergent series which is term by term larger than the series to be tested. If one feels the unknown series is divergent, then one sets out to find a known divergent series that is term by term smaller than the series being tested. If you make the wrong guess, nothing results.

Example 4   Use the comparison tests to determine the convergence of the series $\sum\limits_{i=3}^{\infty} (1/\ln i)$.

Solution   For all integers $i \ge 3$, $\ln i < i$. Therefore, $1/\ln i > 1/i$. Since $\sum\limits_{i=3}^{\infty} (1/i)$ is part of the harmonic series, it is known to be divergent. Therefore $\sum\limits_{i=3}^{\infty} (1/\ln i)$ is term by term larger than a known divergent series, and thus it is divergent.

Notice we had to guess how the series behaved. Such guesses are usually based on a feel for what the series looks like.

Example 5   Test $\displaystyle\sum_{i=1}^{\infty} \frac{1}{(i + 1)(i + 2)}$ for convergence.

Solution $\displaystyle\frac{1}{(i + 1)(i + 2)} = \frac{1}{i^2 + 3i + 2}$

It is reasonable to assume that for i large enough, this behaves like $1/i^2$. Therefore we will try to compare this series with $\displaystyle\sum_{i=1}^{\infty} (1/i^2)$, which is a p series with $p = 2 > 1$. Therefore it is convergent. Fortunately,

$$i^2 + 3i + 2 > i^2$$

thus

$$\frac{1}{i^2 + 3i + 2} < \frac{1}{i^2}$$

and $\displaystyle\sum_{i=1}^{\infty} \frac{1}{(i + 1)(i + 2)}$ is term by term smaller than a known convergent series, and must be convergent.

In summary, here are the important *series* that have been discussed so far:

1. The geometric series $\displaystyle\sum_{i=1}^{\infty} ar^{i-1}$ converges to $a/(1 - r)$ if $|r| < 1$; otherwise it diverges.

2. The harmonic series $\displaystyle\sum_{i=1}^{\infty} (1/i)$ diverges.

3. The p series $\displaystyle\sum_{i=1}^{\infty} (1/i^p)$ converges if $p > 1$ and diverges if $p \leq 1$.

Here are the *tests* that are available to test a series for *convergence* or *divergence*:

1. If $\displaystyle\lim_{n \to \infty} a_n \neq 0$, then $\displaystyle\sum_{i=1}^{\infty} a_i$ diverges.

2. In the Integral Test, $\sum_{i=1}^{\infty} f(i)$ and $\int_n^{\infty} f(x)\, dx$ converge or diverge together.

3. In the comparison tests:

    I. If $a_i \leqslant b_i$ for all i and $\sum_{i=1}^{\infty} b_i$ converges, then $\sum_{i=1}^{\infty} a_i$ converges.

    II. If $a_i \geqslant b_i$ for all i and $\sum_{i=1}^{\infty} b_i$ diverges, then $\sum_{i=1}^{\infty} a_i$ diverges.

9.3   EXERCISES

1. Construct a figure similar to Figure 9.2 to justify the comparison test for divergence.

In Exercises 2–7, establish the convergence or divergence of each series using the Integral Test.

2. $\displaystyle\sum_{i=1}^{\infty} \frac{1}{i^2 + 1}$

3. $\displaystyle\sum_{i=1}^{\infty} \frac{i}{(i^2 + 1)^{3/2}}$

4. $\displaystyle\sum_{i=1}^{\infty} \frac{i}{e^{2i}}$

5. $\displaystyle\sum_{i=1}^{\infty} \frac{1}{\sqrt{i^2 + 1}}$

6. $\displaystyle\sum_{i=1}^{\infty} \frac{1}{(i + 1)(i + 2)}$

7. $\displaystyle\sum_{i=2}^{\infty} \frac{1}{i(\ln i)^2}$

In Exercises 8–13, establish the convergence or divergence of each series using a comparison test.

8. $\displaystyle\sum_{i=1}^{\infty} \frac{1}{(i + 1)(i + 2)}$

9. $\displaystyle\sum_{i=3}^{\infty} \frac{i}{(i - 1)(i - 2)}$

10. $\displaystyle\sum_{i=1}^{\infty} \frac{1}{3^i + 2}$

11. $\displaystyle\sum_{i=1}^{\infty} \frac{1}{3^i - 2}$

12. $\displaystyle\sum_{i=2}^{\infty} \frac{1}{i - 1}$

13. $\displaystyle\sum_{i=1}^{\infty} \frac{3}{2^i + 1}$

In Exercises 14–25, determine if each series is convergent or divergent.

14. $\displaystyle\sum_{i=1}^{\infty} \frac{1}{(i + 1) \ln (i + 1)}$

15. $\displaystyle\sum_{i=1}^{\infty} \frac{1}{\sqrt{i^2 + 2i + 2}}$

16. $\displaystyle\sum_{i=1}^{\infty} \frac{1}{(3i)^2}$

17. $\displaystyle\sum_{i=1}^{\infty} \frac{1}{(3i)^{1/2}}$

18. $\displaystyle\sum_{i=1}^{\infty} \frac{2 + \cos i}{i^2}$

19. $\displaystyle\sum_{i=1}^{\infty} \frac{\ln i}{i}$

20. $\displaystyle\sum_{i=1}^{\infty} \frac{i}{(i + 1)^2}$

21. $\displaystyle\sum_{i=1}^{\infty} \frac{i}{(i + 1)^3}$

22. $\displaystyle\sum_{i=1}^{\infty} \frac{i}{i + 1}$

23. $\displaystyle\sum_{i=1}^{\infty} \frac{i + 1}{(i + 1)3^i}$

24. $\displaystyle\sum_{i=1}^{\infty} i^2 \sin (\pi i)$

25. $\displaystyle\sum_{i=1}^{\infty} \frac{(-1)^i}{\cos [\pi(i + 1)]}$

26. Must two series be in the proper relationship for every term of both series when using the comparison tests? Modify the statements of the comparison tests to reflect your answer.

In Exercises 27–30, note that an alternate form of the comparison test is stated as follows: If $\displaystyle\sum_{i=1}^{\infty} a_i$ and $\displaystyle\sum_{i=1}^{\infty} b_i$ are two positive-termed series, and if

$$\lim_{n\to\infty} \frac{a_n}{b_n} = L \qquad (L > 0 \quad and \quad L\ finite)$$

then

$$\sum_{i=1}^{\infty} a_i \qquad and \qquad \sum_{i=1}^{\infty} b_i$$

both converge or both diverge. Use this version to test the following series.

27. $\displaystyle\sum_{i=1}^{\infty} \frac{i}{i^2 + 1}$

28. $\displaystyle\sum_{i=1}^{\infty} \frac{1}{\sqrt{i^2 + 1}}$

29. $\displaystyle\sum_{i=1}^{\infty} \frac{1}{i^2 + 2i + 3}$

30. $\displaystyle\sum_{i=1}^{\infty} \frac{i^7 - 2i^3}{i^8 + 2i^7 - 4}$

9.4 THE RATIO TEST AND L'HÔSPITAL'S RULES

The integral test and the comparison test relate the terms of a series to other quantities outside of the original series. The Ratio Test analyzes the convergence or divergence of a series by examining how consecutive terms of the series itself relate to each other.

The Ratio Test

The series $\displaystyle\sum_{i=1}^{\infty} a_i$ is convergent (assuming $a_i \geqslant 0$) if

$$\lim_{n \to \infty} \frac{a_{n+1}}{a_n} < 1$$

The series is divergent if

$$\lim_{n \to \infty} \frac{a_{n+1}}{a_n} > 1$$

If

$$\lim_{n \to \infty} \frac{a_{n+1}}{a_n} = 1$$

the Ratio Test fails, that is, no conclusion can be drawn.

To establish this result, consider

$$\lim_{n \to \infty} \frac{a_{n+1}}{a_n} = L$$

Now suppose that $L < 1$. Then there is a number r such that $L < r < 1$, and further, if n is large enough, say greater than some fixed number N, we can be sure that

$$\frac{a_{n+1}}{a_n} < r$$

Thus,

$$a_{n+1} < ra_n$$
$$a_{n+2} < ra_{n+1} < r(ra_n) = r^2 a_n$$
$$a_{n+3} < ra_{n+2} < r(r^2 a_n) = r^3 a_n$$

Then the series $\displaystyle\sum_{i=n+1}^{\infty} a_i$ is term by term less than $\displaystyle\sum_{i=2}^{\infty} r^{i-1} a_n$. Since $\displaystyle\sum_{i=2}^{\infty} r^{i-1} a_n$ is a geometric series, it is convergent when $r < 1$. Thus $\displaystyle\sum_{i=n+1}^{\infty} a_i$ is convergent by the comparison test. If $\displaystyle\sum_{i=n+1}^{\infty} a_i$ converges, so does $\sum_{i=1}^{\infty} a_i$, since they differ only by a finite number of terms. Similarly, if $\lim_{n\to\infty} (a_{n+1}/a_n) = L$ and $L > 1$, the comparison test can be used to establish that the series is divergent. This approach also makes it clear why the Ratio Test fails if $\lim_{n\to\infty} (a_{n+1}/a_n) = 1$. The application of the Ratio Test is simpler than that of the comparison test as we don't have to guess the result in advance.

Example 1   Apply the Ratio Test to the convergent geometric series $\sum_{i=1}^{\infty} (1/3^i)$.

Solution $a_n = \dfrac{1}{3^n}$ and $a_{n+1} = \dfrac{1}{3^{n+1}}$

Thus,

$$\lim_{n\to\infty} \frac{a_{n+1}}{a_n} = \lim_{n\to\infty} \frac{1/3^{n+1}}{1/3^n}$$
$$= \lim_{n\to\infty} \frac{3^n}{3^{n+1}}$$
$$= \lim_{n\to\infty} \frac{1}{3} = \frac{1}{3}$$

Since $\frac{1}{3} < 1$, the Ratio Test indicates that $\sum_{i=1}^{\infty} (1/3^i)$ converges.

Example 2   Consider the series

$$\frac{1}{1} + \frac{1}{2(1)} + \frac{1}{3(2)(1)} + \frac{1}{4(3)(2)(1)} +$$
$$\cdots + \frac{1}{n(n-1)(n-2)\cdots(3)(2)(1)} + \cdots$$

or

$$\sum_{i=1}^{\infty} \frac{1}{i(i-1)\cdots(3)(2)(1)}$$

Test this series for convergence using the Ratio Test.

Solution $a_n = \dfrac{1}{n(n-1)(n-2)\cdots(3)(2)(1)}$

$$a_{n+1} = \frac{1}{(n+1)(n)(n-1)\cdots(3)(2)(1)}$$

$$\frac{a_{n+1}}{a_n} = \frac{\dfrac{1}{(n+1)(n)(n-1)\cdots(3)(2)(1)}}{\dfrac{1}{n(n-1)(n-2)\cdots(3)(2)(1)}}$$

$$= \frac{n(n-1)(n-2)\cdots(3)(2)(1)}{(n+1)(n)(n-1)\cdots(3)(2)(1)}$$

$$= \frac{1}{n+1}$$

Thus,

$$\lim_{n\to\infty} \frac{a_{n+1}}{a_n} = \lim_{n\to\infty} \frac{1}{n+1} = 0 < 1$$

This series has been shown to be convergent by use of the Ratio Test.

Factors like $n(n-1)(n-2)\cdots(3)(2)(1)$ are sufficiently common and useful to warrant a special notation. We use $n!$ (read "n factorial") to represent $n(n-1)\cdots(3)(2)(1)$ for $n \geq 2$ and n an integer. In general,

$$n! = n(n-1)! = n(n-1)(n-2)! \qquad \text{and so forth}$$
$$(n+1)! = (n+1)n!$$

The definition of $n!$ does not reasonably apply if $n = 0$ or $n = 1$. These are defined as special cases, just as $a^0 = 1$ is defined as a special case.

Definition

$$0! = 1! = 1$$

Using factorials in the previous series example we have:

Example 3   Test $\displaystyle\sum_{i=1}^{\infty} (1/i!)$ for convergence.

Solution
$$a_n = \frac{1}{n!}$$

$$a_{n+1} = \frac{1}{(n+1)!}$$

$$\frac{a_{n+1}}{a_n} = \frac{1/(n+1)!}{1/n!} = \frac{n!}{(n+1)!}$$

$$= \frac{n!}{(n+1)n!} = \frac{1}{n+1}$$

$$\lim_{n \to \infty} \frac{a_{n+1}}{a_n} = \lim_{n \to \infty} \frac{1}{n+1} = 0$$

Thus by the Ratio Test the series is convergent.

Example 4

Test the series $\sum_{i=1}^{\infty} (2^i/i!)$ for convergence.

Solution
$$a_n = \frac{2^n}{n!}$$

$$a_{n+1} = \frac{2^{n+1}}{(n+1)!}$$

$$\frac{a_{n+1}}{a_n} = \frac{2^{n+1}/(n+1)!}{2^n/n!} = \frac{2^{n+1}}{2^n} \cdot \frac{n!}{(n+1)!} = \frac{2}{n+1}$$

$$\lim_{n \to \infty} \frac{a_{n+1}}{a_n} = \lim_{n \to \infty} \frac{2}{n+1} = 0 < 1$$

Thus, the series has been shown to be convergent by the Ratio Test.

To aid in finding limits such as the one in Example 4, two very useful rules involving just such cases were developed by L'Hôspital. Guillaume Francois Marquis de L'Hôspital was a French disciple of Leibniz and a student of one of the great names in the development of calculus, Jacob Bernoulli. L'Hôspital was also the author of the first published textbook on differential calculus, *Analyse des infiniments petits pour l'intelligence des lignes courbes.* The following are L'Hôspital's rules.

I. Let f and g be two functions such that

$$\lim_{x \to a} f(x) = \infty \qquad \text{and} \qquad \lim_{x \to a} g(x) = \infty$$

Then

$$\lim_{x \to a} \frac{f(x)}{g(x)} = \lim_{x \to a} \frac{f'(x)}{g'(x)}$$

where a may be either finite or infinite.

II. Let f and g be two functions such that

$$\lim_{x \to a} f(x) = 0 \qquad \text{and} \qquad \lim_{x \to a} g(x) = 0$$

Then

$$\lim_{x \to a} \frac{f(x)}{g(x)} = \lim_{x \to a} \frac{f'(x)}{g'(x)}$$

where a may be either finite or infinite.

In both cases it is assumed that f and g are differentiable functions over the values in question except possibly at $x = a$.

These rules of L'Hôspital, which are really theorems, are stated without proof. However, a simple proof of the following restricted form is given.

Proof   Suppose $f(a) = g(a) = 0$. Then

$$\lim_{x \to a} \frac{f(x)}{g(x)} = \lim_{x \to a} \frac{f(x) - f(a)}{g(x) - g(a)} = \lim_{x \to a} \frac{\dfrac{f(x) - f(a)}{x - a}}{\dfrac{g(x) - g(a)}{x - a}} = \frac{f'(a)}{g'(a)}$$

Example 5   Find $\lim\limits_{x \to 2} \dfrac{x^2 - 4}{x - 2}$.

Solution   Without L'Hôspital's rule:

$$\lim_{x \to 2} \frac{x^2 - 4}{x - 2} = \lim_{x \to 2} \frac{(x - 2)(x + 2)}{(x - 2)} = 4$$

With L'Hôspital's rule:

$$\lim_{x \to 2} \frac{x^2 - 4}{x - 2}$$

is of the required form:

$$\lim_{x \to 2} (x^2 - 4) = 0 = \lim_{x \to 2} (x - 2)$$

Therefore since $\dfrac{d}{dx} (x^2 - 4) = 2x$ and $\dfrac{d}{dx} (x - 2) = 1$,

$$\lim_{x \to 2} \frac{x^2 - 4}{x - 2} = \lim_{x \to 2} \frac{2x}{1} = 4$$

Example 6   Find $\lim\limits_{x \to 0} (\sin x)/x$.

Solution   Since $\lim\limits_{x \to 0} \sin x = 0$, $\lim\limits_{x \to 0} x = 0$, and

$$\frac{d}{dx} \sin x = \cos x \qquad \frac{d}{dx} x = 1 \qquad \lim_{x \to 0} \cos x = 1$$

we have

$$\lim_{x \to 0} \frac{\sin x}{x} = \lim_{x \to 0} \frac{\cos x}{1} = 1$$

Example 7   Test $\displaystyle\sum_{i=1}^{\infty} \frac{2^i}{\ln i^2}$ for convergence.

Solution $a_n = \dfrac{2^n}{\ln n^2} = \dfrac{2^n}{2 \ln n} = \dfrac{2^{n-1}}{\ln n}$

$a_{n+1} = \dfrac{2^n}{\ln (n + 1)}$

$\dfrac{a_{n+1}}{a_n} = \dfrac{2^n/\ln (n + 1)}{2^{n-1}/\ln n} = \dfrac{2^n}{2^{n-1}} \left(\dfrac{\ln n}{\ln (n + 1)} \right) = 2 \dfrac{\ln n}{\ln (n + 1)}$

However,

$$\lim_{n \to \infty} \frac{a_{n+1}}{a_n} = \lim_{n \to \infty} 2 \frac{\ln n}{\ln (n + 1)} = ?$$

Using L'Hôspital's rule,

$$\lim_{x \to \infty} 2 \frac{\ln x}{\ln (x + 1)} = \lim_{x \to \infty} 2 \frac{1/x}{1/(x + 1)}$$

since

$$\frac{d}{dx} \ln x = \frac{1}{x} \qquad \frac{d}{dx} \ln (x + 1) = \frac{1}{x + 1}$$

and

$$\lim_{x \to \infty} \ln x = \lim_{x \to \infty} \ln (x + 1) = \infty$$

Therefore,

$$\lim_{x \to \infty} 2 \frac{\ln x}{\ln (x + 1)} = \lim_{x \to \infty} 2 \frac{x + 1}{x} = \lim_{x \to \infty} 2 \frac{1}{1} = 2$$

since

$$\frac{d}{dx} (x + 1) = 1 \qquad \frac{d}{dx} x = 1$$

and

$$\lim_{x \to \infty} (x + 1) = \infty = \lim_{x \to \infty} x$$

Here L'Hôspital's rules have been applied twice. Since

$$\lim_{n \to \infty} \frac{a_{n+1}}{a_n} = 2$$

we conclude that $\displaystyle\sum_{i=1}^{\infty} \frac{2^i}{\ln i^2}$ is divergent.

Example 8   Use the Ratio Test to test the harmonic series $\displaystyle\sum_{i=1}^{\infty} (1/i)$.

Solution $a_n = 1/n \qquad a_{n+1} = 1/(n + 1)$

$$\lim_{n \to \infty} \frac{a_{n+1}}{a_n} = \lim_{n \to \infty} \frac{1/(n + 1)}{1/n} = \lim_{n \to \infty} \frac{n}{n + 1} = \lim_{n \to \infty} \frac{1}{1} = 1$$

The Ratio Test fails to draw any conclusion.

Example 9   Use the Ratio Test on the series $\displaystyle\sum_{i=1}^{\infty} (1/i^2)$, a convergent p series.

Solution $a_n = \dfrac{1}{n^2}$ $a_{n+1} = \dfrac{1}{(n+1)^2}$

$$\lim_{n \to \infty} \frac{a_{n+1}}{a_n} = \lim_{n \to \infty} \frac{1/(n+1)^2}{1/n^2} = \lim_{n \to \infty} \frac{n^2}{(n+1)^2}$$

$$= \lim_{n \to \infty} \frac{2n}{2(n+1)} = \lim_{n \to \infty} \frac{2}{2} = 1$$

Again the Ratio Test fails.

From the latter two examples, it is clear that with either a convergent or a divergent series the Ratio Test can fail. Also, in both cases, L'Hôspital's rules were used in the evaluation of the limit. The alert reader may have noticed a conceptual problem when applying L'Hôspital's rules in these cases; $(d/dn)n^2$ is not defined, since we are considering n^2 defined only for integer values for n. However, x^2, x a real number, is defined and takes the same values as n^2 when $x = n$. Further, $f(x) = x^2$ is differentiable for all real numbers, not just integers. L'Hôspital's rules can be applied by treating expressions like n^2 or $\ln n$ as functions of a continuous real variable. That is, we may treat the factors as differentiable functions and differentiate using the usual rules. This practice can be formally established in a more complete presentation.

9.4 EXERCISES

In Exercises 1–10, find the indicated limits.

1. $\displaystyle\lim_{x \to 3} \frac{x^2 - 9}{x - 3}$

2. $\displaystyle\lim_{x \to -2} \frac{x^3 + 6x^2 + 12x + 8}{x + 2}$

3. $\displaystyle\lim_{x \to \infty} \frac{2^x}{x^2}$

4. $\displaystyle\lim_{x \to \infty} \frac{x^{12}}{e^x}$

5. $\displaystyle\lim_{x \to \infty} \frac{\ln x}{x}$

6. $\displaystyle\lim_{x \to \infty} \frac{x}{\ln x}$

7. $\displaystyle\lim_{x \to \infty} \frac{\sin(1/x)}{\tan^{-1}(1/x)}$

8. $\displaystyle\lim_{x \to \infty} \frac{\ln(e^{2x}/x)}{\sqrt{x}}$

9. $\displaystyle\lim_{x \to \infty} \frac{1 + \cos 2x}{1 - \sin x}$

10. $\displaystyle\lim_{x \to 3} \frac{x^3 - 3x^2 + 9x - 27}{x - 3}$

In Exercises 11–25, use the Ratio Test to determine if the series are convergent.

11. $\displaystyle\sum_{i=1}^{\infty} \frac{i}{2^i}$

12. $\displaystyle\sum_{i=2}^{\infty} \frac{1}{\ln i^2}$

13. $\displaystyle\sum_{i=1}^{\infty} \frac{i+2}{i(i+1)}$

14. $\displaystyle\sum_{i=1}^{\infty} \sin\left(\frac{\pi}{i}\right)$

15. $\displaystyle\sum_{i=1}^{\infty} \frac{2^i}{i!}$

16. $\displaystyle\sum_{i=1}^{\infty} \frac{1(3)(5)(7)(9)\cdots(2i-1)}{2(4)(6)(8)(10)\cdots(2i)}$

17. $\displaystyle\sum_{i=1}^{\infty} \frac{i!}{2^i}$

18. $\displaystyle\sum_{i=1}^{\infty} \frac{e^i}{i^3}$

19. $\displaystyle\sum_{i=1}^{\infty} \frac{e^{-1}}{i^2}$

20. $\displaystyle\sum_{i=1}^{\infty} \frac{\ln i}{i^3}$

21. $\displaystyle\sum_{i=1}^{\infty} \frac{1}{3^{i-1}}$

22. $\displaystyle\sum_{i=1}^{\infty} \frac{i}{i^2-1}$

23. $\displaystyle\sum_{i=1}^{\infty} \frac{i-1}{i!}$

24. $\displaystyle\sum_{i=1}^{\infty} \frac{i!}{(2i)!}$

25. $\displaystyle\sum_{i=1}^{\infty} \frac{i}{2(i!)}$

26. Suppose $\displaystyle\sum_{i=1}^{\infty} a_i$ can be shown to be convergent by the Ratio Test. Does this mean $\displaystyle\sum_{i=1}^{\infty} (a_i)^2$ is convergent? Justify your answer.

27. Show that $\displaystyle\sum_{i=1}^{\infty} a_i$ may converge, but $\displaystyle\sum_{i=1}^{\infty} \sqrt{a_i}$ may still diverge.

9.5   ALTERNATING SERIES AND ABSOLUTE CONVERGENCE

All of the series considered so far have included only terms with the same sign. The simplest kind of series with terms which differ in algebraic sign is one where the terms alternate in sign. Such series are called *alternating series*.

Example 1

$$\sum_{i=1}^{\infty} \frac{(-1)^{i+1}}{i} = 1 - \frac{1}{2} + \frac{1}{3} - \frac{1}{4} + \frac{1}{5} - \frac{1}{6} + \cdots$$

$$\sum_{i=1}^{\infty} \frac{(-1)^{i+1}}{i^2} = 1 - \frac{1}{4} + \frac{1}{9} - \frac{1}{16} + \cdots$$

$$\sum_{i=1}^{\infty} \frac{(-1)^i}{i!} = \frac{-1}{1} + \frac{1}{2(1)} - \frac{1}{3(2)(1)} + \frac{1}{4(3)(2)(1)} - \cdots$$

The convergence of an alternating series is a much simpler matter to examine.

Test for the Convergence of Alternating Series

Consider the alternating series $\sum_{i=1}^{\infty} (-1)^{i+1}a_i$, where $a_i \geq 0$ for all i. This series converges if $a_{i+1} < a_i$ for all i, and $\lim_{n\to\infty} a_n = 0$.

To establish this test, consider a series meeting the conditions of the test:

$$\sum_{i=1}^{\infty} (-1)^{i+1}a_i \qquad \text{with} \quad \lim_{n\to\infty} a_n = 0 \quad \text{and} \quad a_{i+1} < a_i \quad \text{for all } i$$

Then

$$\sum_{i=1}^{2n} (-1)^{i+1}a_i = a_1 - a_2 + a_3 - a_4 + \cdots - a_{2n}$$
$$= (a_1 - a_2) + (a_3 - a_4) + \cdots + (a_{2n-1} - a_{2n})$$

Each term in parentheses is positive because of the condition that $a_{i+1} < a_i$. Hence, the sum is positive.

On the other hand,

$$\sum_{i=1}^{2n} (-1)^{i+1}a_i = a_1 - (a_2 - a_3) - (a_4 - a_5) - (a_6 - a_7) - \cdots - (a_{2n})$$

Thus, although $\sum_{i=1}^{2n} (-1)^{i+1}a_i$ is positive, it is less than a_1, since it has been shown to be a_1 minus a number of positive terms.

Consider the fact that

$$\sum_{i=1}^{2n} (-1)^{i+1}a_i < \sum_{i=1}^{2n+2} (-1)^{i+1}a_i$$

because

$$\sum_{i=1}^{2n+2} (-1)^{i+1}a_i = \sum_{i=1}^{2n} (-1)^{i+1}a_i + (a_{2n+1} - a_{2n+2})$$

and $a_{2n+1} - a_{2n+2}$ is positive.

$$\sum_{i=1}^{2n} (-1)^{i+1}a_i < \sum_{i=1}^{2n+2} (-1)^{i+1}a_i < \sum_{i=1}^{2n+4} (-1)^{i+1}a_i < \cdots < a_1$$

So

$$\lim_{n\to\infty} \sum_{i=1}^{2n} (-1)^{i+1} a_i$$

gets bigger as n tends to infinity but it is always less than a_1, the first term. Therefore we can conclude that

$$\lim_{n\to\infty} \sum_{i=1}^{2n} (-1)^{i+1} a_i$$

exists, hence

$$\lim_{n\to\infty} \sum_{i=1}^{n} (-1)^{i+1} a_i$$

exists, and the series converges.

Example 2   Does the alternating harmonic series $\displaystyle\sum_{i=1}^{\infty} \frac{(-1)^{i+1}}{i}$ converge?

Solution   Since $a_n = 1/n$, $a_{n+1} = 1/(n + 1)$, and $1/(n + 1) < 1/n$ for all n, and $\lim_{n\to\infty} a_n = \lim_{n\to\infty} (1/n) = 0$, the series is convergent.

Example 3   Does the alternating p series $\displaystyle\sum_{i=1}^{\infty} \frac{(-1)^{i+1}}{i^2}$ converge?

Solution   Here $a_n = 1/n^2$, $a_{n+1} = 1/(n + 1)^2$, and $1/(n + 1)^2 < 1/n^2$ for all positive integers n. Also,

$$\lim_{n\to\infty} a_n = \lim_{n\to\infty} \frac{1}{n^2} = 0$$

and the series is convergent.

The last two examples illustrate an interesting feature of alternating series. Associated with the two convergent alternating series are two regular series found by leaving out the alternating feature. These are

$$\sum_{i=1}^{\infty} \frac{1}{i} \qquad \text{and} \qquad \sum_{i=1}^{\infty} \frac{1}{i^2}$$

The first, $\sum\limits_{i=1}^{\infty} (1/i)$, is the harmonic series, and is divergent. The second, $\sum\limits_{i=1}^{\infty} (1/i^2)$, is a p series with $p > 1$, and is convergent. It would seem that without the alternating feature, some convergent alternating series would fail to converge. For that reason convergent alternating series are classified according to whether or not they require the alternating feature for their convergence.

Definition

Consider a convergent alternating series $\sum\limits_{i=1}^{\infty} (-1)^{i+1} a_i$,   with $a_i \geq 0$ for all i. Associated with this series is a second positive termed series, $\sum\limits_{i=1}^{\infty} a_i$. If $\sum\limits_{i=1}^{\infty} a_i$ converges, then $\sum\limits_{i=1}^{\infty} (-1)^{i+1} a_i$ is said to be *absolutely convergent*. If $\sum\limits_{i=1}^{\infty} a_i$ diverges, then $\sum\limits_{i=1}^{\infty} (-1)^{i+1} a_i$ is said to be *conditionally convergent*.

In other words, if a convergent alternating series does not converge without the alternating feature, it is called a conditionally convergent series. If it would still converge without the alternating signs, it is called absolutely convergent.

Example 4 $\sum\limits_{i=1}^{\infty} \dfrac{(-1)^{i+1}}{i^2}$ is absolutely convergent, because $\sum\limits_{i=1}^{\infty} \dfrac{1}{i^2}$ is convergent.

Example 5 $\sum\limits_{i=1}^{\infty} \dfrac{(-1)^{i+1}}{i}$ is conditionally convergent, because $\sum\limits_{i=1}^{\infty} \dfrac{1}{i}$ is divergent.

Fortunately, if an alternating series is absolutely convergent, it is convergent. This means that a possible first step in testing an alternating series for convergence is to test the associated positive term series for convergence.

Example 6   Check the convergence of $\sum\limits_{i=1}^{\infty} \dfrac{(-1)^{i+1}}{i 3^i}$.

Solution $a_n = \dfrac{1}{n3^n}$ and $\displaystyle\lim_{n\to\infty} a_n = \lim_{n\to\infty} \dfrac{1}{n3^n} = 0$

Using the Ratio Test,

$$a_{n+1} = \frac{1}{(n+1)3^{n+1}}$$

$$\frac{a_{n+1}}{a_n} = \frac{1/(n+1)3^{n+1}}{1/n3^n}$$

$$= \frac{n3^n}{(n+1)3^{n+1}} = \frac{n}{(n+1)(3)}$$

$$\lim_{n\to\infty} \frac{a_{n+1}}{a_n} = \lim_{n\to\infty} \frac{n}{3(n+1)} = \frac{1}{3} < 1$$

Therefore, $\displaystyle\sum_{i=1}^{\infty} (1/i3^i)$ is convergent; thus

$$\sum_{i=1}^{\infty} \frac{(-1)^{i+1}}{i3^i}$$

is absolutely convergent.

Example 7   Test the series $\displaystyle\sum_{i=2}^{\infty} \frac{(-1)^i}{i \ln i}$ for convergence.

Solution $a_n = \dfrac{1}{n \ln n}$

$$\lim_{n\to\infty} a_n = \lim_{n\to\infty} \frac{1}{n \ln n} = 0$$

$$a_{n+1} = \frac{1}{(n+1) \ln (n+1)}$$

$$\frac{a_{n+1}}{a_n} = \frac{1/[(n+1) \ln (n+1)]}{1/(n \ln n)} = \frac{n \ln n}{(n+1) \ln (n+1)}$$

$$\lim_{n\to\infty} \frac{a_{n+1}}{a_n} = \lim_{n\to\infty} \frac{n \ln n}{(n+1) \ln (n+1)}$$

$$= \lim_{n\to\infty} \left(\frac{n}{n+1}\right) \cdot \lim_{n\to\infty} \frac{\ln n}{\ln (n+1)}$$

$$= \lim_{n\to\infty} \frac{1}{1} \cdot \lim_{n\to\infty} \frac{1/n}{1/(n+1)}$$

$$= \lim_{n\to\infty} \frac{n+1}{n} = \lim_{n\to\infty} \frac{1}{1} = 1$$

The Ratio Test fails! Perhaps the Integral Test will work: $f(x) = 1/(x \ln x)$, and the improper integral is

$$\int_2^\infty \frac{1}{x \ln x} \, dx = \lim_{b \to \infty} \int_2^b \frac{1}{x \ln x} \, dx$$

If $u = \ln x$, then $du = (1/x) \, dx$. Therefore,

$$\int \frac{1}{x \ln x} \, dx = \int \frac{1}{u} \, du = \ln u$$

Thus,

$$\lim_{b \to \infty} \int_2^b \frac{1}{x \ln x} \, dx = \lim_{b \to \infty} \left\{ \ln (\ln x) \, \Big]_2^b \right\}$$
$$= \lim_{b \to \infty} [\ln (\ln b) - \ln (\ln 2)] = \infty$$

The integral, and thus the series, diverges. However,

$$\frac{1}{(i + 1) \ln (i + 1)} < \frac{1}{i \ln i}$$

Thus $\sum_{i=2}^\infty \frac{(-1)^i}{i \ln i}$ is conditionally convergent.

9.5 EXERCISES

In Exercises 1–20, test each alternating series for convergence and absolute convergence.

1. $\sum_{i=1}^\infty \frac{(-1)^{i+1}}{i^2 + 1}$

2. $\sum_{i=1}^\infty \frac{(-1)^{i+1} e^i}{i}$

3. $\sum_{i=1}^\infty (-1)^{i+1} \frac{\ln i}{i}$

4. $\sum_{i=1}^\infty \frac{(-1)^{i+1}}{2^{2i}}$

5. $\sum_{i=1}^\infty \frac{(-1)^i}{i!}$

6. $\sum_{i=1}^\infty (-1)^{i+1}$

7. $\sum_{i=1}^\infty \frac{(-1)^i}{i3^i}$

8. $\sum_{i=1}^\infty \frac{(-4)^{i+1}}{i^4}$

9. $\displaystyle\sum_{i=1}^{\infty} \frac{(-1)^{i+1}}{(2i-1)!}$

10. $\displaystyle\sum_{i=1}^{\infty} (-1)^{i+1} \cdot \frac{i^2+1}{i^3}$

11. $\displaystyle\sum_{i=1}^{\infty} (-1)^{i+1} i \left(\frac{2}{3}\right)^i$

12. $\displaystyle\sum_{i=1}^{\infty} \frac{(-1)^i 7^i}{2(4)(6) \cdots (2i)}$

13. $\displaystyle\sum_{i=1}^{\infty} (-1)^{i+1} \frac{7^{i+2}}{5^{i+1}}$

14. $\displaystyle\sum_{i=1}^{\infty} \frac{(-1)^{i+1} 9^{i+4}}{10^{i+3}}$

15. $\displaystyle\sum_{i=1}^{\infty} \frac{(-1)^{i+1}(3+2i)}{7^i}$

16. $\displaystyle\sum_{i=1}^{\infty} \frac{(-1)^{i+1}}{i+1} \cdot i$

17. $\displaystyle\sum_{i=1}^{\infty} \frac{(-1)^{i+1}}{\cos \pi i}$

18. $\displaystyle\sum_{i=1}^{\infty} \frac{(-1)^{i+1}}{\sin \pi i}$

19. $\displaystyle\sum_{i=1}^{\infty} \frac{\sin \pi i}{(-1)^{i+1}}$

20. $\displaystyle\sum_{i=3}^{\infty} \frac{\cos \pi i}{\ln i}$

To do Exercises 21–30, note that it can be shown that the error in approximating the infinite sum of a convergent alternating series with a finite number of terms is no more, in absolute value, than the value of the first omitted term. That is, if one estimates

$$\sum_{i=1}^{\infty} (-1)^i a_i \qquad \textit{as approximately} \qquad \sum_{i=1}^{n} (-1)^i a_i$$

the maximum possible error is a_{n+1}.

In Exercises 21–25, estimate the error in approximating the indicated sum with the first four terms of the series.

21. $\displaystyle\sum_{i=1}^{\infty} \frac{(-1)^{i+1}}{i}$

22. $\displaystyle\sum_{i=1}^{\infty} \frac{(-1)^{i+1}}{i!}$

23. $\displaystyle\sum_{i=1}^{\infty} \frac{(-1)^{i+1}}{i 2^i}$

24. $\displaystyle\sum_{i=1}^{\infty} \frac{(-1)^{i+1}}{(3i+1)^3}$

25. $\displaystyle\sum_{i=1}^{\infty} \frac{(-1)^{i+1} \cdot i}{2^i}$

In Exercises 26–30, determine how many terms one would have to include in order that the estimated sum have no more than 0.001 error.

26. $\displaystyle\sum_{i=1}^{\infty} \frac{(-1)^{i+1}}{i}$

27. $\displaystyle\sum_{i=1}^{\infty} \frac{(-1)^{i+1}}{i!}$

28. $\displaystyle\sum_{i=1}^{\infty} \frac{(-1)^{i+1}}{i^i}$ **29.** $\displaystyle\sum_{i=1}^{\infty} \frac{(-1)^{i+1} i!}{3^i}$

30. $\displaystyle\sum_{i=1}^{\infty} \frac{(-1)^{i+1} \sin \pi i}{i^3}$

9.6 POWER SERIES

Definition

An infinite series of the form

$$\sum_{i=0}^{\infty} a_i(x - c)^i = a_0 + a_1(x - c) + a_2(x - c)^2 + a_3(x - c)^3 + \cdots$$

is called a *power series* in $x - c$, where it is understood that the first term is a_0 rather than $a_0(x - c)^0$. The term $(x - c)^0$ does not present any problems unless $x = c$, in which case $(x - c)^0 = 0^0$, which is undefined.

Example 1   The following four series are power series.

$$\sum_{i=0}^{\infty} \frac{x^i}{i!} = 1 + x + \frac{x^2}{2!} + \frac{x^3}{3!} + \cdots$$

$$\sum_{i=0}^{\infty} i^2(x - 3)^i = 0 + (x - 3) + 2^2(x - 3)^2 + \cdots$$

$$\sum_{i=1}^{\infty} \frac{(-1)^{i+1}}{i} \left(x - \frac{\pi}{2}\right)^i = \left(x - \frac{\pi}{2}\right) - \frac{\left(x - \frac{\pi}{2}\right)^2}{2} + \frac{\left(x - \frac{\pi}{2}\right)^3}{3} - \cdots$$

$$\sum_{i=0}^{\infty} x^i = 1 + x + x^2 + x^3 + \cdots$$

The question of convergence and divergence of such series clearly depends on the actual value assumed by x.

> **Definition**
>
> The set of values of x for which the infinite series
>
> $$\sum_{i=0}^{\infty} a_i(x - c)^i$$
>
> converges is called the *interval of convergence* of the series.

For a given power series, the Ratio Test can usually be relied upon to determine the interval of convergence; however, the endpoints of the interval usually correspond to values of x for which the Ratio Test fails, and as such must be tested by some other means. In applying the Ratio Test, we will have to consider the absolute value of the terms since, unlike the series considered earlier, we cannot guarantee all of the terms of the series are positive.

Example 2   For what values of x does the series $\sum_{i=0}^{\infty} x^i$ converge?

Solution $a_n = |x^n| \qquad a_{n+1} = |x^{n+1}|$

$$\frac{a_{n+1}}{a_n} = \left|\frac{x^{n+1}}{x^n}\right| = |x|$$

Therefore,

$$\lim_{n \to \infty} \frac{a_{n+1}}{a_n} = \lim_{n \to \infty} |x| = |x|$$

Clearly if $|x| < 1$ the series is convergent, and if $|x| > 1$, the series is divergent. If $|x| = 1$,

$$\lim_{n \to \infty} a_n \neq 0$$

and the series is divergent. Hence, $\sum_{i=0}^{\infty} x^i$ is convergent if $-1 < x < 1$.

Let

$$u_n = \left| a_n(x - c)^n \right|$$

where n is the appropriate power for $(x - c)$. Then the Ratio Test would be applied to $\lim\limits_{n \to \infty} (u_{n+1}/u_n)$.

Example 3   Determine the interval of convergence of the series

$$\sum_{i=1}^{\infty} \frac{(x - 2)^{i-1}}{3^i i^2}$$

Solution $u_n = \left| \dfrac{(x - 2)^{n-1}}{3^n n^2} \right| \qquad u_{n+1} = \left| \dfrac{(x - 2)^n}{3^{n+1}(n + 1)^2} \right|$

$$\frac{u_{n+1}}{u_n} = \left| \frac{\dfrac{(x - 2)^n}{3^{n+1}(n + 1)^2}}{\dfrac{(x - 2)^{n-1}}{3^n n^2}} \right| = \left| \frac{(x - 2)^n 3^n n^2}{(x - 2)^{n-1} 3^{n+1}(n + 1)^2} \right|$$

$$= \frac{|x - 2|}{3} \cdot \frac{n^2}{(n + 1)^2}$$

$$\lim_{n \to \infty} \frac{u_{n+1}}{u_n} = \lim_{n \to \infty} \left[\frac{|x - 2|}{3} \cdot \frac{n^2}{(n + 1)^2} \right]$$

$$= \lim_{n \to \infty} \frac{|x - 2|}{3} \cdot \frac{2n}{2(n + 1)}$$

$$= \frac{|x - 2|}{3}$$

If $|x - 2|/3 < 1$, the series is convergent. Thus the series is convergent   if $|x - 2|/3 < 1$, $|x - 2| < 3$, $-3 < x - 2 < 3$,   or $-1 < x < 5$. The series is divergent if $x > 5$ or $x < -1$, as then the ratio would be greater than 1. What happens if $x = -1$ or $x = 5$? In the first case, $x = -1$, the series becomes

$$\sum_{i=1}^{\infty} \frac{(-3)^{i-1}}{3^i i^2} = \sum_{i=1}^{\infty} \frac{(-1)^{i-1}}{3 i^2} = \frac{1}{3} \sum_{i=1}^{\infty} \frac{(-1)^{i-1}}{i^2}$$

which is a convergent alternating p series. In the case $x = 5$, the series becomes

$$\sum_{i=1}^{\infty} \frac{3^{i-1}}{3^i i^2} = \sum_{i=1}^{\infty} \frac{1}{3 i^2} = \frac{1}{3} \sum_{i=1}^{\infty} \frac{1}{i^2}$$

which is also a convergent p series. In either case, the series is convergent; thus the interval of convergence is $-1 \leqslant x \leqslant 5$.

Example 4   Determine the interval of convergence for the series $\displaystyle\sum_{i=0}^{\infty} \frac{x^{2i}}{i!}$.

Solution $u_n = \left| \dfrac{x^{2n}}{n!} \right|$ $u_{n+1} = \left| \dfrac{x^{2(n+1)}}{(n+1)!} \right|$

$$\frac{u_{n+1}}{u_n} = \left| \frac{\dfrac{x^{2n+2}}{(n+1)!}}{\dfrac{x^{2n}}{n!}} \right| = \left| \frac{x^{2n+2}n!}{x^{2n}(n+1)!} \right| = \frac{x^2}{n+1}$$

$$\lim_{n\to\infty} \frac{u_{n+1}}{u_n} = \lim_{n\to\infty} \frac{x^2}{n+1} = 0$$

Therefore, if x is any finite number, the series converges. The interval of convergence is $(-\infty, \infty)$.

Example 5   Find the interval of convergence for $\displaystyle\sum_{i=0}^{\infty} i!x^i$.

Solution $u_n = \left| n!x^n \right|$ $u_{n+1} = \left| (n+1)!x^{n+1} \right|$

$$\frac{u_{n+1}}{u_n} = \left| \frac{(n+1)!x^{n+1}}{n!x^n} \right| = (n+1)\left| x \right|$$

$$\lim_{n\to\infty} \frac{u_{n+1}}{u_n} = \lim_{n\to\infty} (n+1)\left| x \right| = \infty \qquad (x \neq 0)$$

This series converges only if $x = 0$.

9.6 EXERCISES

In Exercises 1–20, determine the interval of convergence for each power series.

1. $\displaystyle\sum_{i=0}^{\infty} \frac{x^i}{i!}$

2. $\displaystyle\sum_{i=1}^{\infty} i(x-2)^i$

3. $\displaystyle\sum_{i=1}^{\infty} \frac{2^i x^i}{i^2}$

4. $\displaystyle\sum_{i=0}^{\infty} \frac{(x-2)^i}{2^i}$

5. $\displaystyle\sum_{i=0}^{\infty} \frac{(x-10)^i}{i!}$

6. $\displaystyle\sum_{i=0}^{\infty} x^i \left(\frac{3}{2}\right)^i$

7. $\displaystyle\sum_{i=1}^{\infty} \frac{(-1)^{i+1}x^i}{3^i i^2}$

8. $\displaystyle\sum_{i=0}^{\infty} \frac{(x+1)^i}{(i+1)}$

9. $\displaystyle\sum_{i=1}^{\infty} \frac{x^i}{\ln(i+1)}$

10. $\displaystyle\sum_{i=1}^{\infty} \frac{(3x-1)^i}{i^2}$

11. $\displaystyle\sum_{i=1}^{\infty} \frac{(2x+1)^i}{(-i)^3}$

12. $\displaystyle\sum_{i=0}^{\infty} \frac{(-1)^i x^{2i+1}}{(2i+1)!}$

13. $\displaystyle\sum_{i=0}^{\infty} \frac{(-1)^i x^{2i}}{(2i)!}$

14. $\displaystyle\sum_{i=0}^{\infty} \frac{(-1)^i(x-8)^i}{i}$

15. $\displaystyle\sum_{i=0}^{\infty} \frac{(-1)^i(2x-1)^i}{(2i)^2}$

16. $\displaystyle\sum_{i=0}^{\infty} \frac{(-1)^i(x+2)^i}{i+1}$

17. $\displaystyle\sum_{i=0}^{\infty} \frac{x^i}{(i+3)^2}$

18. $\displaystyle\sum_{i=0}^{\infty} \frac{x^i}{(i+1)^5}$

19. $\displaystyle\sum_{i=0}^{\infty} x^i i$

20. $\displaystyle\sum_{i=0}^{\infty} \frac{(-1)^{i+1}(x-3)^i i}{i+3}$

In Exercises 21–24, note that for a power series of the form $\displaystyle\sum_{i=0}^{\infty} a_i x^i$ the interval of convergence can often be stated in the form $|x| < R$, $|x| \le R$, $-R \le x < R$, or $-R < x \le R$. In these cases, R is called the radius of convergence of the series. Find the radius of convergence for each of the following series.

21. $\displaystyle\sum_{i=0}^{\infty} \frac{(-1)^i x^{2i}}{(2i)!}$

22. $\displaystyle\sum_{i=0}^{\infty} (-1)^i x^i i$

23. $\displaystyle\sum_{i=0}^{\infty} \frac{(-1)^i x^i}{2i}$

24. $\displaystyle\sum_{i=0}^{\infty} \frac{x^i}{5^i}$

25. Find any power series whose interval of convergence is from -2 to $+2$.

26. Find any power series whose interval of convergence is from 28 to 32.

27. Find any power series whose interval of convergence is from $C - R$ to $C + R$, where C is some real number and R is a given positive real number.

9.7   FUNCTIONS DEFINED BY POWER SERIES

If x is a value within the interval of convergence of a given power series,

$$\sum_{i=0}^{\infty} a_i x^i$$

then we can be sure that the series converges. The actual value to which the series converges depends on the value selected for x. Therefore, it is appropriate to indicate

$$\sum_{i=0}^{\infty} a_i x^i = f(x)$$

Example 1   Represent the function $f(x) = 1/(1 - x)$ as a power series.

Solution   By long division we have

$$\frac{1}{1 - x} = 1 + x + x^2 + \cdots = \sum_{i=0}^{\infty} x^i$$

provided $|x| < 1$, that is, x is within the interval of convergence of the series.

If a power series has a radius of convergence R, then it can be shown that its derivative and its antiderivative will also converge for all x within R. These facts can be used to create new power series.

Example 2   Find a power series to represent $f(x) = \ln(1 - x)$.

Solution   Integrate the power series of the last example.

$$\int \frac{1}{1 - x}\, dx = \int (1 + x + x^2 + \cdots)\, dx$$

$$-\ln(1 - x) = x + \frac{x^2}{2} + \frac{x^3}{3} + \cdots$$

$$= \sum_{i=1}^{\infty} \frac{x^i}{i}$$

$$\ln(1 - x) = -\sum_{i=1}^{\infty} \frac{x^i}{i} \qquad (|x| < 1)$$

Example 3   If $f(x) = \sum\limits_{i=0}^{\infty} \dfrac{x^i}{i!} = 1 + x + \dfrac{x^2}{2!} + \dfrac{x^3}{3!} + \cdots$, then show that $f'(x) = f(x)$.

Solution $f'(x) = 1 + \dfrac{2x}{2!} + \dfrac{3x^2}{3!} + \dfrac{4x^3}{4!} + \cdots$

$\qquad\quad = 1 + x + \dfrac{x^2}{2!} + \dfrac{x^3}{3!} + \cdots$

$\qquad\quad = f(x)$

This suggests that $f(x) = e^x$ because e^x is the only function we have seen such that $f'(x) = f(x)$.

Example 4   We will show later that

$$\sin x = x - \frac{x^3}{3!} + \frac{x^5}{5!} - \frac{x^7}{7!} + \cdots = \sum_{i=0}^{\infty} \frac{(-1)^i x^{2i+1}}{(2i+1)!}$$

Use this fact to find an infinite series for $\cos x$.

Solution $\dfrac{d\sin x}{dx} = \cos x$

$\cos x = \dfrac{d\sin x}{dx} = 1 - \dfrac{3x^2}{3!} + \dfrac{5x^4}{5!} - \dfrac{7x^6}{7!} + \cdots$

$\qquad\quad = 1 - \dfrac{x^2}{2!} + \dfrac{x^4}{4!} - \dfrac{x^6}{6!} + \cdots$

$\qquad\quad = \sum\limits_{i=0}^{\infty} \dfrac{(-1)^i x^{2i}}{(2i)!}$

Example 5   Find an infinite series to represent $\tan^{-1} x$.

Solution $\dfrac{d}{dx}(\tan^{-1} x) = \dfrac{1}{1 + x^2}$

However,

$$\frac{1}{1 + x^2} = 1 - x^2 + x^4 - x^6 + \cdots$$

Therefore

$$\tan^{-1} x = \int \frac{1}{1 + x^2} \, dx$$
$$= \int (1 - x^2 + x^4 - x^6 + \cdots) \, dx$$
$$= x - \frac{x^3}{3} + \frac{x^5}{5} - \frac{x^7}{7} + \cdots$$
$$= \sum_{i=0}^{\infty} \frac{(-1)^i x^{2i+1}}{2i + 1}$$

As a simple variation, consider the following.

Example 6   Find a power series for $(\tan^{-1} x)/x$ valid for $0 < x < 1$.

Solution   Within the specified interval

$$\tan^{-1} x = \sum_{i=0}^{\infty} \frac{(-1)^i x^{2i+1}}{2i + 1}$$

Hence

$$\frac{\tan^{-1} x}{x} = \frac{1}{x} \sum_{i=0}^{\infty} \frac{(-1)^i x^{2i+1}}{2i + 1} = \sum_{i=0}^{\infty} \frac{(-1)^i x^{2i}}{2i + 1}$$

9.7   EXERCISES

In Exercises 1–10, using any of the methods discussed in this section, find a power series representing the indicated function. Determine the interval of convergence in each case.

1. $\dfrac{1}{1 + x}$ **2.** $\dfrac{1}{1 - x^2}$ **3.** $\dfrac{x}{1 + x^3}$

4. $\ln (1 + x)$ **5.** $\ln (1 + x)^2$ **6.** $\ln (1 + x)^x$

7. $\ln (1 + x)^{1/x}$

8. $\tan^{-1} x^2$ [*Hint:*   Use the series for $\tan^{-1} x$, with x replaced by x^2.]

9. $\ln \left(\dfrac{1 + x}{1 - x} \right)$ [*Hint:* $\ln \left(\dfrac{1 + x}{1 - x} \right) = \ln (1 + x) - \ln (1 - x)$.]

10. $\dfrac{1 + x}{1 - x}$

11. Show that $\sum_{i=1}^{\infty} a_i(x - c)^i$ and $\sum_{i=1}^{\infty} \dfrac{a_i(x - c)^{i+1}}{i + 1}$ have the same radius of convergence.

In Exercises 12–17, note that a very useful method of finding power series is the generalized form of the binomial expansion:

$$(a + x)^n = a^n + na^{n-1}x + \frac{n(n - 1)a^{n-2}x^2}{2!}$$
$$+ \frac{n(n - 1)(n - 2)a^{n-3}x^3}{3!} + \cdots$$
$$+ \frac{n(n - 1)(n - 2) \cdots (n - i + 1)a^{n-i}x^i}{i!} + \cdots$$

For example,

$$\frac{1}{\sqrt{1 + x}} = (1 + x)^{-1/2} = 1^{-1/2} + (-\tfrac{1}{2})1^{-3/2}x + \frac{(-\tfrac{1}{2})(-\tfrac{3}{2})1^{-5/2}x^2}{2!}$$
$$+ \frac{(-\tfrac{1}{2})(-\tfrac{3}{2})(-\tfrac{5}{2})1^{-7/2}x^3}{3!} + \cdots$$
$$= 1 - \frac{1}{2} x + \frac{1(3)}{2(2)(2!)} x^2 - \frac{1(3)(5)}{(2)^3(3!)} x^3 + \cdots$$
$$= 1 - \frac{1}{2} x + \frac{1(3)}{2(4)} x^2 - \frac{1(3)(5)}{2(4)(6)} x^3 + \frac{1(3)(5)(7)}{2(4)(6)(8)} x^4 - \cdots$$

12. Determine the interval of convergence for the power series for $1/\sqrt{1 + x}$.

13. By replacing x by $-x$ in the power series for $1/\sqrt{1 + x}$ one gets a series for $1/\sqrt{1 - x}$. Find this series and establish its interval of convergence.

14. By replacing x by $-x^2$ in the power series for $1/\sqrt{1 + x}$ one gets a power series for $1/\sqrt{1 - x^2}$. Find this series and determine its interval of convergence.

15. Use the results of Exercise 14, together with the fact that

$$\int \frac{1}{\sqrt{1 - x^2}} \, dx = \sin^{-1} x$$

to find an infinite power series for $\sin^{-1} x$.

16. By replacing x by x^2 in the series for $1/\sqrt{1 + x}$ find a series for $1/\sqrt{1 + x^2}$, together with its interval of convergence.

17. Since

$$\int \frac{dx}{\sqrt{1 + x^2}} = \ln (x + \sqrt{1 + x^2})$$

use the results of Exercise 16 to find an infinite series for $\ln (x + \sqrt{1 + x^2})$.

18. Assume that

$$\sin x = \sum_{i=0}^{\infty} \frac{(-1)^i x^{2i+1}}{(2i + 1)!}$$

Find a power series for $(\sin x)/x$. Discuss the behavior of the series as $\lim_{x \to 0} [(\sin x)/x]$ is considered.

19. Since $\cot^{-1} x = (\pi/2) - \tan^{-1} x$, find an infinite series for $\cot^{-1} x$.

20. Since $\csc^{-1} x = \sin^{-1} (1/x)$, use the infinite series for $\sin^{-1} x$ to find an infinite series for $\csc^{-1} x$ by replacing x by $1/x$. Is this a power series?

9.8 TAYLOR'S SERIES

Sometimes, if we have been lucky, we have been able to find an infinite series representing a known function. It is natural to wonder if it is possible to find, in a systematic way without some good guess or lucky break, an infinite power series for a given function.

Consider a function f. Since power series are differentiable over and over again, at least within their interval of convergence, we should assume that if the series is an alternative representation of f, f is also differentiable in this manner. Then can we find an infinite power series of the form

$$\sum_{i=0}^{\infty} a_i (x - c)^i$$

which, for a given value of x and within some interval around c, has the property

$$f(x) = \sum_{i=0}^{\infty} a_i (x - c)^i$$

Clearly the problem is one of finding the proper form for the a_i's. If $f(x) = \sum\limits_{i=0}^{\infty} a_i(x - c)^i$, then in expanded form

$$f(x) = a_0 + a_1(x - c) + a_2(x - c)^2$$
$$+ a_3(x - c)^3 + \cdots + a_n(x - c)^n + \cdots$$

Then

$$f'(x) = a_1 + 2a_2(x - c) + 3a_3(x - c)^2 + 4a_4(x - c)^3 + \cdots$$
$$+ na_n(x - c)^{n-1} + \cdots$$
$$f''(x) = 2a_2 + (3)2a_3(x - c) + (4)3a_4(x - c)^2 + \cdots$$
$$+ n(n - 1)a_n(x - c)^{n-2} + \cdots$$
$$f'''(x) = (3)2a_3 + (4)(3)2a_4(x - c) + (5)(4)(3)a_5(x - c)^2 + \cdots$$
$$+ n(n - 1)(n - 2)a_n(x - c)^{n-3} + \cdots$$

In general,

$$f^{(n)}(x) = n(n - 1)(n - 2) \cdots (3)2a_n$$
$$+ (n + 1)(n)(n - 1) \cdots 2a_{n+1}(x - c) + \cdots$$
$$= n!a_n + (n + 1)(n)(n - 1) \cdots 2a_{n+1}(x - c) + \cdots$$

where $f^{(n)}(x) = d^n f(x)/dx^n$ = the nth derivative of $f(x)$. If $x = c$ then each of these equations reduces to a single term of the series, because $(c - c) = 0$. Specifically

$$f(c) = a_0$$
$$f'(c) = a_1$$

$$f''(c) = 2a_2 \qquad \text{or} \qquad a_2 = \frac{f''(c)}{2}$$

$$f'''(c) = 3(2)a_3 \qquad \text{or} \qquad a_3 = \frac{f'''(c)}{3!}$$

or, in general,

$$f^{(n)}(c) = n!a_n \qquad \text{or} \qquad a_n = \frac{f^{(n)}(c)}{n!}$$

The conclusion of this analysis is *if* there is a power series for a given function in terms of powers of $(x - c)$, then the coefficients must be given by

$$a_n = \frac{f^{(n)}(c)}{n!}$$

This result is called *Taylor's Formula,* and the series is called *Taylor's series.* In the special case where $c = 0$, the result is called *Maclaurin's series.*

Example 1   Find the Maclaurin's series representing e^x.

Solution   If $f(x) = e^x$, then

$$f'(x) = e^x$$
$$f''(x) = e^x$$

and in general,

$$f^{(n)}(x) = e^x$$

At $c = 0$,

$$f(c) = e^0 = 1$$
$$f'(c) = e^0 = 1$$
$$f^{(n)}(c) = e^0 = 1$$

Therefore,

$$a_0 = 1 = f(0)$$
$$a_1 = f'(0) = 1$$
$$a_2 = \frac{f''(0)}{2!} = \frac{1}{2!}$$
$$a_3 = \frac{f'''(0)}{3!} = \frac{1}{3!}$$
$$a_n = \frac{f^{(n)}(0)}{n!} = \frac{1}{n!}$$

Thus,

$$e^x = a_0 + a_1x + a_2x^2 + a_3x^3 + \cdots$$
$$= 1 + x + \frac{x^2}{2!} + \frac{x^3}{3!} + \frac{x^3}{4!} + \cdots$$
$$= \sum_{i=0}^{\infty} \frac{x^i}{i!}$$

This was the expected result. The interval of convergence for this series is from $-\infty$ to $+\infty$.

It can be shown that if a Taylor's series representing a given function is truncated after a finite number of terms, say n, the error made by estimating the function with the finite series can be estimated by

$$R_n(x) = \frac{f^{(n+1)}(z)}{(n+1)!}(x-c)^{n+1}$$

where $R_n(x)$ is the remainder term for the series, and z is a number between x and c.

Example 2 Estimate the error made in assuming that e is given by the first 11 terms of its Taylor's series.

Solution Considering 11 terms implies that $n = 10$, then $n + 1 = 11$, and

$$f^{(11)}(0) = e^0 = 1$$

since

$$e^x = \sum_{i=0}^{\infty} \frac{x^i}{i!}$$

and thus

$$e = \sum_{i=0}^{\infty} \frac{1}{i!}$$

If

$$e \approx 1 + 1 + \frac{1}{2!} + \frac{1}{3!} + \frac{1}{4!} + \frac{1}{5!} + \frac{1}{6!} + \frac{1}{7!} + \frac{1}{8!} + \frac{1}{9!} + \frac{1}{10!}$$

the error, $R_{11}(1) = e^z/11!$, where $0 \leqslant z \leqslant 1$. But $e^z/11! \leqslant 3^1/11!$; therefore the error is less than $3/11! \approx 0.000000075$.

Example 3 Find the Maclaurin's series for $\sin x$.

Solution

$$\begin{aligned}
f(x) &= \sin x \\
f'(x) &= \cos x \\
f''(x) &= -\sin x \\
f'''(x) &= -\cos x
\end{aligned}$$

$$f^{(4)}(x) = \sin x = f(x)$$
$$f^{(5)}(x) = \cos x = f'(x)$$
$$f^{(6)}(x) = -\sin x = f''(x)$$
$$f^{(7)}(x) = -\cos x = f'''(x)$$
$$f^{(8)}(x) = \sin x = f(x) = f^{(4)}(x)$$

In general,

$$f^{(4n)}(x) = \sin x$$
$$f^{(4n+1)}(x) = \cos x$$
$$f^{(4n+2)}(x) = -\sin x$$
$$f^{(4n+3)}(x) = -\cos x$$

for $n = 0, 1, 2, 3, \ldots$. If $x = c = 0$,

$$f^{(4n)}(0) = \sin 0 = 0$$
$$f^{(4n+1)}(0) = \cos 0 = 1$$
$$f^{(4n+2)}(0) = -\sin 0 = 0$$
$$f^{(4n+3)}(0) = -\cos 0 = -1$$

Therefore,

$$a_{4n} = 0$$
$$a_{4n+1} = \frac{1}{(4n+1)!}$$
$$a_{4n+2} = 0$$
$$a_{4n+3} = \frac{-1}{(4n+3)!}$$

for $n = 0, 1, 2, 3, 4, \ldots$. In other words, a's with even subscripts are zero, whereas those with odd subscripts alternate in sign. Thus,

$$\sin x = 0 + \frac{x}{1!} + 0 - \frac{x^3}{3!} + 0 + \frac{x^5}{5!} + 0 - \frac{x^7}{7!} + \cdots = \sum_{i=0}^{\infty} \frac{(-1)^i x^{2i+1}}{(2i+1)!}$$

It is a simple matter to verify that this series converges for all real values of x.

Example 4 If

$$\sin x = \sum_{i=0}^{\infty} \frac{(-1)^i x^{2i+1}}{(2i+1)!}$$

find the Maclaurin's series for $\cos x$.

Solution Using term by term differentiation as was done in an example in Section 9.7,

$$\cos x = \sum_{i=0}^{\infty} \frac{(-1)^i x^{2i}}{(2i)!}$$

Would we have obtained the same series directly from Taylor's series? For $f(x) = \cos x$ the verification is left as an exercise. However, it will be informative to verify by an example the fact that Taylor's series agrees with series obtained other ways.

Example 5 Verify by use of Maclaurin's series that

$$\frac{1}{1-x} = 1 + x + x^2 + x^3 + \cdots = \sum_{i=0}^{\infty} x^i \qquad (|x| < 1)$$

Solution Here $c = 0$ and $a_n = f^{(n)}(0)/n!$.

$$f(x) = \frac{1}{1-x}$$
$$f'(x) = \frac{-1}{(1-x)^2}(-1) = \frac{1}{(1-x)^2}$$
$$f''(x) = \frac{-2}{(1-x)^3}(-1) = \frac{2}{(1-x)^3}$$
$$f'''(x) = \frac{+2(-3)}{(1-x)^4}(-1) = \frac{2(3)}{(1-x)^4}$$

In general,

$$f^{(n)}(x) = \frac{n!}{(1-x)^{n+1}}$$

At $x = c = 0$,

$$f^{(n)}(0) = n!$$

Therefore,

$$a_n = \frac{f^{(n)}(0)}{n!} = \frac{n!}{n!} = 1$$

for all n. Here,

$$\frac{1}{1-x} = \sum_{i=0}^{\infty} a_i x^i = \sum_{i=0}^{\infty} x^i$$

When a series is known, or found by using some other method, Taylor's Formula is often useful in evaluating derivatives.

Example 6   Find $\dfrac{d^7}{dx^7} \tan^{-1} x \Big|_{x=0}$

Solution   Using term by term integration, we found that

$$\tan^{-1} x = x - \frac{x^3}{3} + \frac{x^5}{5} - \frac{x^7}{7} + \frac{x^9}{9} - \cdots$$

Using Maclaurin's series,

$$\tan^{-1} x = a_0 + a_1 x + a_2 x^2 + a_3 x^3 + a_4 x^4 + a_5 x^5 + a_6 x^6 + a_7 x^7 + \cdots$$

where

$$a_n = \frac{f^{(n)}(0)}{n!}$$

Comparing the two series, it is apparent that

$$a_7 = \frac{\dfrac{d^7}{dx^7} \tan^{-1} x \Big|_{x=0}}{7!}$$

while, at the same time, $a_7 = -\frac{1}{7}$ from the actual series. Thus,

$$\frac{\dfrac{d^7}{dx^7} \tan^{-1} (x) \Big|_{x=0}}{7!} = -\frac{1}{7}$$

or

$$\frac{d^7}{dx^7} \tan^{-1} (x) \Big|_{x=0} = \frac{-7!}{7} = -6! = -720$$

Actually, the most important applications of infinite series relate to solving equations like

$$\frac{d^2y}{dx^2} - xy = 0$$

If we assume that

$$y = \sum_{i=0}^{\infty} a_i x^i$$

then

$$y' = \sum_{i=1}^{\infty} a_i i x^{i-1}$$

$$y'' = \sum_{i=2}^{\infty} a_i i(i-1) x^{i-2}$$

and

$$xy = \sum_{i=0}^{\infty} a_i x^{i+1}$$

In terms of the power series, the equation becomes

$$\sum_{i=2}^{\infty} i(i-1) a_i x^{i-2} - \sum_{i=0}^{\infty} a_i x^{i+1} = 0$$

or

$$2(1)a_2 + 3(2)a_3 x + 4(3)a_4 x^2 + 5(4)a_5 x^3 +$$
$$\cdots - (a_0 x + a_1 x^2 + a_2 x^3 + \cdots) = 0$$

or

$$2a_2 + (6a_3 - a_0)x + (12a_4 - a_1)x^2 + \cdots = 0$$

To satisfy the equation, all of the coefficients must vanish. Hence,

$$a_2 = 0 \qquad 6a_3 - a_0 = 0 \qquad 12a_4 - a_1 = 0 \qquad \text{and so forth}$$

or

$$a_3 = \frac{a_0}{6} \qquad a_4 = \frac{a_1}{12} \qquad a_5 = 0 \qquad \text{and so forth}$$

Hence,

$$y = a_0 + a_1 x + \frac{a_0}{6} x^3 + \frac{a_1}{12} x^4 + \frac{a_0}{2(3)(5)(6)} x^6 + \cdots$$

Thus, if a solution exists it must be represented by the indicated power series.

In attempting to solve differential equations involving an unknown function and its derivatives as above, it is often possible to find an infinite power series representing the function. It frequently happens in these cases that the function found has no counterpart among the algebraic functions. This leads to a whole class of functions, defined only in terms of a power series, that represent nonelementary functions. Often it is only with functions found in terms of an infinite series that suitable mathematical models of real-world phenomena can be found.

9.8   EXERCISES

1. Use Taylor's formula to find a Maclaurin's series for $\ln |x + 1|$. Determine the radius of convergence.

2. If $f(x) = \dfrac{e^x + e^{-x}}{2}$, find a series in powers of x for f using Taylor's Formula.

3. Using the series developed for e^x find a series for $f(x) = \dfrac{e^x + e^{-x}}{2}$.

4. $f(x) = \dfrac{e^x + e^{-x}}{2}$ is called the *hyperbolic cosine* of x and is usually denoted as $\cosh x = \dfrac{e^x + e^{-x}}{2}$. Verify, using the infinite series representation of $\cosh x$, that $\dfrac{d^2}{dx^2} \cosh x = \cosh x$.

5. Verify by direct calculation that $\dfrac{d^2}{dx^2} \cosh x = \cosh x$.

6. Find a Taylor's series with $c = \pi/2$ for $\cos x$.

7. Find a Taylor's series with $c = \pi/2$ for $\sin x$.

8. Develop the Maclaurin's series for $\tan^{-1} x$.

9. Develop the Maclaurin's series for $\sin^{-1} x$.

10. Find, by any method, a Maclaurin's series for e^{x^2}.

11. Find, using the result of Exercise 10,

$$\frac{d^7}{dx^7}\, e^{x^2}\Bigg|_{x=0}$$

12. Find a Taylor's series for e^x in powers of $(x - 1)$, that is, with $c = 1$.

13. Find a Taylor's series with $c = 1$ for $f(x) = \sqrt{x}$.

14. What problems would one encounter if one tried to find a Maclaurin's series for $f(x) = \sqrt{x}$?

15. Test the validity of the binomial expansion

$$(a + x)^n = a^n + na^{n-1}x$$
$$+ \frac{n(n-1)a^{n-2}x^2}{2!} + \frac{n(n-1)(n-2)a^{n-3}x^3}{3!} +$$
$$\cdots + \frac{n(n-1)(n-2)\cdots(n-i+1)a^{n-i}x^i}{i!} + \cdots$$

by finding the first four terms of the Maclaurin's series for $f(x) = (a + x)^n$, where n is some rational number.

16. Consider $f(x) = (4 - x)^3$. Find the Taylor's series for f. Does this result agree with the ordinary expansion of $(4 - x)^3$ by the usual rules of algebra?

17. Estimate the error made in equating sin 1 to the first four nonzero terms of the Maclaurin's series for $\sin x$, with $x = 1$.

18. Estimate the error one would make if one said that $\cos x$ was exactly equal to the first three nonzero terms of the Maclaurin's series for $\cos x$ provided $0 \leqslant x < \pi/4$.

19. Since $\tan^{-1} 1 = \pi/4$, how many terms would be necessary to estimate $\pi/4$ using the series developed for $\tan^{-1} x$, so that the error would be less than 0.005?

In Exercises 20–23, find the first three nonzero terms of the Maclaurin's series for each of the indicated functions.

20. $f(x) = \sec x$ 21. $f(x) = \ln (\sec x)$

22. $f(x) = e^x \sin x$ 23. $f(x) = x \ln (1 + x)$

24. Find a power series such that

$$y = \sum_{i=0}^{\infty} a_i x^i$$

is a solution to the equation $(dy/dx) - y = 0$. Does the power series you find appear to be one of the known functions? Which one?

NEW TERMS

Arithmetic sequence   Harmonic series
Convergence p-series
Divergence   Sequence
Geometric sequence   Series

REVIEW

1. Define a sequence.
2. Define a series.
3. Define convergence and divergence for a sequence.
4. Define convergence and divergence for a series.
5. What is a sequence of partial sums? How does it relate to the concept of an infinite series?

In Exercises 6–10, determine if the sequences are convergent or divergent. Find a_1, a_2, and a_7 in each case.

6. $\{a_n\}$, where $a_n = \dfrac{\ln(n + 1)}{n}$

7. $\{a_n\}$, where $a_n = 3^n$

8. $\{a_n\}$, where $a_n = \dfrac{n^2 - 1}{(n + 3)^2}$

9. $\{a_n\}$, where $a_n = 3^{-n}$

10. $\{a_n\}$, where $a_n = \dfrac{\ln(n^2)}{2n^3}$

In Exercises 11–13, a sequence $\{a_n\}$ is defined recursively. Find a general form for a_n, and test the sequence for convergence.

11. $a_{n+1} = a_n \left(\dfrac{1}{n + 1}\right)$, $a_1 = 1$ 12. $a_{n+1} = a_n + (n + 1)$, $a_1 = 1$

13. $a_{n+1} = a_n + 1$, $a_1 = 1$

14. State the Integral Test for convergence of a series. What are the principal problems encountered in applying the Integral Test?

In Exercises 15–18, test the following series for convergence using the Integral Test.

15. $\displaystyle\sum_{i=1}^{\infty} \frac{1}{i\sqrt{i}}$

16. $\displaystyle\sum_{i=1}^{\infty} \frac{1}{3i+1}$

17. $\displaystyle\sum_{i=2}^{\infty} \frac{\ln i}{i^2}$

18. $\displaystyle\sum_{i=1}^{\infty} \frac{i^2}{e^i}$

19. State the comparison tests for convergence or divergence. What are the principal problems one encounters in attempting to use the comparison tests?

In Exercises 20–23, use a comparison test to establish the convergence or divergence of the series.

20. $\displaystyle\sum_{i=1}^{\infty} \frac{1}{(3i)^4}$

21. $\displaystyle\sum_{i=1}^{\infty} \frac{|\sin i|}{i^2}$

22. $\displaystyle\sum_{i=1}^{\infty} \frac{1}{2^{1/i}}$

23. $\displaystyle\sum_{i=1}^{\infty} \frac{(i+1)!}{(i+2)!}$

24. State the Ratio Test for the convergence of an infinite series. What problems can one expect to encounter in attempting to apply it?

In Exercises 25–28, using the Ratio Test, test for convergence.

25. $\displaystyle\sum_{i=1}^{\infty} \frac{i!}{2^i}$

26. $\displaystyle\sum_{i=1}^{\infty} \frac{1(3)(5)(7)(9) \cdots (2i+1)}{2(4)(6)(8)(10) \cdots 2i(2i+2)} \left(\frac{2}{3}\right)^i$

27. $\displaystyle\sum_{i=1}^{\infty} \frac{2^i}{i^2}$

28. $\displaystyle\sum_{i=1}^{\infty} \frac{i^2}{2^i}$

29. Under what conditions does an alternating series converge?

30. What are absolute and conditional convergence?

In Exercises 31–34, test for absolute convergence.

31. $\displaystyle\sum_{i=1}^{\infty} \frac{(-1)^{i+1}}{i(i+2)}$

32. $\displaystyle\sum_{i=1}^{\infty} \frac{(2i-1)(-1)^{i+1}}{5i+1}$

33. $\displaystyle\sum_{i=1}^{\infty} \frac{(-1)^i}{i3^i}$

34. $\displaystyle\sum_{i=1}^{\infty} \frac{(-1)^{i+1}}{i+5}$

In Exercises 35–40, test for convergence by any method.

35. $\displaystyle\sum_{i=1}^{\infty} \frac{i}{i^4 + 1}$ **36.** $\displaystyle\sum_{i=1}^{\infty} \frac{1}{(2i + 2)^3}$

37. $\displaystyle\sum_{i=3}^{\infty} \frac{1}{i(\ln i)^{3/2}}$ **38.** $\displaystyle\sum_{i=1}^{\infty} \left[i - \left(\frac{2}{3}\right)^{i+1} \right]$

39. $\displaystyle\sum_{i=1}^{\infty} \frac{2 + 6 \sqrt{i}}{i}$ **40.** $\displaystyle\sum_{i=1}^{\infty} \frac{(-1)(-2)(-3) \cdots (-i)}{(2)(4)(6)(8) \cdots (2i)}$

41. For a given series $\displaystyle\sum_{i=1}^{\infty} a_i$ with S_n the partial sum, $S_n \cdot S_{n+1} < 0$. How can this happen? Can you find an example where this takes place?

42. What is a power series?

43. Does a power series converge for all values of the argument?

In Exercises 44–50, determine the interval of convergence.

44. $\displaystyle\sum_{i=1}^{\infty} i(x - 2)^i$ **45.** $\displaystyle\sum_{i=2}^{\infty} \frac{x^i}{\ln (i)}$

46. $\displaystyle\sum_{i=1}^{\infty} (-1)^i \frac{1(3)(5)(7)(9) \cdots (2i + 1)}{2(4)(6)(8)(10) \cdots (2i)} x^i$

47. $\displaystyle\sum_{i=1}^{\infty} \frac{i!}{i^i} x^i$ **48.** $\displaystyle\sum_{i=1}^{\infty} \frac{(x + 3)^i}{i 2^i}$

49. $\displaystyle\sum_{i=1}^{\infty} \frac{(x + 1)^i}{4^i}$ **50.** $\displaystyle\sum_{i=1}^{\infty} \frac{(x^2 + 2x + 1)^i}{i}$

51. Does the concept of an interval of convergence apply to a series like $\displaystyle\sum_{i=1}^{\infty} \frac{1}{ix^i}$? If it does, what is the interval of convergence?

52. If a function f can be represented by a power series, what advantage does such a series representation have over the more usual and more compact representation?

53. Use the idea of partial sums or term by term differentiation to argue that if

$$\sum_{i=0}^{\infty} a_i x^i \qquad \text{and} \qquad \sum_{i=0}^{\infty} b_i x^i$$

have the same interval of convergence, and

$$\sum_{i=0}^{\infty} a_i x^i = \sum_{i=0}^{\infty} b_i x^i$$

for every x in that interval, then $a_n = b_n$ for all n.

In Exercises 54–60, find by any method a series of the form

$$\sum_{i=0}^{\infty} a_i (x - c)^i$$

for the indicated value of c. Determine the interval of convergence in each case.

54. $f(x) = \sin x$, $c = \dfrac{\pi}{4}$ **55.** $f(x) = \ln\left(\dfrac{1 - x}{1 - x^2}\right)$, $c = 0$

56. $f(x) = e^{-x^2}$, $c = 0$

57. $f(x) = \sinh x = \dfrac{e^x - e^{-x}}{2}$ (sinh x is read "hyperbolic sine of x"), $c = 0$

58. $f(x) = (1 + x)^{1/3}$, $c = 0$ **59.** $f(x) = \dfrac{1}{x}$, $c = 1$

60. $f(x) = \sec x$, $c = 0$

In Exercises 61 and 62, assume

$$e^{\cos x} = e\left(1 - \frac{x^2}{2} + \frac{x^4}{4} - \frac{31 x^6}{720} + \cdots\right)$$

61. Find $\dfrac{d^4}{dx^4} e^{\cos x}\Big|_{x=0}$ **62.** Find $\dfrac{d^5}{dx^5} e^{\cos x}\Big|_{x=0}$

63. If

$$\frac{\ln(1 + x)}{1 + x} = x - \left(1 + \frac{1}{2}\right)x^2 + \left(1 + \frac{1}{2} + \frac{1}{3}\right)x^3$$
$$- \left(1 + \frac{1}{2} + \frac{1}{3} + \frac{1}{4}\right)x^4 + \cdots$$

find

$$\frac{d^{10}}{dx^{10}} \frac{\ln(1 + x)}{1 + x}\Big|_{x=0}$$

64. Estimate $\sin(\pi/4)$ with 4 nonzero terms of the Maclaurin's series for $\sin x$. What error is involved in this estimate?

Appendix

Table I

Values of $\dfrac{1}{\sqrt{2\pi}}\displaystyle\int_0^{z_1} e^{-\frac{1}{2}z^2}\,dz$ or Normal Curve or Gaussian Areas

z_1	0.00	0.01	0.02	0.03	0.04	0.05	0.06	0.07	0.08	0.09
0.0	0.0000	0.0040	0.0080	0.0120	0.0160	0.0199	0.0239	0.0279	0.0319	0.0359
0.1	.0398	.0438	.0478	.0517	.0557	.0596	.0636	.0675	.0714	.0753
0.2	.0793	.0832	.0871	.0910	.0948	.0987	.1026	.1064	.1103	.1141
0.3	.1179	.1217	.1255	.1293	.1331	.1368	.1406	.1443	.1480	.1517
0.4	.1554	.1591	.1628	.1664	.1700	.1736	.1772	.1808	.1844	.1879
0.5	.1915	.1950	.1985	.2019	.2054	.2088	.2123	.2157	.2190	.2224
0.6	.2257	.2291	.2324	.2357	.2389	.2422	.2454	.2486	.2517	.2549
0.7	.2580	.2611	.2642	.2673	.2704	.2734	.2764	.2794	.2823	.2852
0.8	.2881	.2910	.2939	.2967	.2995	.3023	.3051	.3078	.3106	.3133
0.9	.3159	.3186	.3212	.3238	.3264	.3289	.3315	.3340	.3365	.3389
1.0	.3413	.3438	.3461	.3485	.3508	.3531	.3554	.3577	.3599	.3621
1.1	.3643	.3665	.3686	.3708	.3729	.3749	.3770	.3790	.3810	.3830
1.2	.3849	.3869	.3888	.3907	.3925	.3944	.3962	.3980	.3997	.4015
1.3	.4032	.4049	.4066	.4082	.4099	.4115	.4131	.4147	.4162	.4177
1.4	.4192	.4207	.4222	.4236	.4251	.4265	.4279	.4292	.4306	.4319
1.5	.4332	.4345	.4357	.4370	.4382	.4394	.4406	.4418	.4429	.4441
1.6	.4452	.4463	.4474	.4484	.4495	.4505	.4515	.4525	.4535	.4545
1.7	.4554	.4564	.4573	.4582	.4591	.4599	.4608	.4616	.4625	.4633
1.8	.4641	.4649	.4656	.4664	.4671	.4678	.4686	.4693	.4699	.4706
1.9	.4713	.4719	.4726	.4732	.4738	.4744	.4750	.4756	.4761	.4767
2.0	.4772	.4778	.4783	.4788	.4793	.4798	.4803	.4808	.4812	.4817
2.1	.4821	.4826	.4830	.4834	.4838	.4842	.4846	.4850	.4854	.4857
2.2	.4861	.4864	.4868	.4871	.4875	.4878	.4881	.4884	.4887	.4890
2.3	.4893	.4896	.4898	.4901	.4904	.4906	.4909	.4911	.4913	.4916
2.4	.4918	.4920	.4922	.4925	.4927	.4929	.4931	.4932	.4934	.4936
2.5	.4938	.4940	.4941	.4943	.4945	.4946	.4948	.4949	.4951	.4952
2.6	.4953	.4955	.4956	.4957	.4959	.4960	.4961	.4962	.4963	.4964
2.7	.4965	.4966	.4967	.4968	.4969	.4970	.4971	.4972	.4973	.4974
2.8	.4974	.4975	.4976	.4977	.4977	.4978	.4979	.4979	.4980	.4981
2.9	.4981	.4982	.4982	.4983	.4984	.4984	.4985	.4985	.4986	.4986
3.0	.4987	.4987	.4987	.4988	.4988	.4989	.4989	.4989	.4990	.4990

Table II
Natural Logarithms

x	$\ln x$	x	$\ln x$	x	$\ln x$
0.1	-2.303	3.5	1.253	6.9	1.932
0.2	-1.609	3.6	1.281	7.0	1.946
0.3	-1.204	3.7	1.308	7.1	1.960
0.4	-0.916	3.8	1.335	7.2	1.974
0.5	-0.693	3.9	1.361	7.3	1.988
0.6	-0.511	4.0	1.386	7.4	2.001
0.7	-0.357	4.1	1.411	7.5	2.015
0.8	-0.223	4.2	1.435	7.6	2.028
0.9	-0.105	4.3	1.459	7.7	2.041
1.0	0.000	4.4	1.482	7.8	2.054
1.1	0.095	4.5	1.504	7.9	2.067
1.2	0.182	4.6	1.526	8.0	2.079
1.3	0.262	4.7	1.548	8.1	2.092
1.4	0.336	4.8	1.569	8.2	2.104
1.5	0.405	4.9	1.589	8.3	2.116
1.6	0.470	5.0	1.609	8.4	2.128
1.7	0.531	5.1	1.629	8.5	2.140
1.8	0.588	5.2	1.649	8.6	2.152
1.9	0.642	5.3	1.668	8.7	2.163
2.0	0.693	5.4	1.686	8.8	2.175
2.1	0.742	5.5	1.705	8.9	2.186
2.2	0.788	5.6	1.723	9.0	2.197
2.3	0.833	5.7	1.740	9.1	2.208
2.4	0.875	5.8	1.758	9.2	2.219
2.5	0.916	5.9	1.775	9.3	2.230
2.6	0.956	6.0	1.792	9.4	2.241
2.7	0.993	6.1	1.808	9.5	2.251
2.8	1.030	6.2	1.825	9.6	2.262
2.9	1.065	6.3	1.841	9.7	2.272
3.0	1.099	6.4	1.856	9.8	2.282
3.1	1.131	6.5	1.872	9.9	2.293
3.2	1.163	6.6	1.887	10.0	2.303
3.3	1.194	6.7	1.902		
3.4	1.224	6.8	1.917		

Table III

e^x

x	0	1	2	3	4	5	6	7	8	9
0.0	1.0000	1.0101	1.0202	1.0305	1.0408	1.0513	1.0618	1.0725	1.0833	1.0942
0.1	1.1052	1.1163	1.1275	1.1388	1.1503	1.1618	1.1735	1.1853	1.1972	1.2092
0.2	1.2214	1.2337	1.2461	1.2586	1.2712	1.2840	1.2969	1.3100	1.3231	1.3364
0.3	1.3499	1.3634	1.3771	1.3910	1.4049	1.4191	1.4333	1.4477	1.4623	1.4770
0.4	1.4918	1.5068	1.5220	1.5373	1.5527	1.5683	1.5841	1.6000	1.6161	1.6323
0.5	1.6487	1.6653	1.6820	1.6989	1.7160	1.7333	1.7507	1.7683	1.7860	1.8040
0.6	1.8221	1.8404	1.8589	1.8776	1.8965	1.9155	1.9348	1.9542	1.9739	1.9937
0.7	2.0138	2.0340	2.0544	2.0751	2.0959	2.1170	2.1383	2.1598	2.1815	2.2034
0.8	2.2255	2.2479	2.2705	2.2933	2.3164	2.3396	2.3632	2.3869	2.4109	2.4351
0.9	2.4596	2.4843	2.5093	2.5345	2.5600	2.5857	2.6117	2.6379	2.6645	2.6912
1.0	2.7183	2.7456	2.7732	2.8011	2.8292	2.8577	2.8864	2.9154	2.9447	2.9743
1.1	3.0042	3.0344	3.0649	3.0957	3.1268	3.1582	3.1899	3.2220	3.2544	3.2871
1.2	3.3201	3.3535	3.3872	3.4212	3.4556	3.4903	3.5254	3.5609	3.5966	3.6328
1.3	3.6693	3.7062	3.7434	3.7810	3.8190	3.8574	3.8962	3.9354	3.9749	4.0149
1.4	4.0552	4.0960	4.1371	4.1787	4.2207	4.2631	4.3060	4.3492	4.3929	4.4371
1.5	4.4817	4.5267	4.5722	4.6182	4.6646	4.7115	4.7588	4.8066	4.8550	4.9037
1.6	4.9530	5.0028	5.0531	5.1039	5.1552	5.2070	5.2593	5.3122	5.3656	5.4195
1.7	5.4739	5.5290	5.5845	5.6407	5.6973	5.7546	5.8124	5.8709	5.9299	5.9895
1.8	6.0496	6.1104	6.1719	6.2339	6.2965	6.3598	6.4237	6.4883	6.5535	6.6194
1.9	6.6859	6.7531	6.8210	6.8895	6.9588	7.0287	7.0993	7.1707	7.2427	7.3155
2.0	7.3891	7.4633	7.5383	7.6141	7.6906	7.7679	7.8460	7.9248	8.0045	8.0849
2.1	8.1662	8.2482	8.3311	8.4149	8.4994	8.5849	8.6711	8.7583	8.8463	8.9352
2.2	9.0250	9.1157	9.2073	9.2999	9.3933	9.4877	9.5831	9.6794	9.7767	9.8749
2.3	9.9742	10.074	10.176	10.278	10.381	10.486	10.591	10.697	10.805	10.913
2.4	11.023	11.134	11.246	11.359	11.473	11.588	11.705	11.822	11.941	12.061
2.5	12.182	12.305	12.429	12.554	12.680	12.807	12.936	13.066	13.197	13.330
2.6	13.464	13.599	13.736	13.874	14.013	14.154	14.296	14.440	14.585	14.732
2.7	14.880	15.029	15.180	15.333	15.487	15.643	15.800	15.959	16.119	16.281
2.8	16.445	16.610	16.777	16.945	17.116	17.288	17.462	17.637	17.814	17.993
2.9	18.174	18.357	18.541	18.728	18.916	19.106	19.298	19.492	19.688	19.886
3.0	20.086	20.287	20.491	20.697	20.905	21.115	21.328	21.542	21.758	21.977
3.1	22.198	22.421	22.646	22.874	23.104	23.336	23.571	23.807	24.047	24.288
3.2	24.533	24.779	25.028	25.280	25.534	25.790	26.050	26.311	26.576	26.843
3.3	27.113	27.385	27.660	27.938	28.219	28.503	28.789	29.079	29.371	29.666
3.4	29.964	30.265	30.569	30.877	31.187	31.500	31.817	32.137	32.460	32.786
3.5	33.115	33.448	33.784	34.124	34.467	34.813	35.163	35.517	35.874	36.234
3.6	36.598	36.966	37.338	37.713	38.092	38.475	38.861	39.252	39.646	40.045
3.7	40.447	40.854	41.264	41.679	42.098	42.521	42.948	43.380	43.816	44.256
3.8	44.701	45.150	45.604	46.063	46.525	46.993	47.465	47.942	48.424	48.911
3.9	49.402	49.999	50.400	50.907	51.419	51.935	52.457	52.985	53.517	54.055

Table III
e^x (continued)

x	0	1	2	3	4	5	6	7	8	9
4.	54.598	60.340	66.686	73.700	81.451	90.017	99.484	109.95	121.51	134.29
5.	148.41	164.02	181.27	200.34	221.41	244.69	270.43	298.87	330.30	365.04
6.	403.43	445.86	492.75	544.57	601.85	665.14	735.20	812.41	897.85	992.27
7.	1096.6	1212.0	1339.4	1480.3	1636.0	1808.0	1998.2	2208.3	2440.6	2697.3
8.	2981.0	3294.5	3641.0	4023.9	4447.1	4914.8	5431.7	6002.9	6634.2	7332.0
9.	8103.1	8955.3	9897.1	10938	12088	13360	14765	16318	18034	19930
10.	22026									

Table IV

e^{-x}

x	0	1	2	3	4	5	6	7	8	9
0.0	1.00000	.99005	.98020	.97045	.96079	.95123	.94176	.93239	.92312	.91393
0.1	.90484	.89583	.88692	.87810	.86936	.86071	.85214	.84366	.83527	.82696
0.2	.81873	.81058	.80252	.79453	.78663	.77880	.77105	.76338	.75578	.74826
0.3	.74082	.73345	.72615	.71892	.71177	.70469	.69768	.69073	.68386	.67706
0.4	.67032	.66365	.65705	.65051	.64404	.63763	.63128	.62500	.61878	.61263
0.5	.60653	.60050	.59452	.58860	.58275	.57695	.57121	.56553	.55990	.55433
0.6	.54881	.54335	.53794	.53259	.52729	.52205	.51685	.51171	.50662	.50158
0.7	.49659	.49164	.48675	.48191	.47711	.47237	.46767	.46301	.45841	.45384
0.8	.44933	.44486	.44043	.43605	.43171	.42741	.42316	.41895	.41478	.41066
0.9	.40657	.40252	.39852	.39455	.39063	.38674	.38289	.37908	.37531	.37158
1.0	.36788	.36422	.36060	.35701	.35345	.34994	.34646	.34301	.33960	.33622
1.1	.33287	.32956	.32628	.32303	.31982	.31664	.31349	.31037	.30728	.30422
1.2	.30119	.29820	.29523	.29229	.28938	.28650	.28365	.28083	.27804	.27527
1.3	.27253	.26982	.26714	.26448	.26185	.25924	.25666	.25411	.25158	.24908
1.4	.24660	.24414	.24171	.23931	.23693	.23457	.23224	.22993	.22764	.22537
1.5	.22313	.22091	.21871	.21654	.21438	.21225	.21014	.20805	.20598	.20393
1.6	.20190	.19989	.19790	.19593	.19398	.19205	.19014	.18825	.18637	.18452
1.7	.18268	.18087	.17907	.17728	.17552	.17377	.17204	.17033	.16864	.16696
1.8	.16530	.16365	.16203	.16041	.15882	.15724	.15567	.15412	.15259	.15107
1.9	.14957	.14808	.14661	.14515	.14370	.14227	.14086	.13946	.13807	.13670
2.0	.13534	.13399	.13266	.13134	.13003	.12873	.12745	.12619	.12493	.12369
2.1	.12246	.12124	.12003	.11884	.11765	.11648	.11533	.11418	.11304	.11192
2.2	.11080	.10970	.10861	.10753	.10646	.10540	.10435	.10331	.10228	.10127
2.3	.10026	.09926	.09827	.09730	.09633	.09537	.09442	.09348	.09255	.09163
2.4	.09072	.08982	.08892	.08804	.08716	.08629	.08543	.08458	.08374	.08291
2.5	.08208	.08127	.08046	.07966	.07887	.07808	.07730	.07654	.07577	.07502
2.6	.07427	.07353	.07280	.07208	.07136	.07065	.06995	.06925	.06856	.06788
2.7	.06721	.06654	.06587	.06522	.06457	.06393	.06329	.06266	.06204	.06142
2.8	.06081	.06020	.05961	.05901	.05843	.05784	.05727	.05670	.05613	.05558

Table IV

e^{-x} (continued)

x	0	1	2	3	4	5	6	7	8	9
2.9	.05502	.05448	.05393	.05340	.05287	.05234	.05182	.05130	.05079	.05029
3.0	.04979	.04929	.04880	.04832	.04783	.04736	.04689	.04642	.04596	.04550
3.1	.04505	.04460	.04416	.04372	.04328	.04285	.04243	.04200	.04159	.04117
3.2	.04076	.04036	.03996	.03956	.03916	.03877	.03839	.03801	.03763	.03725
3.3	.03688	.03652	.03615	.03579	.03544	.03508	.03474	.03439	.03405	.03371
3.4	.03337	.03304	.03271	.03239	.03206	.03175	.03143	.03112	.03081	.03050
3.5	.03020	.02990	.02960	.02930	.02901	.02872	.02844	.02816	.02788	.02760
3.6	.02732	.02705	.02678	.02652	.02625	.02599	.02573	.02548	.02522	.02497
3.7	.02472	.02448	.02423	.02399	.02375	.02352	.02328	.02305	.02282	.02260
3.8	.02237	.02215	.02193	.02171	.02149	.02128	.02107	.02086	.02065	.02045
3.9	.02024	.02004	.01984	.01964	.01945	.01925	.01906	.01887	.01869	.01850
4.	.018316	.016573	.014996	.013569	.012277	.011109	.010052	$.0^2 90953$	$.0^2 82297$	$.0^2 74466$
5.	$.0^2 67397$	$.0^2 60967$	$.0^2 55166$	$.0^2 49916$	$.0^2 45166$	$.0^2 40868$	$.0^2 36979$	$.0^2 33460$	$.0^2 30276$	$.0^2 27394$
6.	$.0^2 24788$	$.0^2 22429$	$.0^2 20294$	$.0^2 18363$	$.0^2 16616$	$.0^2 15034$	$.0^2 13604$	$.0^2 12309$	$.0^2 11138$	$.0^2 10078$
7.	$.0^3 91188$	$.0^3 82510$	$.0^3 74659$	$.0^3 67554$	$.0^3 61125$	$.0^3 55308$	$.0^3 50045$	$.0^3 45283$	$.0^3 40973$	$.0^3 37074$
8.	$.0^3 33546$	$.0^3 30354$	$.0^3 27465$	$.0^3 24852$	$.0^3 22487$	$.0^3 20347$	$.0^3 18411$	$.0^3 16659$	$.0^3 15073$	$.0^3 13639$
9.	$.0^3 12341$	$.0^3 11167$	$.0^3 10104$	$.0^4 91424$	$.0^4 82724$	$.0^4 74852$	$.0^4 67729$	$.0^4 61283$	$.0^4 55452$	$.0^4 50175$
10.	$.0^4 45400$									

Table V
Trigonometric Functions

Rad.	Deg.	sin	tan	sec	csc	cot	cos	Deg.	Rad.
0.000	0°	0.000	0.000	1.000	—	—	1.000	90°	1.571
.017	1°	.017	.017	1.000	57.30	57.29	1.000	89°	1.553
.035	2°	.035	.035	1.001	28.65	28.64	0.999	88°	1.536
.052	3°	.052	.052	1.001	19.11	19.08	.999	87°	1.518
.070	4°	.070	.070	1.002	14.34	14.30	.998	86°	1.501
.087	5°	.087	.087	1.004	11.47	11.43	.996	85°	1.484
.105	6°	.105	.105	1.006	9.567	9.514	.995	84°	1.466
.122	7°	.122	.123	1.008	8.206	8.144	.993	83°	1.449
.140	8°	.139	.141	1.010	7.185	7.115	.990	82°	1.431
.157	9°	.156	.158	1.012	6.392	6.314	.988	81°	1.414
.175	10°	.174	.176	1.015	5.759	5.671	.985	80°	1.396
.192	11°	.191	.194	1.019	5.241	5.145	.982	79°	1.379
.209	12°	.208	.213	1.022	4.810	4.705	.978	78°	1.361
.227	13°	.225	.231	1.026	4.445	4.331	.974	77°	1.344
.244	14°	.242	.249	1.031	4.134	4.011	.970	76°	1.326
.262	15°	.259	.268	1.035	3.864	3.732	.966	75°	1.309
.279	16°	.276	.287	1.040	3.628	3.487	.961	74°	1.292
.297	17°	.292	.306	1.046	3.420	3.271	.956	73°	1.274
.314	18°	.309	.325	1.051	3.236	3.078	.951	72°	1.257
.332	19°	.326	.344	1.058	3.072	2.904	.946	71°	1.239
.349	20°	.342	.364	1.064	2.924	2.747	.940	70°	1.222
.367	21°	.358	.384	1.071	2.790	2.605	.934	69°	1.204
.384	22°	.375	.404	1.079	2.669	2.475	.927	68°	1.187
.401	23°	.391	.424	1.086	2.559	2.356	.921	67°	1.169
.419	24°	.407	.445	1.095	2.459	2.246	.914	66°	1.152
.436	25°	.423	.466	1.103	2.366	2.145	.906	65°	1.134
.454	26°	.438	.488	1.113	2.281	2.050	.899	64°	1.117
.471	27°	.454	.510	1.122	2.203	1.963	.891	63°	1.110
.489	28°	.469	.532	1.133	2.130	1.881	.883	62°	1.082
.506	29°	.485	.554	1.143	2.063	1.804	.875	61°	1.065
.524	30°	.500	.577	1.155	2.000	1.732	.866	60°	1.047
.541	31°	.515	.601	1.167	1.942	1.664	.857	59°	1.030
.559	32°	.530	.625	1.179	1.887	1.600	.848	58°	1.012
.576	33°	.545	.649	1.192	1.836	1.540	.839	57°	0.995
.593	34°	.559	.675	1.206	1.788	1.483	.829	56°	0.977
.611	35°	.574	.700	1.221	1.743	1.428	.819	55°	0.960
.628	36°	.588	.727	1.236	1.701	1.376	.809	54°	0.942
.646	37°	.602	.754	1.252	1.662	1.327	.799	53°	0.925
.663	38°	.616	.781	1.269	1.624	1.280	.788	52°	0.908
.681	39°	.629	.810	1.287	1.589	1.235	.777	51°	0.890
.698	40°	.643	.839	1.305	1.556	1.192	.766	50°	0.873
.716	41°	.656	.869	1.325	1.524	1.150	.755	49°	0.855
.733	42°	.669	.900	1.346	1.494	1.111	.743	48°	0.838
.750	43°	.682	.933	1.367	1.466	1.072	.731	47°	0.820
.768	44°	.695	0.966	1.390	1.440	1.036	.719	46°	0.803
.785	45°	.707	1.000	1.414	1.414	1.000	.707	45°	0.785
Rad.	Deg.	cos	cot	csc	sec	tan	sin	Deg.	Rad.

Answers to Selected Exercises

CHAPTER 1

1.1 Exercises

1. $-12\frac{2}{9}$°C, 10°C, $22\frac{2}{9}$°C, 260°C, $-35\frac{5}{9}$°C

3. -40°C $= -40$°F

5. 168

7. $\frac{3}{4}$

9. 16

11. 224

13. 76

15. 9

17. -48

19. $\frac{227}{4}$

21. $y = z + 10$

23. $y = z^2 - 4z + 7$

25. $y = z + z^2$

27. $y = \dfrac{2(z^2 - 6z + 9)}{(z^2 - 6z + 6)(z^2 - 6z + 12)}$

29. $y = \dfrac{z^4}{16} + \dfrac{z^3}{4} + \dfrac{3z^2}{4} + \dfrac{7z}{2}$

31. $x + 3$

33. $3x + 4$

35. $(x^2/4) + 8xy + 64y^2$

37. $x^4 + x^2 + 6x + 4$

39. $5(2x + 1)(x - 1)$

41. $[3x - 3y + 2w]^2$

43. $7(x^2 + 9)$

45. $1/(x - 5)^2$

47. $11x/12$

49. $(x + 15)/3$

51. $\dfrac{x^4 + x - x^3y - y + xy}{y(x^3 + 1)}$

53. $\dfrac{2x}{(x - y)(x + y)^2}$

55. $x = 0, y = \frac{4}{3}$

57. $x = \frac{22}{7}$, $y = \frac{23}{7}$

59. $x = 4$, or $x = \frac{2}{3}$

1.2 Exercises

1. $1, 9$

3. $\frac{9}{4}, \frac{121}{4}$

5. $z^2 - 4z + 4, z^2 + 4z + 4$

7. $\frac{3}{8}, -\frac{3}{7}$

9. $\dfrac{x^2}{x^2 - 5}, \dfrac{x^3}{x^3 - 5}$

11. $\sqrt{x}, \sqrt{x - 2}$

13. $|-x - 1|, \sqrt{x - 4}$

15. $\sqrt{5/(x - 5)}, \sqrt{-5/x}$

17. $1/2(y + 1), 1/2(y - 3)$

19. $\dfrac{x - 1}{2(2 - x)}, \dfrac{1}{2(y - 2)}$

21. $-x - 4$

23. $-x - d - 3$

25. 1

27. $a = 1, b = 0$

29. $-\frac{1}{4}, \sqrt{5}/2$

31. $1/(2y + 1), \sqrt{2y/(2y + 1)}$

33. $\sqrt{-1 - x^2}, 1/(2\sqrt{-x^2 - 1} + 1)$

35. 4

1.3 Exercises

1. Real numbers   **3.** Real numbers   **5.** $x \geqslant 6$
7. $x \neq -2$ **9.** $x > 5$
11. Function: domain $= \{2, 3, 4, -1\} =$ range
13. Function: domain, all reals; range, $\{y \mid y \geqslant -1\}$
15. Not a function   **17.** $f(-1) = 6$ **19.** $3a^2 - 2a + 1$
21. $6xh + 3h^2 - 2h$ **23.** 4
25. $\frac{1}{3}(x + 1)^3 - 2(x + 1) + 4$ or $\frac{1}{3}x^3 + x^2 - x + \frac{7}{3}$
27. $x^2 + hx + \frac{1}{3}h^2 - 2$
29. Yes   **31.** Yes

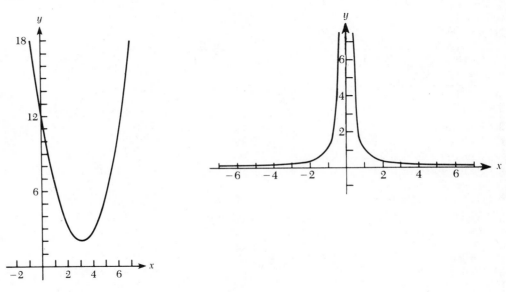

33. Yes   **35.** Yes

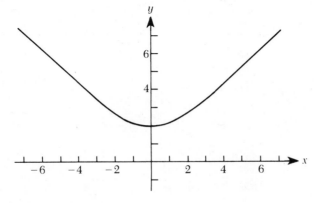
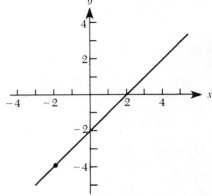

37. $f(1000) = -\$150$
$f(10,000) = \$3000$
$f(100,000) = \$34,500$
$f(x) = 0$ when $x = 10,000/7$
A profit is realized when the 1429th item is produced.

39. Domain: $\{x \mid x \geq 0\}$

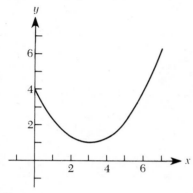

41. Domain: $\{x \mid x \geq 20\}$

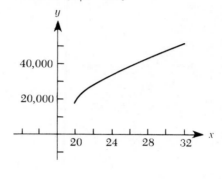

43. Function

45. Nonfunction

47. $f(x) + g(x) = x^2 + 5x + 1$
domain: {all reals}
$f(x) - g(x) = x^2 + x - 1$
domain: {all reals}
$f(x) \cdot g(x) = 2x^3 + 7x^2 + 3x$
domain: {all reals}

$\dfrac{f(x)}{g(x)} = \dfrac{x^2 + 3x}{2x + 1}$

domain: $\{x \mid x \neq -\frac{1}{2}\}$

49. $f(x) + g(x) = \sqrt{x^2 - 1} + x + 2$
domain: $\{x \mid |x| \geq 1\}$
$f(x) - g(x) = \sqrt{x^2 - 1} - x - 2$
domain: $\{x \mid |x| \geq 1\}$
$f(x) \cdot g(x) = (x + 2)\sqrt{x^2 - 1}$
domain: $\{x \mid |x| \geq 1\}$

$\dfrac{f(x)}{g(x)} = \dfrac{\sqrt{x^2 - 1}}{x + 2}$

domain: $\{x \mid |x| \geq 1,\ x \neq -2\}$

51. $f(x) + g(x) = 1 - \sqrt{x} + x^3$
domain: $\{x \mid x \geq 0\}$
$f(x) - g(x) = 1 - \sqrt{x} - x^3$
domain: $\{x \mid x \geq 0\}$
$f(x) \cdot g(x) = (1 - \sqrt{x})x^3$
domain: $\{x \mid x \geq 0\}$

$\dfrac{f(x)}{g(x)} = \dfrac{1 - \sqrt{x}}{x^3}$

domain: $\{x \mid x > 0\}$

53. $f \circ g = 1/(1 - \sqrt{x})$
domain: $\{x \mid x \geq 0,\ x \neq 1\}$

55. $f \circ g = (1 + \sqrt{x})^2$
domain: $\{x \mid x \geq 0\}$

57. $g \circ f = 2(x^2 + 3x) + 1$
$= 2x^2 + 6x + 1$
domain: {all reals}

59. $g \circ f = \sqrt{x^2 - 1} + 2$
domain: $\{x \mid |x| \geq 1\}$

61. $g \circ f = (1 - \sqrt{x})^3$
domain: $\{x \mid x \geq 0\}$

63. -1

65. Undefined

67. $1/(1 - \sqrt{x + h})$

69.

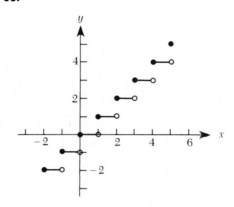

71.   Fare for 4.3 miles = 95¢.

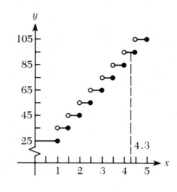

1.4   Exercises

1. $f(x) = \begin{cases} -x - 2, & x \in (-\infty, -1) \\ -x, & x \in [-1, 2) \\ x + 2, & x \in [2, \infty) \end{cases}$

3. 4, 8   **5.** 1, 5, 11, 11   **7.** $[-5, -4]$ **9.** $[-1/2, 1/2]$

11.    **13.** ─┼──┼──┼─●●○─┼──┼─► −200 0 200 400   **15.**

17. The real numbers between -4 and 8 excluding the ends

○●●●●●●●●●●●●●●○
  −4 −2 0 2 4 6 8

19. The real numbers between -3 and 3 including both numbers.

─┼─●●●●●●●●●●●●●●●●●─┼─►
  −4   −2   0   2   4

21. $[-1, \infty)$ **23.** $(0, \infty)$ **25.** $(-\infty, 0]$
27. $[-1, 3)$ **29.** $[-1, 3]$ **31.** $-1 \leqslant x \leqslant 5$
33. $-1 < x \leqslant 5$ **35.** $-1/2 \leqslant x < 0$ or $0 < x \leqslant 1/2$
37. Should be $\infty)$ **39.** $2 \not< -1$ **41.** $3, 3, x^2$
43. All values   **45.** $[-1, 5]$

47.

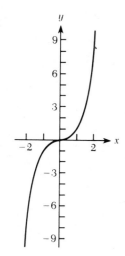

49.

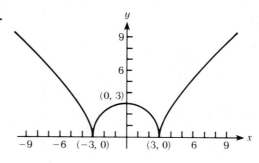

51.

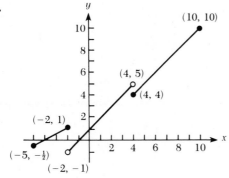

53. $1, 1, -2, -4$ **55.**

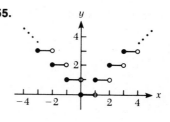

1.5   Exercises

1. Function, $f^{-1}(x) = x$

3. Function, $f^{-1}(x) = (x - 5)/4$

5. Function, $f^{-1}(x) = (2 - x)/6$

7. $f^{-1}(x) = 1/(2 - x)$

9. Function, $f^{-1}(x) = 2(2 - x)/(1 - x)$

11. $f^{-1}(x) = (3x - 7)/(2 - x)$

13. $f^{-1}(x) = x^2 - 25$

15. $f^{-1}(x) = x^{2/3}$

17. $f^{-1}(x) = \pm \sqrt{x} + 5$, not a function

19. $y = (x + 2)^2, x \geqslant -2$

21. $y = x^4, x \geqslant 0$

23.

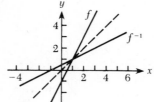

25.

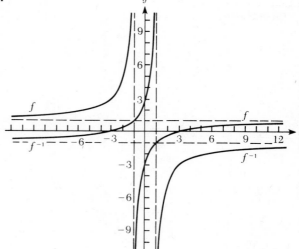

27.

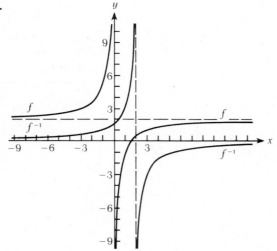

29.

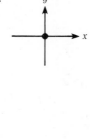

31. $f^{-1}(x) = \begin{cases} x, & x \in [-4, 0] \\ \sqrt{x}, & x \in (0, 16] \end{cases}$

domain $= [-4, 16]$

35. No inverse function

39. No inverse function

41. $f^{-1}(x) = (x - 2)/3, f(f^{-1}(x)) = f^{-1}(f(x)) = x$

33. $h^{-1}(x) = \begin{cases} \sqrt{-x}, & x \in (-\infty, 0] \\ -x, & x \in (0, \infty) \end{cases}$

domain $= \{$all reals$\}$

37. Inverse function exists

1.6 Exercises

1. 9   **3.** $1/\sqrt{2}$ **5.** 9   **7.** 10^4 **9.** 10^3
11. 3   **13.** 10   **15.** -5 **17.** $-1/3$ **19.** -5
21. 2   **23.** 1000   **25.** b **27.** 10^{-3} **29.** 10^{-1}
31. #2   **33.** #6   **35.** #4   **37.** #2   **39.** #2
41. 3.90   **43.** 1.16   **45.** 8.64   **47.** $\dfrac{\log 8.1}{\log 2}$ **49.** $\dfrac{\log 8}{\log e} \approx 2.08$

1.7 Exercises

1. 0.2   **3.** 14.25   **5.** $2ah + h^2$
7. $6x(\Delta x) + 3(\Delta x)^2 + 2(\Delta x)$ **9.** $3x^2h + 3xh^2 + h^3$ **11.** 2
13. $2\sqrt{17} - 8$ **15.** $2ax + ah + b$ **17.** $6a^2 + 6ah + 2h^2$
19. 144   **21.** 96.016   **23.** 15.3
25. 225/14
27. \$45
29. Marginal cost is the average rate of change of cost taken over the interval of the one additional unit of production.

Review

1. $98\frac{3}{5}$ or 98.6°F, $64\frac{2}{5}$°F, -4°F, 32°F, $413\frac{3}{5}$°F
2. 2.78°C, -7.78°C, -28.89°C, -17.78°C, 100°C
3. -130
4. 1
5. $\frac{15}{14}$
6. $\frac{101}{3}$
7. 162
8. -9.7
9. 1
10. 1
11. $y = z^2$
12. $y = (z^2 + 2)^2 - 2(z^2 + 2) + 2$
13. $y = \dfrac{2z}{(z-1)(z+1)}$
14. $y = \dfrac{z^2 - z - 4}{z^2 - z - 6}$
15. $\frac{1}{2}$, 0, 2
16. $-6, -6, 0$
17. $\dfrac{x+2}{x+3}, \dfrac{x-2}{x-1}, \dfrac{x^2+x}{x^2+x+1}$
18. $x^2 + 3x - 4$, $x^2 - 5x$, $x^4 + 2x^3 - x - 6$
19. $0, x^4 - x^2 - 6, 0, \dfrac{x^4 - x^2 - 6}{x^4 - x^2 - 5}$
20. $\frac{3}{4}, \dfrac{x^2}{x^2+1}, -\frac{99}{16}, \left(\dfrac{x^2}{x^2+1}\right)^2 - \left(\dfrac{x^2}{x^2+1}\right) - 6$
21. Function, $x \neq 0$
22. Function, all reals
23. Function, $x \in (-3, 3)$
24. Nonfunction

25. Function

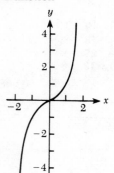

26. Function

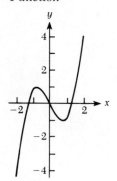

27. Nonfunction

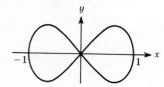

28. Function

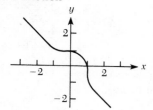

29. 1, 3, 9, 5

30. 12, 1, 1

31.

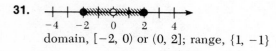

domain, $[-2, 0)$ or $(0, 2]$; range, $\{1, -1\}$

32.

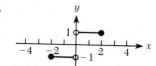

33. Function, all reals, range $\{7\}$

34. 7, 7, 7, 7, 7

35. 8, 5

36. $x^{3/2} - 2x^{1/2} + 4$, $(x + h)^3 - 2(x + h) + 4$

37. $3x^2 + 3xh + h^2 - 2$

38. $(f + g)(x) = 3x^2 + 2x$, all reals

39. $(f - g)(x) = 3x^2 - 4x - 4$, all reals

40. $(f \cdot g)(x) = (3x^2 - x - 2)(3x + 2)$, all reals

41. $\left(\dfrac{f}{g}\right)(x) = \dfrac{3x^2 - x - 2}{3x + 2}$, $x \neq -\dfrac{2}{3}$

42. $f \circ g(x) = 3(3x + 2)^2 - (3x + 2) - 2$, all reals

43. $3(3x^2 - x - 2) + 2$, all reals

44. Not a function

45. $g^{-1}(x) = (x - 2)/3$

46. $f^{-1}(x) = 3/(2 - x)$

47. $y = (x + 7)/3$

48.

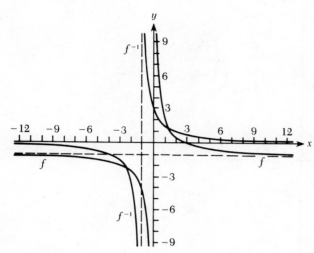

49. No inverse function

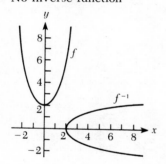

50.

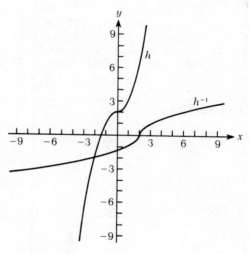

51.

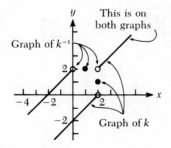

52. $\frac{1}{256}$ **53.** $2\sqrt{2}$
54. 10^{1000} **55.** 1000
56. -3 **57.** $-\frac{1}{3}$
58. 3   **59.** None exists
60. 10   **61.** $2x\,\Delta x + (\Delta x)^2 - 3\Delta x$
62. $3x^2 + 3x + 1$ **63.** 3
64. $-8b - 4h$

CHAPTER 2

2.1   Exercises

1. 0.4, 0.6, 0.8, 1.4, 1.2   **3.** $\lim_{x\to 3} f(x) = 1$ **5.** 0

7. 2   **9.** 132   **11.** 2
13. -1 **15.** -7 **17.** 12

19. Undefined   **21.** $\dfrac{\sqrt{a} - a}{a - a^2} = \dfrac{1}{\sqrt{a} + a}$ **23.** $2\sqrt{a}$

25. 11   **27.** 6   **29.** $3x^2$
31. Continuous   **33.** Continuous
35. Discontinuous since $f(2)$ is undefined
37. Discontinuous since $f(2)$ is undefined
39. 1   **41.** 6   **43.** 1, 2

2.2   Exercises

1. 0   **3.** $\frac{1}{2}$ **5.** $\frac{3}{5}$ **7.** -3
9. ∞ **11.** 0   **13.** $-\frac{1}{3}$ **15.** 2
17. $-\frac{5}{4}$ **19.** 0   **21.** $-\frac{2}{5}$ **23.** Undefined
25. -1 **27.** 1

2.3   Exercises

1. $\frac{5}{4}$ **3.** $-\frac{5}{7}$ **5.** Undefined
7. $-3, 5$ **9.** $\frac{4}{3}, \frac{7}{3}$ **11.** 0, 4
13. $y = x$ **15.** $3x - 4y = -13$ **17.** $y = 2$
19. $y = x$ **21.** $7x + 4y = -1$ **23.** $y = -5$
25. $x = -5$ **27.** m = marginal cost
                                      b = fixed cost

29. **a.** $90   **31.** **a.** $650   **33.** **a.** $7000
     b. $9   **b.** $8.125   **b.** $58.33
     c. $40   **c.** $10   **c.** $1000
     d. $50   **d.** $640   **d.** $6000
     e. $5   **e.** $8   **e.** $50

2.4 Exercises

1. $dy/dx = 2x$ **3.** $dy/dx = -6$ **5.** $D_x y = -2x$ **7.** $D_x y = -2/x^3$
9. $f'(3) = -\frac{1}{9}$ **11.** $h'(-1) = -5$ **13.** $f'(0) = 0$ **15.** $-\frac{1}{36}$
17. 1   **19.** 0.99   **21.** $6I - 2$
23. Yes, $\lim\limits_{x \to 1^+} f(x) = \lim\limits_{x \to 1^-} f(x) = 1 = f(1)$ **25.** No, $\lim\limits_{x \to -2^-} f(x) = 4$ and $\lim\limits_{x \to -2^+} f(x) = -1$
27. Yes, $\lim\limits_{x \to -1^+} f(x) = \lim\limits_{x \to -1^-} f(x) = \frac{1}{2} = f(-1)$

2.5 Exercises

1. $3x^2$ **3.** $35x^4$ **5.** $-36x^{-4}$ **7.** $1/(2\sqrt{x})$
9. $15x^{2/3}$ **11.** $6x - 6$ **13.** $6x - 4$ **15.** $2ax + b$
17. -20 **19.** -4
21. $\dfrac{3x^5 - 2}{x^3} = 3x^2 + \dfrac{2}{x^3}$ **23.** $\dfrac{5x^2 + 9x + 1}{2\sqrt{x}} = \dfrac{5}{2}x^{3/2} + \dfrac{9}{2}x^{1/2} + \dfrac{1}{2}x^{-1/2}$
25. $3x^2 + 8x + 3$ **27.** $y' = -\dfrac{2}{x^2}$
29. $y' = \dfrac{30x - 50}{x^3} = \dfrac{30}{x^2} - \dfrac{50}{x^3}$

2.6 Exercises

1. $30x + 11$ **3.** $3t^2 + 10t + 4$
5. $24x^3 + 30x^2 + 24x + 20$ **7.** $48t^3 + 42t^2 + 28t - 12$
9. $108x^5 + 165x^4 + 114x^2 + 38x + 2$ **11.** $3t^2 + 10t + 3$
13. $3x^2 - 1$ **15.** $-\dfrac{2}{(x - 1)^2}$ **17.** $\dfrac{1 - t^2}{(t^2 + 1)^2}$ **19.** $\dfrac{x^2 + 2x - 5}{(x + 1)^2}$
21. $\dfrac{-4(t + 1)}{(t - 1)^3}$ **23.** $\dfrac{-3x^2}{(x^3 - 1)^2}$ **25.** $-\dfrac{1}{3t^{2/3}(t^{1/3} - 1)^2}$ **27.** $-\dfrac{1}{\sqrt{x}(1 + \sqrt{x})^2}$
29. $-\dfrac{29}{625}$

2.7 Exercises

1. $6(x + 5)^5$ **3.** $(2x + 1)^{-1/2}$ **5.** $x(x^2 + 1)^{-1/2}$ **7.** $\dfrac{-2}{3(x + 4)^{5/3}}$
9. $\dfrac{-(3x^2 + 2)}{3(x^3 + 2x)^{4/3}}$ **11.** $\dfrac{9x + 4}{2\sqrt{3x + 2}}$ **13.** $\dfrac{5x^3 + 90x}{(x^2 + 9)^{3/2}}$
15. $\frac{1}{3}(x + \sqrt{x})^{-2/3}\left(1 + \dfrac{1}{2\sqrt{x}}\right)$ **17.** $\dfrac{2x - 5}{(x + 2)^2 \sqrt{x^2 + 5}}$
19. $\dfrac{-x^2 - x}{(x^2 - 1)^{3/2}(x^3 + 1)^{2/3}}$ **21.** $\dfrac{2\sqrt{x^2 + 1} + x}{2\sqrt{x^2 + 1}\sqrt{2x + \sqrt{x^2 + 1}}}$
23. Undefined   **25.** $\sqrt{2}\,x^{\sqrt{2} - 1}$

2.8   Exercises

1. 0

3. 4

5. $6x$

7. $2 + (6/x^4)$

9. $\dfrac{4}{(x-1)^3}$

11. $\dfrac{4x}{(x^3+2)^{5/3}}$

13. $12x^2 - 4$

15. $\dfrac{1-2x}{(x+1)^{5/2}(x-1)^{3/2}}$

17. $\frac{35}{4}x^{3/2}$

19. $\dfrac{3x^4 + 6x^2 - 1}{4(x^3+x)^{3/2}}$

21. 24

23. $-6/(x-1)^4$

25. $6/(x+1)^4$

27. 0

29. $101!\,x^{-102}$

31. When $n < a$, $a(a-1)(a-2)\cdots(a-n+1)x^{a-n}$
When $n > a$, the nth derivative of x^a is zero

33.

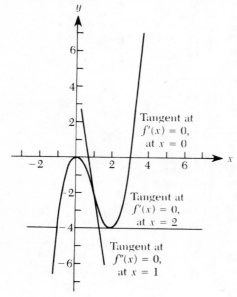

Tangent at $f'(x) = 0$, at $x = 0$

Tangent at $f'(x) = 0$, at $x = 2$

Tangent at $f''(x) = 0$, at $x = 1$

Review

1. 20

2. 210

3. 3

4. \$3.29

5. 230

6. $6x + 5 - \dfrac{2}{x^3} - \dfrac{3}{2x^{3/2}}$

7. $-1/x^{3/2}$

8. $-1/(x+2)^2$

9. $\dfrac{4t^4 + 9t^2 + 3}{\sqrt{t^2+1}}$

10. -1

11. $-\dfrac{2+x^2}{x^3\sqrt{x^2+1}}$

12. $\dfrac{1}{2}\sqrt{\dfrac{x+\sqrt{x^2+1}}{x^2+1}}$

13. $\frac{5}{2}x^{3/2}$

14. $\dfrac{4x^2+3x}{2\sqrt{x^2+x}}$

15. $\dfrac{-2}{(x+\sqrt{x^2+1})^2\sqrt{x^2+1}}$

16. $4/(x^2 + 4)^{3/2}$

17. $\dfrac{6x^2 + 3}{\sqrt{x^2 + 1}}$

18. $-\dfrac{5 \cdot 7 \cdot 9}{2^3} x^{-11/2}$

19. $-3x/(x^2 + 1)^{5/2}$

20. $499/10,000 = \$0.0499$

21. $3\frac{9}{16}$ grams per hour

CHAPTER 3

3.1 Exercises

1. If $x_1 = 1$, then $x_3 = 1.75$

3. If $x_1 = 2.5$, then $x_3 = 2.16$

5. $1.5, -2.67$

7. No real roots

9. $7.32, 1.37, -0.70$

11. 1.41

13. 2.83

15. If $x_1 = 0$, then $x_3 = -0.452$
If $x_1 = -1$, then $x_3 = -0.430$

17. 2.83

19. -0.3557

23. 0.7937

25. 1.2589

3.2 Exercises

1. Decreasing, concave upward

3. Decreasing, concave upward

5. Neither, concave upward

7. Increasing, neither

9. Not defined at $x = 0$

11. Increasing, concave upward

13. Increasing: $x > -1$
Decreasing: $x < -1$
Concave up: for all x

15. Increasing: $t < \frac{5}{2}$ or $t > 5$
Decreasing: $\frac{5}{2} < t < 5$
Concave up: $t > \frac{15}{4}$
Concave down: $t < \frac{15}{4}$

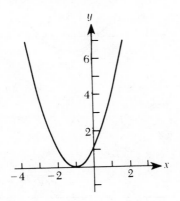

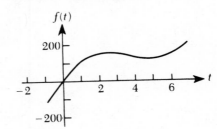

17.   Increasing: never
Decreasing: $x > 0$ or $x < 0$
Concave up: $x > 0$
Concave down: $x < 0$

19. $(0, 0)$

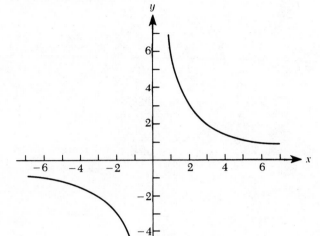

21. $(0, 0)$

23.

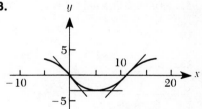

3.3   Exercises

1. $(-2, -5)$ min

3.   None

5.   None

7. $(1, -8)$ min
$(-1, 8)$ max

9. $(3, -77)$ min
$(-2, 48)$ max

11. $(1, 11)$ min
$(-1, 11)$ min
$(0, 12)$ max

13. No max or min   **15.** (1, 2) min   **17.** $(\frac{1}{8}, -\frac{3}{4})$ min

(−1, −2) max

19. (1, 0) min   **21.** $(-\frac{8}{27}, \frac{4}{27})$ max, (0, 0) min

---

3.4 Exercises

1. 15, 15

3. 5, 25

5. 90.7 cubic inches

7. 160 × 160

9. 6 × 6 × 15

11. $(5\sqrt{2} + 4) \times (10\sqrt{2} + 8)$

or 11.1 × 22.1

13. $r = 2, h = \dfrac{16}{\pi}$

15. Right angles

17. \$275

19. 175

21. $s = 20\sqrt{5} \approx 44.7$ miles per hour

23. $\left(10 - \dfrac{8\sqrt{3}}{3}\right)$ miles ≈ 5.4 miles from A

25. 250 items

3.5 Exercises

1. $dy = 16x\, dx$

3. $dy = \dfrac{x\, dx}{\sqrt{x^2 + 25}}$

5. $dy = (4x^3 + 6x)\, dx$

7. $dy = \dfrac{dx}{(x + 1)^2}$

9. 0.14

11. 3×10^{-5}

13. $dy = 2, \Delta y = 2.04$

15. $dy = 1.2, \Delta y = 1.261$

17. $\dfrac{142}{1331}$

19. −39

21. $V \approx 226.8$ cubic inches

23. $\sqrt{100 + dx} = 10 + \dfrac{dx}{20}$

25. $2\sqrt{\dfrac{3}{\pi}}$

3.6 Exercises

1. $\dfrac{dy}{dx} = \dfrac{x}{y}$

3. $\dfrac{dy}{dx} = \dfrac{-1 - y}{x}$

5. $\dfrac{dy}{dx} = \dfrac{y - 3x^2}{2y - x}$

7. $\dfrac{dy}{dx} = \dfrac{-y}{2(x + 1)}$

9. $\dfrac{dy}{dx} = \dfrac{2xy^2 - y - 1}{-2x^2y + x + 1}$

11. $\dfrac{dy}{dx} = \dfrac{-y^{1/3}}{x^{1/3}}$

13. $\frac{4}{3}$

15. $\frac{1}{3}$

17. $\frac{6}{5}$

19. $-1/y^3$

21. $-\dfrac{16}{9y^3}$

23. 0

25. $\dfrac{3x^2 + 4xy + y^2}{(y + 2x)^3} = 0$

3.7 Exercises

1. \$3 per hour

3. \$0 per week

5. \$26,000 per day

7. \$1400 per day

9. $1600 per day

11. -12π square inches per minute, -2π square inches per minute

13. $1/20\pi$ feet per second
17. approx. 27 feet per minute
21. $13 per week

15. -14 square feet per second
19. $230/\sqrt{61} \approx 29.4$ knots

3.8 Exercises

1. $f^{-1}(x) = \dfrac{x-2}{3}, \ \frac{1}{3}$

3. $f^{-1}(x) = (x-2)^{1/5}, \ \frac{1}{5}(x-2)^{-4/5}$

5. $f^{-1}(x) = \dfrac{2x+4}{x-1}, \ \dfrac{-6}{(x-1)^2}$

7. $f^{-1}(x) = \dfrac{x+5}{x-2}, \ \dfrac{-7}{(x-2)^2}$

9. $f^{-1}(x) = \dfrac{5x-3}{2(1+2x)}, \ \dfrac{11}{2(1+2x)^2}$

11. $\frac{1}{3}$

13. $1/(5y^4)$
19. $(5-4y)^2/22$
25. $1/(3y^2-2)$

15. $-(y-2)^2/6$
21. $1/[3(y-3)^2]$
27. $-y^3/2$

17. $-(y-1)^2/7$
23. $-y^4/3$
29. $(y^2+1)^2/(1-y^2)$

3.9 Exercises

1. $c = \frac{7}{2}$

3. $c = \dfrac{4-\sqrt{13}}{3} \approx 0.13$

5. $c = \sqrt{14} \approx 3.74$

7. $\dfrac{f(3)-f(0)}{3-0} = 0$

$d = 2$

$0 < 2 < 3$

9. $\dfrac{f(3.5)-f(3.1)}{3.5-3.1} = -20$

$d = 3 + \dfrac{1}{\sqrt{20}} \approx 3.22$

$3.1 < 3 + \dfrac{1}{\sqrt{20}} < 3.5$

11.

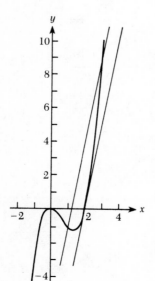

13. $f(3) = 12 \neq 0$
15. Not continuous at $x = 1$
17. Not continuous at $x = 0$
19. Works fine and satisfies mean value theorem

Review

1. 1.74 and -5.74

3. Increasing: $x > -2$
Decreasing: $x < -2$
Always concave upward
Minimum: $(-2, -3)$
No max or inflection pts.

2. -0.34

4. Always increasing
Concave upward: $x > 0$
Concave downward: $x < 0$
No max or min; inflection pt.: $(0, 0)$

5. $\dfrac{dy}{dx} = \dfrac{3x}{4y}$

6. $\dfrac{dy}{dx} = -\sqrt{\dfrac{y}{x}}$

7. $\dfrac{dy}{dx} = \dfrac{3x^2 - 4y + 2x}{4x}$

8. $\dfrac{dy}{dx} = \dfrac{2y^{1/3}}{x^{1/3}(3y^{1/3} - 2)}$

9. $\dfrac{dy}{dx} = \dfrac{(2x + y)}{1 - x - 2y}$

10. $y' = \dfrac{-2(1 + y)}{1 + 2x}$

11. $y'' = -\dfrac{4}{y^3}$

12. $y'' = 0$

13. $y'' = \dfrac{2}{(1 - 2y)^3}$

14. x

15. 20 monetary units per hour

16. ≈ 21.6 cubic inches

17. $dy = (6x + 8)\,dx$
$\Delta y = 6x(\Delta x) + 3(\Delta x)^2 + 8(\Delta x)$

18. 18 square feet per minute

19. 125×125 feet

20. 50 passengers

21. 911 feet from A

22. Let $c = 3$

23. $\dfrac{-1 + \sqrt{19}}{3} = c$

CHAPTER 4

4.1 Exercises

1. $\dfrac{3x^2}{2} + C$

3. $\dfrac{x^4}{4} + \dfrac{x^3}{3} - x + C$

5. $\dfrac{3x^{4/3}}{4} + C$

7. $-\dfrac{1}{x} + C$

9. $\frac{1}{3}(x + 2)^3 + C$

11. $-\dfrac{2}{x} - \dfrac{3}{2x^2} + C$

13. $\frac{1}{3}(x^2 + 5)^{3/2} + C$

15. $-\sqrt{1 - x^2} + C$

17. $\frac{1}{10}(6x + 9)^5 + C$

19. $-\dfrac{x^5}{5} + \dfrac{5x^3}{3} - 4x + C$

21. $\frac{1}{4}(4x^2 + 2x + 1)^4 + C$

23. $\dfrac{-1}{2(2x^2 - 3x + 1)^2} + C$

25. $-\dfrac{1}{x} - \dfrac{2}{\sqrt{x}} + C$

27. $\frac{1}{2}(x^4 - 1)^{1/2} + C$

29. $x^3 + \dfrac{x^2}{2} - 3x + C$

31. $-\dfrac{4}{x} - \dfrac{2}{3\sqrt{x}} + C$

33. $x^3 - 4x; C = 0$

35. $\frac{3}{4}x^{4/3} - 2; C = -2$

37. $\dfrac{1}{x^2 - 4}$

39. $x^2 - x; x^2 - x + C$

4.2 Exercises

1. 30

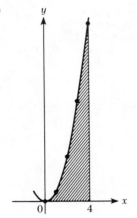

3. 28

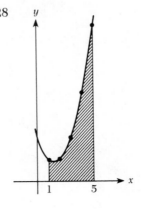

5. 13

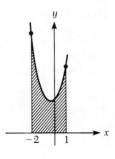

7. 13 (see figure for Exercise 5)

9. 30.875

11. $\displaystyle\int_1^3 x^2\, dx$

13. 6.75
19. 1.30

15. 7.69

17. 1.22

4.3 Exercises

1. $-1 + 1 + 3 + 5 + 7 = 15$

3. $1 + \frac{1}{2} + \frac{1}{3} + \frac{1}{4} + \frac{1}{5} + \frac{1}{6} = 2.45$ (or $2\frac{9}{20}$)

5. $(1 - \frac{1}{2}) + (\frac{1}{2} - \frac{1}{3}) + (\frac{1}{3} - \frac{1}{4}) + (\frac{1}{4} - \frac{1}{5}) = \frac{4}{5}$

7. $\dfrac{n^3 + 6n^2 + 8n}{3}$

9. $\dfrac{4n^3 + 27n^2 + 59n}{6}$

11. $\frac{64}{3}$

13. $\frac{14}{3}$

15. $\frac{27}{2}$

17. 0

19. $\frac{5}{6}$

21.

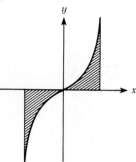

23. 4

25. 0

27. \$73,200

4.4   Exercises

1. $c = \dfrac{4\sqrt{3}}{3} \approx 2.31$

3. $c = \dfrac{2\sqrt{3}}{3} \approx 1.15$

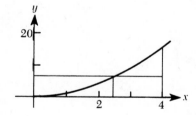

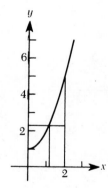

5. $c = \dfrac{-1 + \sqrt{31}}{6} \approx 0.76$ or $c = \dfrac{-1 - \sqrt{31}}{6} \approx -1.09$

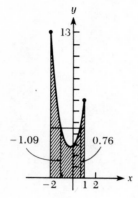

7. $\frac{8}{3}$

9. $-\frac{7}{3}$

11. $\frac{17}{12}$

13. 0

15. $-\frac{1}{12}$

17. $\frac{2438}{15}$

19. $\frac{7666}{15}$

21. No, not continuous at $x = 1$

23. Applies

25. Applies

27. $4\frac{1}{3}$

29. $37\frac{1}{3}$

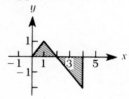

4.5   Exercises

1. $[a, b]$ cannot contain 0

3. Any interval $[a, b]$

5. $[a, b]$ cannot contain 1 or -2

7. $10\frac{2}{3}$

9. $37\frac{1}{3}$

11. $\frac{7}{27}$

13. $\frac{14}{3}$

15. 2

17. $\dfrac{4a^{5/2}}{15}$

19. $3\frac{7}{12}$

21. 0

23. $71\frac{2}{3}$

25. $15\frac{1}{15}$

27. 52

29. $\frac{2}{3}$

4.6   Exercises

1. Form 7
$$2\left[\frac{1}{4(6)}(2x + 3)^6 - \frac{3}{4(5)}(2x + 3)^5\right] + C$$
$$= \frac{(2x + 3)^6}{12} - \frac{3(2x + 3)^5}{10} + C$$

3. Form 5
$$\frac{1}{4}\ln|4x + 5| + C$$

5. Form 42

$\frac{1}{4} e^{4x} + C$

7. Form 24

$\frac{-2}{2x + 10} + C = \frac{-1}{x + 5} + C$

9. Form 11

$\frac{1}{3} \left(\frac{1}{\sqrt{5}} \ln \left| \frac{\sqrt{2x + 5} - \sqrt{5}}{\sqrt{2x + 5} + \sqrt{5}} \right| \right) + C$

11. Form 7

$\frac{1}{28} (2x + 5)^7 - \frac{5}{24} (2x + 5)^6 + C$

13. Form 43

$\frac{1}{4} x^2 e^{4x} - \frac{1}{8} xe^{4x} + \frac{1}{32} e^{4x} + C$

15. Form 48

$\ln \left| \ln x \right| + C$

17. Form 23

$\frac{1}{\sqrt{5}} \ln \left| \frac{2x + 5 - \sqrt{5}}{2x + 5 + \sqrt{5}} \right| + C$

19. Form 53

$3! = 3 \cdot 2 \cdot 1 = 6$

21. Form 7; 220

23. Form 5

$\frac{1}{2}[\ln 11 - \ln 7] = \frac{1}{2} \ln \frac{11}{7} \approx 0.23$

25. Form 2 with $u = x^2 + 3$

$\frac{2}{3} (7 \sqrt{7} - 8) \approx 7.01$

4.7 Exercises

1. 22   **3.** 14.5   **5.** 15.29   **7.** 60   **9.** 1.01   **11.** 0.75

Review

1. a. $\frac{15}{4}$
b. $\frac{11}{4}$
c. $\frac{8}{3}$

2. a. $\frac{39}{2}$
b. 18
c. 18

3. a. 28
b. $22\frac{1}{2}$
c. 22

4. a. $22\frac{1}{4}$
b. $18\frac{3}{4}$
c. $18\frac{2}{3}$

5. a. $6\frac{1}{2}$
b. 6
c. 6

6. $\frac{32}{3}$

7. 12

8. $C = 4/\sqrt{3} \approx 2.31$

9. $\frac{x^3}{3} + 2x + C$

10. $\frac{x^6}{2} + \frac{x^4}{2} - \frac{x^3}{3} + x + C$

11. $-\frac{1}{x} + C$

12. $2 \sqrt{x} + C$

13. $\frac{2}{5}(x + 3)^{5/2} + C$

14. $\frac{2}{5}x^{5/2} + \frac{2}{3}x^{3/2} - 3x + C$

15. $\frac{3}{8}(2x - 1)^{4/3} + C$

16. $-\frac{3}{8}(1 - 2x)^{4/3} + C$

17. $\frac{3}{4}(2x - 1)^{2/3} + C$

18. $\sqrt{2x - 1} + C$

19. $\sqrt{x^2 - 1} + C$

20. $-\sqrt{1 - x^2} + C$

21. $\frac{1}{15}(x^3 + 3x + 4)^5 + C$

22. $-\frac{1}{4(x^3 + 2x - 4)^4} + C$

23. $\frac{x^2}{2} + \frac{4}{3} x^{3/2} + x + C$

24. $-\frac{2}{3}$

25. $18\frac{2}{3}$

26. 28

27. 12

28. $37\frac{11}{12}$

29. 48

30. $1077\frac{1}{6}$

31. $\frac{49}{12}$

32. $\frac{5}{14}$

33. $1\frac{1}{10}$

34. $2 \sqrt{5} - 4$

35. $\frac{2}{3}$

36. Form 8; $\dfrac{2}{15(4)} (2x + 3)^{3/2} (6x - 6) + C = \dfrac{1}{5} (2x + 3)^{3/2} (x - 1) + C$

37. Forms 43 and 51; $x^2 e^x - 2e^x(x - 1) + C$

38. Form 2: $u = 4x^2 + 1$; $\frac{3}{40} (4x^2 + 1)^5 + C$

39. Form 23; $\dfrac{1}{\sqrt{12}} \ln \left| \dfrac{2x - 6 - \sqrt{12}}{2x - 6 + \sqrt{12}} \right| + C$

40. Form 5; $\frac{1}{3} \ln |3x + 7| + C$

CHAPTER 5

5.1 Exercises

1. $270°$	**3.** $120°$	**5.** $210°$	**7.** $-90°$
9. $-420°$	**11.** $3\pi/2$	**13.** $7\pi/4$	**15.** $-2\pi/3$
17. $-25\pi/6$	**19.** $\sqrt{3}/2$	**21.** $1/\sqrt{3}$	**23.** 1
25. $\sqrt{3}$	**27.** -2	**29.** 1	**31.** 0
33. Undefined	**35.** Undefined	**37.** Undefined	**39.** Undefined
41. 1	**43.** $\frac{1}{2}$	**45.** $\sqrt{3}/2$	**47.** 1.221
49. 5.759	**51.** 0.554	**53.** 0.999	**55.** -0.087
57. -1.006	**59.** -0.819	**61.** 0.839	

5.2 Exercises

1. $\alpha \neq 0, \pi/2, \pi, 3\pi/2, 2\pi$	**3.** $\alpha \neq 0, \pi, 2\pi$	**5.** $\alpha \neq 0, \pi, 2\pi$
7. $\alpha \neq \pi/2, 3\pi/2$	**9.** $\dfrac{1 - \sqrt{3}}{2\sqrt{2}}$	**11.** $\dfrac{1 + \sqrt{3}}{2\sqrt{2}}$
13. $-\sin\theta$	**15.** $-\sin\theta$	**17.** $\tan\theta$
19. $\sqrt{3}/2$	**21.** $1/\sqrt{2}$	

5.3 Exercises

1. 2	**3.** $2/\pi$	**5.** $-\infty$	**7.** $+\infty$
9. 0	**11.** 1	**13.** No limit	**15.** 3
17. n	**19.** No limit	**21.** 1	**23.** 1
25. 2	**27.** 0	**29.** No limit	

5.4 Exercises

1. $5\cos 5x$	**3.** $-4\sin 4x$
5. $3\cos t$	**7.** $24\sin 6t$
9. $\cos x + \sin x$	**11.** $-2\cos x \sin x = -\sin 2x$
13. $\cos x^2 - 2x^2 \sin x^2$	**15.** $\cos^2 x - \sin^2 x = \cos 2x$
17. $\dfrac{\cos x}{2\sqrt{\sin x}}$	**19.** $\dfrac{x\cos x - \sin x}{x^2}$
21. $\dfrac{-2\cos t}{(1 + \sin t)^2}$	**23.** $\dfrac{-\cos\sqrt{x}\sin\sqrt{x}}{\sqrt{x}}$

25. $-\pi^2$

29. $-4 \sin 2x$

33. $0, y = 1$

27. $\approx 10^3(3(3.14)^2) \approx \$29,600$ per year

31. $-3 \cos x + x \sin x$

35. $\pi/6, \pi/2, 5\pi/6, 7\pi/6, 3\pi/2, 11\pi/6$

5.5 Exercises

1. $3 \sec 3x \tan 3x$

5. $-2x \csc^2 x^2$

9. $2 \tan t \sec^2 t$

13. $2t \csc \sqrt{t} - \dfrac{t^2 \csc \sqrt{t} \cot \sqrt{t}}{2 \sqrt{t}}$

17. $\dfrac{2 \tan x \cos 2x - \sin 2x \sec^2 x}{\tan^2 x} = -2 \sin 2x$

21. $2 \sec^2 x \tan x$

25. $2x^2 \sec^2 x \tan x + 4x \sec^2 x + 2 \tan x$

29. $18x \csc^2 3x \cot 3x - 6 \csc^2 3x$

33. π

3. $6 \sec^2 2x$

7. $8x \csc x^2 \cot x^2$

11. $6t + 7 + 2t \sec t^2 \tan t^2$

15. $\cos t \sec^2 (\sin t)$

19. $\dfrac{2 \sec^2 x}{\sec^2 x + 2 \tan x}$

23. $\sec^3 x + \sec x \tan^2 x$

27. $\sec^4 x + 2 \sec^2 x \tan^2 x$

31. 2

35. $4 \sec^2 x \tan^2 x + 2 \sec^4 x$

5.6 Exercises

1. $\pi/3$

7. $3\pi/4$

13. $\frac{1}{2}$

19. -1

25. 1.030

31. $[-1, 1]$

37. $\frac{3}{5}, \frac{4}{5}, \frac{5}{4}, \frac{3}{4}, \frac{4}{3}, \frac{5}{3}$

3. $\pi/2$

9. 0

15. Undefined

21. 0.681

27. 1.065

33. All real numbers

5. $-\pi/3$

11. $-\pi/6$

17. $\pi/2$

23. 0.890

29. 0.140

35. $x \geq 1, x \leq -1$

5.7 Exercises

5. $\dfrac{2x}{\sqrt{1 - x^4}}$

9. $2x \cot^{-1} (2x) - \dfrac{2x^2}{1 + 4x^2}$

13. $\dfrac{64x \tan^{-1} x^2}{1 + x^4}$

17. $\dfrac{2x}{(1 + x^2)^2}$

21. $\frac{1}{4}$

7. $\dfrac{3 - 12x^2}{1 + (3x - 4x^3)^2}$

11. 1

15. $\dfrac{-x}{(1 - x^2)^{3/2}}$

19. $\dfrac{|x|}{x} \left(\dfrac{2x^2 - 1}{x^2 (x^2 - 1)^{3/2}} \right), |x| > 1$

23. 40π miles per minute

5.8 Exercises

1. $\frac{1}{2} \sin 2x + C$

5. $\frac{1}{4} \tan^4 x + C$

9. $\frac{1}{3} \sin^3 \theta + C$

13. $-\tan^{-1} x + C$

3. $2 \tan (\theta/2) + C$

7. $-\cot x + C$

11. $(1/\sqrt{3}) \sin^{-1} (\sqrt{3} x) + C$

15. $\frac{1}{6} \tan^{-1} \left(\dfrac{3x}{2} \right) + C$

17. $\sin^{-1}\left(\dfrac{x-1}{\sqrt{2}}\right) + C$

19. $\tan^{-1}(\sin x) + C$

21. $-\sec\left(\dfrac{1}{x}\right) + C$

23. $-\tan^{-1}(\cos x) + C$

25. $-\frac{1}{2}\cos^4(x/2) + C$

27. $\cos(1/x) + C$

29. $\pi/6$

31. 2

33. -1

35. $\dfrac{x}{2} + \dfrac{\sin 2x}{4} + C$

37. $\dfrac{x}{2} - \dfrac{\sin 6x}{12} + C$

39. $\dfrac{1}{2}\left(\dfrac{\sin 2x}{2} - \dfrac{\sin 8x}{8}\right) + C$

41. $x/2\,\sqrt{9-x^2} + \frac{9}{2}\sin^{-1}(x/3) + C$

43. $\sqrt{x^2-4} - 2\sec^{-1}(x/2) + C$

45. $\frac{1}{2}\sin 2x - \frac{1}{6}\sin^3 2x + C$

47. $-2\sqrt{1-\sin x} + C$

Review

1. $-\sqrt{3}/2$

2. $1/\sqrt{2}$

3. 1

4. Undefined

5. 1.010

6. $-1/\sqrt{3}$

7. $(1+\sqrt{3})/(2\sqrt{2})$

8. 0

11. $-\infty$

12. 2

13. 0.14

14. $\frac{5}{2}$

15. ∞

16. $-\infty$

17. $\dfrac{x\cos x - 2\sin x}{x^3}$

18. $4x\tan x^2\sec^2 x^2$

19. $6\sec^2 3x \tan 3x$

20. $-2x\csc x^2 \cot x^2$

21. 0

22. $\sin^{-1} x + \dfrac{x}{\sqrt{1-x^2}}$

23. $\dfrac{1}{2\sqrt{x+1}\,(x+2)}$

24. 1

25. $\dfrac{2}{|x|\,\sqrt{x^4-1}}$

26. $\tan^{-1}(1/x) - x/(1+x^2)$

27. $2\sec^2 x \tan x$

28. $\dfrac{|x|}{x}\left(\dfrac{2x^2-1}{x^2\,(x^2-1)^{3/2}}\right),\ |x| > 1$

29. $\sin x$

30. $\sin x$

31. 0.714 gram per hour

32. $-\frac{1}{3}\cos 3x + C$

33. $2\sin(x/2) + C$

34. $\frac{1}{2}\tan(2x+3) + C$

35. $\frac{1}{3}\sin^3 x + C$

36. $\dfrac{1}{2\cos^2 x} + C = \dfrac{1}{2}\sec^2 x + C$

37. $\frac{1}{4}\tan^4 x + C$

38. $\frac{1}{2}\sin x^2 + C$

39. $\frac{1}{2}\sin^3 x^2 + C$

40. $\dfrac{1}{\sqrt{2}}\tan^{-1}(x\sqrt{2}) + C$

41. $\sin^{-1}(x/\sqrt{2}) + C$

42. $-\sin^{-1}(\cos x) + C$

43. $\sin^2(x/2) + C$
or $-\frac{1}{2}\cos x + C$

44. $-\dfrac{1}{2}\left[\dfrac{\cos(x/4)}{1/4} + \dfrac{\cos(3x/4)}{3/4}\right] + C$

45. $\dfrac{x}{2} - \dfrac{\sin 6x}{12} + C$

46. $\dfrac{2x + 1}{4} + \dfrac{\sin (4x + 2)}{8} + C$

47. $\dfrac{2}{\sqrt{3}} \tan^{-1} \left(\dfrac{2x + 1}{\sqrt{3}} \right) + C$

48. $-\frac{1}{3} \sin^2 x \cos x - \frac{2}{3} \cos x + C$ or $\frac{1}{3} \cos^3 x - \cos x + C$

49. $\frac{1}{3} \sec^{-1} (x/3) + C$

50. $-\frac{1}{4} \cos (2x^2) + C$

51. 1

52. $\pi/4$

53. $(\pi/2) - 1$

54. $\pi/4$

55. $\pi/4$

CHAPTER 6

6.1 Exercises

1. $3/(3x + 2)$

3. $4x/(x^2 + 3)$

5. $-14x/[(x^2 + 3)(2x^2 - 1)]$

7. $1 + \ln x$

9. $x/(x^2 + 3)$

11. $1/\sqrt{x^2 + 1}$

13. $(x^2 + 1)/(x^3 + 3x)$

15. $-2/[(2x + 1)(2x - 1)]$

17. $(1 - 2 \ln x)/x^3$

19. $1/[x \ln x]$

21. $f'(x) = 1/x$, which is positive for all positive values of x

23. 1

25. $\eta = \frac{3}{2}$; demand is elastic

27. $\eta = \frac{1}{2}$; demand is inelastic

6.2 Exercises

1. $\ln|x + 2| + C$

3. $\ln|3x^2 - 1| + C$

5. $-\frac{1}{2} \ln|3 - x^2| + C$

7. $2 \ln|\ln x| + C$

9. $\dfrac{x^3}{3} + \dfrac{x^2}{2} - 3x + 9 \ln|x + 3| + C$

11. $-\frac{1}{15} \ln|1 - 5x^3| + C$

13. $\frac{1}{4} \ln|2x^2 + 3| + C$

15. $\frac{1}{2} \ln|x^2 + 2x + 5| + C$

17. $\dfrac{x^3}{3} + \dfrac{x^2}{2} + x + \ln|x - 1| + C$

19. $\frac{1}{3} \ln|x^3 + 3x| + C$

21. $\ln 5 \approx 1.61$

23. $45 - \frac{1}{2} \ln 21 \approx 43.48$

25. $2 + \frac{2}{3} \ln 2 \approx 2.46$

27. $\ln 2 \approx 0.69$

6.3 Exercises

1. $2xe^{x^2}$

3. $\frac{1}{2}(e^x + e^{-x})$

5. $-2xe^{-x^2}$

7. ex^{e-1}

9. 1

11. $1/x$

13. $2x3^{x^2} \ln 3$

15. $2/(x \ln 10)$

17. $e^x + ex^{e-1}$

19. $2(5e)^{2x}(\ln 5 + 1)$

21. $1 + \ln x$

23. $e^x [\ln x + (1/x)]$

25. $\frac{1}{3}e^{3x} + C$

27. $\ln (1 + e^x) + C$

29. $2^{3x+2}/(3 \ln 2) + C$

31. $-e^{-x} + C$

33. $\frac{1}{2}(e^x + e^{-x}) + C$

35. $\frac{1}{5}e^{x^5} + C$

37. $-e^{1/x} + C$

39. $\ln|x| + C$

41. $\frac{1}{3}(e^{12} - e^6)$

43. $(3^8 - 1)/(4 \ln 3)$

45. $(e - 1)/2e \approx 0.32$

47.

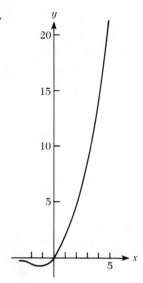

49. $\frac{15}{2}$

6.4   Exercises

1. $(e^x/2)\,(\sin x - \cos x) + C$

3. $(x^3/3)\,(\ln x - \frac{1}{3}) + C$

5. $-e^{-x}(x^2 + 2x + 2) + C$

7. $x \ln |x + 1| - x + \ln |x + 1| + C$

9. $-x^2 \cos x + 2x \sin x + 2 \cos x + C$

11. $\dfrac{3x^2}{2} \tan^{-1} x - \dfrac{3x}{2} + \dfrac{3}{2} \tan^{-1} x + C$

13. $\dfrac{x^2}{4} + \dfrac{x \sin 2x}{4} + \dfrac{\cos 2x}{8} + C$

15. $\dfrac{-xe^{-ax}}{a} - \dfrac{e^{-ax}}{a^2} + C$

17. 1

19. $\dfrac{-5}{e^4} + \dfrac{2}{e}$

21. $\frac{9}{2} \ln 3 - 2$

23. $(\pi^2 + 4)/16$

25. 1

6.5   Exercises

1. $\frac{1}{2}(\sin^{-1} x + x \sqrt{1 - x^2}) + C$

3. $\frac{1}{2}[x \sqrt{4 - x^2} + 4 \sin^{-1} (x/2)] + C$

5. $-\sqrt{9 - x^2} + \sin^{-1} (x/3) + C$

7. $\frac{1}{2}(\tan \theta - \theta) + C$

9. $\sqrt{x^2 + 1} - \ln \left| \dfrac{\sqrt{x^2 + 1} + 1}{x} \right| + C$

11. $\frac{1}{4} \ln \left| \dfrac{3 + x}{1 - x} \right| + C$

13. $x - \tan^{-1}x + C$

15. $\frac{1}{2}(x \sqrt{x^2 - 1} - \ln |x + \sqrt{x^2 - 1}|) + C$

17. $\pi/4$

19. $(\pi + 2)/8$

21. $\frac{52}{9}$

23. $\frac{1}{4} \ln 3$

6.6 Exercises

1. 2
3. Diverges
5. Diverges
7. Diverges
9. 2
11. $2\sqrt{7}$
13. $\pi/2$
15. Diverges
17. $-\frac{1}{2}$
19. 2

21.
$$\int_0^\infty e^{-ax}\,dx = \lim_{b\to\infty}\int_0^b e^{-ax}\,dx$$
$$= -\frac{1}{a}\lim_{b\to\infty}\frac{1}{e^{ax}}\Big]_0^b$$
$$= -\frac{1}{a}\lim_{b\to\infty}\left(\frac{1}{e^{ab}}-1\right) = \frac{1}{a}$$

23. Yes, when $n \le 0$.
25. Divergent
27. 2
29. $\frac{3}{2}(\sqrt[3]{4}-1)$

Review

1. $2xe^{x^2+1}$
2. $-1/(2\sqrt{e^x})$
3. $e^{x^2} + 2x^2e^{x^2}$
4. $2(3^{2x+5})\ln 3$
5. $-10^{2-x}\ln 10$
6. $\dfrac{2x+1}{x^2+x+1}$
7. $1 + \ln x$
8. $\dfrac{2x}{x^2+1} - \dfrac{3}{3x-2}$
9. $\dfrac{4x^3}{3(x^4+1)}$
10. $e^x\left(-\dfrac{1}{x^2} + \dfrac{1}{x}\right)$
11. $2/(x\ln 5)$
12. $3/[(3x+1)\ln 2]$
13. $\frac{1}{3}e^{3x} + C$
14. $\frac{1}{2}e^{x^2} + C$
15. $\frac{1}{2}(\ln x)^2 + C$
16. $\ln|x+4| + C$
17. $\frac{1}{2}\ln|x^2+4| + C$
18. $\frac{1}{2}\tan^{-1}(x/2) + C$
19. $x\ln x - x + C$
20. $\dfrac{3^{x^2}}{2\ln 3} + C$
21. $\frac{1}{5}e^{2x}(2\sin x - \cos x) + C$
22. $2\sin^{-1}\left(\dfrac{x}{2}\right) + \dfrac{x\sqrt{4-x^2}}{2} + C$
23. $\frac{1}{4}\ln\left|\dfrac{2+x}{2-x}\right| + C$
24. $-2\cos\sqrt{x} + C$
25. $x\sin x + \cos x + C$
26. $x\sin^{-1}x + \sqrt{1-x^2} + C$
27. $\dfrac{x^2}{2}[\ln(x^2) - 1] + C = x^2\ln x - \frac{1}{2}x^2 + C$
28. $\dfrac{x^2}{2}(\ln x - \frac{1}{2}) + C$
29. $\dfrac{1}{\sqrt{2}}\tan^{-1}\left(\dfrac{x+2}{\sqrt{2}}\right) + C$
30. $\sin^{-1}\left(\dfrac{x+2}{\sqrt{5}}\right) + C$
31. 184
32. 42

33. $(2e^4 - 10)/e^2$

34. $8 \ln 4 - \frac{15}{4}$

35. $7/\ln 2$

36. $\pi/4$

37. $-\dfrac{\pi}{4} + \dfrac{1}{2} \ln 2$

38. $\pi/4$

39. $\ln 2$

40. $(\pi/2) + 1$

41. 8

42. Diverges

43. $\pi/4$

44. $\frac{1}{2}$

45. Diverges

46. $\min (-\pi/4, -1/(e^{\pi/4} \sqrt{2}))$
$\max (3\pi/4, e^{3\pi/4}/\sqrt{2})$

47. Decreasing $(-\pi, -\pi/4)$, $(3\pi/4, \pi)$
Increasing $(-\pi/4, 3\pi/4)$
Concave up $(-\pi/2, \pi/2)$
Concave down $(-\pi, -\pi/2)$ $(\pi/2, \pi)$

48.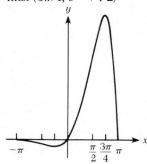

49. $\dfrac{e^\pi - e^{-\pi}}{2}$; no, part of the area is below the x axis

CHAPTER 7

7.1 Exercises

1. $\frac{65}{4}$

3. $\frac{22}{3}$

5. $\frac{3}{4}$

7. $\frac{5}{4}$

9. $\frac{1}{6}$

11. 4

13. $\frac{32}{3}$

15. 2

17. $\frac{3}{10}$

19. $4 - e \ln 4$

21. $(3 \sqrt{3} - \pi)/3$

23. 29.23

25. No area defined

27. $2 \displaystyle\int_0^8 (4 - x^{2/3})^{3/2} \, dx$

7.2 Exercises

1. For a sufficiently large number of 70-year-old men, 4 out of 9 may expect to reach their 80th birthday.

3. No; there are two red aces

5. $\sqrt{2}/2$

7. $\frac{1}{2}$

9. $\frac{5}{14}$

11. $\frac{1}{2}$

13. $\frac{1}{6}$

15. $\approx \dfrac{1}{4} + \dfrac{1.25}{\pi}$

17. 4

19. $\pi/6$

21. $\frac{20}{27}$

23. $\approx 43\%$

25. $P(x) = \frac{3}{8}x^2;\ \frac{7}{8}$

27. $P(x) = 3x^2/(x^3 + 1)^2;\ \frac{1}{4}$

29. $P(x) = 1/(2x^{3/2});\ \frac{2}{3}$

31. $P(x) = \dfrac{1}{\pi} \cdot \dfrac{1}{x^2 + 2x + 2}; \dfrac{1}{2}$

35. 1

33. $\pi/6$

37. $a = \pi/3, b = \pi/2, c = 2\pi/3$

7.3   Exercises

1. $\frac{45}{14}$

7. $5/(1 - e^{5/2})$

13. $\mu = 3/2, \sigma^2 = 9/20$

19. $\mu = 1/(2e\ \sqrt{2\pi})$

3. 0

9. 0

15. $\mu = 5, \sigma^2 = 27/5$

5. $\frac{53}{34}$

11. $\mu = 3/5, \sigma^2 = 12/175$

17. $\mu = 0$

7.4   Exercises

1. 0.4772

7. 0.8185

13. Too small to measure

17. 0.7062 (see Figure 7.11) Each toss has a width of 1, therefore to include 190 we must include $189\frac{1}{2} - 190\frac{1}{2}$

19. 0.6736

25. **a.** 0.84
b. 0.67

3. 0.4772

9. 0.0228

15. 95.99%

21. 0.5394

27. **a.** Not possible
b. Not possible

5. 0.5328

11. 95.44%

23. **a.** 1.44
b. -0.92

7.5   Exercises

1. $\$6000(\sqrt[3]{4} - 1)$

9. $\approx\$79,700$

17. $\$32,000$

3. $\$9333\frac{1}{3}$

11. $\approx\$5000$

19. $\$3600$

5. $\$500 \ln\left(\frac{18}{13}\right)$

13. $\approx\$12,360$

7. $\$3547.67$

15. $\approx\$10,100$

7.6   Exercises

1. The integral of marginal revenue is total revenue.

3. $R(x) = 1000x - 4x^2$

7. $\$15,573.75$

11. $\$2721.75$

15. $\frac{4}{3}$

19. 5627 tons

5. 200

9. $\$463.33$

13. $\frac{3}{4}(10^{4/3}) - \frac{1}{3}(10^{3/2}) + 30 \approx \$35.62$

17. 100

7.7   Exercises

1. $c = \displaystyle\int_b^d g(y)\, dy$, where $g(d) = 0$

5. 5.6

9. $4 - \ln 5$

13. 19,800

17. Selling price = 3 units, $C = \frac{1}{3}, P = \frac{1}{3}$

3. $60\frac{3}{4}$

7. 60.75

11. $7e^3 - 1$

15. Selling price = 9 units, $C = \frac{2}{3}, P = 4$

7.8 Exercises

1. $\dfrac{20}{\sqrt[100]{2}}$ grams, $\dfrac{5}{256}$ grams, $\dfrac{20}{2^{100}} = \dfrac{5}{2^{98}}$ grams

3. $\approx 74{,}300$ years old (too old for accurate measurement—see Exercise 4)

5. 7.6×10^8 **7.** 2112 A.D.

9. 1992 A.D. Human population growth is limited and dependent on factors which must be accounted for

11. 2114 A.D.   **13.** 1.71 hours

15. $k = \dfrac{-\ln 2}{t}$ for growth, whereas $k = \dfrac{\ln 2}{t_h}$ for half-life when $(-)$ is included in expression.

17. ≈ 14 days   **21.** Never

7.9 Exercises

1. 1   **3.** $3 - 2 \ln 2$ **5.** 2
7. $\ln (1.5)$ **9.** $\sqrt{2}/2 - \cos 1$ **11.** 8π
13. 39π **15.** $81\pi/2$ **17.** $\pi^2/2$
19. $3\pi/8$ **21.** $\pi/2$ **23.** Undefined

25. $\pi \displaystyle\int_c^d [f(y)]^2 \, dy$ **27.** $40\pi/3$ **29.** $256\pi/15$

7.10 Exercises

1. $y = \dfrac{(x + 3)^3}{3} + C$ **3.** $y = \dfrac{x^3}{3} - x^2 + 4x + C$

5. $y^2 = \dfrac{1}{Cx} - 1$ **7.** $y = -\ln |xe^{-x} + e^{-x} + C|$

9. $y = (3x^{1/2} + C)^{2/3}$ **11.** $y = Ce^{\sin^{-1} x}$
13. $y^2 = 2(x^2 - 1)$ **15.** $y = -(\sin 2x)/2 + 1$
17. $y = e^{-x} + Ce^{-2x}$ **19.** $y = \sin x \cos x + C \cos x$

21. $y = \dfrac{x}{2} + \dfrac{C}{x}$ **23.** $y = e^{3x} + e^{2x}$

25. 0.05   **27.** $\approx 23.3°C$

Review

1. 39   **2.** $\dfrac{67}{6}$ **3.** $\frac{1}{3}$

4. $\frac{1}{2}(\ln 2)^2$ **5.** $\dfrac{2 - \sqrt{3}}{2}$ **6.** $\frac{4}{3}$

7. $\dfrac{3\pi^2}{32} + \dfrac{\pi}{4} + \dfrac{\sqrt{2}}{2}$ **8.** $\dfrac{\pi}{4} - \frac{1}{2} \ln 2 + e - 1$

11. At least one possible result will happen 100% of the time

12. $\dfrac{\sqrt{2}}{4}$

13. 0.3794

14. $\frac{1}{4}$

15. $\dfrac{1}{\pi}\left[\tan^{-1}\frac{1}{3} - \tan^{-1}\left(-\frac{1}{2}\right)\right]$

16. $\approx 63\%$

17. $\approx 0.7\%$

18. $P(x) = \dfrac{3x^2}{(x^3 + 1)^2}$

19. $\mu = \frac{4}{7}$, $\sigma^2 = \frac{18}{245}$

20. $\mu = -1$, $\sigma^2 = \frac{12}{5}$

21. 0.7938

22. 0.2257

23. 0.5468

24. 0.5000

25. 0.3050

26. 1.34

27. 0.36

28. \$150

29. $1000 \ln 5 \approx \$1609$

30. $1500 \tan^{-1} 6 \approx \$2108.47$

31. \$5000

32. $\approx \$91,000$

33. $\approx \$13,600$

34. $\approx \$9600$

35. $\approx \$13,350$

36. $\approx \$49,280$

37. $1000 \ln 6 \approx \$1790$

38. Price = 3 units
$P = 2$, $C = 16/3$

39. Price = 15 units
$P = 25$, $C = 83\frac{1}{3}$

40. Price = 2 units
$P = \dfrac{\pi}{4}$, $C = 1$

41. Price = 36 units
$P = 213\frac{1}{3}$, $C = 341\frac{1}{3}$

42. $37\frac{1}{3}$

43. $\dfrac{577e^6 - 1}{4e^6} \approx 144.25$

44. $\dfrac{23e^4 - 1}{2e^4} \approx 11.49$

45. $8\sqrt[4]{10} - \frac{25}{6}\sqrt[4]{10} + 3$

46. 10

47. ≈ 510 years

48. 2.80×10^8 years

49. $k = 0.67\%$ or $\frac{2}{3}$ of 1\%

50. $\ln 4 - \frac{3}{4}$

51. $2e^3 - e^2$

52. $\frac{1}{2}$

53. $\ln |\sec 1|$

54. $\pi(\ln 4 + \frac{3}{2})$

55. $\pi(\ln 4 + \frac{145}{6})$

56. $\dfrac{\pi e^2}{2}(e^2 - 1)$

57. $2\pi e^2$

58. \$0.59

59. approx. $2\frac{1}{2}$ pounds

60. $2y^3 = 3x^2 + C$

61. $3e^{-y^2} = -2x^3 + C$

62. $y = 1 + Ce^{-x^2}$

63. $y = x^3/2 + xC$

CHAPTER 8

8.1 Exercises

3. $f(0, 0) = 2$
$f(1, 1) = \sqrt{\dfrac{5}{2}}$
$f(2, 2) = 2\sqrt{\dfrac{2}{3}}$
$f(4, 1) = \sqrt{10}$
$f(2x, 3) = \sqrt{x^2 + 1}$
$f(u, v) = \dfrac{\sqrt{u^2 + 4}}{\sqrt{v + 1}}$
$f(x^2 + y^2, x) = \dfrac{\sqrt{(x^2 + y^2)^2 + 4}}{\sqrt{x + 1}}$

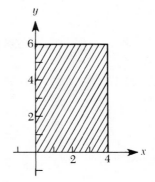

5. $f(0, 0)$ meaningless
$f(1, 1) = 1$
$f(2, 2)$ meaningless
$f(4, 1)$ meaningless
$f(2x, 3) = \dfrac{1}{36x^2}$, where $6x = 1$

$f(u, v) = \dfrac{1}{u^2v^2}$, where $uv = 1$

$f(x^2 + y^2, x) = \dfrac{1}{(x^2 + y^2)^2 x^2}$, where $(x^2 + y^2)x = 1$

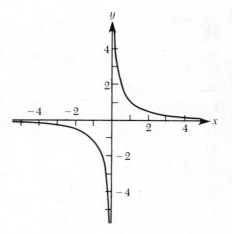

7. $f(0, 0) = 0, f(1, 1) = 8, f(2, 2) = 64, f(4, 1) = 125$
$f(2x, 3) = (2x + 3)^3, f(u, v) = (u + v)^3, f(x^2 + y^2, x) = (x^2 + y^2 + x)^3$
9. $f(0, 0)$ meaningless, $f(1, 1)$ meaningless, $f(2, 2)$ meaningless, $f(4, 1)$ meaningless, $f(2x, 3)$ meaningless

$f(u, v) = \cos\left(\pi u - \dfrac{v}{2}\,\pi\right)$, where $u = 2, v^2 \leq 1$

$f(x^2 + y^2, x) = \cos\left[\pi(x^2 + y^2) - \dfrac{x}{2}\,\pi\right]$, where $x^2 + y^2 = 2, x^2 \leq 1$

11. $D = \{(x, y) \mid x \neq -2, y \neq 3\}$

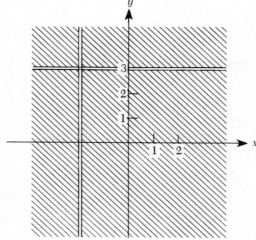

13. $D = \{(x, y) \mid x, y \text{ real numbers}\}$
15. $D = \{(x, y) \mid x, y \text{ real numbers}\}$
17. $D = \{(x, y) \mid xy \geq 0\}$; graph all of quadrants I and III, and axes
19. No values of x and y make this a real valued function
21. $D = \{(x, y) \mid xy \neq n\pi \text{ for any integer } n\}$
23. $D = \{(x, y) \mid x \text{ and } y \text{ are real numbers}\}$

25. 27

27. $\frac{1}{2}$

29. $\frac{4}{9}$

31. $\frac{9}{2}$

33. 1

35. 1

37. $\frac{1}{2}$

39. 1

41. $2/\sqrt{3}$

43. 1

45. 2

47. 1

8.2 Exercises

1. $\sqrt{41}$

3. $2\sqrt{2}$

5. 2

7.

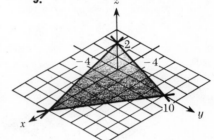

9.

11.

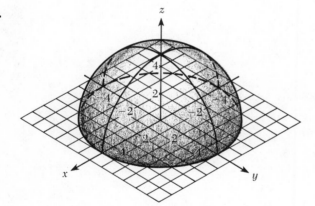

13.

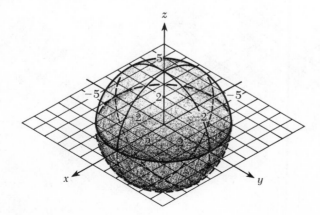

15.

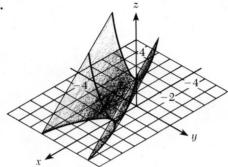

17.

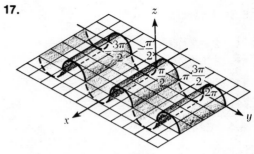

19.

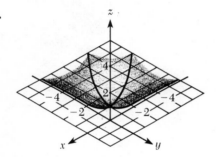

21. $(0, 0, 4), (0, 0, -4)$

23.

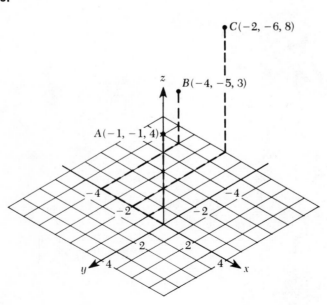

8.3 Exercises

1. $z = 34$

3. $P = 19$

5. $z = 30$, max; $z = -150$, min

7. $I = 22$, max; $I = -138$, min

9. 22

11. No vitamin A; 10 pounds vitamin B

13. \$83.33

15. 33 bags of A and 67 bags of B

17. 34 quarts Scotch, 20 quarts beer

8.4 Exercises

1. $2x + 2y$, $2x + 2y$

3. $\frac{1}{2}, \frac{1}{2}$

5. $\frac{1}{5} x^{-4/5} - \frac{\sqrt{y}}{2\sqrt{x}}$, $\frac{1}{5} y^{-4/5} - \frac{\sqrt{x}}{2\sqrt{y}}$

7. $-\frac{(y^2 + 2)}{(x^2 - 5x + 8)^2} (2x - 5)$, $\frac{2y}{x^2 - 5x + 8}$

9. $-4(x - y)^3$, $4(x - y)^3 + 4(y - 1)^3$

11. $2 + y$, $x + 2y + 3$

13. $e^{x^2 + 2xy}(2x + 2y)$, $e^{x^2 + 2xy}(2x)$

15. $\cos(x^2 + 2\cos y)(2x)$, $\cos(x^2 + 2\cos y)(-2\sin y)$

17. $\frac{\partial^2 z}{\partial x^2} = \frac{6}{x^4 y^2}$, $\frac{\partial^2 z}{\partial y^2} = \frac{6}{x^2 y^4}$

$\frac{\partial^2 z}{\partial x\,\partial y} = \frac{4}{x^3 y^3}$, $\frac{\partial^2 z}{\partial y\,\partial x} = \frac{4}{x^3 y^3}$

19. $\frac{\partial^2 z}{\partial x^2} = \frac{2(y^2 + 2)(3x^2 - 15x + 17)}{(x^2 - 5x + 8)^3}$, $\frac{\partial^2 z}{\partial y^2} = \frac{2}{x^2 - 5x + 8}$

$\frac{\partial^2 z}{\partial x\,\partial y} = \frac{-2y(2x - 5)}{(x^2 - 5x + 8)^2}$, $\frac{\partial^2 z}{\partial y\,\partial x} = \frac{-2y(2x - 5)}{(x^2 - 5x + 8)^2}$

21. $\frac{\partial^2 z}{\partial x^2} = 2$, $\frac{\partial^2 z}{\partial y^2} = 2$

$\frac{\partial^2 z}{\partial y\,\partial x} = 1$, $\frac{\partial^2 z}{\partial x\,\partial y} = 1$

23. $f_{yyy}(x, y) = \frac{4}{y^3}$, $f_{xyx}(x, y) = 0$

25. $f_{xxy}(x, y) = ye^{xy}(xy + 2)$
$f_{xyx}(x, y) = ye^{xy}(xy + 2)$
$f_{yxx}(x, y) = ye^{xy}(xy + 2)$

27. $4y$

31. $\left(\frac{4}{3}, \frac{2}{3}\right)$

33. $\sin(xy) - x^5 + y^3 + C$

35. $xy + y\ln(xy) + C$

8.5 Exercises

1. $f(2, 1) = -8$, min

3. $f(2, -1) = -1$, min

5. $f(2, 0) = -16$, min

7. $f(3, 3) = -27$, min

9. $f(-\frac{1}{2}, 4) = -6$, max

11. $f(-3, 2k\pi) = 36$, min; $f(3, (2k + 1)\pi) = 36$, min

13. $f\left(\frac{4}{3}, \frac{4}{3}\right) = \frac{64}{27}$, max

15. No max or min

17. $f\left(\frac{\sqrt{2}}{2}, \frac{\sqrt{2}}{2}\right) = 1 + \ln 2$

19. \$200 million on welfare, \$100 million on prisons

21. $V = 14{,}815$ max at \$66,667 in both Los Angeles and San Francisco

23. max at $(1, \frac{4}{3})$; that is, \$100,000 on television and \$133,333.33 on magazines

25. $y = \frac{3}{4}x + \frac{1}{6}$

27. $y = -\frac{15}{38}x + \frac{33}{38}$

8.6 Exercises

1. $\dfrac{\partial z}{\partial u} = 2x \ln v + 2y \dfrac{v}{u}$

$\dfrac{\partial z}{\partial v} = 2x \dfrac{u}{v} + 2y \ln u$

3. $\dfrac{\partial z}{\partial u} = \dfrac{1}{u}(e^{x+y} + e^{-x-y})$

$= v + \dfrac{1}{u^2 v}$

$\dfrac{\partial z}{\partial v} = \dfrac{1}{v}(e^{x+y} + e^{-x-y})$

$= u + \dfrac{1}{uv^2}$

5. $\dfrac{\partial z}{\partial u} = \dfrac{1}{y} e^{x/y}\left(1 - \dfrac{2x}{y}\right)$

$\dfrac{\partial z}{\partial v} = \dfrac{1}{y} e^{x/y}\left(2 + \dfrac{x}{y}\right)$

7. $\dfrac{dz}{dt} = \dfrac{2t}{y} - \dfrac{x}{y^2} + 1 = 2$

9. $\dfrac{dw}{dx} = (2u + ve^u)(\sin x + v) + e^u(\cos x - u)$

11. $dz = 2(x + 2y)\,dx + 4(x + 2y)\,dy$

13. $dz = (2x + 2y + 1)\,dx + (2x + 4y + 1)\,dy$

15. $dz = 0$

19. 0.2296

21. 28 ± 0.115 square centimeters

23. $\$0.575\pi \pm 0.0264\pi$

8.7 Exercises

1. 36   **3.** 0   **5.** $\frac{3125}{3}$ **7.** $-\frac{207}{20}$

9. 1   **11.** $\frac{8}{15}$ **13.** 12   **15.** 0

17. $\frac{16}{5}$ **19.** $(3\pi/8) - 1$ **21.** No.   **23.** Yes

25. Notation is meaningless as outer limits must be constants

8.8 Exercises

1. 8

3. $\frac{8}{3}$

5. $\frac{1}{2}$

7. $(e^4 - 1)^2 \approx 2872.8$

9. 45

11. $\dfrac{\pi}{12} + \dfrac{\sqrt{3}}{8}$

13. $(4 \ln 4) - 8 + e$

15. $e^2 - 3$

17. $4\pi/3$

19. $\frac{33}{250}$

21. $-\frac{243}{4}$

23. $25\pi/2$

25. $\frac{21}{2}$

27. $\displaystyle\int_0^1 \int_{\sqrt[3]{x}}^{\sqrt[3]{x}} dy\,dx$

29. $\displaystyle\int_{-1}^{1} \int_x^1 \sqrt{1 + y^2}\,dy\,dx$

31. $\displaystyle\int_{-2}^{0} \int_{-1}^{x/2} e^{-x^2}\,dy\,dx + \int_0^8 \int_{x/2}^4 e^{-x^2}\,dy\,dx$

33. $\frac{9}{2}$

35. 1

37. 2

39. $\frac{1}{8}$

8.9 Exercises

1. $\dfrac{\partial f}{\partial x} = 2xy^3 z^4,\ \dfrac{\partial^2 f}{\partial x^2} = 2y^3 z^4,\ \dfrac{\partial f}{\partial y} = 3x^2 y^2 z^4,\ \dfrac{\partial^2 f}{\partial x\,\partial y} = 6xy^2 z^4$

3. $\dfrac{\partial f}{\partial x} = \dfrac{y^2}{z}, \dfrac{\partial f}{\partial y} = \dfrac{2xy}{z}, \dfrac{\partial f}{\partial z} = \dfrac{-xy^2}{z^2}, \dfrac{\partial^2 f}{\partial z^2} = \dfrac{2xy^2}{z^3}$

5. $\dfrac{\partial f}{\partial x} = 2x \ln y, \dfrac{\partial^2 f}{\partial x^2} = 2 \ln y, \dfrac{\partial f}{\partial z} = \dfrac{-y}{z} + y, \dfrac{\partial^2 f}{\partial x \, \partial y} = \dfrac{2x}{y}$

7. $\dfrac{\partial f}{\partial x} = \tfrac{1}{2}e^{(x+y+z)/2}, \dfrac{\partial f}{\partial y} = \tfrac{1}{2}e^{(x+y+z)/2}, \dfrac{\partial f}{\partial z} = \tfrac{1}{2}e^{(x+y+z)/2}, \dfrac{\partial^3 f}{\partial x \, \partial y \, \partial z} = \tfrac{1}{8}e^{(x+y+z)/2}$

9. $\dfrac{\partial f}{\partial u} = \dfrac{2u}{w}, \dfrac{\partial f}{\partial v} = \dfrac{-2v}{w}, \dfrac{\partial f}{\partial w} = \dfrac{v^2 - u^2}{w^2}, \dfrac{\partial^3 f}{\partial w^3} = \dfrac{6(v^2 - u^2)}{w^4}$

11. $\dfrac{\partial f}{\partial x} = \dfrac{1}{x}, \dfrac{\partial f}{\partial y} = \dfrac{1}{y}, \dfrac{\partial f}{\partial z} = \dfrac{1}{z}, \dfrac{\partial f}{\partial w} = \dfrac{1}{w}, \dfrac{\partial^2 f}{\partial x \, \partial y} = 0$

13. $\dfrac{\partial f}{\partial x_2} = x_1 \cos(x_1 x_2) + 2x_2, \dfrac{\partial^2 f}{\partial x_3^2} = 2, \dfrac{\partial^2 f}{\partial x_1 \, \partial x_3} = 0$

15. $\tfrac{15}{4}$ **17.** $2^{12}/15$ **19.** 0 **21.** 0 **23.** 3^n

25. $2xyz^3(2u^2v) + x^2 z^3 \left(\dfrac{-u}{v^2}\right) + 3x^2 yz^2 \left(\dfrac{2v}{u}\right)$

27. $2u^2 x \cos(x^2) \sin y \cos z - \dfrac{u\pi}{v^2} \sin(x^2) \cos y \cos z + \dfrac{v \sin(x^2) \sin y \sin z}{\sqrt{u^2 - v^2}}$

29. $\dfrac{\partial z}{\partial y_3} = 9y_3^2 z, \dfrac{\partial z}{\partial y_5} = 5y_5^4 z$

Review

2. The domain is all real numbers for which the function is defined

3. $D = \{(x, y) \mid x^2 + y^2 \leq 25\}$

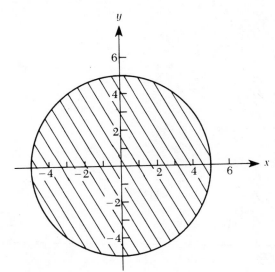

4. $D = \{(x, y) | xy > 0\}$

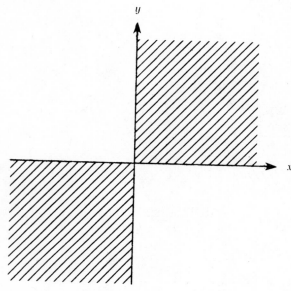

5. $D = \{(x, y) | x \text{ and } y \text{ are real}\}$
6. $D = \{(x, y) | x \neq -y\}$

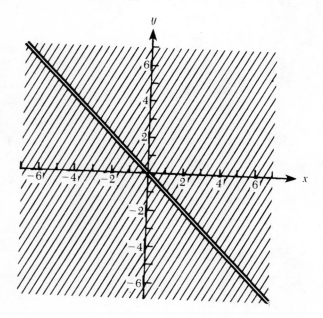

7. $D = \{(x, y) \,|\, y \neq -x\}$

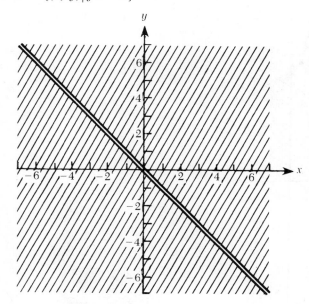

8. $f(1, 1) = 2$

$f(3, \ln 2) = \dfrac{e^3}{2}\,[9 + (\ln 2)^2]$

$f(0, 1) = \dfrac{1}{e}$

$f(-1, -2) = 5e$

$f(x, x) = 2x^2$

9. $f\left(0, \dfrac{2}{\pi}\right) = e^{2/\pi}$

$f\left(\dfrac{4}{\pi}, \dfrac{2}{\pi}\right) = \tfrac{1}{2}e^{6/\pi}$

$f(1, -1)$ meaningless

10. $f(1, 1) = 0$

$f(2, 1)$ meaningless

$f(0, 0) = 0$

$f(x, x) = 0$

$f(w, w) = 0$

11.

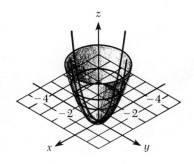

12.

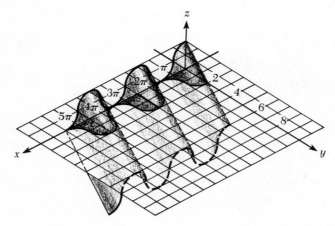

13.

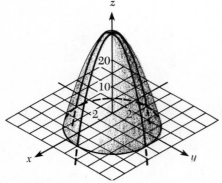

14.

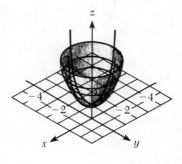

15.

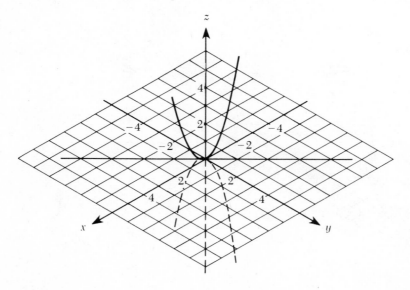

16.

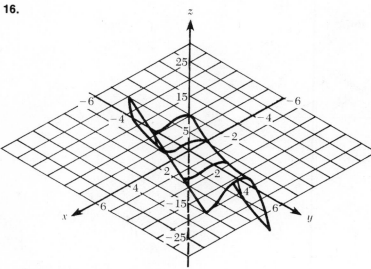

17.

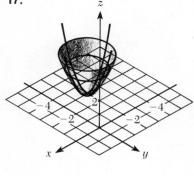

18.

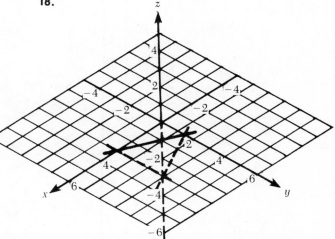

19.

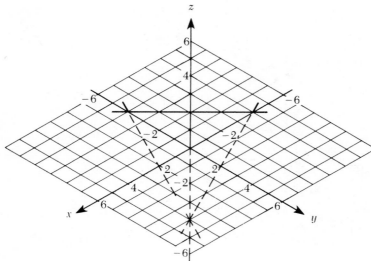

20.

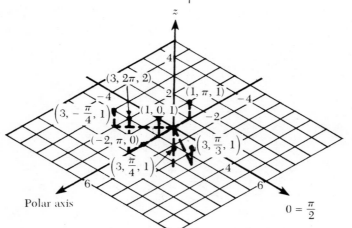

21. Right circular cylinder, radius 5, z axis as center line

22. All points, $P(r, \theta, z)$ for which $\theta = \pi/4$ describe a plane containing the z axis and intersecting the xy plane in a line that makes an angle of $\pi/4$ with the polar axis

23. Hemisphere above the xy plane; center at origin, radius 5

24. $z = \sqrt{r^2(\sin^2 \theta - \cos^2 \theta) + 10r \cos \theta}$

25. $\dfrac{\partial z}{\partial x} = e^y + ye^x$, $\dfrac{\partial z}{\partial y} = xe^y + e^x$, $\dfrac{\partial^2 z}{\partial x\, \partial y} = e^y + e^x$

26. $\dfrac{\partial^3 z}{\partial x\, \partial y^2} = \dfrac{8x(x^4 - 8x^2y^2 + 3y^4)}{(x^2 + y^2)^4}$

27. $\dfrac{\partial^4 z}{\partial x^4} = 0$

28. $\dfrac{\partial^2 z}{\partial u^2} = 0$

29. $\dfrac{\partial z}{\partial u} = 2x \cos u \cos v - 2y \sin u \cos v = 0$,

$\dfrac{\partial z}{\partial v} = -2 \cos v \sin v$

30. $\dfrac{\partial z}{\partial u} = 0,\ \dfrac{\partial z}{\partial v} = -\cos(\cos v)\sin v$

34. $\dfrac{dy}{dx} = \dfrac{y - 2x}{2y - x}$

35. $\dfrac{dy}{dx} = -1$

36. No max or min

37. $f\left(\pi, \dfrac{\pi}{2}\right) = -2,\ \text{min}$

$f\left(0, \dfrac{3\pi}{2}\right) = 2,\ \text{max}$

$f\left(2\pi, \dfrac{3\pi}{2}\right) = 2,\ \text{max}$

38. No max or min

39. $f(1, 1) = -7,\ \text{min}$

40. $f(2, \frac{1}{2}) = 0,\ \text{min};\ f(-2, -\frac{1}{2}) = 0;\ \text{min}$

41. $f(2, 0) = -4,\ \text{min}$

42. $-\frac{1}{3}$

43. $-\sqrt{3}/6$

44. $-\frac{10}{3}$

45. $\dfrac{\pi}{2}\ln 2$

46. 68

47. $\frac{33}{5}$

48. $2(2 - e)$

49. $\dfrac{e^2 - 3}{2}$

50. 0

51. $\frac{3}{20}$

52. $\frac{29}{42}$

53. $\frac{9}{2}$

54. $\dfrac{\partial^2 w}{\partial z^2} = 0,\ \dfrac{\partial^2 w}{\partial x^2} = -\dfrac{2}{x^2},\ \dfrac{\partial^2 w}{\partial y^2} = 0$

55. $\dfrac{\partial u}{\partial v} = 2v\,\dfrac{\partial u}{\partial x} + r^3\,\dfrac{2u}{\partial y} + \dfrac{2v}{r^2}\,\dfrac{\partial u}{\partial z} - \dfrac{2}{v^3}\,\dfrac{\partial u}{\partial w},$

$\dfrac{\partial u}{\partial r} = \dfrac{2}{r^3}\,\dfrac{\partial u}{\partial x} + 3vr^2\,\dfrac{\partial u}{\partial y} - \dfrac{2v^2}{r^3}\,\dfrac{\partial u}{\partial z}$

56. $\displaystyle\sum_{i=1}^{100} \dfrac{i^2 - 1}{i^2}\,x^{i2 - 2}$

57. $\pi/4$

58. $5^5/24$

59. 0

60. 200 boxes tomatoes, 400 boxes lettuce, 200 boxes onions

61. 36 pages news, 15 pages advertising, 9 pages photos

CHAPTER 9

9.1   Exercises

1. $a_1 = 4,\ a_2 = \frac{5}{4},\ a_3 = \frac{2}{3},\ a_{20} = \frac{23}{400}$
Convergent

3. $a_1 = 2,\ a_2 = (\frac{3}{2})^{1/2},\ a_3 = (\frac{4}{3})^{1/3},\ a_{20} = (\frac{21}{20})^{1/20}$
Convergent

5. $a_1 = \dfrac{1}{\sqrt{2} - 1},\ a_2 = \dfrac{1}{\sqrt{3} - \sqrt{2}}$

$a_3 = \dfrac{1}{2 - \sqrt{3}},\ a_{20} = \dfrac{1}{\sqrt{21} - \sqrt{20}}$
Divergent

7. $a_1 = \frac{1}{2},\ a_2 = \frac{23}{18},\ a_3 = \frac{10}{7},$

$a_{20} = \dfrac{3(20)^3 - 1}{2(20)^3 + 2} = \dfrac{23,999}{16,002}$
Convergent

9. $a_1 = \frac{1}{4},\ a_2 = \frac{1}{4},\ a_3 = \frac{1}{4},\ a_{20} = \frac{1}{4}$
Convergent

11. $a_n = \dfrac{1}{4^{n-1}}$
Convergent

13. $a_n = 2n - 1$
Divergent

15. $a_n = 1 + \displaystyle\sum_{i=2}^{n} \frac{1}{i}$
Divergent (this will be shown to be divergent in Section 9.2)

17. $a_n = \dfrac{4}{(3n + 4)(3n + 1)}$
Convergent

19. $a_n = n$
Divergent

21. $a_n = (\frac{1}{2})^{n-1}$
Convergent

23. $a_n = 4$
Convergent

25. $a_n = \dfrac{1}{n(n + 1)}$
Convergent

27. $|r| < 1$

29. Convergent

31. Convergent

33. Divergent

35. Divergent

9.2 Exercises

1. $\displaystyle\lim_{n\to\infty} \frac{n}{n + 1} = 1 \neq 0$
Divergent

3. $\displaystyle\lim_{n\to\infty} \frac{1}{4^n} = 0$
May or may not converge

5. $\displaystyle\lim_{n\to\infty} (\tfrac{3}{2})^n \neq 0$
Divergent

7. Convergent

9. Divergent

11. Divergent

13. $a_n = \dfrac{n^2 + n - 1}{n(n + 1)}$

15. $a_n = \dfrac{5}{(4n + 1)(4n - 3)}$

17. $a_n = \dfrac{1}{n(n + 1)}$

19. $a_n = 2(3)^{n-1}$

21. $\frac{1}{3}$

23. $\frac{22}{7}$

25. $10\frac{371}{999}$

27. 0

29. 500 feet

31. Yes; $\displaystyle\sum_{i=1}^{\infty} c_i = \sum_{i=1}^{\infty} a_i + \sum_{i=1}^{\infty} b_i$, where $c_i = a_i + b_i$

33. Unknown

35. Could be either

37. If any of the series converge, the rest do also

39. $2915

9.3 Exercises

3. Convergent

5. Divergent

7. Convergent

9. Divergent

11. Convergent

13. Convergent

15. Divergent

17. Divergent

19. Divergent

21. Convergent

23. Convergent

25. Divergent

27. Divergent

29. Convergent

9.4 Exercises

1. 6

3. Divergent

5. 0

7. 1

9. No limit

11. Convergent

13. Ratio test fails

15. Convergent

17. Divergent

19. Test fails

21. Convergent

23. Convergent

25. Convergent

27. Let $a_i = 1/i^2$; then $\sqrt{a_i} = 1/i$ and $\sum\limits_{i=1}^{\infty} (1/i^2)$ converges, whereas $\sum\limits_{i=1}^{\infty} (1/i)$ diverges

---

9.5 Exercises

1. Absolutely convergent   **3.** Conditionally convergent
5. Absolutely convergent   **7.** Absolutely convergent
9. Absolutely convergent   **11.** Absolutely convergent
13. Divergent   **15.** Absolutely convergent
17. Divergent   **19.** Absolutely convergent
21. $\frac{1}{5}$ **23.** $\frac{1}{160}$
25. $\frac{5}{32}$ **27.** 7
29. Divergent series

9.6 Exercises

1. $(-\infty, \infty)$ **3.** $[-\frac{1}{2}, \frac{1}{2}]$ **5.** $(-\infty, \infty)$
7. $[-3, 3]$ **9.** $[-1, 1)$ **11.** $[-1, 0]$
13. $(-\infty, \infty)$ **15.** $[0, 1]$ **17.** $[-1, 1]$
19. $(-1, 1)$ **21.** $R = \infty$ **23.** $R = 1$

25. $\sum\limits_{i=1}^{\infty} \left(\frac{x}{2}\right)^i$ **27.** $\sum\limits_{i=1}^{\infty} \frac{(x - C)^i}{R^i}$

9.7 Exercises

1. $1 + \sum\limits_{i=1}^{\infty} (-x)^i, |x| < 1$ **3.** $\sum\limits_{i=1}^{\infty} x^{3i-2} (-1)^{i+1}, |x| < 1$

5. $2 \sum\limits_{i=1}^{\infty} \frac{(-1)^{i-1} x^i}{i}, R = 1, -1 < x \leqslant 1$ **7.** $\sum\limits_{i=1}^{\infty} \frac{(-1)^{i-1} x^{i-1}}{i}, -1 < x \leqslant 1$ and $x \neq 0$

9. $2 \sum\limits_{i=1}^{\infty} \frac{x^{2i-1}}{2i - 1}, R = 1, -1 < x < 1$

13. $\dfrac{1}{\sqrt{1 - x}} = 1 + \dfrac{x}{2} + \dfrac{1(3)}{2(4)} x^2 + \dfrac{1(3)(5)}{2(4)(6)} x^3 + \cdots, R = 1$

15. $\sin^{-1} x = x + \dfrac{1}{2} \cdot \dfrac{x^3}{3} + \dfrac{1(3)}{2(4)} \cdot \dfrac{x^5}{5} + \dfrac{1(3)(5)}{2(4)(6)} \cdot \dfrac{x^7}{7} + \cdots, R = 1$

17. $\ln (x + \sqrt{1 + x^2}) = x - \dfrac{1}{2} \cdot \dfrac{x^3}{3} + \dfrac{1(3)}{2(4)} \cdot \dfrac{x^5}{5} - \dfrac{1 \cdot 3 \cdot 5}{2 \cdot 4 \cdot 6} \cdot \dfrac{x^7}{7} + \cdots, R = 1$

19. $\cot^{-1} x = \dfrac{\pi}{2} - x + \dfrac{x^3}{3} - \dfrac{x^5}{5} + \cdots, R = 1$

9.8 Exercises

1. $\ln(1+x) = x - \dfrac{x^2}{2} + \dfrac{x^3}{3} - \dfrac{x^4}{4} + \cdots + (-1)^{n-1}\dfrac{x^n}{n} - \cdots, R = 1$

3. $\displaystyle\sum_{i=0}^{\infty} \dfrac{x^{2i}}{(2i)!}$

7. $1 - \dfrac{\left(x - \dfrac{\pi}{2}\right)^2}{2!} + \dfrac{\left(x - \dfrac{\pi}{2}\right)^4}{4!} - \dfrac{\left(x - \dfrac{\pi}{2}\right)^6}{6!} + \cdots$

9. $\sin^{-1} x = x + \dfrac{x^3}{2(3)} + \dfrac{3x^5}{2(4)(5)} + \dfrac{1(3)(5)x^7}{2(4)(6)(7)} + \cdots$

11. 0

13. $1 + \dfrac{(x-1)}{2} + \displaystyle\sum_{i=2}^{\infty} (-1)^{i+1} \dfrac{1(3)(5)\cdots(2i-3)}{2^i\, i!} (x-1)^i$

17. $\dfrac{1}{9!}$

19. 100 nonzero terms

21. $\ln(\sec x) = \dfrac{x^2}{2} + \dfrac{x^4}{12} + \dfrac{x^6}{45}$

23. $x \ln(x+1) = x^2 - \dfrac{x^3}{2} + \dfrac{x^4}{3}$

Review

6. $a_1 = \ln 2, a_2 = \dfrac{\ln 3}{2}, a_7 = \dfrac{\ln 8}{7}$
Convergent

7. $a_1 = 3, a_2 = 9, a_7 = 3^7$
Divergent

8. $a_1 = 0, a_2 = \frac{3}{25}, a_7 = \frac{12}{25}$
Convergent

9. $a_1 = \frac{1}{3}, a_2 = \frac{1}{9}, a_7 = \dfrac{1}{3^7}$
Convergent

10. $a_1 = 0, a_2 = \dfrac{\ln 4}{16}, a_7 = \dfrac{\ln 49}{686}$
Convergent

11. $a_n = \dfrac{1}{n!}$
Convergent

12. $a_n = \dfrac{n(n+1)}{2}$
Divergent

13. $a_n = n$
Divergent

15. Convergent

16. Divergent

17. Convergent

18. Convergent

20. Convergent

21. Convergent

22. Divergent

23. Divergent

25. Divergent

26. Convergent

27. Divergent

28. Convergent

31. Absolutely convergent

32. Divergent

33. Absolutely convergent

34. Conditionally convergent

35. Convergent

36. Convergent

37. Convergent

38. Divergent

39. Divergent

40. Convergent

41. An alternating divergent series

43. No

44. $(1, 3)$

45. $[-1, 1)$

46. $(-1, 1]$

47. $(-e, e)$

48. $[-5, 1)$

49. $(-5, 3)$

50. $(-2, 0)$

51. Yes; $x > 1$ or $x \le -1$

54. $\dfrac{1}{\sqrt{2}} \left[1 + \left(x - \dfrac{\pi}{4} \right) - \dfrac{1}{2!} \left(x - \dfrac{\pi}{4} \right)^2 - \dfrac{1}{3!} \left(x - \dfrac{\pi}{4} \right)^3 + \cdots \right], (-\infty, \infty)$

55. $-x + \dfrac{x^2}{2} - \dfrac{x^3}{3} + \dfrac{x^4}{4} - \cdots, (-1, 1]$

56. $\displaystyle\sum_{i=0}^{\infty} \dfrac{(-1)^i x^{2i}}{i!}, (-\infty, \infty)$

57. $\displaystyle\sum_{i=0}^{\infty} \dfrac{x^{(2i+1)}}{(2i+1)!}, (-\infty, \infty)$

58. $1 + \dfrac{x}{3} - \dfrac{2x^2}{3(6)} + \dfrac{2(5)}{3(6)(9)} x^3 - \cdots, (-1, 1]$

59. $1 - (x - 1) + (x - 1)^2 - (x - 1)^3 + (x - 1)^4 - \cdots, (0, 2)$

60. $1 + \dfrac{x^2}{2} + \dfrac{5x^4}{24} + \dfrac{61x^6}{720} + \cdots, \left(-\dfrac{\pi}{2}, \dfrac{\pi}{2} \right)$

61. $3!e$

62. 0

63. $-10! \left(\displaystyle\sum_{i=1}^{10} \dfrac{1}{i} \right)$

64. $\sin \dfrac{\pi}{4} \approx \dfrac{\pi}{4} - \dfrac{1}{3!} \left(\dfrac{\pi}{4} \right)^3 + \dfrac{1}{5!} \left(\dfrac{\pi}{4} \right)^5 - \dfrac{1}{7!} \left(\dfrac{\pi}{4} \right)^7, E < \dfrac{\pi^9}{9!4^9}$

Index

INDEX OF APPLICATIONS

SOCIAL SCIENCE

STATISTICS